Small AC Generator

DISCARD

SERVICE MANUAL ■ 1ST EDITION

Small AC Generator Manufacturers

■ Coleman   ■ Deere   ■ Homelite   ■ Mitsubishi   ■ Porter Cable

■ Dayton   ■ Generac   ■ Honda   ■ North Star   ■ Robin

PRIMEDIA
Business Magazines & Media Inc.
P.O. Box 12901 ■ Overland Park, KS 66282-2901
Phone: 800-262-1954 Fax: 800-633-6219

621. 3134

Cover photo courtesy of Mitsubishi Generators and Pumps.

II

CONTENTS

GENERATOR BASICS

GENERATOR SAFETY

When servicing portable generators and electrical systems, pay close and careful attention to all published precautions. Where care should be exercised, there are four levels of notification:

1. DANGER – You WILL BE killed or seriously injured if you don't follow instructions.

2. WARNING – You MIGHT BE killed or seriously injured if you don't follow instructions.

3. CAUTION – You might be hurt if you don't follow instructions.

4. NOTE – A helpful hint to prevent equipment damage and/or to make the job easier.

Human skin offers somewhat of a barrier to the flow of electricity – to a point. Below about 48 volts, the hazard is small. However, at the 120 or 240 volt level produced by portable generators:

A – One milliampere (.001 Amp – 1/1000th of an amp) produces a tingling sensation in most people. This is referred to as perception current.

B – Shock level is at five milliamperes (.005 amp). This is only 1/43 of the current required to operate a 25-watt light bulb, and will produce a violent muscle reaction, normally causing the individual to drop or be thrown back from the item causing the shock.

C – Let-go current is about twice shock level, or 10 milliamperes (.010 amp). Above this level, the individual loses the ability to let go of the shocking item. If no help is available to separate the victim from the power source, serious injury or death may occur.

D – Electrocution caused by ventricular fibrillation occurs at about 100 milliamperes (1/10th of an amp – less than half the power needed to operate a 25-watt light bulb), when the heart muscle fibers lose control and the heart is no longer able to pump blood. Observe the following precautions when servicing portable generators:

- NEVER work on a wet generator or a generator in a wet or rainy area. Moisture creates an excellent short-circuit ground.
- ALWAYS use tools and test equipment with insulated handles. Whenever possible, use electrically insulated gloves.
- NEVER allow uninsulated wires and terminals to hang loosely from circuits being power-tested.
- ALWAYS use proper service procedures.
- NEVER service a standby generator connected to a transfer switch without first isolating the generator from the power grid.
- ALWAYS ground the generator properly prior to test-loading.

- NEVER operate the generator in confined spaces or in any area without sufficient fresh-air ventilation.
- ALWAYS make sure that the generator is on secure footing so it cannot move around while running.
- NEVER allow unqualified personnel to service or operate the generator.
- ALWAYS make sure that the switches on power tools and appliances are Off prior to plugging them into the generator.
- NEVER contact hot generator or engine components, or rotating parts.
- ALWAYS use three-prong extension cords for tools and appliances which are not double-insulated.

Fig. GB101 – Displaying magnetic lines of force (flux) with iron filings and a horseshoe magnet.

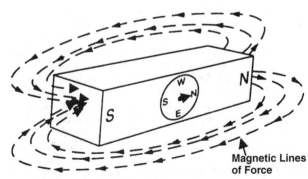

Fig. GB102 – Sketch showing the flow of magnetic flux from North to South on the outside of a bar magnet.

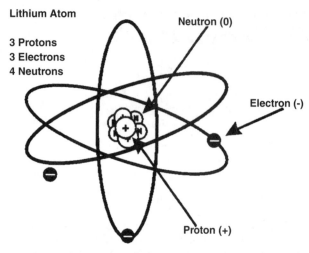

Lithium Atom

3 Protons
3 Electrons
4 Neutrons

Fig. GB103 – Diagram of a Lithium atom showing balanced protons and electrons.

PRINCIPLES OF OPERATION

MAGNETISM AND ELECTRICITY

Electricity and magnetism are inseparable. Electricity cannot be produced without magnetism, and the production of electricity creates additional magnetism.

Magnetic lines of force, although invisible, can be readily observed by placing a piece of paper over the poles of a horseshoe magnet, then sprinkling iron filings onto the paper. The results are seen in Fig. GB101, as the filing positions are determined by the flux, the magnetic lines of force. Flux is more concentrated around the poles and moves outside the magnet from the north magnetic pole to the south magnetic pole (Fig. GB102). To help understand magnetism, let's take a simple look at atomic structure.

An atom is made up of a center core (nucleus) which contains the neutral charges (neutrons) and the positive charges (protons) in a tight grouping. The negative charges (electrons) are constantly orbiting the nucleus.

Atoms, by nature, strive to be neutral – to have their negative charges balance their positive charges. This is the basis for the "opposite or unlike charges attract; like charges repel" principle of magnetism. Fig. GB103, for example, shows a two-dimensional view of the lithium atom. It is possible to break an electron loose from its orbit around the nucleus. If this happens to an atom with an equal number of protons and electrons, then the atom is left with a positive (+) net charge and is referred to as a positive ion. If a neutral atom receives a stray electron, it then becomes negatively charged, thereby becoming a negative ion.

The natural tendency of a positive ion is to attract an electron from a neighbor atom. The neighbor then becomes a positive ion, and a chain reaction begins in which each atom, in turn, borrows an electron from its neighbor. This electron-borrowing creates a flow of current that continues until all the atoms have achieved a state of balance. Electron transfer – the "flow" of electrons – is electricity.

Science has proven that electrical current tends to flow from points of high potential (areas with a surplus of electrons) to points of lower potential (areas with an electron deficiency), or, in other words, from negative to positive (- to +): This is called the Electron Theory. However, many years

prior to the discovery of the electron theory, contemporary convention chose current flow to be from positive to negative, because it was thought that positive ions travelled to seek out electrons. Although incorrect and confusing, electricity today is still governed by the "conventional theory", and the electron theory is still relegated to theory.

Some materials such as silver and copper will readily transfer electrons from atom to atom – these materials are called *conductors*. Other materials such as rubber, glass, and plastic hold their electrons very tightly, having what is referred to as "bound" electrons – these materials are *non-conductors*, more commonly known as *insulators*.

Electrical force – the pressure which causes electricity to "flow" – is called *voltage*, and refers to the potential for electrical current to flow from one point to another. For this reason, voltage is referred to as *electrical potential*. The unit of measurement for voltage is the *volt*, and the reference symbol is *V*. Voltage is sometimes also referred to as *electromotive force*, or EMF, with a symbol of *E*.

Electrical volume, the amount of electricity flowing through a conductor or a system, is known as current, and is measured in *amperes* or, more commonly, *amps*. The symbol for amperes is *I*.

Electrical resistance, the degree to which a conductor will not allow current to flow, is called just that, resistance. It is measured in units called *ohms*. The symbol for ohms is the Greek letter Omega, Ω.

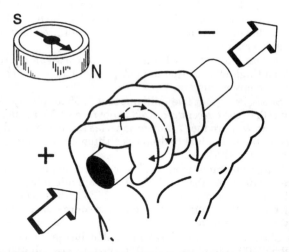

Fig. GB104 – The "Right-hand Rule" of magnetism for finding magnetic North in an electromagnetic conductor.

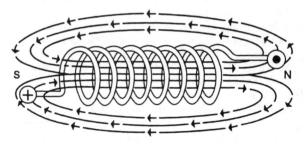

Fig. GB105 – Sketch showing the magnetic flux around an electromagnetized coil of wire.

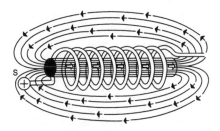

Fig. GB106 – Sketch showing a stronger magnetic field around the Fig. GB105 coil by placing an iron core inside the wire coil.

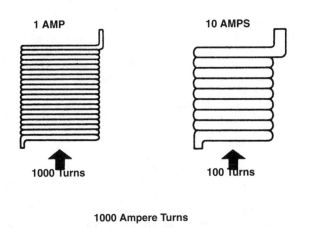

1 AMP   10 AMPS

1000 Turns   100 Turns

1000 Ampere Turns

Fig. GB107 – Sketch showing the relationship between the number of wire coils, the current being fed through the coil, and the measurement of ampere-turns.

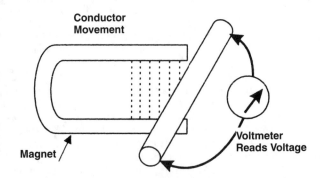

Conductor Movement

Voltmeter Reads Voltage

Magnet

Fig. GB108 – Creating electromagnetic induction by moving a wire conductor through a magnetic field.

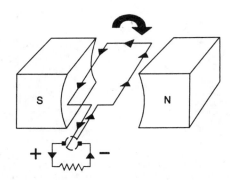

Fig. GB109 – Sketch of a simple revolving-armature generator.

Work performed by an electrical current is called *power*, and is measured in units called watts. The symbol for watts is *W*.

Electromagnetism

A magnetic field can also be created where there is none by feeding electrical current through a conductor or wire. When this is done, each end of the wire becomes a pole, and the movement of electrons inside the wire causes a magnetic field around the wire.

To determine the direction of the magnetic lines of force around the conductor use the "right-hand rule." Place your right hand around the conductor with the thumb pointing in the direction of current flow (Fig. GB104). The fingers will then be pointing in the direction of the flux or, in other words, to the North magnetic pole.

To strengthen this electrically created magnetic field, the wires can be formed into a coil (Fig. GB105). With lines of magnetic force entering the coil at one end and departing the other, the coil ends form the poles, with the flux concentrated inside the coil.

To determine polarity in a coil, again use the "right-hand rule." Grasp the coil with the fingers pointing in the direction of current flow through the coil. The thumb then points to the coil's north pole.

To further strengthen this *electromagnetic* field, the coil can be wound around a core of magnetizable material, such as iron or certain steel alloys (Fig. GB106). Since air is a poor conductor of flux and since ferrous metals are good conductors, using such a core greatly increases magnetic strength.

Magnetic-field strength is determined by the amount of current being fed into the coil and the number of turns in the coil; it is measured in "ampere turns." Fig. GB107 shows the first 1000-turn coil being fed 1 amp and the second 100-turn coil being fed 10 amps. The strength of their magnetic fields will be the same – 1000 ampere-turns.

AC Generator Theory

If the iron filing paper were removed from the magnet in Fig. GB101, a wire conductor could be moved through the pole-to-pole magnetic field. If a sensitive voltmeter were connected to the wire, a voltage could be read, indicating electron flow through the conductor (Fig. GB108). This is the principle of electromagnetic induction, the movement of the conductor through the magnetic field *induces* the transfer and movement of electrons within the conductor. Three things are necessary to generate voltage.
1. A magnetic field.
2. A conductor.
3. Motion of the conductor through the magnetic field or of the magnetic field past the conductor.

Taking the simple generator one step farther, the conductor wire could be formed into a loop and rotated inside two wider-spaced poles (Fig. GB109). By taking output from the wire ends through a rotational connection, current would flow whenever the wire was rotated.

If the Fig. GB109 conductor was rotated continuously in the same direction, one side of the wire would first pass one pole, then the other. This would cause current to change direction each half revolution, and is exactly what happens in a generator which produces *alternating* current.

By connecting a meter to Fig. GB109 and starting out with conductor Leg A in the 0° or twelve o'clock position (Fig. GB110), the conductor can be rotated clockwise to indicate on the meter the intensity and direction of current flow.

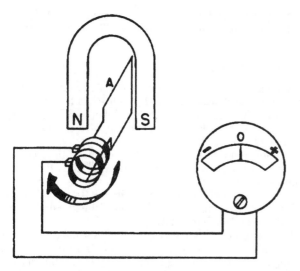

Fig. GB110 – Connecting a meter to the Fig. GB109 generator allows current flow to be measured.

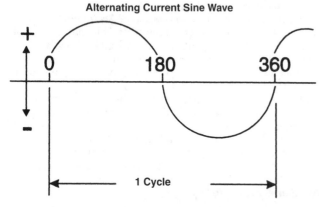

Fig. GB111 – Graphic sine wave representation of one cycle (one complete armature revolution) of alternating current.

- With Leg A midway between the S and N poles (0°), both poles exert an equal force on the wire, and no electrons flow. This results in a no-current reading on the meter.
- As the loop begins to rotate clockwise, the S pole flux begins to exert more influence on the electrons in Leg A, causing the current to flow in the wire and the meter to begin registering (+).
- When the loop is rotated 90° to align Leg A with the S pole, the current reaches its maximum (+) value.
- As the loop continues to rotate, Leg A moves farther from the S pole flux concentration, causing decreasing electron and current flow and a movement of the meter needle back toward zero.
- At the 180° (six o'clock) position, Leg A is again equidistant from either pole, resulting in no current flow and no meter reading.
- As Leg A continues to rotate up to the 270° (nine o'clock) position, it comes under the influence of the flux from the N pole and begins moving electrons in the

opposite direction, causing the meter to begin reading current flow in the (-) direction.
- With Leg A in the 270° position, the N pole exerts its strongest force on the loop, causing the meter to read maximum (-).
- As Leg A continues to rotate back to twelve o'clock, current flow decreases and the meter needle registers decreasing (-) current flow until Leg A is again at its twelve o'clock starting point and zero current flow.

On paper, this rise and fall of current from positive to negative during one revolution is shown as Fig. GB111, and is referred to as the AC sine wave. It shows one "cycle" of alternating current. Rotating the loop 60 times per second gives us an alternating current frequency of 60 cycles per second which, until recently, was abbreviated "cps". However, electrical terminology changed somewhat, and cycles per second is now referred to as Hertz, abbreviated "Hz". The United States of America, as well as most of North America, uses 60 Hz electricity. Many other nations use a lower 50 Hz frequency.

Since one wire loop will produce measurable, but not usable, electricity, there are three methods of increasing output:
1. Using a stronger magnetic field.
2. Using more loops of wire.
3. Increasing the speed with which we cut the lines of force.

Since speed, or generator RPM, has just been shown to be directly proportional to Hertz, and since Hz must be kept at 60 to remain compatible with the equipment the generator will be powering, Option 3 can be eliminated. The following section will show how Options 1 and 2 are used to allow generators to produce the necessary output.

GENERATOR CONSTRUCTION

With the basic background in the preceding sections, we can now understand how most portable generators produce electric power by rotating a magnetic field (rotor) inside of, or around, a set of stationary windings (stator). Depending on the generator model, this field could be one of three styles: 1) brush type; 2) brushless; or 3) rare earth (permanent) magnet.

In brush-type or brushless rotors, ever-present residual magnetism in the rotor's metal core induces a small amount of current flow in the stator to begin producing electricity upon initial start-up. This current is fed back into the rotor, increasing the magnetic field and induced current to the point of saturation. At this point, the generator is operating at its rated capacity.

In permanent magnet rotors, the rare earth magnets are normally always fully magnetized, thereby naturally producing the full-output magnetic field. However, due to the loads placed on a rotor with a ring of permanent magnets

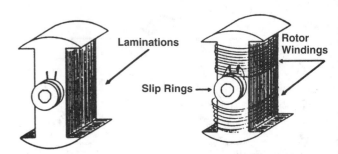

Fig. GB112 – Sketch showing the laminated alloy-steel plates used as the rotor's core, with the rotor windings being started around the core.

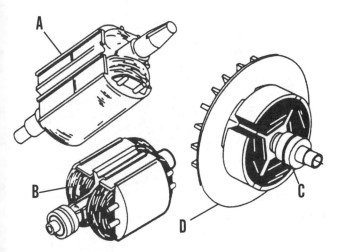

Fig. GB113 – View of typical 2-pole and 4-pole generator rotors.

A. 2-pole brushless rotor
B. 2-pole brush-style rotor
C. Slip rings on brush-style rotor
D. 4-pole rotor

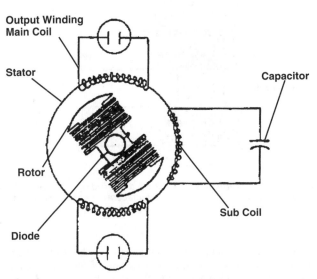

Fig. GB115 – Typical end view of the winding circuitry of a brushless generator. Refer to the text for a functional explanation.

attached, and the resulting rotational limitations, most permanent magnet generators are of a smaller capacity, usually under 3kW.

Rotors

The rotor core is constructed of a stack of steel-alloy plates called the laminations (Fig. GB112). The core is laminated in order to reduce the magnetic turbulence (eddy currents) that would be present in a large one-piece core. Each plate has its own N and S pole, thereby better aligning the core magnetism.

All rotors start out with a small amount of inherent magnetism. Some rotors retain residual magnetism in the special alloy laminations, while others have small permanent magnets fastened to the laminations.

Wires wound around the core carry the current needed to produce an electromagnetic field in the rotor. The strength of the magnetic field is determined by how much current is sent through the coil windings.

Most portable generator rotors have one North and one South pole (Fig. GB113). Since one revolution of a two-pole magnet (rotor) produces one cycle of AC (one Hz), and since we operate on a 60 Hz system, the two-pole rotor must rotate 60 times per second to produce the proper electrical frequency. With 60 seconds in a minute, the rotor must then rotate at 60×60, or 3600 RPM.

Some generator manufacturers produce four-pole rotors which have two north and two south poles (Fig. GB113). By doubling the number of poles that induce voltage into the stator, the rotor must only rotate at half the speed of a two-pole rotor to still produce 60 Hz electricity.

Brush-style Rotors

In the brush-style rotor, the two ends of the coil of wire that is wound around the rotor core are connected to the rotor's slip rings. The brushes contact the slip rings, smooth rings of conducting material, while the rotor turns. As the rotor begins turning, residual magnetism in the rotor induces AC voltage in the excitation windings. A group of

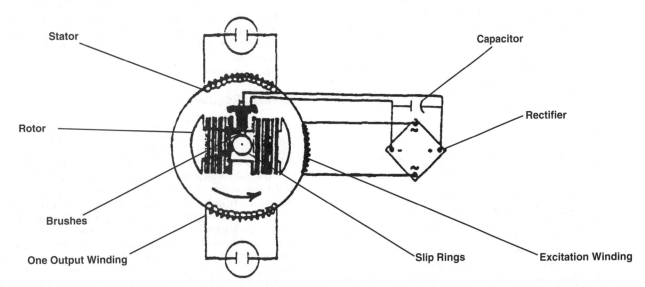

Fig. GB114 – Typical end view of the winding circuitry of a brush-style generator. Refer to the text for a functional explanation.

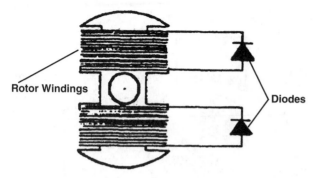

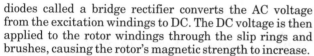

Fig. GB116 – Typical end view of the winding circuitry of a brushless rotor showing the diode placement in the windings.

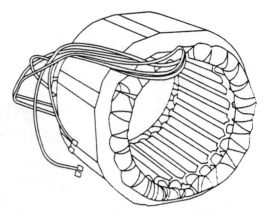

Fig. GB117 – Sketch of a typical stator assembly.

diodes called a bridge rectifier converts the AC voltage from the excitation windings to DC. The DC voltage is then applied to the rotor windings through the slip rings and brushes, causing the rotor's magnetic strength to increase.

The rotor's increased magnetism then induces current into the stator windings, resulting in a voltage increase up to the rated output. Refer to Fig. GB114 for an illustration of the brush-style generator.

Brushless Rotors

Some rotors do not use brushes to increase rotor magnetism. Fig. GB115 shows an illustration of the brushless system used by most manufacturers.

Brushless generators have a "sub coil" winding in the stator. This winding is equivalent to the excitation winding in a brush-style generator. When the rotor starts to turn and the rotor windings pass the stator's sub coil winding, the residual magnetism in the rotor's core causes the sub coil to produce a voltage. This voltage is applied to the capacitor (condenser) connected in series to the sub coil winding. The capacitor builds a charge to a predetermined value, then releases it.

This charge causes current to flow in the sub coil, creating a strong magnetic field just as the rotor coils approach the sub coil. The sub coil's magnetic field induces an AC voltage which is rectified to DC by two diodes in the rotor windings (Fig. GB116). The DC then feeds through the rotor windings, increasing the rotor magnet's strength, subsequently increasing output to the rated voltage.

When a load is applied to the generator, rotor current magnetizes the main coil. Since the main coil and the sub coil share a common core, the main coil induces a current flow in the sub coil. This current flow increases the strength of the sub coil's magnetic field, thereby inducing a stronger field in the rotor. This increases the generator's output to match the load.

Stators

The stator is so named because it is the stationary magnetic component of the generator. Fig. GB117 is a sketch of a typical stator.

Like the rotor, the stator is made up of a stack of laminated alloy steel plates which form the core. The winding wire is wound through slots in the core. Electromagnetic induction from the spinning rotor then produces current flow through the windings, creating output.

Most stators have multiple windings:

• A 120-volt generator will have one output winding and one excitation winding.

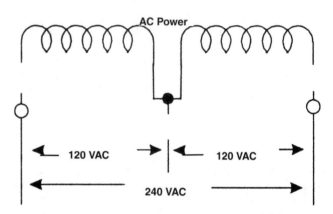

Fig. GB118 – Schematic view of a dual-voltage AC stator winding showing the series connection between the two 120V windings in order to produce 240 volts.

• A 120/240-volt generator will have two output windings and one excitation winding.
• A 120/240-volt generator with a DC output/battery charge circuit will have two output windings, one excitation winding, and one DC output winding.

Figure GB118 shows how a typical 120/240V stator output winding pair is pictured on a wiring diagram. Browsing through the wiring diagrams of the generators in this manual will provide excellent examples of all different styles of stators.

Voltage Regulators

Most portable brush-style generators utilize one of two types of regulation to control voltage output: inherent or electronic. Inherent voltage regulation, as the name implies, uses qualities built into the generator design to keep the output voltage within limits. The excitation coil wound onto the stator (Fig. GB119) produces AC power that is bridge-rectified to DC, then filtered by a capacitor. This DC voltage then feeds directly into the rotor through the slip rings. Under no-load conditions, the excitation winding is energized solely by the rotor. As the main windings take on load, extra flux is produced by the load current flowing through the windings. This tends to boost the excitation winding output, thereby increasing inductive output to match the load.

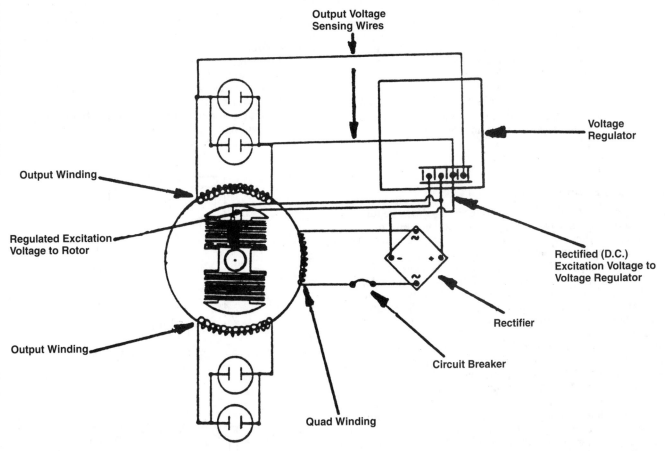

Fig. GB119 – *Typical end view of the winding circuitry of a brush-style generator showing the electronic voltage regulation system.*

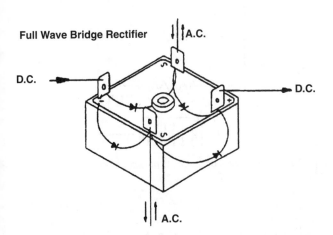

Fig. GB120 – *Skeleton view of a typical full-wave bridge rectifier showing internal diode placement and connections as well as external input and output connections.*

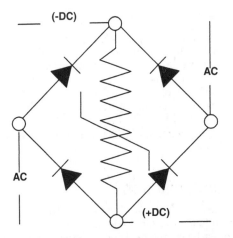

Fig. GB121 – *Schematic view of a late-style Fig. GB120 full-wave bridge rectifier showing the internal "varistor" (variable resistor) in the circuit to protect the diodes against voltage surges.*

The only disadvantage of this regulation method is that the range of regulation is usually between 15%-20%. This results in no-load voltages up to 145 volts and full-load voltages of 110 volts on a 120-volt system.

With electronic voltage regulation (EVR), the bridge rectified DC voltage from the excitation winding, instead of going directly to the rotor brushes, proceeds to the solid-state voltage regulator (Fig. GB119). When the generator receives a load, AC output voltage decreases slightly. The regulator immediately detects this voltage drop by way of its connections to the receptacles, sending more DC voltage to the rotor windings and increasing the rotor's magnetic strength.

Some regulators also have a bypass circuit to ease generator start-up by allowing the residual voltage to feed into the generator field unobstructed until output voltage has reached its no-load rating.

It should be noted that the circuit breaker in the line between the stator excitation winding and the bridge rectifier on this type system protects the components inside the generator, not the load.

The advantage to the EVR system is that voltage regulation is held to much closer tolerances, usually ±2%, contributing to better tool performance and longer life.

Components

Full-Wave Bridge Rectifier

In order to better understand rotor and voltage regulator function, as discussed in previous sections, an understanding of the component which makes DC current possible – the bridge rectifier – is helpful.

A functional sketch of a bridge rectifier is shown in Fig. GB120. Within the bridge rectifier are four diodes, solid-state components that allow current to pass in one direction only. The diodes are arranged and connected so that when AC is fed into the two AC terminals, only the positive half of the sine wave passes through the diodes, thereby making the other two terminals DC+ and DC-. Tests on specific generator models later in this manual will show that, when any two adjacent terminals are continuity-tested, current will pass in one direction but not in the opposite direction.

Some bridge rectifiers also have a "varistor" (a variable resistor) incorporated into the internal circuit, as the Fig. GB121 schematic shows. The purpose of a varistor is to permit normal voltage to pass through the diodes but to shunt excessively-high voltage to ground. This protects the bridge-rectifier's diodes from voltage surges or spikes.

Resistors

Resistors limit current flow by providing a resistance in the circuit. They can be either fixed or variable. Externally-variable resistors are commonly referred to as rheostats or potentiometers, "pots" for short. Internally-variable resistors are usually called "varistors" and have no external adjustment. They are load-sensitive and load-adjusted.

The resistance value of most resistors is marked on the resistor, itself.

ELECTRICAL CIRCUITS

A "circuit" is a path for electricity to take as it provides power to a load, and for the load to receive the power, the circuit must be complete. There are two basic types of circuits which can supply an electrical load: Parallel or Series.

In a series circuit (Fig. GB122), electricity has only one path to take. A break in one part of the circuit will prevent current from flowing in any part of the circuit. In a series circuit:

- Current flow (amps) is the same in every part of the circuit.
- The total resistance (ohms) is the sum of the individual loads.
- The total voltage across all resistances is the sum of the voltages across the individual resistances.

In a parallel circuit (Fig. GB123), electricity has more than one path to take. A break in one part of the circuit will

not affect every part of the circuit, unless the break is immediately beyond the power-supply connections. In a parallel circuit:

- Total current flow (amps) is the sum of the current flow through all the branches.
- Total resistance is less than the resistance of any individual branch.
- The voltage applied to each branch is the same as the source voltage.

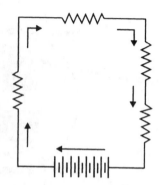

Fig. GB122 – Diagram of a simple series circuit.

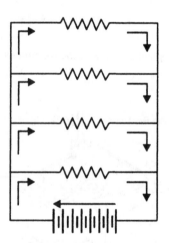

Fig. GB123 – Diagram of a simple parallel circuit.

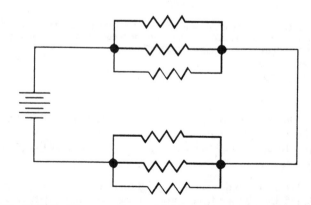

Fig. GB124 – Diagram of a combination series-parallel circuit.

Some loads can be combined into a series-parallel circuit (Fig. GB124).

There are two basic types of circuit faults: Short Circuit or Open Circuit.

In a short circuit, electricity returns to ground through an unintentional path caused by a decrease in resistance through that part of the circuit. Since electricity seeks the path of least resistance, current will tend to flow across the "short" circuit rather than the "long" circuit as the system was originally designed.

An open circuit is, as the name states, open somewhere along the circuit; it is an incomplete circuit, one which does not have a complete path for current to travel from supply to load and back to supply.

SERVICE HINTS AND HELPS

Portable generator performance is affected by altitude above sea level: Engine-horsepower output decreases approximately 3½% for every 1,000 ft. rise in elevation.

- GFCI (Ground Fault Circuit Interrupter) receptacles must be tested "Hot". To test a GFCI on a non-functioning generator in preparation for troubleshooting, disconnect the GFCI generator leads and connect a wall-socket jumper to the GFCI feed terminals.
- When testing diodes, determine test-meter polarity by performing the first test as suggested in the Troubleshooting/Test Procedures sections, then compare the results to the specifications listed. If the test results are the reverse of those specified, then the meter has reverse polarity from the meter used to determine the test procedures. To obtain correct test results, either reverse the meter test leads or the test readings.
- Always drain engine crankcase oil hot, but allow a generator to cool down before disassembly in order to prevent burned skin and stripped aluminum threads.
- When handling wire-wound components such as stators and rotors, exercise caution so that the winding insulation (varnish) will not be damaged; missing insulation causes shorts.

- When servicing brushes on brush-style generators, always observe proper brush polarity. Reversing the brush leads will damage the generator.
- When attempting to diagnose an overheat problem, make sure that the cooling-air inlet and outlet passages are not covered or obstructed.
- Most Idle-Control systems will not detect loads below one amp (1 A). For this reason, if a load of less than 120 watts is connected to a 120-watt generator, or 240 watts on a 240-watt generator, make sure that the idle-control switch is OFF so the unit will run at rated speed to produce proper output.

GROUNDING

Always use the ground terminal and wing nut on the generator frame to connect the generator to a proper ground. The generator-to-ground connection should be made with a minimum #8 gauge solid-conductor wire.

A metal underground water pipe in direct contact with the earth for at least 10 feet is the ideal ground. If this is not available, use at least an 8-foot length of 3/4-inch diameter pipe with a non-corrosive surface, or a 5/8-inch diameter steel or iron rod. The pipe or rod should be driven into the gorund to a depth of eight feet or, if this is not possible, buried in a trench.

All electrical tools operated from the generator must be properly grounded by use of a third wire or be double insulated.

OVERLOADS AND UNDERLOADS

In many instances of generator problems, the trouble comes from improper use or application rather than a fault in the generator, itself.

Extension cords are the primary cause of underloads. Using a cord with too small a wire gauge or too long a length will, because of resistance from the small size or long length, cause a voltage drop between the generator and the load. Full voltage, therefore, cannot reach the load, usually resulting in equipment burnout. Refer to the following chart for the correct extension-cord length for the load applied. Loads are listed for 120-volt output; the load can be doubled for 240 volts.

Maximum Power Switch Parallels 120V Windings

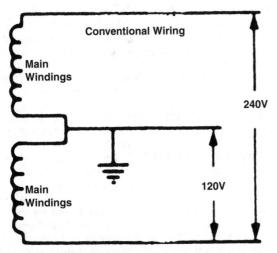

Fig. GB125 – Schematic view of the two main windings in a dual-voltage generator series-connected so both 120V and 240V loads can be supplied. Only half the rated generator output is available at 240 volts in this mode, thereby reducing the 120V output by half.

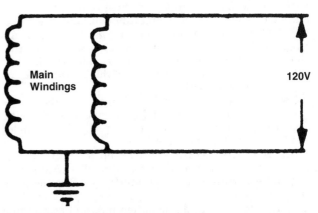

Fig. GB126 – Schematic view of the two main windings in a dual-voltage generator parallel-connected so the generator can deliver full-rated output at 120 volts. This type connection and output is an option offered by some manufacturers, and is not standard equipment on all dual-voltage generators.

LOAD	Maximum Allowable Wattage Cable Length in Feet, by Gauge				
	8#	10#	12#	14#	16#
300		1000	600	375	250
600		500	300	200	125
900		350	200	125	100
1200		250	150	100	50
1800		150	100	65	
2400	175	125	75	50	
3000	150	100	60		
3600	125	65			
4800	90				

Overloads can be caused by:

A. Having too many loads connected to the generator

B. Having a proper-sized load connected to the generator, but one which has an internal fault or short.

C. Having a proper-sized running load, but one which requires excessive starting output to get running.

A generator with 120/240-volt capability can also be overloaded by running more than 50% of rated capacity as a 120-volt load unless the generator is equipped with a switch which connects the two stator windings in parallel. This connection then allows full-rated output to be available as 120V output, while eliminating the availability of 240V output. These switches go by a variety of names. such as Max Power, Full Power, or Dual Voltage. Consult the generator for this option. Figure GB125 shows a generator's main windings connected in series, allowing both 120V and 240V output – note that the load must be split to obtain both outputs. Figure GB126 shows this same generator's main windings now parallel-connected after switching, with 100% of output available as 120V.

GENERATOR TERMINOLOGY

AC: Acronym and abbreviation for alternating current.

ADAPTER, Engine Housing or Front Housing: The part which fits between the engine and the generator components mounted to the engine.

ALTERNATING CURRENT: The constantly-reversing flow of electrons in a conductor (wire). A good example of AC is household electricity.

ALTERNATOR: A device for converting mechanical energy into AC electrical energy. It can also be called an AC generator.

AMMETER: An instrument to measure electric current flow.

AMPERAGE: The strength or intensity of an electric current, measured in Amperes.

AMPERE: Commonly referred to as Amps, it is the unit of current flow through a conductor; a measurement of electrons flowing past a given point at a given time. One ampere will flow when one Volt overcomes one Ohm of resistance.

ARMATURE: The conductor-holding part of a generator or dynamo, normally a core of laminated steel, around which are wound the conductors. A rotor is a rotating armature; a stator is a stationary armature.

BRUSH: Usually a carbon, graphite, and/or copper alloy conductor which maintains electrical contact between a stationary component and a rotating component.

COMMUTATOR: A grouping of conductor segments which makes selective connections to rotating armature coils through brushes. In a DC generator, the brushes and commutator make a rotating switch which converts armature-induced AC to DC.

CONDUCTOR: A wire or cable designed for the flow of electricity. Conductors can be either solid or stranded. Solid conductors are normally used for AC, while stranded conductors work better for DC.

CONTINUOUS LOAD: Any load up to the maximum full-rated load which a generator is capable of supplying indefinitely, within the generator's specified temperature-rise limits.

CORE: The assembled lamination pieces of a generator which constitute the magnetic structure.

CRADLE: The framework supporting and protecting the generator unit.

CURRENT: The rate of flow of electricity. See AMPERE.

CYCLE: One complete reversal of an alternating current or voltage, from zero to a positive maximum back to zero to a negative maximum and back to zero. The number of cycles-per-second is the frequency, and is measured in Hertz (Hz).

DC: Acronym and abbreviation for direct current.

DIODE: A solid-state component which allows electricity to pass in only one direction through a conductor.

DIRECT CURRENT: DC is the flow of electrons through a conductor in only one direction. A battery is a good example of DC.

DYNAMO: Another name for a generator, usually used in countries other than the U.S.

EDDY CURRENT: Inductive magnetic turbulence, as opposed to current flow. These currents create generator-efficiency losses; they are reduced by the use of special thin steel laminates, such as those in armatures.

EFFICIENCY: The ratio of a generator's useful power output to its total power production, expressed as a percent. See RATED POWER.

ELECTRICITY: A form of energy produced by the flow of electrons.

ELECTROMAGNET: A device with a core of magnetic material, usually soft iron, which is surrounded either partially or completely by an embedded coil of wire. When electricity is fed through the wire, the device becomes completely magnetized.

ELECTROMAGNETIC FIELD: A magnetic field around a conductor; this field is generated by the flow of current through the conductor.

ELECTROMAGNETIC INDUCTION: The flow of electricity through a conductor by placing that conductor near another conductor which is already electrified.

ELECTROMOTIVE FORCE: The difference in electrical charge between the component consuming electrical power and the unit producing the power.

ELECTRON: The negatively-charged part of an atom.

ELECTRON THEORY: This theory, based on the magnetic premise that like charges repel and opposite charges attract, holds that the negatively-charged electron is attracted to a positive particle, thereby resulting in a negative-to-positive flow. This contradicts the commonly-accepted conventional theory which holds that current flows from positive to negative.

EMF: Acronym and abbreviation for electromotive force.

ENDBELL: The outermost generator component which positions the rotor bearing, thereby supporting the end of the rotor shaft. It is also referred to as the end housing or rear housing.

FIELD: An area under the influence of magnetic lines of force.

FIELD COIL: A conductor wound around a field pole.

FIELD CURRENT: Electrical flow through the field coils and windings which creates a magnetic field.

FIELD WINDINGS: The sum total of all field coils.

FLASHING THE FIELD: Popular term for restoring lost residual magnetism in a generator field, usually the rotor of a portable generator.

FLUX: Magnetic lines of force.

FLUX DENSITY: Magnetic lines of force per unit of area.

FREQUENCY: In electricity, the number of AC cycles per second, measured in Hertz (Hz).

GAUGE: The size of a conductor, measured by its diameter. The smaller the gauge number, the larger the gauge size.

GENERATOR: A device for converting mechanical energy into electrical energy.

GROUND: A connection, either intentional or accidental, between an electric circuit and the earth or some object or conducting body serving in place of, or as an electrical path **to, the earth.**

HERTZ: Abbreviated Hz, it is a unit of frequency equal to one cycle per second (cps), a cycle being one positive electrical pulse and one negative electrical pulse. Pulses are produced by the passing of both the north and south poles of the magnetic field over the conductor.

HIGH-CYCLE GENERATOR: One designed to produce 180 Hz, 3-phase output. Some manufacturers subdivide this output to allow the generator to be used for conventional needs.

HOUSE CURRENT: The electricity delivered to a building by a public utility.

HOUSING: See ADAPTER or ENDBELL.

HZ: Abbreviation for Hertz.

IDLE CONTROL: A system for lowering the engine speed to idle during no-load periods.

INDUCTION: The causing of a flow of current in a conductor by passing that conductor through a magnetic field.

INSULATOR: A non-conducting material used for preventing the flow of electricity.

KILOWATT: 1,000 watts.

KILOWATT-HOUR: An electrical measurement derived from the consumption of 1,000 watts per hour.

kW: Abbreviation for kilowatt.

MAGNETO: A permanent magnet alternator. Most magnetos are used to supply ignition spark for internal combustion engines or as detonators for explosives.

NO-LOAD: A state of generator operation where the generator is running but is not supplying power to any equipment.

NON-CONDUCTOR: An insulator; a material which will not transmit electricity.

OHM: The amount of resistance necessary for one amp of current to flow in a conductor when one volt of electromotive force is applied.

OHM'S LAW: E (volts) = I (amps) × R (ohms); also I = E/R, or E = I/R.

OUT-OF-PHASE: A condition in which the AC voltage waves of two generating systems do not coincide.

OVERLOAD: A load applied to the generator in excess of the output the generator is capable of producing. Also see SURGE OUTPUT.

PARALLEL CONNECTION: An electrical-connection system in which the input and output connections of multiple elements are joined, thereby providing more than one path for current flow.

PERMANENT MAGNET: A material which retains induced magnetic properties, all properly aligned, after it is removed from a magnetic field.

PF: Acronym and abbreviation for Power Factor.

PHASE: The number of complete voltage and/or current sine waves generated per 360 electrical (usually rotational) degrees. Each phase requires its own set of windings.

POLE: Either one of a pair of areas which are positioned at opposite ends or sides of a preset, fixed distance.

POWER FACTOR: Expressed as a decimal or percent, it is the ratio of true power (watts) to the volt-amps (VA) of an AC circuit. When amps and volts are in phase, the power factor is 1.0. When the current wave precedes the voltage wave, the PF is higher than 1.0; when the voltage wave precedes the current wave (which is normally the case), the PF is below 1.0.

RATED OUTPUT: The net amount of guaranteed continuous output which the generator is designed to provide, expressed in watts. See SURGE OUTPUT.

RATED SPEED: The revolutions per minute (RPM) at which the generator is designed to produce its rated output.

RATED VOLTAGE: The voltage at which the generator is designed to operate.

REAL POWER: The term, expressed in kW, derived by multiplying voltage times current times power factor, then dividing by 1000.

RECTIFIER: A device, usually made up of multiple diodes, for converting AC to DC.

RELAY: An electrically-operated switch with low-amperage primary-circuit contacts; it is usually used in control circuits.

RESIDUAL VOLTAGE: Voltage generated with zero field current.

RESISTANCE: Opposition to the flow of current.

ROTOR: The moving, or rotating, part of an electrical generator, usually composed of an iron-based core wound with wire. The flow of current through its windings creates a magnetic field around the windings. The strength of the magnetic field can be increased by (A) forming the wires into coils, (B) increasing the wire size, or (C) increasing current flow through the wires.

RPM: Acronym and abbreviation for revolutions per minute, the speed at which a rotating item's central shaft turns.

SERIES CONNECTION: An electrical-connection system of multiple elements in which the output terminal of one element is connected to the input terminal of the next element, thereby providing only one path for electrical flow.

SERIES WOUND: A type of generator winding in which all current output passes through the field windings.

SHORT CIRCUIT: A normally-unintentional electrical contact between current-carrying components which results in the current being passed through an undesirable path.

SHUNT CONNECTION: A connection in which the respective terminals of two or more devices are connected together; a synonym for parallel connection.

SHUNT WOUND: A type of generator winding in which the field current is supplied from the armature potential.

SINGLE PHASE: A single-voltage AC system in which cycle reversals occur at the same time and are of the same alternating polarity throughout the system.

STATOR: The non-moving, or stationary, part of an electrical generator. Like the rotor, it is also composed of an iron-based core wound with wire. The greater the number of turns of wire in the stator winding, the greater the induced electromotive force (EMF) when magnetic lines of force (flux) cut through the windings.

SURGE OUTPUT: The maximum amount of wattage a generator is capable of producing for a brief period of time without causing damage to the generator. It is usually 10%-20% above rated output and is helpful for motor-starting loads which decrease once the motor is running.

THREE PHASE: An output of three complete, separate sine waves spaced 120 electrical degrees apart.

TRANSFORMER: A component consisting of two or more conductor coils joined only by magnetic induction and used to transfer electrical energy from one circuit to another. This transfer does not change frequency but normally does change voltage and current.

UNITY POWER FACTOR: A power factor of 1.

UNIVERSAL MOTOR: An electric motor which can operate on either AC or DC voltage.

VIBRATION MOUNT: A rubber-like device fastened between the engine/generator and cradle frame to reduce operational vibration.

VOLT: The unit used to measure electromotive force; the amount of EMF necessary to force a current (electron) flow of one amp through a resistance of one ohm. It is the difference in electrical potential or electrical balance – the difference between the number of protons and the number of electrons in the circuit – with current flow being the attempt to regain that balance.

VOLTAGE: Electric potential or potential difference expressed in volts.

VOLTAGE DROP: Voltage reduction caused by current flowing through a resistance.

WATT: The amount of electrical power equal to one amp of current flow under pressure of one volt. It takes exactly 746 watts of electrical power to equal one horsepower (746 W = 1 HP). One kilowatt equals 1,000 watts. For mathematical calculations, Watts = Volts × Amps. The rule of thumb for engine power requirements for portable generators, taking both internal and external losses into account, is two horsepower per kilowatt.

WINDING: A generator-armature conductor coil; the wiring of a stator or rotor.

ATI

Active Technologies, Inc.
1117 LaVelle Rd.
Alamogordo, NM 88310

In the early 1990s, ATI manufactured lightweight portable generators producing less than 1000 watts. These units were named LIGHTNING CHARGERS (Fig. ATI). In the summer of 1995, ATI was purchased by Coleman Powermate, Inc. The following ATI generators are similar to Coleman POWERMATE 1000 models:

GEN-50926
GEN-50939
GEN-50967
GEN-50969

GEN-50998
GEN-50999
LC12
LC12/24
12/115

The main difference between ATI and Coleman is that most ATI generators produce 120 volts DC instead of AC.

Refer to the Coleman POWERMATE 1000 section in this manual for available service procedures.

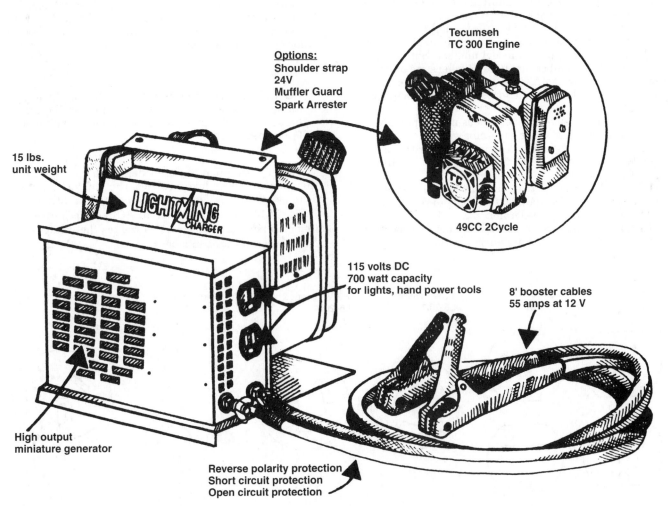

**Tecumseh
TC 300 Engine**

49CC 2Cycle

Options:
Shoulder strap
24V
Muffler Guard
Spark Arrester

15 lbs.
unit weight

115 volts DC
700 watt capacity
for lights, hand power tools

8' booster cables
55 amps at 12 V

High output
miniature generator

Reverse polarity protection
Short circuit protection
Open circuit protection

Fig. ATI

COLEMAN POWERMATE, INC.

125 Airport Rd.
Kearney, NE 68848

INTRODUCTION

The Coleman numbers specify design series and normal rated continuous-run wattage. A model number in the MAXA name series, for example, could be 52-2000, PM522000, or PM0522000.01.

Coleman generators are identified using the six numbers immediately to the left and the two numbers to the right of the decimal dot. In model numbers with no decimal, the last six numbers are used. Hyphens are unimportant. In the previous MAXA example, for instance:

52 indicates the construction series and name series ("MAXA").

20 indicates a 2000-watt continuous-run output; and

00 indicates a first-series generator with no options.

The 01 to the right of the decimal specifies revisions. In this particular model of the MAXA series, it identifies a brushless generator.

The two-digit construction series prefix, 52 in this example, determines which specification chart to refer to when testing or servicing the generator.

The three letters which prefix most newer generators (e.g. PMO) are for marketing purposes; they identify the type of Coleman product, the market it is produced for, and the usual engine powering the unit.

Chart 1 lists model number prefixes with corresponding model name series and continuous-run wattages. For assistance with generator identification cross-referencing, Chart 2 lists consecutive model numbers matched with model names and rated outputs, and Chart 3 lists model names alphabetically, matched with corresponding model numbers.

NOTE: The Powermate registered trademark name immediately following the Coleman trademark name on the generator label does not constitute a Powermate model name.

The only exception to the model-numbering system at this time is in the industrial series. The 1 of the 61 prefix is also the number designating wattage in the 10,000 range.

Fig. CL101 shows the different AC receptacles used on Coleman portable generators as well as the voltages and NEMA identification numbers.

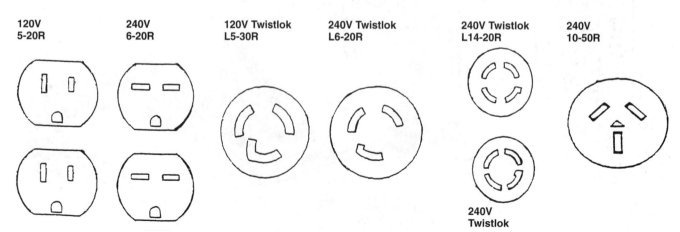

| 120V
5-20R | 240V
6-20R | 120V Twistlok
L5-30R | 240V Twistlok
L6-20R | 240V Twistlok
L14-20R | 240V
10-50R |

240V
Twistlok

Fig. CL101 – Various receptacles used on Coleman portable generators, including voltages and NEMA identification numbers.

Chart 1

Model Number Prefix	Model Name Series	Continuous-Run Wattage
20	Ultimite	N/A - DC only
30	Powermate 1000 w/2-cycle engine	800-1000
34	Voltswagon	2250
40	Pulse, Sport, or Powermate	1000-2000
42	Ultra	2000
44	PM	500-2800
45	Commercial or Industrial	4000-7000
47	Vantage	3000-8000
48	Contractor	3500-6000
50	Powerhouse or Magna Force	2500-8000
51	Powerpartner	4000-6000
52	Maxa or Contractor	2000-7000
53	I/C, Powerbase, or Pro-Gen	3000-5000
54	Powerbase	2000-7000
55	Vertex or Commercial	2000-7500
56	HP (Honda-Powered)	3500-4000
61	Industrial	10,000-12,000
20, 25	Powerweld	

Chart 2

Model Number	Model Name	Rated Output Watts	Volts
200900	Ultimite	N/A	12VDC
201000	Ultimite	N/A	12VDC
201100	Ultimite	N/A	12/24VDC
201101	Ultimite	N/A	12/24VDC
204000	Powerweld 150	50 Amp	
255203	Powerweld 175	175 Amp	
300800	Powermate 1000	850AC	120VAC, 12VDC
300850	Powermate 1000	850AC	120VAC, 12VDC
300910	Powermate 1000	850AC	120VAC, 12VDC
300915	Powermate 1000	850AC	120VAC, 12/24VDC
301010	Powermate 1000	850AC	120VAC, 12VDC
301015	Powermate 1000	850AC	120VAC, 12/24VDC
342250	Voltswagon	2000	120
342250.01	Voltswagon	2000	120
401000	Pulse 1000 (Brushless)	800AC	120VAC, 12VDC
401600	Sport 1600 (Brushless)	1250AC	120VAC, 12VDC
401750	Pulse 1750 (Brushless)	1400AC	120VAC, 12VDC
401755	Pulse Plus 1750 (Brushless)	1400AC	120VAC, 12VDC
401805	Powermate 1850 (Brushless) or Pulse 1850 (Brushless)	1500AC	120VAC, 12VDC
401850	Pulse 1850 (Brushless)	1500AC	120VAC, 12VDC
401851	Sport 1800 Or 1850 (Brushless)	1500AC	120VAC, 12VDC
422200	Ultra (Brushless)	2000	120
422201	Ultra (Brushless)	2000	120
422500	Ultra (Brushless)	2000	120
422502	Ultra (Brushless)	2000	120
422505	Ultra (Brushless)	2000	120
440500	Pm 600 (Brushless)	500AC	20VAC, 12VDC
440750	Pm 800 (Brushless)	750AC	120VAC, 12VDC
441250	Pm Ohv1600 & Ohv 1650	1250AC	120VAC, 12VDC
441500	Pm 1500 (Brushless)	1500AC	120VAC, 12VDC

Chart 2 (continued)

442000	Pm Ohv2800	2000AC	120VAC, 12VDC
454002	Commercial	4000	120/240
454003	Commercial	4000	120/240
454022	Commercial	4000	120/240
454024	Industrial	4000	120/240
454025	Industrial	4000	120/240
455002	Commercial	5000	120/240
455022	Commercial	5000	120/240
455023	Commercial	5000	120/240
455024	Industrial	5000	120/240
455025	Industrial	5000	120/240
457022	Commercial	7000	120/240
473003	Vantage	3000AC	120VAC, 12VDC
473503	Vantage	3500AC	120VAC, 12VDC
474203	Vantage	4000	120
474603	Vantage	4000	120/240
474603.01	Vantage	4000	120/240
475003	Vantage	5000	120/240
476503	Vantage	6000	120/240
477022	Vantage	7000	120/240
477023	Vantage	7000	120/240
478022	Vantage	8000	120/240
483503	Contractor	3500	120/240
483508	Contractor	3500	120/240
485203	Contractor	5000	120/240
486223	Contractor	6000	120/240
486228	Contractor	6000	120/240
486623	Contractor	6000	120/240
502500	Powerhouse	2500	120
503300	Powerhouse	3000	120
504202	Powerhouse	4000	120/240
504222	Powerhouse	4000	120/240
504602	Powerhouse	4500	120/240
505602	Powerhouse	5000	120/240
505622	Powerhouse	5000	120/240
505622.01	Magna Force	5000	120/240
508622	Powerhouse	8000	120/240
514001.01	Powerpartner		
516000.01	Powerpartner		
522000	Maxa	2250	120
522000.01	Maxa (Brushless)	2000	120
522004	Maxa (Brushless)	2250	120
522400	Maxa Plus	2250	120
523000	Maxa	3000	120
523001	Maxa	3000	120
523001.01	Maxa (Brushless)	3000	120
523008	Contractor	3000	120
523300	Maxa Er Plus	3000	120
524000	Maxa	4000	120/240
524020	Maxa	4000	120/240
524052	Maxa	4000	120/240
524202	Maxa/Er	4000	120/240
524208	Contractor	4000	120/240
524302	Maxa/Er Plus	4250	120/240
524602	Maxa Er Plus	4250	120/240

Chart 2 (continued)

524604	Maxa/Er Plus	4250	120/240
524702	Maxa/Er	4250	120/240
524702.01	Maxa Er	4250	120/240
525000	Maxa	5000	120/240
525020	Maxa	5000	120/240
525202	Maxa/Er	5000	120/240
525202.02	Maxa	5000	120/240
525208	Contractor	5000	120/240
525302.01	Maxa/Er Plus	5000	120/240
525302.02	Maxa Er Plus	5000	120/240
525302.03	Maxa Er Plus	5000	120/240
525402	Maxa	5000	120/240
534000	I/C	4000	120/240
534202.01	Powerbase	4000	120/240
534502	Pro-Gen	4500	120/240
534805	I/C (Brushless)	4500	120/240
535000.03	Pro-Gen (Brushless)	5000	120/240
535202	Pro-Gen	5000	120/240
536503.17	Pro-Gen	6000	120/240
542000	Powerbase	2000	120
542000.01	Powerbase (Brushless)	2000	120
542004	Powerbase (Brushless)	2000	120
542200	Powerbase	2000	120
542200.01	Powerbase (Brushless)	2000	120
542400	Powerbase	2000	120
543800	Powerbase	3000	120
544000	Powerbase	4000	120/240
544003	Powerbase	4000	120/240
544020	Powerbase	4000	120/240
544200	Powerbase	4000	120/240
544202	Powerbase	4000	120/240
544202.01	Powerbase	4000	120/240
544202.02	Powerbase	4000	120/240
544208	Powerbase	4000	120/240
544222	Powerbase	4000	120/240
544222.01	Powerbase	4000	120/240
544302	Powerbase	4000	120/240
544302.01	Powerbase	4000	120/240
544502	Powerbase	4000	120/240
545000	Powerbase	5000	120/240
545021	Powerbase	5000	120/240
545202	Powerbase	5000	120/240
545202.01	Powerbase	5000	120/240
545208	Powerbase	5000	120/240
545212	Powerbase	5000	120/240
545212.01	Powerbase	5000	120/240
545212.02	Powerbase	5000	120/240
545222	Powerbase	5000	120/240
545222.01	Powerbase	5000	120/240
545303	Powerbase	5000	120/240
545305	Powerbase	5000	120/240
545305.01	Powerbase	5000	120/240
545305.02	Powerbase	5000	120/240
545305.03	Powerbase	5000	120/240
545305.17	Powerbase	5000	120/240

Chart 2 (continued)

545323	Powerbase	5000	120/240
545402	Powerbase	5000	120/240
552400	Commercial	2000	120
555200	Commercial	5000	120/240
555503	Vertex	5500	120/240
555523	Vertex	5500	120/240
555523.01	Vertex	5500	120/240
557523	Vertex	7500	120/240
557523.01	Vertex	7500	120/240
557823	Vertex	7500	120/240
558023	Vertex	8000	120/240
563503	Hp	3500AC	120/240vac, 12vdc
563505	Hp	3500	120/240
564002	Hp	4000	120/240
564202	Hp	4000	120/240
564302	Hp	4000AC	120/240vac, 12vdc
564303	Hp	4000AC	120/240vac, 12vdc
612023	Industrial	12,000	120/240

Chart 3

Model Name	Model Number
Commercial	454002
Commercial	454003
Commercial	454022
Commercial	455002
Commercial	455022
Commercial	455023
Commercial	457022
Commercial	552400
Commercial	555200
Contractor	483503
Contractor	483508
Contractor	485203
Contractor	486223
Contractor	486228
Contractor	486623
Contractor	523008
Contractor	524208
Contractor	525208
Hp	563503
Hp	563505
Hp	564002
Hp	564202
Hp	564302
Hp	564303
I/C (Brushless)	534805
Industrial	454024
Industrial	454025
Industrial	455024
Industrial	455025
Industrial	610023
Industrial	612023
Magna Force	505622.01
Maxa	522000
Maxa (Brushless)	522000.01

Chart 3 (continued)

Maxa	522004
Maxa Plus	522400
Maxa	523000
Maxa	523001
Maxa	523001.01
Maxa Er Plus	523300
Maxa	524000
Maxa	524020
Maxa	524052
Maxa Er	524202
Maxa Er Plus	524302
Maxa Er Plus	524602
Maxa Er Plus	524604
Maxa Er	524702
Maxa Er	524702.01
Maxa	525000
Maxa	525020
Maxa Er	525202
Maxa	525202.02
Maxa Er Plus	525302.01
Maxa Er Plus	525302.02
Maxa Er Plus	525302.03
Maxa	525402
Pm 600	440500
Pm 800	440750
Pm 1500	441500
Pm Ohv1600 & Ohv1650	441250
Pm Ohv2800	442000
Powerbase	534202.01
Powerbase	542000
Powerbase (Brushless)	542000.01
Powerbase	542004
Powerbase	542200
Powerbase (Brushless)	542200.01
Powerbase	542400
Powerbase	543800
Powerbase	544000
Powerbase	544003
Powerbase	544020
Powerbase	544200
Powerbase	544202
Powerbase	544202.01
Powerbase	544202.02
Powerbase	544208
Powerbase	544222
Powerbase	544222.01
Powerbase	544302
Powerbase	544302.01
Powerbase	544502
Powerbase	545000
Powerbase	545021
Powerbase	545202
Powerbase	545202.01
Powerbase	545208
Powerbase	545212

Chart 3 (continued)

Powerbase	545212.01
Powerbase	545212.02
Powerbase	545222
Powerbase	545222.01
Powerbase	545303
Powerbase	545305
Powerbase	545305.01
Powerbase	545305.02
Powerbase	545305.03
Powerbase	545305.17
Powerbase	545323
Powerbase	545402
Powerhouse	502500
Powerhouse	503300
Powerhouse	504202
Powerhouse	504222
Powerhouse	504602
Powerhouse	505602
Powerhouse	505622
Powerhouse	508622
Powermate 1000	300800
Powermate 1000	300850
Powermate 1000	300910
Powermate 1000	300915
Powermate 1000	301010
Powermate 1000	301015
Powermate 1850	401805
Powerpartner	514001.01
Powerpartner	516000.01
Powerweld 150	204000
Powerweld 175	255203
Pro-Gen	534502
Pro-Gen (Brushless)	535000.03
Pro-Gen	535202
Pro-Gen	536503.17
Pulse	401000
Pulse 1750	401750
Pulse Plus 1750	401755
Pulse 1850	401805
Pulse 1850	401850
Sport	401600
Sport	401851
Ultimite	200900
Ultimite	201000
Ultimite	201100
Ultimite	201101
Ultra	422200
Ultra	422201
Ultra	422500
Ultra	422502
Ultra	422505
Vantage	473003
Vantage	473503
Vantage	474203
Vantage	474603

Chart 3 (continued)

Vantage	474603.01
Vantage	475003
Vantage	476503
Vantage	477022
Vantage	477023
Vantage	478022
Vertex	555503
Vertex	555523
Vertex	555523.01
Vertex	557523
Vertex	557523.01
Vertex	557823
Vertex	558023
Voltswagon	342250
Voltswagon	342250.01

Wiring Color Codes

All Generators Manufactured By Coleman Powermate, Inc. Use The Following Standardized Wiring Color-Coding:

Power windings	B–Black
	B/W–Black With White Tracer Stripe
	R–Red
Excitation	Y–Yellow (Diodes)
	Blu–Blue (Capacitor)
Ground	G–Green
Neutral	W–White
	W/B–White With Black Tracer Stripe
Battery charger	O–Orange
Idle control	O–Orange

COLEMAN

BRUSHLESS GENERATORS, 500-5000 WATT
SERIES 40, 42, 44, 52, 53, & 54

IDENTIFICATION

Identify Coleman brushless generators in the 500-5000W series by removing the snap-fit rectangular cover on the generator endbell and inspecting the area under the cover. Only the rotor bearing should be visible un-

der the cover. If there are two screw-mounted terminals, one on each side of the bearing, those are the brush connectors, and it is a brush-type generator.

Coleman brushless generators powered by four-cycle engines can be found in Series 40, 42, 44, 52, 53, or 54. Fig. CL301 shows typical Series 40 components; Fig. CL302

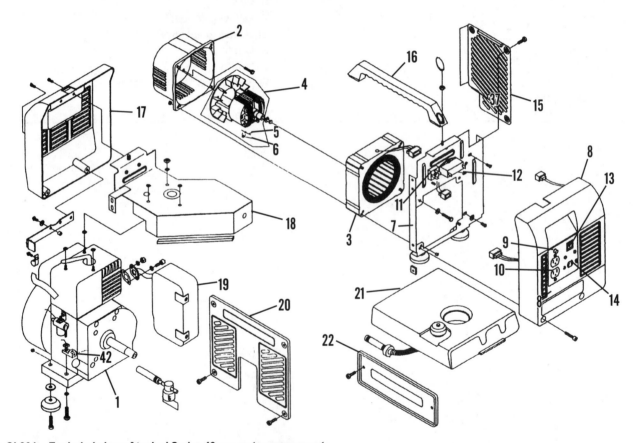

Fig. CL301 – Exploded view of typical Series 40 generator components.

1. Engine
2. Engine adapter/front housing
3. Stator assembly
4. Rotor assembly
5. Rotor diode
6. Rotor varistor
7. Stator bracket assembly

8. Endbell/rear housing assembly
9. AC circuit breaker
10. AC receptacle
11. Diode bridge rectifier
12. Capacitor
13. Lighted rocker switch
14. DC receptacle

15. Muffler shield
16. Handle
17. Cover
18. Engine shroud
19. Muffler
20. Side cover
21. Fuel tank
22. Side panel

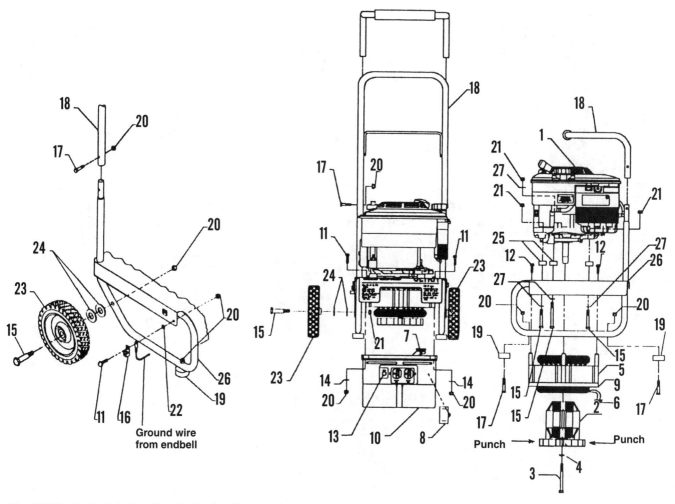

Fig. CL302 – Exploded view of typical Series 42 generator components.

1. Engine
2. Rotor assembly
3. Rotor bolt
4. Rotor bolt washer
5. Stator assembly
6. Stator harness connector
7. Endbell harness connector
8. Capacitor
9. Foam gasket
10. Endbell/rear housing assembly
11. Endbell and ground lug bolts
12. Stator bolts
13. Circuit breaker
14. Flat washer
15. Axle bolt
16. Ground lug terminal
17. Frame bolt
18. Handle
19. Foot pad
20. Nylok nut
21. Hex nut
22. Ground wire star washer
23. Wheel
24. Wheel spacer washer
25. Engine-to-mainframe spacer
26. Main frame
27. Lock washer

shows typical Series 42 components; Fig. CL303 shows typical Series 44 components; and Fig. CL304 shows typical Series 52-54 components.

NOTE: On PULSE and SPORT models equipped with a DC battery charger, the DC receptacle is used for recharging weak batteries only. Do NOT use it to boost-start dead batteries or run power tools.

CAUTION: The PM500/PM600 models (440500) are not equipped with an automatic DC circuit breaker and should only be used to "quick-charge" batteries. Extended charging periods or attempting to charge fresh batteries can result in overcharging.

WIRING COLOR CODES

All portable generators manufactured by Coleman Powermate, Inc. use the following standardized wire color-coding:

Power Windings	B–Black
	B/W–Black with white tracer stripe
	R–Red
Excitation	Y–Yellow (diodes)
	Blu–Blue (capacitor)
Ground	G–Green
Neutral	W–White
	W/B–White with black tracer stripe
Battery Charger	O–Orange
Idle Control	O–Orange

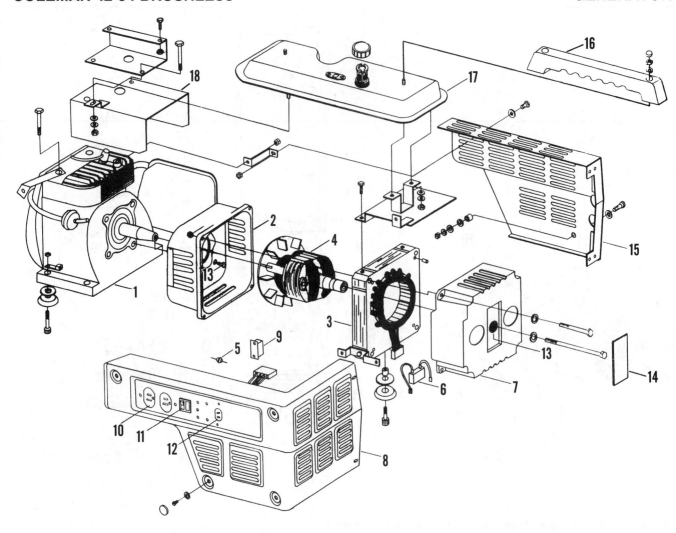

Fig. CL303 – Exploded view of typical Series 44 generator components.

1. Engine
2. Engine adapter
3. Stator
4. Rotor
5. Diode
6. Capacitor

7. Endbell
8. Cover
9. Circuit breaker
10. AC receptacle
11. Switch
12. DC receptacle

13. Bearing
14. Cover
15. Side panel
16. Handle
17. Fuel tank
18. Engine shroud

MAINTENANCE

The endbell bearing and race should be inspected, cleaned, and lubricated any time the endbell is removed. If the bearing feels hot to the touch, replacement may be necessary.

TROUBLESHOOTING

No-load voltages on 60Hz units should be 128-143 volts from the 120V receptacles and 258-278 volts from the 240V receptacles. If little or no output is generated, the following items could be potential causes:

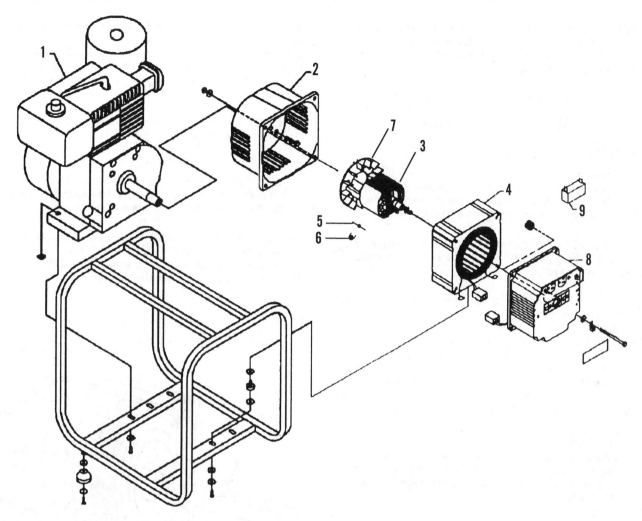

Fig. CL304 – Exploded view of typical Series 52, 53, and 54 generator components

1. Engine
2. Engine adapter
3. Rotor
4. Stator
5. Rotor diode
6. Rotor varistor
7. Fan
8. Endbell
9. Capacitor

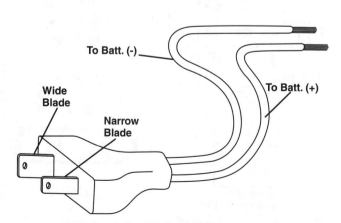

Fig. CL305 – Homemade jumper tool for remagnetizing rotors.

1. Engine RPM: Engine must maintain 3600 RPM under load, or 3650-3800 RPM no-load. Engine RPM could also be set too high, causing overvoltage. Refer to the appropriate engine section for service.

2. Loose component-mounting fasteners (low or no voltage).

3. Circuit breakers (low or no voltage).

4. Insecure wiring connections (low or no voltage).

5. Faulty receptacle (low or no voltage).

6. Capacitor (low or no voltage).

7. Diodes (low or no voltage).

8. Stator windings (low voltage – shorted; no voltage – open).

9. Rotor:

    a. Open or shorted windings (no or low voltage, respectively).

    b. Loss of residual magnetism (no voltage).

NOTE: Residual magnetism can be restored without disassembling the generator, and is the recommended first step if the generator has no output. Refer to RESIDUAL MAGNETISM.

10. Insufficient cooling system ventilation (overheating).

GENERATOR REPAIR

RESIDUAL MAGNETISM

Unlike brush-type generators, there is no test procedure for residual magnetism on brushless generators. If low residual magnetism is suspected as the cause of no output, remagnetize the rotor using the following procedure:

WARNING: Do not attempt to re-excite a PM500/PM600 (440500) unit or the control module will be destroyed.

1. Turn the engine switch off and remove the spark-plug lead from the spark plug.
2. Construct a jumper tool (Fig. CL305) using a male two-prong 110-volt household plug and two 16-gauge minimum wire leads. It must be of sufficient length to reach from the generator's 110V outlet to a nearby fully charged DC power source (minimum 6VDC; 24VDC maximum), such as a 12V automotive battery. Alligator clips on the jumper wire ends would be helpful.

NOTE: Mark the jumper wires so that the wire coming from the narrow spade on the 110V plug connects to the positive battery terminal; the wire from the wide spade must connect to the negative battery terminal.

3. With the jumper tool plugged into one of the generator's 120VAC outlets, connect the narrow spade jumper wire to the positive battery terminal (+), and connect the wide spade jumper wire to the negative terminal (−).
4. Immediately pull the engine starter cord seven times, then disconnect the jumper tool from the battery and generator.
5. Reconnect the spark-plug lead, turn the engine switch on, then start the engine and observe the generator output.
6. Remagnetizing can be attempted up to three times. If, after the third try, the generator still has no output, proceed with further testing and repair.

GENERATOR DISASSEMBLY

All Except 42 Series

Refer to Figs. CL301, CL303, or CL304 for component identification:
1. Place a support under the engine adapter.
2. Remove the four long bolts holding the endbell and stator to the engine adapter. Remove the endbell (endbell/outlet panel assembly on 40 Series) and disconnect the wiring connector blocks.

NOTE: With the endbell removed, the capacitor, rotor, and stator can be tested without further disassembly.

3. Remove any stator support brackets, fasteners, and/or rubber mounts.

4. On 40 and 44 Series models, remove the top handle and sheet metal shields.
5. On 40 Series, remove the stator end-mount bracket.
6. Remove the stator by prying between the stator and engine adapter with a flat blade pry tool. Note orientation of the stator wires and connector.

NOTE: The rotor cannot be removed without first removing the stator.

CAUTION: The plastic rotor fan is part of the rotor assembly and cannot be replaced separately. Handle the rotor carefully to prevent fan damage.

7. Loosen the rotor through-bolt, then apply penetrating oil through the rotor onto the crankshaft taper.
8. Remove the rotor using a suitable puller (Fig. CL 306). Place the puller feet under the bottom of the laminations,

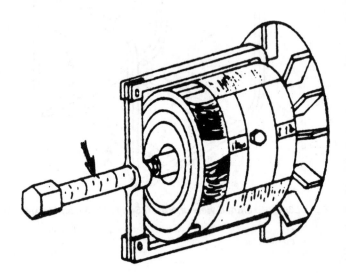

Fig. CL306 – Recommended rotor removal method using puller. See text for NOTE.

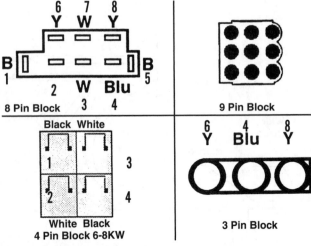

Fig. CL307 – Views of endbell wiring connectors used on Coleman generators.

then place the puller center bolt on the head of the rotor throughbolt.

NOTE: Use extreme care to prevent the puller legs and feet from contacting the rotor windings to prevent damage to winding insulation.

An alternate rotor removal method would be to remove the rotor through-bolt and after applying penetrating oil through the rotor onto the crankshaft taper, firmly but carefully tap the end bearing race with a soft-tip rubber mallet (not a metal hammer) while holding the rotor to prevent it from falling.

42 Series

Refer to Fig. CL302 for component identification:
1. Remove the three bolts (11) holding the endbell (10) to the main frame (26).
2. Disconnect the endbell-to-stator wiring connector (6, 7) and remove the endbell.
3. Remove the rotor bolt and washer (3, 4) from the rotor (2).
4. Carefully insert a brass punch between the rotor fins and against the rotor shaft. Firmly hit the brass punch with a hammer, being careful not to allow the rotor to fall when loosened. If the rotor does not loosen on the first hit,

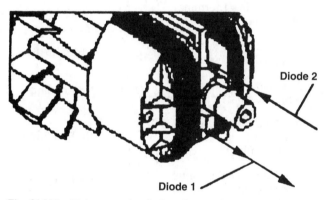

Fig. CL308 – Note correct polarity when replacing diodes.

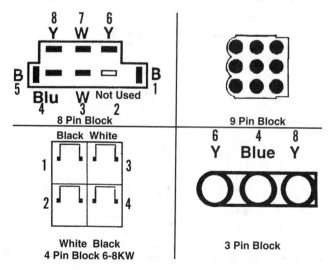

Fig. CL309 – Views of stator wiring connectors used on Coleman generators

remove the punch, rotate the rotor a quarter turn, then repeat the process until the rotor is loose.
5. Remove the three engine mount bolts (15) and remove the engine (1).

NOTE: Do not lose the three engine-to-mainframe spacers (25).

6. Remove the four stator mount bolts (12) from the stator posts and remove the stator.

CAPACITOR TESTING

The capacitor can be tested without removing it from the endbell, provided the stator connector plug is disconnected.

NOTE: Even with the generator stopped, the capacitor should retain a charge if it is working properly. Do not touch the capacitor terminals, as electric shock may result. Always discharge the capacitor by shorting across the terminals with a screwdriver or similar tool with an insulated handle.

To test the capacitor:
1. Set your VOM ohmmeter scale to Rx10,000.
2A. On spade-terminal capacitors, disconnect the harness wires from the capacitor, noting polarity orientation, then touch the VOM leads to the capacitor terminals. The red VOM lead should always connect to the positive (+) capacitor terminal. Connect the negative (−) VOM lead to any bare metal ground in the endbell.
2B. On solder terminal capacitors, connect the positive (+) VOM lead to the #4 terminal of the endbell connector block, as shown in Fig. CL307. Connect the negative (−) VOM lead to any bare metal ground in the endbell.
3. The meter should first indicate low resistance, gradually increasing resistance toward infinity. No meter reading or constant continuity indicates a faulty capacitor, and it must be replaced.
4. On solder terminal capacitors, a failed test could be the result of faulty solder joints. Repeat the test holding the VOM leads directly against the capacitor terminals. Use caution to observe proper polarity. If the test is now successful, one or both of the terminal-to-capacitor wires may be faulty, or the capacitor pigtail solder joint is defective; correct as necessary.

NOTE: If replacing the capacitor, note capacitor polarity prior to removal. Polarity must be correct upon installation.

ROTOR

Diode Test

To test diodes on applicable units:
1. Unsolder one side of the diode, making sure to remove the connected resistor/varistor, if so equipped.

NOTE: Use caution not to break the diode lead.

2. Test the diode using the R × 10K scale on the VOM. A good diode shows resistance in only one direction.
3. If the diode is faulty, replace it. Make sure the polarity is correct (Fig. CL308).

ROTOR WINDING RESISTANCE

Test rotor without removing it from the generator. If hypot testing, the stator should be removed to access the laminations.

NOTE: On diode-equipped units, all resistance tests must be performed with the diodes disconnected from the rotor windings.

To test resistance, use a VOM set on the RX1 scale.

NOTE: All resistance readings are approximate and may vary depending on the winding temperature and the test-equipment accuracy.

Refer to the following chart for rotor resistance readings.

Series	Wattage	Min. Ohms	Max. Ohms
40	1000-2000	13.1	14.1
42	2000	13.5	13.7
44	500-800	9.7	10.2
	1000-1500	10.4	11.4
52 & 54	2000-2250	11.9	12.9
	3000	14	15.4
53	5000	3.6 –each winding–	4.0

Hypot testing is the best way to test rotor windings. A rotor might pass VOM testing and still be faulty. Always follow the tester manufacturer's instructions. The recommended hypot test voltage is 1250 volts; less voltage may not show faults, and higher voltage may damage the rotor. To test the rotor, hypot each winding individually from the winding lead to lamination (ground). The failure of either winding denotes a faulty rotor.

STATOR WINDING RESISTANCE

To test the stator windings:
1. Set the VOM to the R × 1 scale;
2. Read the resistance between the stator connector terminals illustrated in Fig. CL309.
3. Compare the resistance readings to the values shown in the following chart. If no reading is obtained upon each initial test, the connector block and stator leads should be checked for broken wires or faulty terminals. Repair as necessary, if possible, then retest.

NOTE: All resistance readings are approximate and may vary depending on winding temperature and test-equipment accuracy.

Series	Wattage	Field	Min. Ohms	Max. Ohms
40	1000-2000	Black to White	1.05	1.25
		Yellow to Yellow	9.3	9.7
		Orange to Orange	0.215	0.235
42	2000	Black to White	0.60	0.80
		Yellow to Yellow	6.4	6.7
44	500-800	Black to White	2.42	2.44
		Yellow to Yellow	10.95	11.05
		Orange to Orange	0.270	0.275
	1000-1500	Black to White	1.64	1.66
		Yellow to Wellow	9.7	9.8
52 & 54	2000	Black to White	0.51	0.52
		Yellow to Yellow	6.8	6.9
52	3000	Black to White	0.675	0.695
		Yellow to Yellow	5.4	5.6
53	5000	Black to White	0.5	0.6
		Red to Red	2.02	2.1

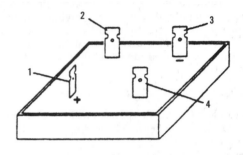

Digital VOM OHM Reading

Pos lead on #1 terminal Neg lead on #2 terminal	1.675-1.731
Pos lead on #1 terminal Neg lead on #3 terminal	3.215-3.301
Pos lead on #1 terminal Neg lead on #4 terminal	1.588-1.612
Pos lead on #2 terminal Neg lead on #3 terminal	1.612-1.616
Pos lead on #2 terminal Neg lead on #1 or #4 terminal	No Reading
Pos lead on #3 terminal Neg lead on #1, #2, or #4 terminal	No Reading
Pos lead on #4 terminal Neg lead on #1 or #2 terminal	No Reading
Pos lead on #4 terminal Neg lead on #3 terminal	1.863-1.871

Fig. CL310 – Bridge Diode Test Chart using Digital VOM.

Analog VOM Testing

	Pos. No.	Black Test Lead (-)			
		1 (+)	2 (AC)	3 (-)	4 (AC)
Red Test Lead (+)	1 (+)		Conducting	Conducting	Conducting
	2 (AC)	Non Conducting		Conducting	Non Conducting
	3 (-)	Non Conducting	Non Conducting		Non Conducting
	4 (AC)	Non Conducting	Non Conducting	Conducting	
	Resistance Readings				

Fig. CL311 – Bridge Diode Test Chart using Analog VOM.

CIRCUIT BREAKER

To test the circuit breaker on units so equipped:
1. Ensure the breaker is set. If tripped, reset.
2. Remove the wires from the rear of the breaker.
3. Using the R × 1 scale on the VOM, test for continuity across the breaker terminals.
4. If no continuity, or if the breaker fails to reset, the breaker is faulty and must be replaced.

BATTERY CHARGER

1. To test the charger windings in the stator on units so equipped, refer to STATOR section; use *Resistance Chart* and test the orange wires.
2. To test the diodes in the charger circuits on units so equipped, refer to the *ROTOR Diode Test* section.
3. To test bridge-diode rectifiers, refer to Figs. CL310 and CL 311 *Bridge Diode Test* Charts.

> **NOTE: For accurate readings, always zero the ohm-meter before testing.**

GENERATOR REASSEMBLY

All Except 42 Series

Refer to respective Figs. CL301, CL303, or CL304 for component identification:
1. If the generator is mounted in a cradle-style frame, place a support under the engine adapter.
2. Ensure that the matching tapers on the engine crank-shaft and inside the rotor shaft are clean and dry. Install the rotor onto the crankshaft. Install the rotor bolt and torque it to 180 in.-lb. (205 N•m).

> **CAUTION: The plastic rotor fan is part of the rotor assembly and cannot be replaced separately. Handle the rotor carefully to prevent fan damage.**

3. Recalling stator wire and connector orientation, align the locator pins and mount the stator tightly to the engine adapter. Using a soft mallet may help seat the stator snugly against the adapter.
4. On 40 Series, install the stator end-mount bracket.
5. On 40-44 Series, install the sheet metal shields and top handle.

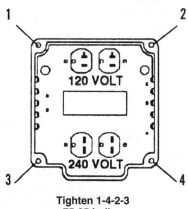

**Tighten 1-4-2-3
75-95 in-lbs.**

Fig. CL312 – Typical torque sequence and values for endbell reassembly.

6. Connect the endbell wiring connector blocks and install the endbell (endbell/outlet panel assembly on 40 Series).

> **CAUTION: When installing the four long endbell bolts, always ensure that flat washers are positioned between the bolt head and the endbell to prevent crushing the endbell flanges. Normally, for proper external grounding, the ground lug (12 – Fig. CL303) mounts on the #4 bolt position as shown in Fig. CL312. When installing the #4 bolt with the ground lug, place the star washer against the head of the bolt, followed by the ground lug, then the flat washer. Another star washer should be installed between the engine adapter flange and the Nylok nut.**

7. Torque the endbell bolts and nuts as shown in Fig. CL312.
8. Reinstall any stator support brackets and/or rubber mounts that were removed during disassembly.
9. Remove the temporary adapter support installed in Step 1.

42 Series

Refer to Fig. CL302 for component identification.
1. Fasten the stator (5) to the main frame (26) using the four stator-mount bolts (12).
2. Correctly position the engine spacers (25) onto the main frame. Position the engine (1) onto the spacers, aligning the engine-mount holes with the spacer holes. Tighten the mount bolts making sure that the three engine-mount bolts (15) have lockwashers against the bolt heads and flat washers under their respective Nylok nuts as necessary.
3. Ensure that matching tapers on the engine crankshaft and inside rotor (2) are clean and dry. Install the rotor onto the crankshaft. Install the rotor bolt assembly (3) and tighten the bolt to 180 in.-lb. (205 N•m).

> **NOTE: After tightening the rotor, remove the spark plug from the engine, then rotate the engine crank-shaft to check for stator-to-rotor clearance. If there is interference (rubbing or scraping), loosen and reposition the engine and/or the stator as necessary to obtain interference-free rotation.**

4. Reconnect the endbell-to-stator wiring connector (6, 7) and position the endbell (10) against the stator, making sure the endbell and stator wires are not pinched. Install the three endbell-to-mainframe mount bolts (11) and tighten securely.

ENGINE

Refer to the appropriate *ENGINE* section in this manual for engine service.

WIRING DIAGRAMS

1. Refer to Fig. CL313 for Series 40 wiring diagram.
2. Refer to Fig. CL314 for Series 42 wiring diagram.
3. Refer to Fig. CL315 for Series 44 wiring diagram.
4. Refer to Fig. CL316 for Series 52 wiring diagram.
5. Refer to Fig. CL317 for Series 53 wiring diagram for units with 120V Twistlok receptacle.
6. Refer to Fig. CL318 for Series 53 wiring diagram for units with 12V DC battery charger option.
7. Refer to Fig. CL319 for Series 54 wiring diagram.

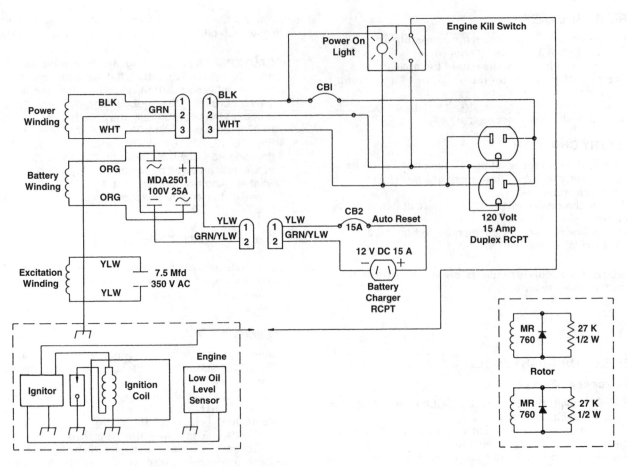

Fig. CL313 – *Wiring Schematic for Series 40 brushless generators, showing optional oil-level sensor.*

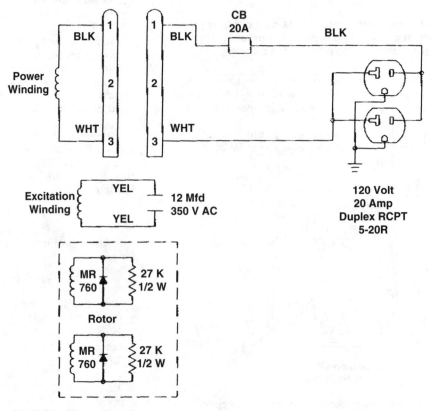

Fig. CL314 – *Wiring Schematic for Series 42 brushless generators.*

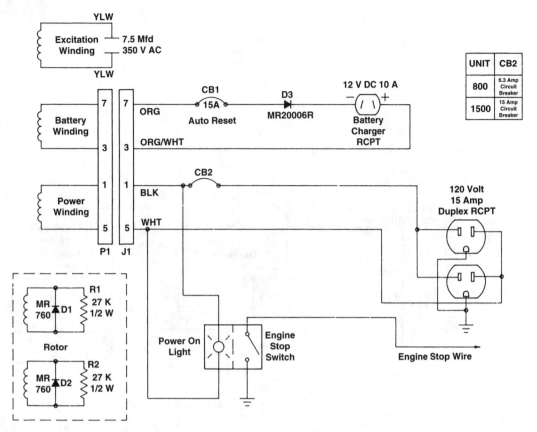

Fig. CL315 – Wiring Schematic for 500-1500W Series 44 brushless generators.

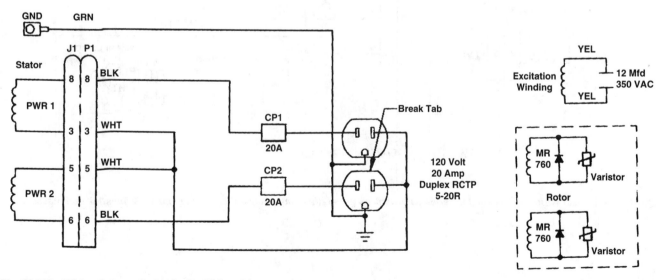

Fig. CL316 – Wiring Schematic for Series 52 brushless generators.

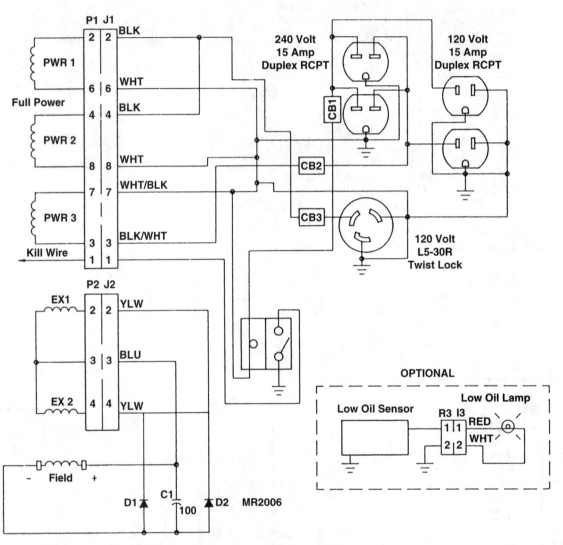

Fig. CL317 – *Typical Wiring Schematic for Series 53 brushless generators with 120/240V capability and 120V Twistlok receptacle.*

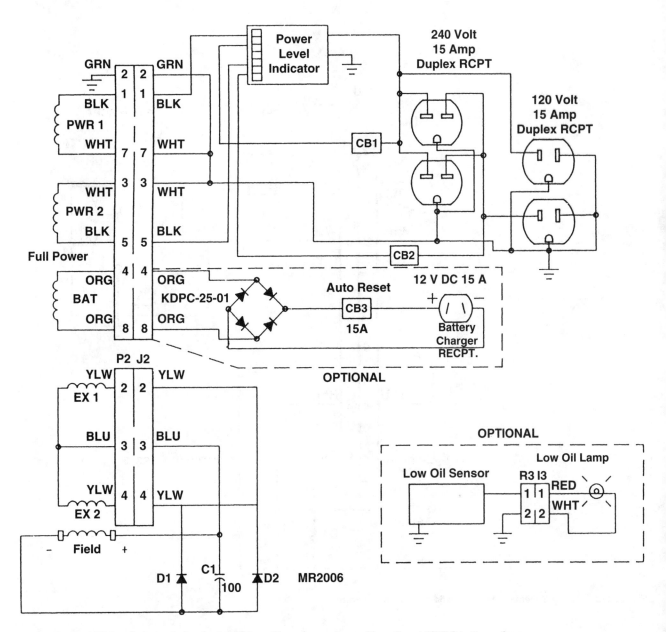

Fig. CL318 – Typical Wiring Schematic for Series 53 brushless generators with optional 12VDC battery charger.

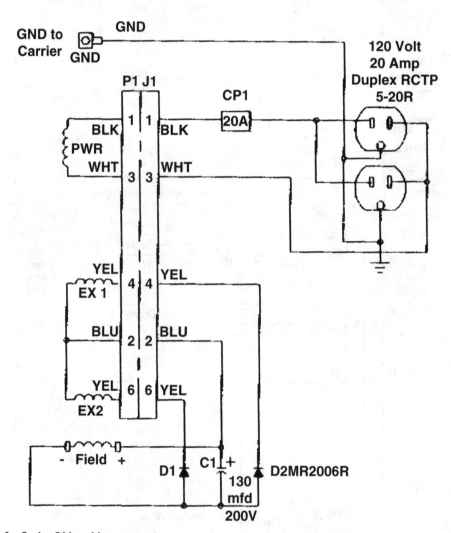

Fig. CL319 – *Wiring Schematic for Series 54 brushless generators.*

COLEMAN

BRUSH-STYLE GENERATORS

HORIZONTAL-ROTOR
1600-8000 WATT Series 34, 44, 45, 47, 48, 50, 52, 53, 54, 55 COMMERCIAL, and 56

IDENTIFICATION

Coleman brush-style generators can be identified by removing the snap fit rectangular cover on the generator endbell (6- Fig. CL401) and inspecting the area under the cover. If two screw mounted terminals are visible, one on each side of the rotor bearing in the center of the rectangle, those are the brush connectors, and it is a brush-type generator. If the only component visible under the cover is the rotor bearing, it is a brushless generator.

Refer to the following list to locate the exploded-view figure of the brush-style generator being serviced:

Series	Wattage	Figure
34	All	CL401
44	All	CL402
45	4-5KW	CL403
45	7KW	CL404
47	3-5KW	CL405
47	8KW	CL406
48	All	CL407
50	2-3KW	CL408
50	4-5KW	CL409
		or
		CL410
50	8KW	CL406
52	2-5KW	CL411
		or
		CL412
53	4-5KW	CL409
		or
		CL410
53	4KW "I/C"	CL413
54	All	CL412
		CL414
		or
		CL415
55	COMMERCIAL	CL416
56	3500W	CL417
56	4-5KW	CL409
		or
		CL410

WIRING COLOR CODES

All portable generators manufactured by Coleman Powermate, Inc. use the following standardized wire color-coding:

Power Windings	B–Black
	B/W–Black with white tracer stripe
	R–Red
Excitation	Y–Yellow (diodes)
	Blu–Blue (capacitor)
Ground	G–Green
Neutral	W–White
	W/B–White with black tracer stripe
Battery Charger	O–Orange
Idle Control	O–Orange

MAINTENANCE

The endbell bearing and race should be inspected, cleaned, and relubricated any time the endbell is removed. If the bearing feels hot to the touch, replacement may be necessary.

Brushes must be at least ¼ inch long. Shorter brushes require replacement. When reinstalling the brushes, do not overtighten the brush screws (see Reassembly section).

TROUBLESHOOTING

No-load voltages on 60Hz units should be 128-143 volts from the 120V receptacles and 258-278 volts from the 240V receptacles. If little or no output is generated, the following items could be potential causes:

1. Engine RPM. Engine must maintain 3600 RPM under load, or 3650-3800 RPM no-load. Engine RPM could also be set too high, causing overvoltage. Refer to the appropriate engine section for engine service.

2. Loose component-mounting fasteners (low or no voltage).

3. Circuit breakers (low or no voltage).

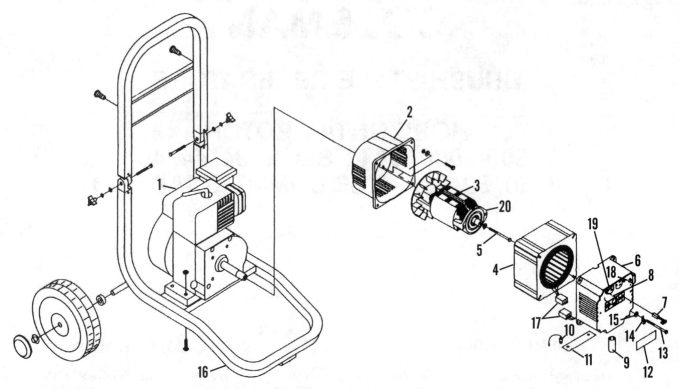

Fig. CL401 – Exploded view of VOLTSWAGON Series 34 generator

1. Engine
2. Engine adapter/front housing
3. Rotor assembly
4. Stator assembly
5. Rotor throughbolt and washer
6. Endbell/rear housing
7. Brush (2)
8. Rotor bearing
9. Capacitor
10. Diode (2)
11. Diode heat sink plate
12. Brush and bearing cover
13. Endbell mount/stator throughbolt
14. Ground lug terminal
15. Flat washer
16. Frame
17. Endbell harness
18. AC circuit breaker
19. AC receptacle
20. Slip rings

4. Insecure wiring connections (low or no voltage).
5. Faulty receptacle (low or no voltage).
6. Capacitor (low or no voltage).
7. Diodes (low or no voltage).
8. Brushes (no voltage).
9. Rotor.
   a. Loss of residual magnetism (no voltage). NOTE: Residual magnetism can be restored without generator disassembly, and is the recommended first step if the generator has no output. Refer to RESIDUAL MAGNETISM.
   b. Open or shorted windings (no or low voltage, respectively).
   c. Dirty, broken, or disconnected slip rings (no voltage).
10. Stator windings (low voltage-shorted; no voltage-open).
11. Insufficient cooling system ventilation (overheating).

GENERATOR REPAIR

RESIDUAL MAGNETISM

If a loss of residual magnetism is suspected as the cause of no output, the rotor can be tested and remagnetized. To test for residual magnetism:

1. Remove the rectangular snap-fit brush cover from the endbell.
2. Start the generator, verifying proper no-load RPM.

3. Using a VOM set on the 250V DC scale, connect the positive (+) VOM lead to the right or top brush, and connect the negative (–) VOM lead to the left or bottom brush. Note VOM reading.
4. If residual magnetism excitation is present, the VOM should read at least the approximate values listed below for the respective wattage range of the unit:

Wattage	DC Voltage
1250	85
2000	78
3000	81
4500	95
6000	144
7000	150

5A. A low voltage reading indicates shorted rotor and/or stator windings or a faulty capacitor or diode(s). For rotor, stator, capacitor, or diode service, refer to the respective repair section.

5B. A no-voltage reading indicates either a loss of residual magnetism or an open in the excitation circuit. To test for an open in the excitation circuit, refer to the Yellow to yellow and/or Yellow to Blue resistance tests in the STATOR section.

6. To remagnetize the rotor, proceed as follows:
   a. Obtain a fully-charged DC power source (6VDC minimum, maximum 24VDC) such as an automotive battery.

b. Start the generator.

c. Connect the positive (+) DC lead to the right or top generator brush.

d. Momentarily (2-3 seconds) touch the negative (-) DC lead to the left or bottom brush.

e. Disconnect the DC leads and retest for recommended voltage or receptacle output. Repeat if necessary, up to three times. If, after the third attempt, the rotor still fails to remagnetize, proceed with further testing and repair.

DISASSEMBLY

Refer to respective Figs. CL401-CL417 for component identification.

1. On units with panel-mounted receptacles, check for generator-to-panel clearance. If necessary, disconnect the receptacle panel wiring connector, then remove the panel. On hard wired units without connectors, carefully set the receptacle panel aside.

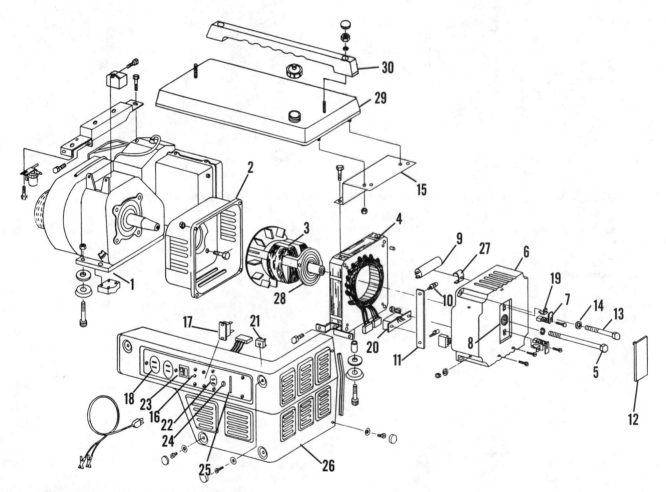

Fig. CL402 – Exploded view of PM Series 44 brush-style generator.

1. Engine
2. Engine adapter/front housing
3. Rotor assembly
4. Stator assembly
5. Rotor through-bolt
6. Endbell/rear housing
7. Brush (2)
8. Rotor bearing
9. Capacitor
10. Diode (2)
11. Diode heat sink plate
12. Brush and bearing cover
13. Endbell/stator through-bolt (4)
14. Flat washer (4)
15. Heat shield
16. AC circuit breaker
17. DC circuit breaker
18. AC receptacle
19. Brush terminal (2)
20. Voltage regulator
21. Diode bridge rectifier
22. DC receptacle
23. Rocker switch
24. Indicator lamp
25. Power-level indicator
26. Cover panel
27. Capacitor holder
28. Slip rings
29. Fuel tank
30. Handle

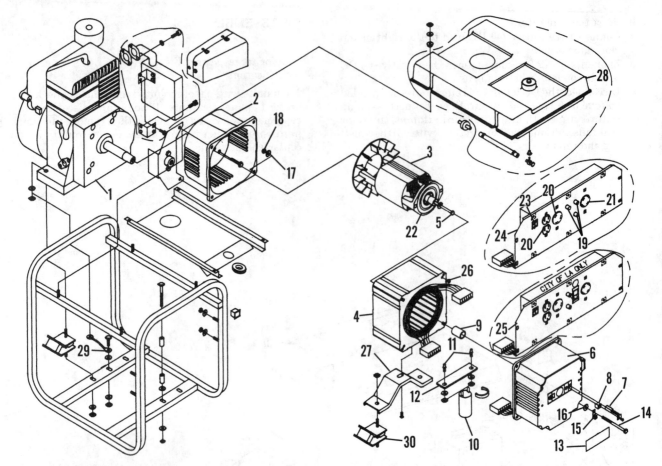

Fig. CL403 – Exploded view of typical COMMERCIAL or INDUSTRIAL Series 45 4000-5000W generator components.

1. Engine
2. Engine adapter/front housing
3. Rotor assembly
4. Stator assembly
5. Rotor through bolt and washer
6. Endbell/rear housing
7. Brush (2)
8. Brush terminal (2)
9. Rotor bearing
10. Capacitor

11. Diode (2)
12. Diode heat sink
13. Brush and bearing cover
14. Endbell/stator through-bolt (4)
15. Ground lug terminal
16. Flat washer (4)
17. Star washer
18. Nylok nut (4)
19. AC circuit breaker
20. 120V AC receptacle

21. 240V AC receptacle
22. Slip rings
23. Rocker switch
24. Control panel, all except L.A. units
25. Control panel, L.A. California units
26. Locator dowel (8)
27. Stator mount bracket
28. Fuel tank
29. Ground wire
30. Isolator

2. Remove the rectangular snap-fit brush cover from the endbell. Remove the brush screws and brushes. Brushes worn to ¼ inch or less in length require replacement.

3. Remove the four long bolts holding the endbell and stator to the engine adapter. Remove the endbell and disconnect the wiring connector blocks.

NOTE: With the endbell removed, the diodes, capacitor, rotor and stator can be tested without further disassembly.

4. Place a support under the engine adapter, then remove any stator support brackets, mounts, and fasteners.

5. Remove the stator by prying between stator and engine adapter with a flat blade pry tool. Note the orientation of the stator wires and connectors.

NOTE: The rotor cannot be removed without first removing the stator.

CAUTION: The plastic rotor fan is part of the rotor assembly and cannot be replaced separately. Handle the rotor carefully to prevent fan damage.

6. Loosen the rotor through-bolt, then apply penetrating oil through the rotor, allowing it to flow onto the crankshaft taper.

7. Remove the rotor using a suitable puller (Fig. CL418). Place the puller feet under the bottom of the laminations, then place the puller center bolt on the head of the rotor through bolt.

NOTE: Use extreme care to prevent the puller legs and feet from contacting the rotor windings to prevent damage to the winding insulation.

An alternate rotor removal method is to remove the rotor through-bolt. After squirting penetrating oil through the rotor onto the crankshaft taper, firmly but carefully tap the end bearing race with a soft-tip rubber mallet (not a metal hammer) while holding the rotor to prevent it from falling.

CAPACITOR TESTING

The capacitor can be tested without removing it from the endbell, provided the brushes are removed and the stator connector plug is disconnected.

NOTE: Even with the generator stopped, the capacitor should retain a charge if it is working properly. Do not touch the capacitor terminals, as electric shock may result. Always discharge the capacitor by shorting across the terminals with a screwdriver or similar tool with an insulated handle.

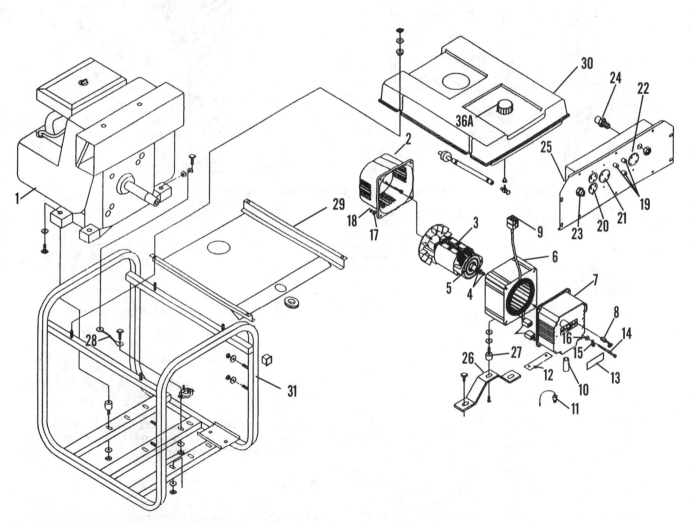

Fig. CL404 – Exploded view of typical COMMERCIAL Series 45 7000W generator components.

1. Engine
2. Engine adapter/housing
3. Rotor assembly
4. Rotor throughbolt
5. Sliprings
6. Stator assembly
7. Endbell/rear housing
8. Brush (2)
9. Rotor bearing
10. Capacitor

11. Diode (2)
12. Diode heat sink
13. Brush and bearing cover
14. Endbell/stator throughbolt
15. Ground lug terminal
16. Flat washer
17. Star washer
18. Nylok nut
19. AC circuit breaker
20. 120V AC receptacle

21. 120V Twistlok receptacle
22. 240V Twistlok receptacle
23. Engine ignition switch
24. Strain relief
25. Control panel assembly
26. Stator mount bracket
27. Stator isolator
28. Ground wire
29. Heat shield
30. Fuel tank
31. Generator frame

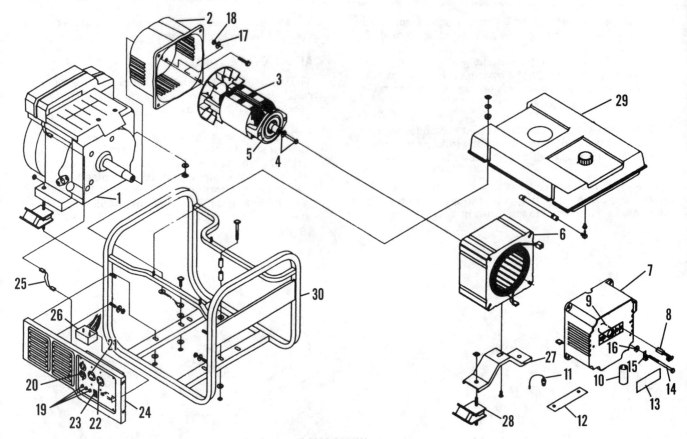

Fig. CL405 – Exploded view of typical VANTAGE Series 47 3000-5000W generator components.

1. Engine	11. Diode (2)	21. 120V AC Twistlok receptacle
2. Engine adapter/front housing	12. Diode heat sink	22. 240V AC Twistlok receptacle
3. Rotor assembly	13. Brush and bearing cover	23. Switch
4. Rotor throughbolt	14. Endbell/stator throughbolt	24. Control panel
5. Slip rings	15. Ground lug terminal	25. Engine harness
6. Stator assembly	16. Flat washer	26. Idle control module
7. Endbell/rear housing	17. Star washer	27. Stator mount bracket
8. Brush (2)	18. Nylok nut	28. Stator isolator
9. Rotor bearing	19. AC circuit breaker	29. Fuel tank
10. Capacitor	20. 120V AC receptacle	30. Generator frame

To test capacitor:

1. Set the VOM ohmmeter scale to R × 10,000.

2A. On spade terminal capacitors, disconnect the harness wires from the capacitor, noting polarity orientation, then touch the VOM leads to the capacitor terminals. The positive (+) VOM lead (usually red in color) should always connect to the positive (+) capacitor terminal.

2B. On solder-terminal capacitors, connect the positive (+) VOM lead (usually red in color) to the #4 terminal of the endbell connector block as shown in Fig. CL419. Connect the negative (-) VOM lead to any bare metal ground in the endbell.

3. The meter should first indicate low resistance, gradually increasing resistance toward infinity. No meter reading or constant continuity indicates a faulty capacitor; replace the capacitor.

4. On solder-terminal capacitors, a failed test could be the result of faulty solder joints. Repeat the test, holding the VOM leads directly against the capacitor terminals, using caution to observe proper polarity. If the test is now successful, one or both of the terminal-to-capacitor wires may

be faulty or the capacitor-pigtail solder joint is defective; correct as necessary.

NOTE: If replacing the capacitor, note the capacitor polarity prior to removal; correct polarity must be observed upon reinstallation.

DIODES

The diodes in brush-style generators can be initially tested without removing them from the endbell or disconnecting any wires, provided the endbell connector is disconnected from the stator. To test either diode:

1. Set the VOM to the Rx1 scale.

2. Connect the positive (+) VOM lead to one of the diode endbell connector terminals. Positions 6 and 8 in Fig. CL419 are diode terminals.

3. Connect the negative (-) VOM lead to the base of the diode or to any bare metal (ground) section of the endbell.

4. Observe the VOM reading.

5. Reverse the VOM leads and observe the reading again. A good diode will show a reading in only one direction of the

two tests. If a reading is observed in both directions, the diode is faulty and should be replaced.

6. If no reading is observed in either direction, trace the yellow wire from the non-reading terminal to the diode pigtail. Repeat the tests, holding one VOM lead directly against the tip of the pigtail.

a If there is still no VOM reading in either direction, diode is faulty and should be replaced.

b If there is now a VOM reading in one direction, either the terminal-to-diode wire is faulty or the diode-pigtail solder joint is defective. Correct as necessary.

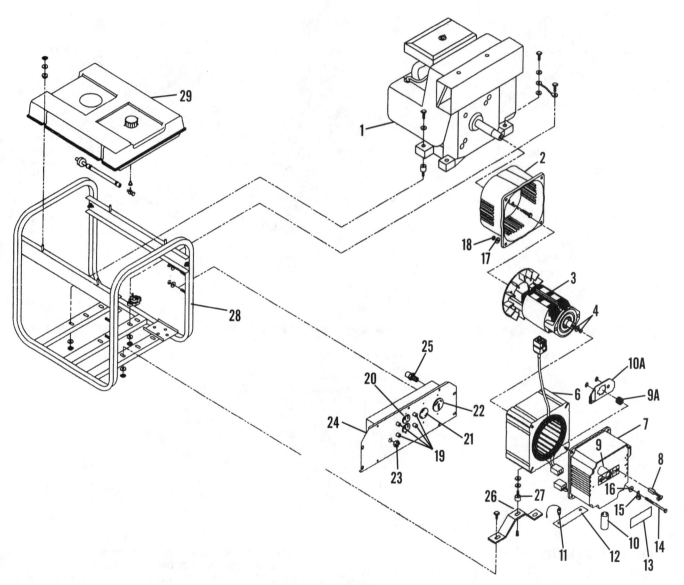

Fig. CL406 – Exploded view of typical VANTAGE Series 47 and POWERHOUSE Series 50 8000W generator components. Series 47 uses a side-mounted receptacle panel; Series 50 uses an end-mounted receptacle panel.

1. Engine
2. Engine adapter/front housing
3. Rotor assembly
4. Rotor throughbolt
5. Slip rings
6. Stator assembly
7. Endbell/rear housing
8. Brush (2)
9. Rotor bearing
9A. Tolerance ring
10. Capacitor

10A. Capacitor shield w/retainers
11. Diode (2)
12. Diode heat sink plate
13. Brush and bearing cover
14. Endbell/stator throughbolt
15. Ground lug terminal
16. Flat washer
17. Star washer
18. Nylok nut
19. AC circuit breaker
20. 120V AC receptacle

21. 120V AC Twistlok receptacle
22. 240V receptacle
23. Engine ignition switch
24. End-mount control panel
25. Strain relief
26. Stator mount bracket
27. Stator isolator
28. Frame
29. Fuel tank

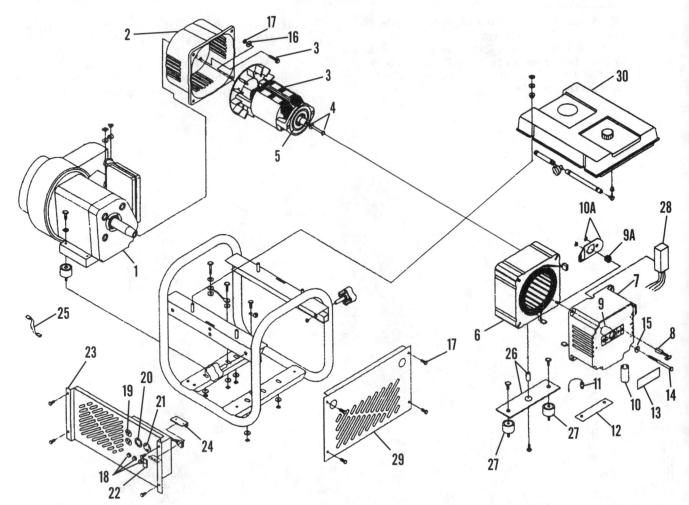

Fig. CL407 – Exploded view of typical CONTRACTOR Series 48 generator components. Idle Control (41) is only used on 6000W units.

1. Engine
2. Engine adapter/front housing
3. Rotor assembly
4. Rotor throughbolt
5. Slip rings
6. Stator assembly
7. Endbell/rear housing
8. Brush (2)
9. Rotor bearing
9A. Tolerance ring
10. Capacitor

10A. Capacitor shield w/retainers
11. Diode (2)
12. Diode heat sink plate
13. Brush and bearing cover
14. Endbell/stator throughbolt
15. Flat washer
16. Star washer
17. Nylok nut
18. AC circuit breakers
19. 120V AC receptacle
20. 120V AC Twistlok receptacle

21. 240V AC Twistlok receptacle
22. Engine switch
23. Control panel
24. Idle-control module
25. Engine harness
26. Stator mount plate and spacer
27. Stator isolators (2)
28. Idle-down control unit
29. End panel
30. Fuel tank

CIRCUIT BREAKERS

Individual circuit breakers are provided for each 120V circuit (a two-plug receptacle is one "circuit"). On most 120/240V units, the circuit breaker protecting a 120V circuit also protects one leg of a 240V receptacle. Units with a 120V Twistlok receptacle have a separate 30A breaker for that circuit.

To test the circuit breaker:

1. Ensure the breaker is set. Reset the breaker if tripped.
2. Remove the wires from the rear of the breaker.
3. Using the R × 1 scale on the VOM, test for continuity across the breaker terminals.
4. If the VOM reads no continuity, or if the breaker fails to reset, the breaker is faulty and must be replaced.

NOTE: If a two-plug 120V receptacle has been replaced and the circuit breaker trips with no load connected to the unit, check to make sure that the tab is broken off between the lead connecting screws on the hot (brass) side of the receptacle.

ROTOR

Rotor Slip Rings

The slip rings are an integral part of the rotor assembly and they cannot be replaced.

Dirty slip rings should be cleaned with a suitable solvent, then lightly sanded with 240-grit sandpaper.

Rough slip rings should be lightly sanded with 240-grit sandpaper.

NOTE: Do not use crocus cloth to sand slip rings.

Cracked, chipped, broken, or deeply-grooved slip rings are not repairable; in these cases, the rotor must be replaced.

Rotor Winding Resistance

The rotor can be tested without removing it from the generator. If hypot testing, the stator should be removed to access the laminations.

To test resistance, use a VOM set on the R × 1 scale. If no reading is observed upon initial testing, inspect the winding to slip ring connection and solder joint. If the solder joint is faulty, and if there is sufficient wire length, resolder the joint and retest.

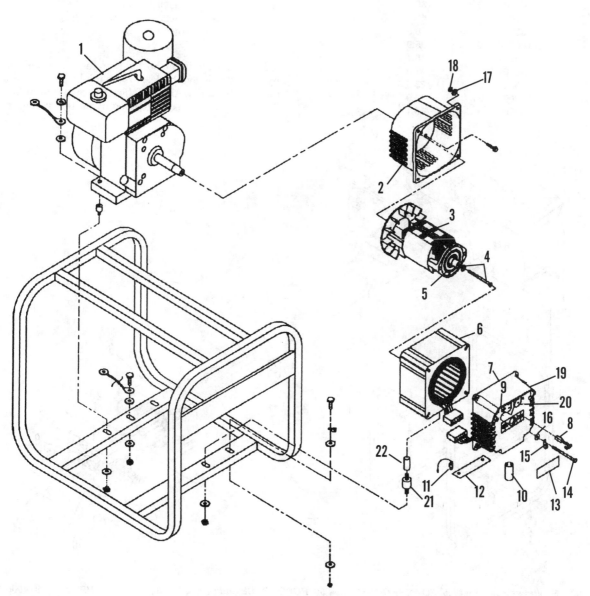

Fig. CL408 – Exploded view of typical POWERHOUSE Series 50 2000-3000W generator components.

1. Engine
2. Engine adapter/front housing
3. Rotor assembly
4. Rotor through bolt and washer
5. Sliprings
6. Stator assembly
7. Endbell/rear housing
8. Brush (2)
9. Rotor bearing
10. Capacitor
11. Diode (2)
12. Diode heat sink plate
13. Brush and bearing cover
14. Endbell/stator through-bolt (4)
15. Ground lug terminal
16. Flat washer (4)
17. Star washer
18. Nylok nut (4)
19. AC circuit breaker
20. 120V AC receptacle
21. Stator isolator
22. Isolator spacer

NOTE: Always reapply high temperature RTV silicone sealer to the inner slip ring solder joint to prevent shorting the wire against the rotor shaft.

NOTE: All resistance readings are approximate and may vary depending on the winding temperature and the accuracy of the test equipment.

Refer to the following chart for the rotor resistance readings:

Series	Wattage	Min. Ohms	Max. Ohms
44	1250-2800	24.0	24.9
45	4000-5000	35.9	36.7
	7000	46.1	48.5
47	3500	28.5	29.9
	4600-5000	35.9	36.7
	7000	46.1	48.5
	8000	47.0	47.4
48	3500	28.5	29.9
	5000	35.9	36.7
	6000	46.1	48.5
50	2500	26.3	27.5
	3000	26.3	27.5
	4000	30.5	34.9
	4500	33.5	34.9
	5000	35.9	36.7
	8000	47.0	47.4
52 & 54	2000-2500	24.0	24.9
	4000	30.0	31.0
	5000	35.9	36.7
	7000	46.1	48.5
53	4000	30.5	34.9
	5000	35.9	36.7
55 COMMERCIAL	2400	26.3	27.5
	5000	35.9	36.7
56	3500	28.5	29.9
	4000	30.5	34.9
	4600	33.5	34.9

Hypot testing is the best way to test rotor windings, if a hypot tester is available. A rotor might pass VOM testing and still be faulty. Always follow the tester manufacturer's instructions. The recommended hypot test voltage is 1250

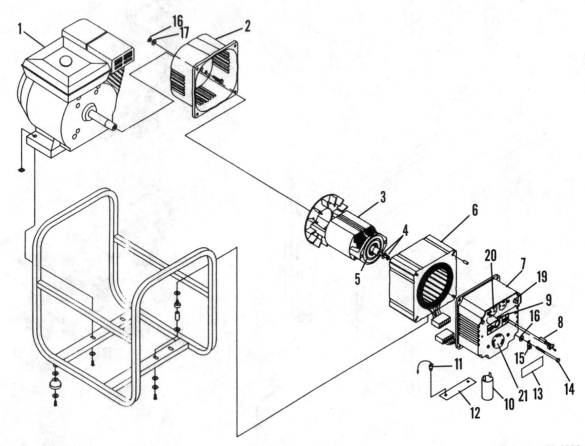

Fig. CL409 – Exploded view of typical *POWERHOUSE Series 50, POWERBASE and PRO-GEN Series 53, and HP Series 56 4000-5000W* generators using endbell-mounted receptacles.

1. Engine
2. Engine adapter/front housing
3. Rotor assembly
4. Rotor bolt and washer
5. Slip rings
6. Stator assembly
7. Endbell/rear housing
8. Brush (2)
9. Rotor bearing
10. Capacitor
11. Diode (2)
12. Diode heat sink plate
13. Brush and bearing cover
14. Endbell/stator through-bolt (4)
15. Ground lug terminal
16. Flat washer (4)
17. Star washer
18. Nylok nut (4)
19. AC circuit breaker
20. 120V AC receptacle
21. 120V AC Twistlok receptacle
22. Stator isolator
23. Isolator spacer

volts; less voltage may not show faults, and higher voltage may damage the rotor. To test the rotor, hypot from each slipring to the lamination (ground). The failure of either test indicates a faulty rotor.

STATOR WINDING RESISTANCE

To test the stator windings:
1. Set the VOM to the R × 1 scale.
2. Read the resistance between the stator connector terminals.

3. Compare the readings to the values shown in the winding resistance chart. If no reading is obtained upon each initial test, check the connector block and stator leads for broken wires or faulty terminals. Repair as necessary, then retest.

NOTE: All resistance readings are approximate and may vary depending on the winding temperature and accuracy of the test equipment.

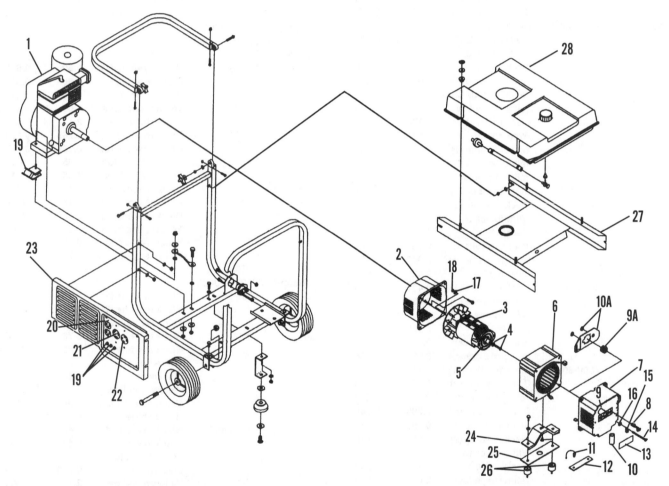

Fig. CL410 – Exploded view of typical POWERHOUSE and MAGNA FORCE Series 50, POWERBASE and PRO-GEN Series 53, and HP Series 56 4000-5000W generators using panel-mounted receptacles. MAGNA FORCE and some PRO-GEN Series 50 5000W generators use an end-mount receptacle panel; others use a side mount panel. Automatic Voltage Regulator is only used on PRO-GEN units with PC model-number prefix.

1. Engine
2. Engine adapter/front housing
3. Rotor assembly
4. Rotor bolt and washer
5. Slip rings
6. Stator assembly
7. Endbell/rear housing
8. Brush (2)
9. Rotor bearing
9A. Tolerance ring
10. Capacitor

10A. Capacitor shield w/retainers
11. Diode (2)
12. Diode heat sink plate
13. Brush and bearing cover
14. Endbell/stator throughbolt
15. Ground lug terminal
16. Flat washer
17. Star washer
18. Nylok nut
19. AC circuit breakers
20. 120V AC receptacle

21. 120V AC Twistlok receptacle
22. 240V AC Twistlok receptacle
23. Side-mount control panel
24. Stator mount bracket
25. Mount reinforcement, POWERHOUSE models
26. Stator isolators
27. Heat shield
28. Fuel tank

STATOR WINDING RESISTANCE

Series	Wattage	Field	Min. Ohms	Max. Ohms
44	1250 & 2000	Black to white	0.47	0.49
		Yellow to blue (1)	1.4	1.5
		Yellow to blue (2)	1.4	1.5
		Yellow to yellow	2.8	3.0
	1600	Black to white (1)	0.835	0.855
		Black to white (2)	0.835	0.855
		Yellow to blue (1)	2.2	2.3
		Yellow to blue (2)	2.2	2.3
		Yellow to yellow	4.4	4.6
		Orange to orange	0.125	0.130
	2800	Black to white (1)	0.47	0.49
		Black to white (2)	0.47	0.49
		Yellow to blue (1)	1.4	1.5
		Yellow to blue (2)	1.4	1.5
		Yellow to yellow	2.8	3.0
		Orange to orange	0.090	0.095
45	4000-5000	Black to white (1)	0.415	0.435
		Black to white (2)	0.415	0.435
		White/black to black/white	0.265	0.285
		Yellow to blue (1)	1.05	1.15
		Yellow to blue (2)	1.05	1.15
		Yellow to yellow	2.10	2.30
	7000	Black to green	0.170	0.190
		White to red	0.170	0.190
		Yellow to blue (1)	1.0	1.1
		Yellow to blue (2)	1.0	1.1
		Yellow to yellow	2.0	2.2
47	3500	Black to white (1)	0.73	0.75
		Black to white (2)	0.73	0.75
		White/black to black/white	1.1	1.25
		Yellow to yellow	2.03	2.23
		Orange to orange	0.086	0.106
		Blue to blue	0.42	0.46
	4600-5000	Black to white (1)	0.435	0.455
		Black to white (2)	0.435	0.455
		White/black to black/white	0.265	0.285
		Yellow to yellow	2.1	2.3
		Orange to orange	0.050	0.055
47	7000-8000	Black to White (1)	0.17	0.19
		Black to white (2)	0.17	0.19
		Yellow to yellow	2.0	2.2
		Orange to orange	0.285	0.340
48	3500	Black to white (1)	0.698	0.718
		Black to white (2)	0.698	0.718
		White/black to black/white	0.578	0.588
		Yellow to yellow	2.03	2.23
		Orange to orange	0.395	0.435
	5000	Black to white (1)	0.435	0.455
		Black to white (2)	0.435	0.455
		White/black to black/white	0.265	0.285
		Yellow to yellow	2.1	2.3
		Orange to orange	0.050	0.055
	6000	Black to white (1)	0.17	0.19
		Black to white (2)	0.17	0.19
		Yellow to yellow	2.0	2.2
		Orange to orange	0.285	0.340
50, 52, 53, 54, &	2000-3000	Black to white	0.35	0.37
55 COMMERCIAL		Yellow (1) to blue	1.25	1.35
		Yellow (2) to blue	1.25	1.25
		Yellow to yellow	2.5	2.7
50	4000-4500	Black to white (1)	0.46	0.48
		Black to white (2)	0.46	0.48
		White/black to black/white	0.295	0.315
		Yellow to yellow	2.1	2.3
		Orange to orange	0.07	0.08
	5000	Black to white (1)	0.435	0.455
		Black to white (2)	0.435	0.455
		White/black to black/white	0.265	0.285
		Yellow to yellow	2.1	2.3
	8000	Black to white (1)	0.147	0.167
		Black to white (2)	0.147	0.167
		Yellow to yellow	2.08	2.28
		Yellow to blue (1)	1.04	1.14
		Yellow to blue (2)	1.04	1.14
52-54	2000-2500 (Refer to 50: 2000-3000)			
52, 53, & 54	4000	Black to white (1)	0.505	0.525
		Black to white (2)	0.505	0.525
		Yellow (1) to blue	1.05	1.15
		Yellow (2) to blue	1.05	1.15
		Yellow to yellow	2.1	2.3
	5000	Black to white (1)	0.265	0.285
		Black to white (2)	0.265	0.285
		Yellow (1) to blue	1.05	1.15
		Yellow (2) to blue	1.05	1.15
		Yellow to yellow	2.1	2.3
	7000	Black to white (1)	0.16	0.18
		Black to white (2)	0.16	0.18
		Yellow (1) to blue	0.85	0.95
		Yellow (2) to blue	0.85	0.95
		Yellow to yellow	1.7	1.9
55	2400 (Refer to 50: 2000-3000)			
55	5000	Black to white (1)	0.265	0.285
		Black to white (2)	0.265	0.285
		Yellow to yellow	2.1	2.3
56	3500	Black to white (1)	0.73	0.75
		Black to white (2)	0.73	0.75
		White/black to black/white	1.10	1.25
		Yellow to yellow	2.03	2.23
		Orange to orange	0.086	0.186
		Blue to blue	0.42	0.46
	4000-4600	Black to white (1)	0.46	0.48
		Black to white (2)	0.46	0.48
		White/black to black/white	0.295	0.315
		Yellow to yellow	2.1	2.3
		Orange to orange	0.07	0.08

IDLE CONTROL

Operation

The idle control allows the engine to idle down when there is no load on the generator. A rocker switch located on the receptacle panel activates the idle control. With the switch on, the generator will not produce output-load voltage.

NOTE: If less than a 250W load is applied at 120V (500W at 240V), the idle control must be in the off position so the unit will run at 3600 RPM to produce

proper voltage. **The idle control will not detect loads below these amounts.**

Troubleshooting

If the unit does not idle with the idle control switch on, perform the following tests and checks.

With the generator shut down, make sure the electromagnetic flapper on the throttle moves freely.

With the generator running, check the following:

1. Check to make sure the generator is producing its rated 120V or 120/240V. Refer to the TROUBLESHOOTING and GENERATOR REPAIR sections as necessary.

2. Make sure the load is properly plugged into the receptacle and that the plug-in cord or the load itself is not faulty.

3. Check the stator output to the idle control module. The module requires 20VAC to operate. The stator-to-module output can be checked as follows:

a. Turn the Idle Control switch off.

b. Unplug the two orange wire harness from the back of the module connection on the control panel.

c. Using a VOM set on the 50VAC scale, test the two orange wire terminals.

d. If the test is successful, turn the switch back on.

4. Check the idle control module output. The idle control magnet on the engine throttle requires 20-24 V DC to operate. To test the module output, disconnect the B/W wire harness from the back of the throttle electromagnet, then use a VOM set on the 25VDC scale. The black wire terminal is positive (+).

5. Check the throttle idle control magnet. With the idle control switch OFF, loosely hold the handle of a flat blade screwdriver so that the blade is approximately ¼ inch (6 mm) away from the electromagnet tip where it pulls the throttle shaft flapper. Turn the switch ON. The magnet should attract the screwdriver blade.

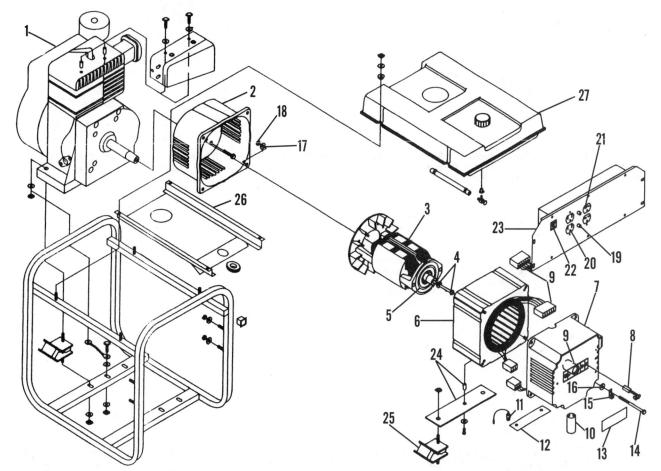

Fig. CL411 – Exploded view of typical MAXA Series 52 2000-5000W generators with end-panel mounted receptacles. Series 52 2000W and 3000W units have only one 120V receptacle (40C) and breaker (40A), and have fewer wires in receptacle connector (9).

1. Engine	10. Capacitor	19. AC circuit breakers
2. Engine adapter/front housing	11. Diode (2)	20. 120V AC receptacles
3. Rotor assembly	12. Diode heat sink plate	21. 240V AC receptacles
4. Rotor through-bolt and washer	13. Brush and bearing cover	22. Lighted rocker switch
5. Sliprings	14. Endbell/stator through-bolt (4)	23. Control panel
6. Stator assembly	15. Ground lug terminal	24. Stator mount plate and spacer
7. Endbell/rear housing	16. Flat washer (4)	25. Stator isolator (2)
8. Brush (2)	17. Star washer	26. Heat shield
9. Rotor bearing	18. Nylok nut (4)	27. Fuel tank

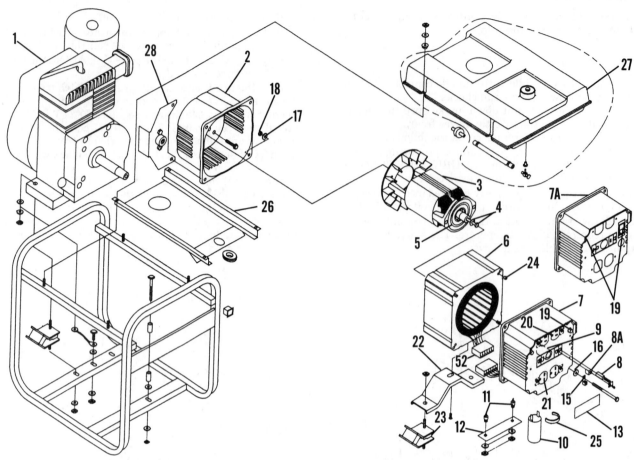

Fig. CL412 – Exploded view of typical MAXA Series 52 2000-5000W and POWERBASE Series 54 4000-5000W generators with endbell-mounted receptacles. Units with PL model-number prefix use rocker-style circuit breakers (56, 57). Series 52 units do not use the pyramid-style stator mount bracket (34).

1. Engine	9. Rotor bearing	19. AC circuit breakers
2. Engine adapter/front housing	10. Capacitor	20. 120V AC receptacle
3. Rotor assembly	11. Diode (2)	21. 240V AC receptacle
4. Rotor bolt and washer	12. Diode heat sink plate	22. Stator mount bracket
5. Sliprings	13. Brush and bearing cover	23. Stator isolator (2)
6. Stator assembly	14. Endbell/stator through-bolt (4)	24. Locator dowel (8)
7. Endbell/rear housing, except L.A. units	15. Ground lug terminal	25. Clamp
7A. Endbell/rear housing, for L.A., California units	16. Flat washer (4)	26. Heat shield
8. Brush (2)	17. Star washer	27. Fuel tank
8A. Brush terminal (2)	18. Nylok nut (4)	28. Electric start switch

6. Make sure the throttle electromagnet is correctly positioned. Adjust as necessary for 1800 RPM idle speed.

7. Make sure the idle mixture and idle speed adjustments on the carburetor are properly calibrated. Refer to the appropriate ENGINE SERVICE section in this manual.

GENERATOR REASSEMBLY

Refer to Figs. CL401-CL417 for component identification.

1. If the engine is mounted in a cradle-style frame, place a support under the engine adapter.

2. Ensure the matching tapers on the engine crankshaft and inside the rotor shaft are clean and dry. Install the rotor onto the crankshaft. Then install the rotor bolt and torque it to 180 in.-lb. (205 N•m).

CAUTION: The plastic rotor fan is part of the rotor assembly and cannot be replaced separately. Handle the rotor carefully to prevent fan damage.

3. Recall the stator wire and connector orientation. Align the locator pins, then mount the stator tightly to the en-

gine adapter. Using a soft mallet may help fit the stator snugly against the adapter.

4. Connect the endbell wiring connector blocks and install the endbell.

CAUTION: Ensure that the wires between the stator and the endbell are not pinched.

CAUTION: When installing the four long endbell bolts, always ensure that flat washers are positioned between the bolt heads and the endbell to prevent crushing the endbell flanges. For proper external grounding, the ground lug (12 – Fig. CL401) mounts on the #4 bolt position (Fig. CL420). When installing the #4 bolt with the ground lug, place the star washer against the head of the bolt, followed by the ground lug, then the flat washer. Install another star washer between the engine adapter flange and the Nylok nut.

5. Torque the endbell bolts and nuts as shown in Fig. CL420.

6. Reinstall any stator support brackets and/or rubber mounts removed during disassembly.

7. Remove the temporary adapter support installed in Step 1.

8. Install the brushes and the brush screws, then install the snap fit brush cover.

CAUTION: The brush screws thread directly into the plastic endbell. Install the brush screws carefully until the screw head bottoms, then twist an additional 1/8 turn only. Overtightening the brush screws will cause stripping of the endbell threads, thereby requiring replacement of the endbell.

9. If the generator has panel-mounted receptacles and the panel was removed for service access, reinstall and reconnect the panel.

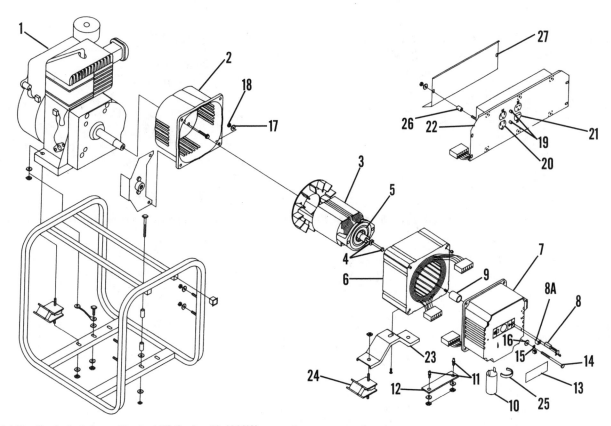

Fig. CL413 – Exploded view of typical I/C Series 53 4000W generator components.

1. Engine
2. Engine adapter/front housing
3. Rotor assembly
4. Rotor through bolt and washer
5. Sliprings
6. Stator assembly
7. Endbell/rear housing
8. Brush (2)
8A. Brush terminal (2)
9. Rotor bearing

10. Capacitor
11. Diode (2)
12. Diode heat sink plate
13. Brush and bearing cover
14. Endbell/stator through-bolt (4)
15. Ground lug terminal
16. Flat washer (4)
17. Star washer
18. Nylok nut (4)
19. AC circuit breakers

20. 120V AC receptacle
21. 240V AC receptacle
22. Control panel
23. Stator mount bracket
24. Stator isolator (2)
25. Clamp
26. Spacer
27. Heat shield

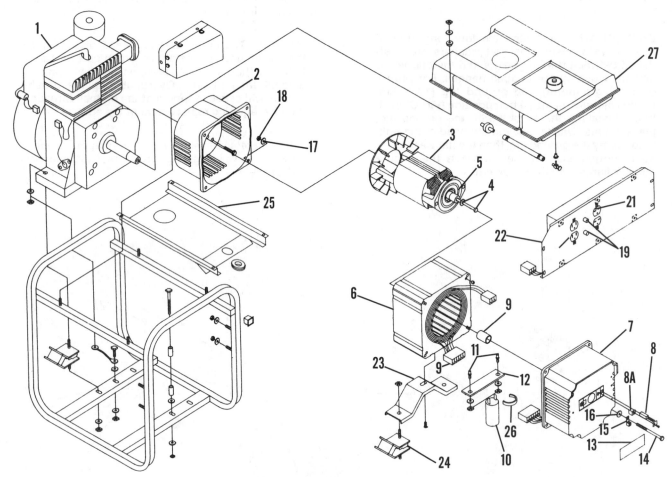

Fig. CL414 – Exploded view of typical POWERBASE Series 54 4000-5000W generators with end panel mounted receptacles, and without Twistlok 240V receptacle.

1. Engine
2. Engine adapter/front housing
3. Rotor assembly
4. Rotor throughbolt
5. Slip rings
6. Stator assembly
7. Endbell/rear housing
8. Brush (2)
8A. Brush terminal (2)
9. Rotor bearing

10. Capacitor
11. Diode (2)
12. Diode heat sink plate
13. Brush and bearing cover
14. Endbell/stator throughbolt
15. Ground lug terminal
16. Flat washer
17. Star washer
18. Nylok nut
19. AC circuit breakers

20. 120V AC receptacle
21. 240V AC receptacle
22. Control panel
23. Stator mount bracket
24. Stator isolator (2)
25. Heat shield
26. Clamp
27. Fuel tank

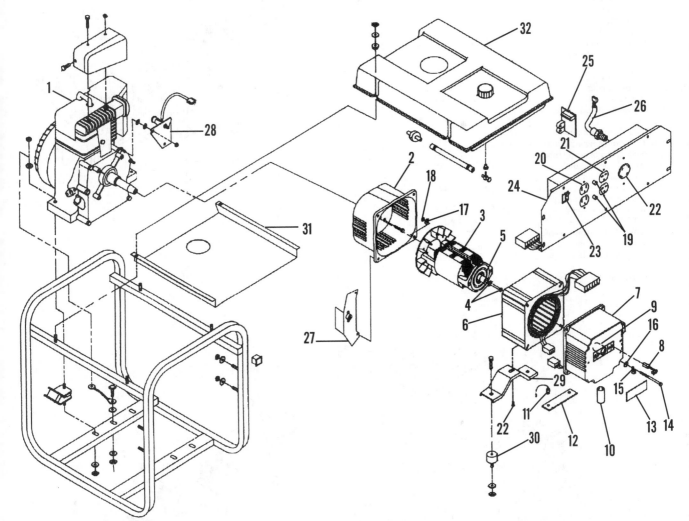

Fig. CL415 – Exploded view of typical POWERBASE Series 54 4000-5000W generators with end panel mounted receptacles and Twistlok 240V receptacle.

1. Engine
2. Engine adapter/front housing
3. Rotor assembly
4. Rotor through bolt and washer
5. Sliprings
6. Stator assembly
7. Endbell/rear housing
8. Brush (2)
9. Rotor bearing
10. Capacitor
11. Diode (2)

12. Diode heat sink plate
13. Brush and bearing cover
14. Endbell/stator through-bolt (4)
15. Ground lug terminal
16. Flat washer (4)
17. Star washer
18. Nylok nut (4)
19. AC circuit breakers
20. 120V AC receptacle
21. 240V AC receptacle
22. 240V AC Twistlok receptacle

23. Lighted rocker switch
24. Control panel
25. Idle-control module
26. Strain relief w/cord
27. Electric-start switch with bracket
28. Idle-control solenoid with bracket
29. Stator mount bracket
30. Stator isolator (2)
31. Heat shield
32. Fuel tank

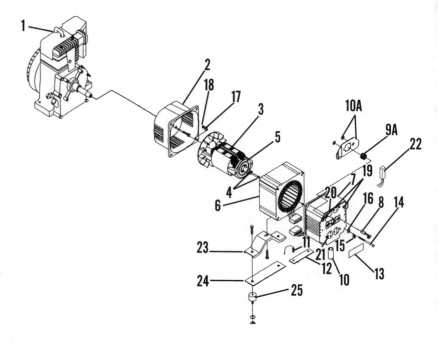

Fig. CL416 – Exploded view of typical COMMERCIAL Series 45 2000-5000W generator components. Only one pair of 120V receptacles is used on 2000W units. Automatic voltage regulator (AVR; 10K) is used on units with PC model-number prefix.

1. Engine
2. Engine adapter/front housing
3. Rotor assembly
4. Rotor throughbolt
5. Slip rings
6. Stator assembly
7. Endbell/rear housing
8. Brush (2)
9. Rotor bearing
9A. Tolerance ring
10. Capacitor
10A. Capacitor shield w/retainers
11. Diode (2)
12. Diode heat sink plate
13. Brush and bearing cover
14. Endbell/stator throughbolt
15. Ground lug terminal
16. Flat washer
17. Star washer
18. Nylok nut
19. AC circuit breakers
20. 120V AC receptacle
21. 120V AC Twistlok receptacle
22. Automatic voltage regulator
23. Stator mount plate
24. Carrier
25. Stator isolator

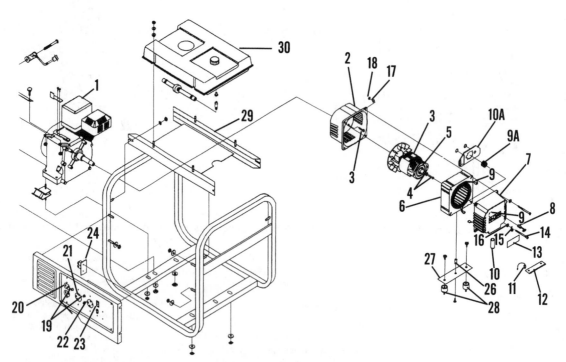

Fig. CL417 – Exploded view of typical HP Series 56 generator components.

1. Engine
2. Engine adapter/front housing
3. Rotor assembly
4. Rotor throughbolt
5. Slip rings
6. Stator assembly
7. Endbell/rear housing
8. Brush (2)
9. Rotor bearing
9A. Tolerance ring
10. Capacitor

10A. Capacitor shield w/retainers
11. Diode (2)
12. Diode heat sink plate
13. Brush and bearing cover
14. Endbell/stator throughbolt (4)
15. Ground lug terminal
16. Flat washer (4)
17. Star washer
18. Nylok nut (4)
19. AC circuit breakers
20. 120V AC receptacle

21. 120V AC Twistlok receptacle
22. 240V AC Twistlok receptacle
23. Idle-control switch
24. Idle-control module
25. Control panel
26. Stator mount spacer
27. Stator mount plate
28. Stator isolator (2)
29. Heat shield
30. Fuel tank

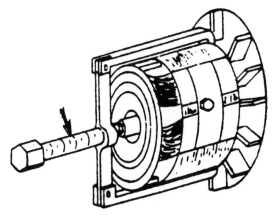

Fig. CL418 – Recommended rotor removal method using puller.

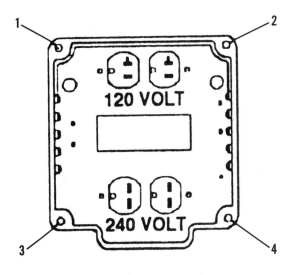

Tighten 1-4-2-3
75-95 in-lbs
(8.4-10.7 N•m)

Fig. CL420 – Typical torque sequence and values for endbell re-assembly.

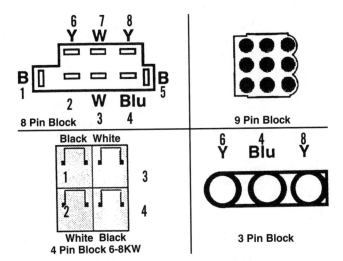

Fig. CL419 – Views of endbell wiring connectors used on Coleman generators.

WIRING DIAGRAMS

1. Refer to Fig. CL421 for Series 44 wiring diagram.
2. Refer to Fig. CL422 for Series 45 4KW & 5KW wiring diagram.
3. Refer to Fig. CL423 for Series 45 7KW wiring diagram.
4. Refer to Fig. CL424 for non-series specific single voltage output generator wiring diagram.
5. Refer to Fig. CL425 for non-series specific dual voltage output generator wiring diagram.
6. Refer to Fig. CL426 for Series 52 and 54 2KW generators.
7. Refer to Fig. CL427 for Series 52 and 54 3-7KW generators.
8. Refer to Fig. CL428 for non-series specific single voltage output generator with idle control wiring diagram.
9. Refer to Fig. CL429 for non-series specific dual voltage output generator with idle control wiring diagram.

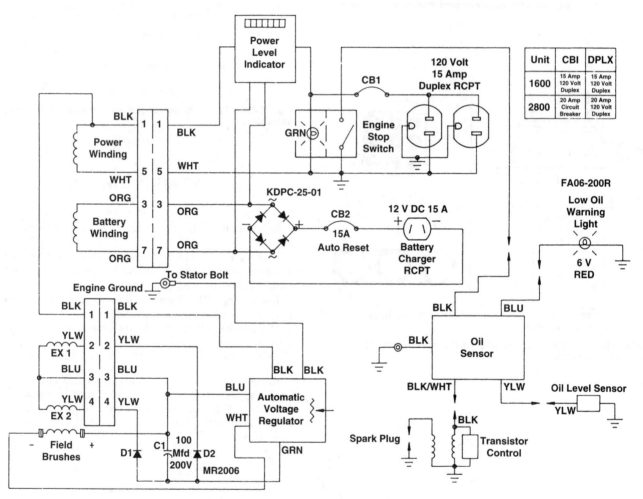

Fig. CL421 – Wiring schematic for Series 44 generators, including battery charger used on some models.

Unit	CBI	DPLX
1600	15 Amp 120 Volt Duplex	15 Amp 120 Volt Duplex
2800	20 Amp Circuit Breaker	20 Amp 120 Volt Duplex

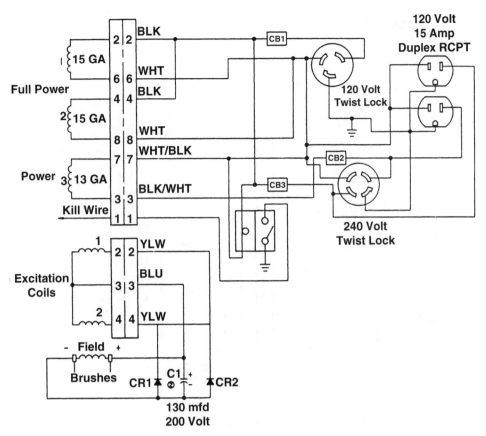

Fig. CL422 – Wiring schematic for Series 45 4000W and 5000W generators.

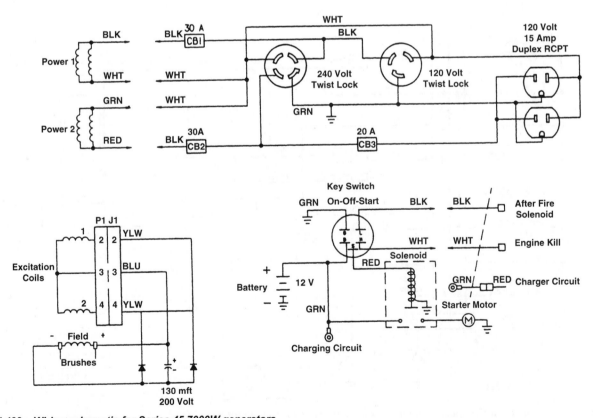

Fig. CL423 – Wiring schematic for Series 45 7000W generators.

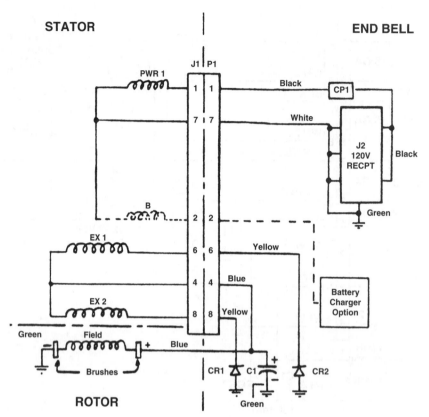

Fig CL424 – *Typical non series-specific wiring schematic for single voltage output generators.*

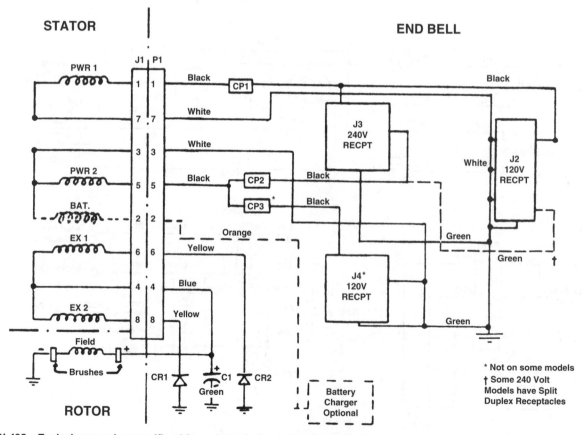

Fig. CL425 – *Typical non series specific wiring schematic for dual voltage output generators.*

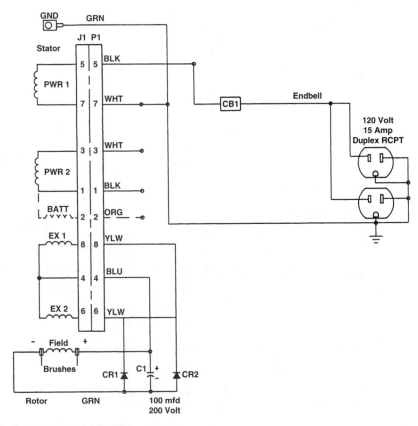

Fig. CL426 – Wiring schematic for Series 52 and 54 2000W generators.

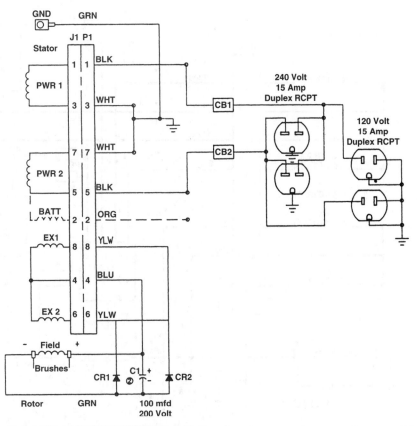

Fig. CL427 – Wiring schematic for Series 52 and 54 3000-7000W generators.

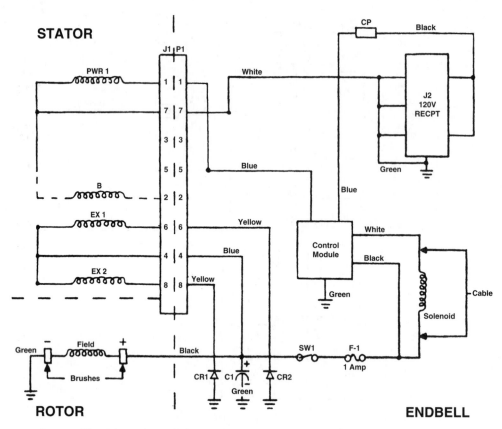

Fig. CL428 – *Typical non series specific wiring schematic for single voltage output generators with Idle Control.*

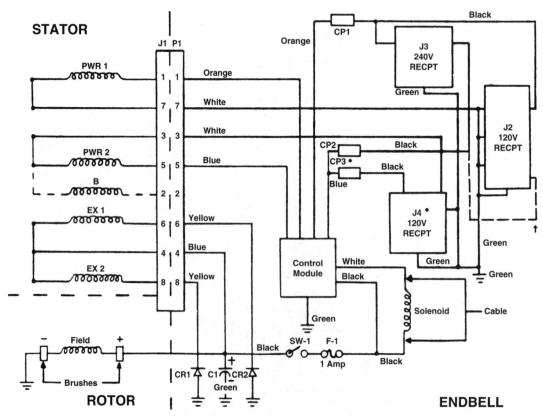

Fig. CL429 – *Typical non series specific wiring schematic for dual voltage output generators with idle control.*

COLEMAN

SERIES 55 "VERTEX"
VERTICAL ROTOR 5500-7500 WATT

IDENTIFICATION

Coleman VERTEX generator components are shown in Figs. CL501, CL502, and CL503. Fig. CL501 displays the components relative to their position on a functioning generator. Fig. CL502 shows the endbell, stator, and related components in their bench-testing position. Fig. CL503 identifies the three different control panels.

WIRING COLOR CODES

All portable generators manufactured by Coleman Powermate, Inc. use the following standardized wire color-coding:

Power Windings	B–Black
	B/W–Black with white tracer stripe
	R–Red
Excitation	Y–Yellow (diodes)
	Blu–Blue (capacitor)
Ground	G–Green
Neutral	W–White
	W/B–White with black tracer stripe
Battery Charger	O–Orange
Idle Control	O–Orange

MAINTENANCE

The endbell bearing and race should be inspected, cleaned, and relubricated any time the endbell is removed. If the bearing runs hot to the touch, replacement may be necessary.

Brushes must be at least ¼ inch long; shorter brushes require replacement. When reinstalling the brushes, do not overtighten the brush screws.

TROUBLESHOOTING

No-load voltages on 60Hz units should be 128-143 volts from the 120V receptacles and 258-278 volts from the 240V receptacles. If little or no output is generated, the following items could be potential causes:

1. Engine RPM – Engine must maintain 3600 RPM under load, or 3650-3800 RPM no-load. Engine RPM could also be set too high, causing overvoltage. Refer to the appropriate engine section for engine service.

2. Loose component-mounting fasteners (low or no voltage).
3. Circuit breakers (low or no voltage).
4. Insecure wiring connections (low or no voltage).
5. Faulty receptacle (low or no voltage).
6. Capacitor (low or no voltage).
7. Diodes (low or no voltage).
8. Brushes (no voltage).
9. Rotor:
   a. Loss of residual magnetism (no voltage).

NOTE: Residual magnetism can be restored without disassembling the generator, and is the recommended first step if the generator has no output.

   b. Open or shorted windings (no or low voltage, respectively).
   c. Dirty, broken, or disconnected slip rings (no voltage).
10. Stator windings (low voltage – shorted; no voltage – open).
11. Insufficient cooling-system ventilation (overheating).

GENERATOR REPAIR

RESIDUAL MAGNETISM

If there is no output due to the loss of residual magnetism, test and remagnetize the rotor. To test for residual magnetism:

1. Remove the rectangular snap-fit brush cover from the endbell.

2. Start the generator, verifying proper no-load RPM.

3. Using a VOM set on the 250VDC scale, connect the positive (+) VOM lead to the right brush and connect the negative (–) VOM lead to the left brush. Right and left brush positions are determined by looking at the generator, while facing the engine spark plug.

4. If residual magnetism excitation is present, the VOM should read approximately 125V for the 5500W generator and 155V for the 7500W generator.

5A. A low-voltage reading indicates shorted rotor and/or stator windings or a faulty capacitor or diode(s). For rotor, stator, capacitor, or diode service, refer to the appropriate repair section.

5B. A no-voltage reading indicates either a loss of residual magnetism or an open in the excitation circuit. To test for an open in the excitation circuit, refer to the yel-

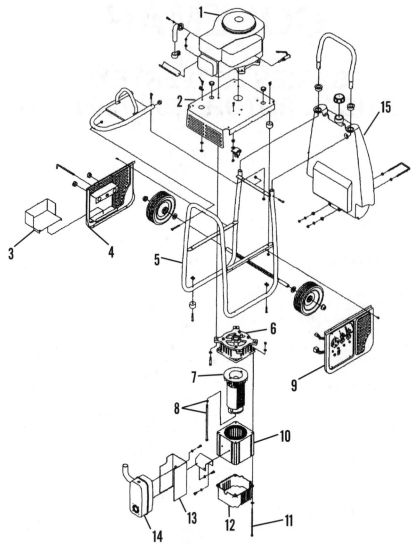

Fig. CL501 – Exploded view of Series 55 VERTEX generator components.

1. Engine
2. Engine to generator mount platform
3. Battery tray
4. Battery panel
5. Main frame
6. Engine adapter/front housing

7. Rotor assembly
8. Rotor through-bolt and washer
9. Control panel
10. Stator assembly
11. Endbell/stator through-bolt (4)
12. Endbell/rear housing

13. Heat shield
14. Muffler
15. Fuel tank

low-to-yellow and/or yellow-to-blue resistance tests in the *stator* section.

6. To remagnetize the rotor, proceed as follows:
   a. Obtain a fully-charged DC power source (6VDC minimum, 24VDC maximum) such as an automotive battery.
   b. Start the generator.
   c. Connect the positive (+) DC lead to the right generator brush.
   d. Momentarily touch the negative (-) DC lead to the left brush.

   e. Disconnect the DC leads and retest for the recommended voltage or receptacle output. Repeat up to three times, if necessary. If, after the third attempt, the rotor still fails to remagnetize, proceed with further testing and repair.

GENERATOR DISASSEMBLY

Refer to Figs. CL501-CL503 for component identification.

1. Remove the fuel tank, drain the engine crankcase oil and remove the battery from electric start units.

2. Lay the generator on its back (fuel tank side).

3. Disconnect the stator leads, starter solenoid harness, and the two red resistor wires from the control panel. Then remove the control panel and the battery/side panel.

4. Remove the muffler assembly.

5. Remove the rectangular snap-fit brush cover from the endbell. Then remove the brush screws and brushes. Brushes worn to ¼ inch or less in length require replacement.

6. Remove the four long bolts holding the endbell and stator to the engine adapter. Remove the endbell and disconnect the wiring connector blocks.

NOTE: With the endbell removed, the diodes, capacitor, rotor and stator can be tested without further disassembly.

7. Remove the stator by prying between the stator and engine adapter with a flat bladed pry tool. Note the orientation of the stator wires and connectors.

NOTE: The rotor cannot be removed without first removing the stator.

CAUTION: The plastic rotor fan is part of the rotor assembly and cannot be replaced separately. Handle the rotor carefully to prevent fan damage.

8. Loosen the rotor throughbolt, then apply penetrating oil through the rotor, allowing it to flow onto the crankshaft taper.

9. Remove the rotor using a puller (Fig. CL504). Place the puller feet under the bottom of the laminations, then place the puller center bolt on the head of the rotor throughbolt.

NOTE: Do not allow the puller legs and feet to contact the rotor windings to prevent damage to the winding insulation.

An alternate rotor removal method is to remove the rotor throughbolt and apply oil through the rotor onto the crank-

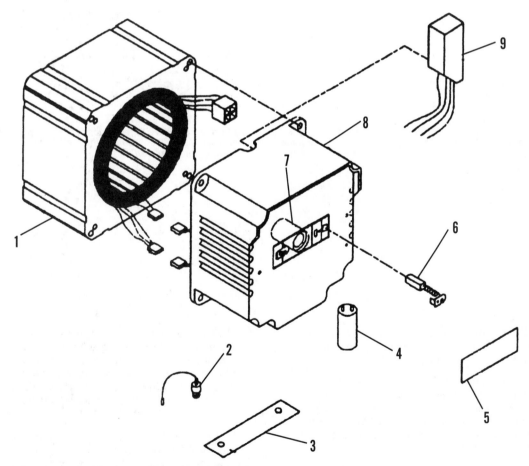

Fig. CL502 – Exploded view of Series 55 VERTEX generator endbell, stator, and related components.

1. Stator assembly	4. Capacitor	7. Rotor bearing
2. Diode (2)	5. Brush and bearing cover	8. Endbell/rear housing
3. Diode heat sink plate	6. Brush (2)	9. Automatic voltage regulator (AVR)

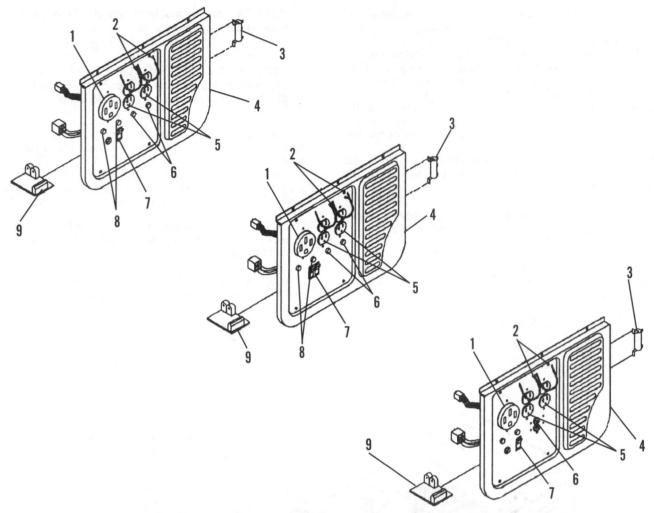

Fig. CL503 – Views of the three different styles of control panels used on Series 55 VERTEX generators: 1. The top panel is used on 5500W units with the PM model-number prefix; 2. The center panel is used on 7500W and 8000W units with the PM model-number prefix; 3. The bottom panel is used on both 5500W and 7500W units with the "PC" model-number prefix.

1. 240V receptacle	4. Control panel	7. Switch
2. Cord bracket	5. 120V receptacle	8. Circuit breaker
3. Resistor	6. Circuit breaker	9. Module

shaft taper. Then firmly tap the end-bearing race with a soft-tip rubber mallet (not a metal hammer) while holding the rotor to prevent it from falling.

CAUTION: Do not separate the two pieces of the engine adapter. These pieces are factory-assembled and machined. If these two pieces are separated after machining, the air-gap alignment between the rotor and the stator will be destroyed, and the adapter must be replaced.

CAPACITOR TESTING

Test the capacitor without removing it from the endbell so long as the brushes are removed and the stator connector plug is disconnected.

NOTE: Even with the generator stopped, the capacitor should retain a charge if it is working properly. Do not touch the capacitor terminals, as electric shock may result. Always discharge the capacitor by shorting across the terminals with a screwdriver or similar tool with an insulated handle.

To test the capacitor:

1. Set the VOM scale to R × 10,000 ohms.

2A. On spade-terminal capacitors, disconnect the harness wires from the capacitor, noting polarity orientation. Touch the VOM leads to the capacitor terminals. The positive (+) VOM lead should always connect to the positive (+) capacitor terminal.

2B. On solder terminal capacitors, connect the positive (+) VOM lead to the blue wire terminal of the endbell connector block. Connect the negative (–) VOM lead to any bare metal ground in the endbell.

3. The meter should first indicate low resistance, gradually increasing resistance toward infinity. Replace the capacitor if there is no meter reading or constant continuity.

4. On solder-terminal capacitors, a failed test could be the result of faulty solder joints. Repeat the test, holding the VOM leads directly against the capacitor terminals, using caution to observe proper polarity. If the test is now successful, one or both of the terminal to capacitor wires may be faulty, or the capacitor pigtail solder joint is defective. Correct as necessary.

NOTE: If replacing the capacitor, note the capacitor polarity before removal. Polarity must be correct upon reinstallation.

DIODES

The diodes in brush-style generators can be initially tested without removing them from the endbell or disconnecting any wires, provided the endbell connector is disconnected from the stator. To test either diode:

1. Set the VOM to the R × 1 scale.
2. Connect the positive (+) VOM lead to the applicable yellow wire diode endbell connector terminal.
3. Connect the negative (-) VOM lead to the base of the diode or to any bare metal (ground) section of the endbell.
4. Read the VOM.
5. Reverse the VOM leads and read the meter again. A good diode shows a reading in only one direction of the two tests. If there is a reading in both directions, the diode is faulty and must be replaced.
6. If there is no reading in either direction, trace the yellow wire from the non-reading terminal to the diode pigtail. Repeat the tests, holding one VOM lead directly against the tip of the pigtail.
   a. If there is still no VOM reading in either direction, replace the faulty diode.
   b. If a VOM reading is now observed in one direction, either the terminal-to-diode wire is faulty or the diode-pigtail solder joint is defective; correct as necessary.

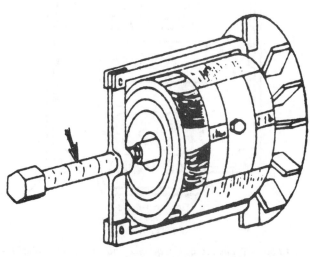

Fig. CL504 – Recommended puller method of rotor removal.

CIRCUIT BREAKERS

Individual circuit breakers are provided for each 120V circuit (a two-plug receptacle is one circuit). On most 120/240V units, the circuit breaker protecting a 120V circuit also protects one leg of a 240V receptacle. Units with a 120V Twistlok receptacle have a separate 30A breaker for that circuit.

To test the circuit breaker:

1. Ensure the breaker is set. Reset the breaker if it is tripped.
2. Remove the wires from the rear of the breaker.
3. Using the R × 1 scale on the VOM, test for continuity across the breaker terminals.
4. If there is no continuity reading or if the breaker fails to reset, the breaker is faulty and must be replaced.

NOTE: If a two-plug 120V receptacle has been replaced and the circuit breaker trips with no load connected to the unit, make sure that the tab is broken off between the lead connecting screws on the hot (brass) side of the receptacle.

ROTOR

Rotor Slip Rings

The slip rings are an integral part of the rotor assembly and cannot be replaced.

Dirty slip rings should be cleaned with a suitable solvent and lightly sanded with 240-grit sandpaper.

Rough slip rings should be lightly sanded with 240-grit sandpaper.

NOTE: Do not use crocus cloth to sand slip rings.

If the slip rings are cracked, chipped, broken, or deeply grooved, replace the rotor.

Rotor Winding Resistance

The rotor can be tested without removing it from the generator. If hypot testing, the stator should be removed to access the laminations.

To test resistance, use a VOM set on the R × 1 scale:
   a. A 5500W rotor should test between 38.8-42.1 ohms.
   b. A 7500W rotor should test between 47.4-47.8 ohms.

If no reading is observed upon initial testing, inspect the winding-to-slip ring connection and solder joint. If the solder joint is faulty, and if there is sufficient wire length, resolder the joint and retest.

NOTE: Always reapply high temperature RTV silicone sealant to the inner slip ring solder joint to prevent shorting the wire against the rotor shaft.

NOTE: All resistance readings are approximate and may vary depending on the winding temperature and the accuracy of the test equipment.

Hypot testing is the best way to test rotor windings. A rotor might pass VOM testing and still be faulty. Always follow the tester manufacturer's instructions. The recommended hypot test voltage is 1250 volts; less voltage may not show faults, and higher voltage may damage the rotor. To test the rotor, hypot from each slip ring to the lamination (ground). If either test fails, the rotor is faulty.

STATOR

Winding Resistance

To test the stator windings:
1. Set the VOM to the R × 1 scale.
2. Using the color-coded stator connector wires as shown in the following chart, read the resistance between each pair of stator terminals.
3. Compare the readings to the values shown in the chart. If no reading is obtained upon each initial test, check the connector block and stator leads for broken wires or faulty terminals. Repair, if possible, then retest.

NOTE: All resistance readings are approximate and may vary depending on the winding temperature and the accuracy of the test equipment.

STATOR WINDING RESISTANCE

Wattage	Field Test	Min. Ohms	Max. Ohms
5500	Black to white (1)	0.282	0.302
	Black to white (2)	0.282	0.302
	Yellow to blue (1)	1.12	1.14
	Yellow to blue (2)	1.12	1.14
	Yellow to yellow	2.24	2.28
	Orange to orange*	2.76	2.96
7500	Black to white (1)	0.40	0.42
	Black to white (2)	0.40	0.42
	Yellow to blue (1)	1.05	1.40
	Yellow to blue (2)	1.05	1.40
	Yellow to yellow	2.1	2.8
	Orange to orange*	2.76	2.96

*Early models

IDLE CONTROL

Operation

The idle control allows the engine to idle down when there is no load on the generator. It is activated by a rocker switch located on the receptacle panel. With the switch on, the generator will not produce output-load voltage.

NOTE: If less than a 250W load is being applied at 120V (500W at 240V), the idle control must be OFF so the unit will run at 3600 RPM to produce proper voltage. The idle control will not detect loads below these amounts.

Troubleshooting

If the unit will not idle with the idle control switch on, perform the following tests and checks:

With the Generator Shut Down:

Make sure the electromagnetic flapper on the throttle shaft moves freely. Repair any tightness or binding before proceeding.

With the Generator Running:

1. Check to make sure the generator is producing its rated 120V or 120/240V. Refer to TROUBLESHOOTING and GENERATOR REPAIR sections as necessary.
2. Make sure that the load is properly plugged into the receptacle and that the plug-in cord or the load itself is not faulty.

3. Check the stator output to the idle control module; 20V AC is necessary to operate the module. The stator-to-module output can be checked as follows:
   a. Turn the Idle Control switch off.
   b. Unplug the two orange wire harness from the back of the module connection on the control panel.
   c. Using a VOM set on the 50VAC scale, test the two orange wire terminals.
   d. If the test is successful, turn the switch back on.
4. Check the idle control module output; 20-24V DC is needed to operate the idle control magnet on the engine throttle. To test the module output, disconnect the black/white wire harness from the back of the throttle electromagnet, then use a VOM set on the 25VDC scale. The black wire terminal is positive (+).
5. Check the throttle idle control magnet: With the idle control switch off, loosely hold the handle of a flat-bladed screwdriver so that the blade is approximately ¼ inch (6 mm) away from the electromagnet tip where it pulls the throttle shaft flapper. Turn the switch on. The magnet should attract the screwdriver blade.
6. Make sure the throttle electromagnet is correctly positioned. Adjust as necessary for 1800 RPM idle speed.
7. Make sure the idle-mixture and idle-speed adjustments on the carburetor are properly calibrated.

REASSEMBLY

Refer to Figs. CL501-CL503 for component identification.
1. Lay the generator frame and engine assembly on its back (fuel tank side).
2. Ensure that the matching tapers on the engine crankshaft and inside the rotor shaft are clean and dry. Install the rotor onto the crankshaft. Install the rotor bolt and torque it to 180 in.-lb. (205 N•m).

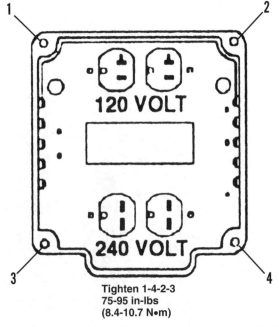

Tighten 1-4-2-3
75-95 in-lbs
(8.4-10.7 N•m)

Fig. CL505 – Typical torque sequence and values for endbell reassembly.

CAUTION: The plastic rotor fan is part of the rotor assembly and cannot be replaced separately. Handle the rotor carefully to prevent fan damage.

3. Note the stator wire and connector orientation. Align the locator pins and mount the stator tightly to the engine adapter. Using a soft mallet may help seat the stator snugly against the adapter.

4. Connect the endbell wiring connector blocks and install the endbell.

CAUTION: Ensure that the wires between the stator and the endbell are not pinched.

CAUTION: When installing the four long endbell bolts, always make sure that flat washers are positioned between the bolt heads and the endbell to prevent crushing the endbell flanges.

5. Torque the endbell bolts and nuts as shown in Fig. CL505.

6. Install the brushes and the brush screws, then install the snap-fit brush cover.

CAUTION: The brush screws thread directly into the plastic endbell. Install the brush screws carefully until the screw head bottoms, then twist an additional 1/8 turn only. Overtightening the brush screws causes stripping of the endbell threads and require replacement of the endbell.

7. Install the muffler assembly.

8. Install the control panel and the battery/side panel. Connect the stator leads, starter-solenoid harness, and the two red resistor wires to the control panel.

9. Stand the generator up into its normal run position, then install the fuel tank and, on electric-start units, the battery. Fill the engine crankcase with oil.

WIRING AND HARNESS DIAGRAMS

Refer to Fig. CL506 for wiring harness detail on 555503 generators.

Refer to Fig. CL507 for wiring harness detail on 555523 generators.

Refer to Fig. CL508 for wiring harness detail on 557523 & 558023 generators.

Refer to Fig. CL509 for wiring diagram on 555523 generators.

Refer to Fig. CL510 for wiring diagram on 557523 and 558023 generators.

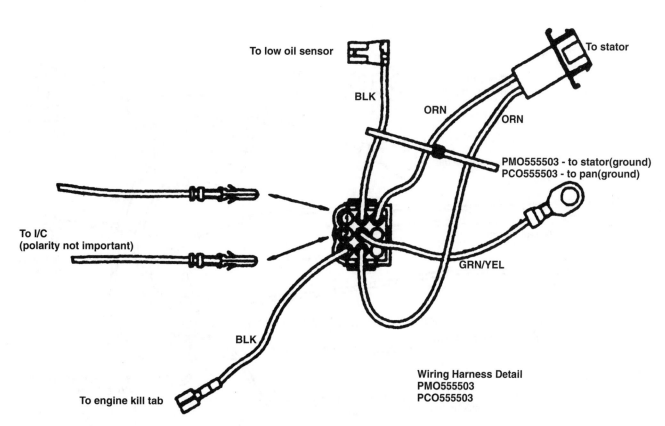

To low oil sensor

To stator

BLK

ORN

ORN

PMO555503 - to stator(ground)
PCO555503 - to pan(ground)

To I/C
(polarity not important)

GRN/YEL

BLK

**Wiring Harness Detail
PMO555503
PCO555503**

To engine kill tab

Fig. CL506 – Wiring harness detail on Model 555503 generators.

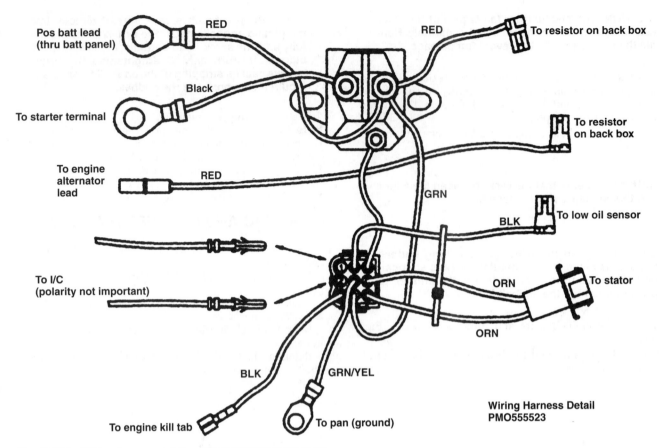

Pos batt lead
(thru batt panel)

RED

RED

To resistor on back box

Black

To starter terminal

To resistor
on back box

To engine
alternator
lead

RED

GRN

BLK

To low oil sensor

To I/C
(polarity not important)

ORN

To stator

ORN

BLK

GRN/YEL

To engine kill tab

To pan (ground)

Wiring Harness Detail
PMO555523

Fig. CL507 – Wiring harness detail on Model 555523 generators.

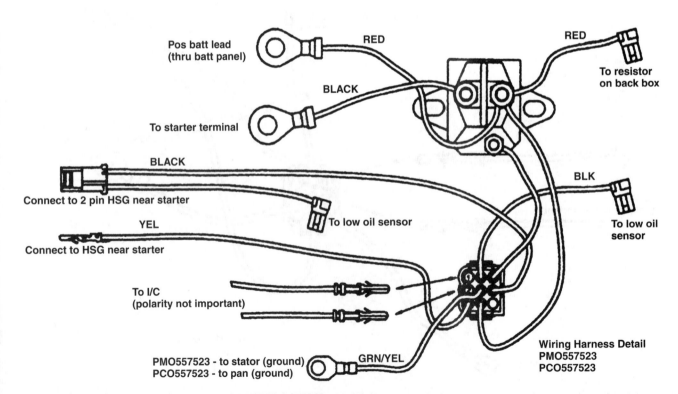

Pos batt lead
(thru batt panel)

RED

RED

To resistor
on back box

BLACK

To starter terminal

BLACK

Connect to 2 pin HSG near starter

To low oil sensor

BLK

YEL

To low oil
sensor

Connect to HSG near starter

To I/C
(polarity not important)

PMO557523 - to stator (ground)
PCO557523 - to pan (ground)

GRN/YEL

Wiring Harness Detail
PMO557523
PCO557523

Fig. CL508 – Wiring harness detail on Model 557523 & 558023 generators.

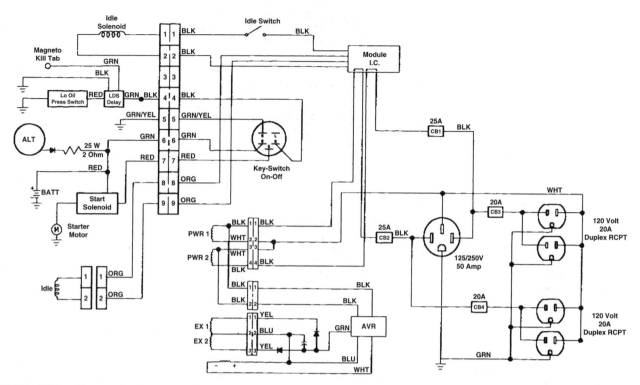

Fig. CL509 – Wiring diagram for Model 555523 generators.

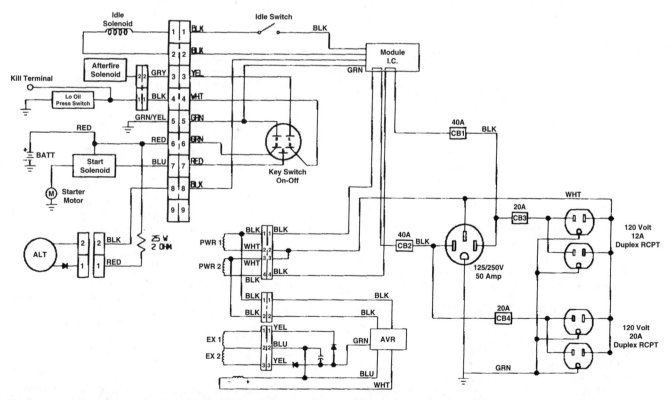

Fig. CL510 – Wiring diagram for Models 557523 and 558023 generators. On models built prior to 02/16/99, the wire from Terminal 6 to the starter solenoid is green, and the wire from Terminal 7 to the starter solenoid is red.

COLEMAN
SERIES INDUSTRIAL
10,000 & 12,000 WATT

IDENTIFICATION

Components for the Coleman 61 series industrial generators are shown in Fig. CL601. These generators are horizontal-rotor brushless-design units powered by Briggs & Stratton Vanguard V-twin engines. The 610023 models use 18 hp; the 612023 use 23 hp.

WIRING COLOR CODES

All portable generators manufactured by Coleman Powermate, Inc. use the following standardized wire color-coding:

Power Windings	B–Black
	B/W–Black with white tracer stripe
	R–Red
Excitation	Y–Yellow (diodes)
	Blu–Blue (capacitor)
Ground	G–Green
Neutral	W–White
	W/B–White with black tracer stripe
Battery Charger	O–Orange
Idle Control	O–Orange

MAINTENANCE

The rotor bearing is a lubricated, sealed, maintenance-free bearing. If it feels excessively hot to the touch, it should be inspected and replaced, if necessary.

TROUBLESHOOTING

No-load voltage at the 240V receptacle should be 245V/62.5Hz at full throttle and 100V/44Hz at idle-control speed. If little or no output is generated, the following circumstances could be potential causes:
1. Engine RPM – The engine must maintain 3600 RPM under load, or 3650-3800 RPM no-load. Engine RPM could also be set too high, causing overvoltage.
2. Loose component-mounting fasteners (low or no voltage).
3. Circuit breakers (low or no voltage).
4. Insecure wiring connections (low or no voltage).
5. Faulty receptacle (low or no voltage).
6. Capacitor (low or no voltage).
7. Diodes (low or no voltage).
8. Stator windings (low voltage – shorted; no voltage – open).
9. Rotor:
   a. Open or shorted windings (no or low voltage, respectively)
   b. Loss of residual magnetism (no voltage).

NOTE: Residual magnetism can be restored without disassembling the generator. This is the recommended first step if generator has no output. Refer to RESIDUAL MAGNETISM.

10. Insufficient cooling system ventilation (overheating)

GENERATOR REPAIR

RESIDUAL MAGNETISM

Unlike brush-type generators, there is no test procedure for residual magnetism on brushless generators. If low residual magnetism is suspected as the cause of no-output, the rotor can be remagnetized as follows:
1. Turn the engine switch off and remove the leads from the spark plugs.
2. Construct a jumper tool using a male two-prong 110-volt household plug and two 16-gauge minimum wire leads. The wires must be long enough to reach from the generator's 110V outlet to a nearby DC power source (minimum 6VDC; 24VDC maximum) such as a 12V automotive battery. An alligator clip on the narrow spade jumper wire end would be helpful.

NOTE: Mark the jumper wires so that the wire coming from the narrow spade on the 110V plug will connect to the positive (+) battery terminal; the wire

from the wide spade must connect to the negative (-) battery terminal.

3. With the jumper tool plugged into one of the generator's 120V AC outlets, connect the narrow spade jumper wire to the battery positive (+) terminal.

4. While using the electric starter to turn the engine over, **momentarily** (1-2 seconds) touch the wide spade jumper wire to the battery negative (-) terminal.

5. Disconnect the jumper tool from the generator and battery. Reconnect the spark plug leads. Turn the engine switch on. Start the engine and observe the generator output.

6. Attempt remagnetizing up to three times. If, after the third try, the generator still has no output, proceed with further testing and repair.

GENERATOR DISASSEMBLY

Refer to Fig. CL601 for component identification:

1. Remove the generator head cover and the louvered end panel and side panels.

2. Disconnect the idle control wires from the idle control magnet and the engine.

3. Disconnect the control panel wires from the generator head connectors, then remove the control panel from the generator carrier frame.

NOTE: At this time, the capacitors and the stator windings can be tested without further disassembly.

4. Remove the muffler.

5. Disconnect the battery ground cable from the bottom of the generator head. Note its position for proper reassembly.

6. Remove the three 5/16-18 Nylok nuts from the bottom studs of the rubber isolators on the stator bracket assembly.

7. Position a support under the engine adapter, making sure the stator is lifted enough so the stator bracket isolator studs clear the generator carrier frame.

8. Remove the four stator-to-engine adapter bolts, and remove the rotor bolt.

NOTE: The rotor cannot be removed without first removing the stator.

9. To remove the stator:

   a. Use a 3-hook puller, placing the three hook arms on the rear rotor bearing/stator housing;

   b. With the puller bolt centered on the rotor shaft, turn the puller down until a 1/8 inch gap opens between the stator housing and the engine adapter. Make sure the exhaust heat shield is not in the way of the stator assembly.

   c. Continue turning down the puller until the rotor bearing journal is free of the rear stator housing and bearing.

10. To remove the rotor, apply penetrating oil to the rotor shaft/engine crankshaft mating area, then tap the bearing journal on the rotor with a rubber or plastic mallet to unlock the rotor from the engine crankshaft.

CAPACITOR TESTING

The capacitors can be tested without removing them from the generator provided the capacitor harnesses are disconnected.

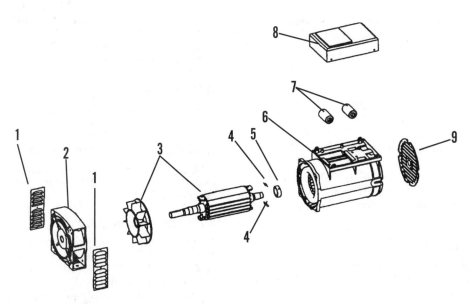

Fig. 601 —Exploded view of generator components.

1. Side panels	4. Diodes	7. Capacitor
2. Engine adapter	5. Bearing	8. Head cover
3. Rotor and fan	6. Stator assembly	9. End panel

NOTE: Even with the generator stopped, the capacitors should retain a charge if they are working properly. Do not touch the capacitor terminals, as electric shock may result. Always discharge the capacitors by shorting across the terminals with a screwdriver or similar tool with an insulated handle.

To test each capacitor:

Set the VOM ohmmeter scale to R × 10,000;

1. On spade terminal capacitors, disconnect the harness wires from the capacitor, noting polarity orientation, then touch the VOM leads to the capacitor terminals. The red VOM lead should always connect to the positive (+) capacitor terminal. Connect the negative (–) VOM lead to any bare metal ground in the endbell.

2. On solder terminal capacitors, connect the positive (+) VOM lead to the blue wire terminal of the endbell. Connect the negative volt lead to ground

3. The meter should first indicate low resistance, gradually increasing resistance toward infinity. No meter reading or constant continuity indicates a faulty capacitor; replace the capacitor.

4. On solder-terminal capacitors, a failed test could be the result of faulty solder joints. Repeat the test, holding the VOM leads directly against the capacitor terminals and using caution to observe proper polarity. If the test is now successful, one or both of the terminal-to-capacitor wires may be faulty or the capacitor-pigtail solder joint is defective; correct as necessary.

NOTE: If replacing the capacitor, note capacitor polarity prior to removal. Polarity must be correct upon reinstallation.

CIRCUIT BREAKER

To test the circuit breaker on units so equipped:

1. Ensure the breaker is set. If it has been tripped, reset it.
2. Remove wires from the rear of the breaker.
3. Using the R × 1 scale on the VOM, test for continuity across the breaker terminals.
4. If no continuity reading is noted, or if the breaker fails to reset, the breaker is faulty and must be replaced.

ROTOR

Diode Test

To test diodes on diode-equipped units:
1. Disconnect one side of the diode.

NOTE: Do not break the diode lead.

2. Test the diode using the Rx10K scale on the VOM. A good diode shows resistance in only one direction.
3. If the diode is faulty, replace the diode, always observing proper polarity.

Rotor Winding Resistance

The rotor can be tested without removing it from the generator. If hypot testing, the stator should be removed to access the laminations.

NOTE: On diode-equipped units, all resistance tests must be performed with the diodes disconnected from the rotor windings.

To test resistance, use a VOM set on the R × 1 scale. Each rotor winding should show a resistance of 5.9-6.0 Ohms.

NOTE: All resistance readings are approximate and may vary depending on the winding temperature and the accuracy of the test equipment.

Hypot testing is the best way to test rotor windings. A rotor might pass VOM testing and still be faulty. Always follow the tester manufacturer's instructions. To test the rotor, hypot each winding individually from the winding lead to lamination (ground). The failure of either winding indicates a faulty rotor.

STATOR WINDING RESISTANCE

To test the stator windings:
1. Set the VOM to the R × 1 scale;
2. Read the resistance between the applicable stator wires as shown in the following chart, then compare the resistance readings to the values shown. If no reading is obtained upon each initial test, check the leads for broken wires or faulty terminals. Repair as necessary, then retest.

NOTE: All resistance readings are approximate and may vary depending on winding temperature and the accuracy of the test equipment.

Field	Ohms
Black to white	0.3
Red to red	1.4
Green to green	1.4

IDLE CONTROL

Operation

The idle control allows the engine to idle down when there is no load on the generator. It is activated by a rocker switch located on the receptacle panel. With the switch on, the generator will not produce output load voltage.

NOTE: If less than a 250W load is being applied at 120V (500W at 240V), the idle control must be in the off position so the unit will run at 3600 RPM to produce proper voltage. The idle control will not detect loads below these amounts.

Troubleshooting

If the unit does not idle with the idle control switch on, perform the following tests and checks.

With the Generator Shut Down

Check the freedom of movement of the throttle shaft electromagnet flapper. Repair any tightness or binding before proceeding.

With the Generator Running

1. Check to make sure the generator is producing its rated 120V or 120/240V. Refer to TROUBLESHOOTING and GENERATOR REPAIR sections as necessary.
2. Make sure that the load is properly plugged into the receptacle and that the plug-in cord or the load itself is not faulty.
3. Check the stator output to the idle control module. The module requires 20VAC to operate. The stator-to-module output can be checked as follows:

a. Turn the idle control switch off.

b. Unplug the two-orange wire harness from the back of the module connection on the control panel.

c. Using a VOM set on the 50VAC scale, test the two orange wire terminals.

d. If the test is successful, turn the switch back on.

4. Check the idle control module output. The idle control magnet on the engine throttle requires 20-24VDC. To test the module output, disconnect the black and white wire harness from the back of the throttle electromagnet, then use a VOM set on the 25VDC scale. The black wire terminal is positive (+).

5. Check the throttle idle control magnet. With the idle control switch off, loosely hold the handle of a flat blade screwdriver so that the blade is approximately ¼ inch (6 mm) away from the electromagnet tip where it pulls the throttle shaft flapper. Turn the switch on. The magnet should attract the screwdriver blade.

6. Make sure the throttle electromagnet is correctly positioned. Adjust, as necessary, for 1800 RPM idle speed.

7. Make sure the idle mixture and idle speed adjustments on the carburetor are properly calibrated.

GENERATOR REASSEMBLY

1. Position a support under the engine adapter.

2. Ensure that the mating surfaces on the rotor and the engine crankshaft are clean and dry. Install the rotor onto the crankshaft, then install the rotor throughbolt finger-tight at this time.

3. Correctly position the stator assembly over the rotor, making certain that the engine is tilted enough to allow the stator bracket isolator studs to clear the generator carrier frame. Also ensure that the exhaust heat shield has sufficient clearance to allow the stator to fit up to the engine adapter.

4. Using a rubber or plastic mallet, drive the stator assembly to fit tightly against the engine adapter. Install and tighten the four stator-to-engine adapter bolts.

5. Tighten the rotor bolt, and reconnect the battery ground cable to its correct position on the bottom of the stator housing as noted during disassembly.

6. Remove the engine adapter support, allowing the three stator-adapter isolator studs to fit into their respective carrier frame holes. Install and tighten the three 5/16-18 isolator stud nuts.

7. Turn the engine over by hand to verify correct rotor-to-stator clearance. If there is any dragging or rubbing, realign as necessary before proceeding.

8. Install the muffler.

9. Install the control panel onto the generator carrier frame, then reconnect the control panel-to-generator head wires, routing the wires through the hole in the generator head cover. Install the cover.

10. Reconnect the yellow idle control wires to the idle control magnet, and the red idle control wire to the white idle control power supply wire coming from the engine.

11. Install the louvered end panel and side panels, insuring that the louver openings face down.

GENERAC
Power Systems, Inc.

1 Generac Way
P.O. Box 239
Jefferson, WI 53549-0239

Generator Model	Rated (Surge) Watts	Output Volts
Home & Away	1500 (1800) AC/120 DC	120 AC/12 DC
SV2400	2400 (2900)	120
SVT4200	4200 (5250)	120/240
SVT5000	5000 (6250)	120/240
SVP(E)5000	5000 (6250)	120/240
3250XE	3250 (4000)	120/240
3500XEP	3500 (4375) AC/120 DC	120/240 AC/12 DC
3500XL	3500 (4375) AC/120 DC	120/240 AC/12 DC
4000XL	4000 (5000) AC/120 DC	120/240 AC/12 DC
5500(E)XL	5500 (6875) AC/120 DC	120/240 AC/12 DC
6500(E)XL	6500 (8125) AC/120 DC	120/240 AC/12 DC
8000(E)XL	8000 (10000) AC/120 DC	120/240 AC/12 DC
10000(E)XL	10000 (12500) AC/120 DC	120/240 AC/12 DC
MC4000	4000 (5000) AC/120 DC	120/240 AC/12 DC
MC5500	5500 (6875) AC/120 DC	120/240 AC/12 DC
MC6500	6500 (8125) AC/120 DC	120/240 AC/12 DC
MC8000	8000 (10000) AC/120 DC	120/240 AC/12 DC
MC10000	10000 (12500) AC/120 DC	120/240 AC/12 DC

NOTE: "(E)" Model-number suffix denotes electric start option.

IDENTIFICATION

These generators are direct-drive, 60 Hz, single-phase units.

Some are brush-style; others are brushless. Brush-style units have rounded endbells with a separate stator lamination-core assembly; brushless units have a square endbell, with a one-piece endbell/stator assembly.

Receptacles are either endbell-mounted or panel-mounted.

MAINTENANCE

ROTOR BEARING

The brush-end rotor bearing is prelubricated and sealed. Inspect and replace as necessary. If a seized bearing is en-

countered, also inspect the rotor journal and the end-housing bearing cavity for scoring.

BRUSHES (IF USED)

The brushes should be replaced when the brush length has worn excessively. New brush length is 10.5 mm (0.414 in.) protruding from the brush holder on 30.0 mm (1.181 in.) free length outside the holder. The positive (+) brush is always the brush closest to the rotor bearing; the negative (-) brush is always the brush closest to the rotor windings.

GENERATOR REPAIR

TOOLS

Besides basic hand tools, the following tools may be needed.

a. Digital VOA meter (VOM).

b. AC frequency meter capable of being plugged into a 120V receptacle.

c. A fully-charged, 12V test battery.

d. Generac Rotor Removal Kit #41079.

If an AC frequency meter is not available, use an accurate engine tachometer as a substitute, since frequency is directly proportional to engine RPM.

TROUBLESHOOTING NOTES

1. Before testing any electrical components, check the integrity of the terminals, connectors, and fuse(s).

2. While testing individual components, also check the wiring integrity.

3. Perform operational testing with the engine running at 3690-3810 RPM, unless specific instructions state otherwise.

4. Wire reference numbers in parentheses refer to wire locator numbers in the wiring diagrams.

5. Always load-test the generator after repairing defects discovered during troubleshooting.

Brushless Units

To troubleshoot the generator:

1. Circuit breakers must be *on*. To test either the AC or DC (if equipped) breakers, access the breaker wiring and continuity-test the breaker terminals with the generator shut down. The breaker is faulty and must be replaced if it displays any of the following conditions:

a. Will not stay set.

b. Shows continuity in the OFF, position.

c. Does not show continuity in the ON position.

2. Receptacles must be functioning. To test the receptacle(s):

a. Access the receptacle connections.

b. Disconnect the return feed wire from the receptacle being tested.

c. Install a jumper wire between the hot and return outlet terminals.

d. Test for continuity across the feed terminals; no continuity indicates a faulty receptacle.

3. Start the generator, allowing it to reach operating temperature.

4. Test the AC voltage at one of the 120V receptacles; output should be 130-140 volts.

5. Stop the generator.

6. If Step 4 output was zero voltage, attempt to restore residual rotor magnetism by flashing the fields:

a. Construct an energizing (test) cord (Fig. GC101).

b. Ensure that the generator is completely assembled.

c. Plug one end of the test cord into one 120V generator receptacle.

d. Start the generator.

Fig. GC101 – To flash the field, construct an energizing cord with two male plugs and two light bulbs. Refer to the text for the correct flashing procedure.

e. Plug the other test cord end into a 120V wall receptacle for 10 seconds.

f. Stop the generator.

g. Unplug both test cord ends.

h. Restart the generator and check AC voltage. If voltage meets the specifications in Step 4, the generator is good. If voltage is still zero, proceed to Step 7.

i. Stop the generator.

7. Test the excitation capacitor.

a. Access the capacitor inside the generator end cover.

b. Noting the capacitor wiring polarity, disconnect the capacitor terminals.

c. Test the capacitor and compare the results with the specification listed on the capacitor. If the capacitor meets specification, proceed with Step 8.

8. Test the rotor winding resistance.

a. Carefully disconnect/unsolder one end of each rotor diode.

b. With the VOM set on the R × 1 scale, test the resistance of the rotor windings (Fig. GC102); compare the readings to the specifications at the end of this section.

c. Test rotor winding resistance between each winding connection to a good ground on the rotor shaft. Resistance should test infinity/no-continuity.

d. If the rotor tested to specification, proceed with Step 9. If it did not meet specification, the rotor is faulty and must be replaced.

9. Test the rotor diodes:

a. Carefully disconnect/unsolder one end of each diode.

b. With the VOM set on the R × 1 scale, test the resistance of the diode, then reverse the meter leads and retest. The diode should show continuity in one direction only. Replace any diode which shows continuity or infinity in both directions.

10. Test the rotor capacitor:

a. Carefully disconnect/unsolder one capacitor lead.

b. Test the capacitance against the value marked on the capacitor. If beyond specification, replace the rotor.

11. Test the stator windings:

a. To test the single-voltage or the first half of dual-voltage AC power windings, disconnect the stator winding wires. Set the VOM to the R × 1 scale. Read resistance between wires #11 and #22 (red and blue wires on unnumbered wire single-voltage units). Compare the reading to the specification at the end of this section.

NOTE: Do not confuse the blue and red stator wires with the blue and red excitation wires. The excitation wires connect to the bridge rectifier or the voltage regulator.

b. Individually test resistance between each wire and any good, clean ground on the stator core. Resistance should read infinity/no-continuity.

c. To test the dual-voltage AC power windings, test the first winding as in Step A. Repeat Step A with wires #22 and #44 (dual-voltage units have numbered wires).

d. With the VOM still on the R × 1 scale, test resistance between the #2 and #6 excitation wires (blue and red wires on single-voltage unnumbered wire units). Compare the reading to the specification at the end of this section.

NOTE: Do not confuse the blue and red stator wires with the blue and red excitation wires. The excitation wires connect to the bridge rectifier or the voltage regulator.

e. Test the resistance between any one excitation winding wire and any one power winding wire. Resistance should read infinity/no-continuity.

f. Test resistance between any excitation-winding wire and any good, clean ground on the stator core. Resistance should read infinity/no-continuity.

g. Any reading in the above tests that do not meet specification indicates a faulty stator.

Brush-Style Direct Excited Units

NOTE: The positive (+) brush is always the brush closest to the rotor bearing; the negative (–) brush is always the brush closest to the rotor windings.

To troubleshoot the generator:

1. Circuit breakers must be on. To test either the AC or DC (if equipped) breakers, access the breaker wiring and continuity-test the breaker terminals with the generator shut down. If the breaker

   a. will not stay set,

   b. shows continuity in the off position, or

   c. does not show continuity in the on position, the breaker is faulty.

2. Receptacles must be functioning. To test the receptacle(s):

   a. Access the receptacle connections.

   b. Disconnect the return feed wire from the receptacle being tested.

   c. Install a jumper wire between the hot and return outlet terminals.

   d. Test for continuity across the feed terminals. No continuity indicates a faulty receptacle.

3. Start the generator, allowing it to reach operating temperature.

4. Test the AC voltage at one of the 120V receptacles; output should be 130-140 volts.

5. Stop the generator.

6. If Step 4 output was less than 6 volts or zero voltage, attempt to restore residual rotor magnetism by using a 12V test battery to flash the fields:

   a. Check the wiring diagram to see if the unit has a field boost assembly in the circuit. If there is field boost, the field boost unit is faulty. Replace the field boost unit and repeat Steps 3-5. Test-battery flashing is not required on generators equipped with field boost.

   b. If the generator is equipped with a combination brush/bridge-rectifier assembly (Fig. GC103):

   • Remove the blue (2) and red (6) wires from brush/rectifier terminals J1 and J2, respectively.

   • Start the engine, bringing it up to rated no-load speed.

   • Apply 12V DC from the test battery to terminals J1 & J2 (polarity unimportant) for approximately five seconds.

   • Disconnect the test battery, then stop the engine.

   • Reconnect the blue (2) wire and red (6) wire to terminals J1 and J2, respectively.

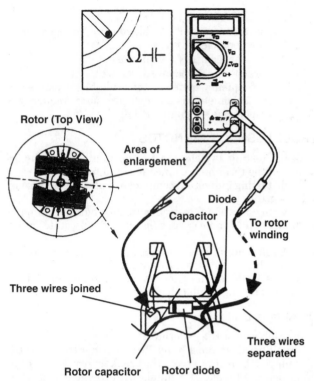

Fig. GC102 – Test points for checking rotor winding resistance. Refer to the text for the correct test procedure.

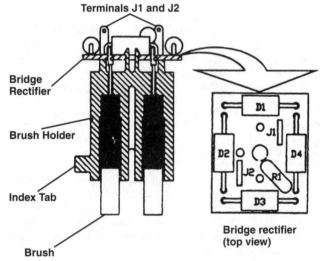

Fig. GC103 – View of combination bridge rectifier/brush assembly.

   • Restart the generator and test AC output.

   c. If the generator is equipped with brushes only, and not a brush/bridge-rectifier assembly:

   • Start the engine, bringing it up to rated no-load speed.

   • Leaving the brush wires connected, connect the positive (+) test battery terminal to the positive (+) brush terminal (closest to the rotor bearing).

- For approximately five seconds, connect the negative (–) test battery terminal to the negative (-) brush terminal (closest to the rotor windings).
- Disconnect the test battery.
- Test AC voltage output.
- Stop, then restart, the engine.
- Retest AC voltage output.

7. If AC voltage now meets specification, the rotor works correctly. If it failed to meet specification, proceed with Step 8.

8. To test the rotor:

The slip rings are an integral part of the rotor assembly; they are not replaceable.

Clean dirty or rough slip rings with a suitable solvent, then lightly sand with ScotchBrite or sandpaper. Do not use steel wool, emery cloth, or crocus cloth.

Cracked, chipped, broken, or grooved slip rings cannot be repaired. In these cases, the rotor must be replaced.

a. Noting the wiring polarity, disconnect the brush wire connectors.

b. With the VOM set on the R × 1 scale, measure field coil winding resistance through the brushes. Compare the reading to the specification at the end of this section.

c. Remove the brush assembly.

d. Measure field coil winding resistance at the rotor slip rings. Resistance should be the same as substep b. On some units, limited access inside the rear bearing carrier housing may require removing the power cable grommet to test the slip rings.

e. Measure field coil winding resistance from each slip ring to a good ground on the rotor shaft or laminations. Resistance should be infinity/no-continuity.

f. If test substeps b and d met specification, the rotor and brushes are OK.

g. If test substep b failed specification but substep d met specification, the rotor is good. Inspect the brushes and test each brush for continuity through its terminal. If faulty, replace.

h. If both test substeps b and d failed specification, the rotor is faulty. Replace the rotor, then inspect and test the brushes.

i. The rotor can also be dynamically tested by performing a current draw test. To perform this test:

- With the VOM on the DCV scale, accurately measure the 12V test battery voltage.
- With the VOM on the R × 1 scale, note the rotor winding resistance across the slip rings.
- Using Ohm's Law applicable to this test (I=V÷R), calculate the current flow (I) by dividing the rotor winding resistance (R) into the battery voltage (V). Note the calculated amperage.
- Disconnect the brush feed wires. Isolate the brushes from the voltage regulator or bridge rectifier.
- Connect the positive (+) terminal of the test battery to the positive (+) side of the VOM/ammeter.
- Connect the ammeter negative (–) lead to the positive (+) brush terminal (wire #4).
- Connect a test wire from the negative (–) brush terminal (wire #1) to the negative (–) battery terminal.
- Note the field winding current draw on the ammeter.
- Start the engine; with the engine running at rated no-load RPM, note the amperage draw on the ammeter.
- Compare the engine running ammeter reading to both the engine stopped ammeter reading and the calculated amperage.

- If the running amperage is significantly higher than the static or calculated amperage, the rotor has a running short. Replace the rotor.
- If there is NO amperage draw during the running test, the rotor has an open winding; replace the rotor.
- If the running amperage is approximately the same as the static and the calculated amperage, the rotor is good.
- Stop the engine and disconnect the test equipment.

NOTE: If a rotor fails, the resulting momentary voltage surge can damage the voltage regulator or bridge rectifier. If the rotor requires replacement, also test these items plus the power regulator board, where used.

9. To test the stator:

NOTE: Compare all readings taken in the following tests to the specifications listed at the end of this section.

1. Disconnect the voltage regulator or bridge rectifier wires, and completely isolate them from the circuit.

2. Using the 12V test battery, connect battery positive (+) to the positive (+) brush (the brush closest to the rotor bearing), and connect battery negative (-) to the negative (-) brush.

3. Start the generator.

4. Measure AC voltage output:

a. On voltage regulator units, measure output at sensing wires (11) and (22).

b. On bridge rectifier units, measure voltage at the 120V receptacle.

c. Across blue (6) and red (2) excitation winding wires.

A minimum of 60VAC should be produced; a slightly higher output is acceptable.

5. Stop the generator.

6. Set the VOM to ohms for the following tests.

7. Unplug the stator-winding connectors.

8. On single-voltage and dual-voltage units, test resistance across the blue (11) to red (22) AC power winding terminals. Note the reading. Do not confuse the blue and red power winding wires with the blue and red excitation winding wires.

9. On dual voltage units, also test resistance across the gray (44) to red (22) AC power winding terminals. Note the reading.

10. Individually test resistance from each blue (11), gray (44), and red (22) power winding terminal to any good, clean ground on the stator laminations. Resistance should read infinity/no-continuity.

11. Test resistance across the blue (2) to red (6) excitation winding terminals. Note the reading.

12. Individually test resistance from each blue (2) and red (6) excitation winding terminal to any good, clean ground on the stator laminations. Resistance should read infinity/no-continuity.

13. Test resistance between the red (22) power winding terminal and the red (6) excitation winding terminal. Resistance should read infinity/no-continuity.

14. On single voltage and dual voltage units, test resistance across the black (55) to brown (66) battery charge winding terminals. Note the reading.

15. On dual-voltage units, also test resistance across the black (55) to brown (77) battery charge winding terminals. Note the reading.

16. Repeat Steps 14 and 15 with terminals (55) to (66A) and (55) to (77A).

17. Individually test resistance from each (55), (66), (66A), (77), and (77A) terminal to any good, clean ground on the stator laminations. Resistance should read infinity/no-continuity.

18. If the stator tests lower than specification in Step 4, tests significantly beyond specification in Steps 8, 9, 11, 14, or 15, or tests any continuity in Steps 10, 12, 13, or 17, the stator is faulty and must be replaced.

19. If all preceding dynamic and static component tests meet specification, the voltage regulator/bridge rectifier is faulty; replace the regulator or rectifier.

Brush Style "Two-Board" Units

NOTE: The positive (+) brush is always the brush closest to the rotor bearing; the negative (–) brush is always the brush closest to the rotor windings.

To troubleshoot the generator:

1. Circuit breakers must be on. To test either the AC or DC (if equipped) breakers, access the breaker wiring and continuity test the breaker terminals with the generator shut down. If the breaker will not stay set, has continuity in the OFF position, or does not show continuity in the ON position, the breaker is faulty.

2. Receptacles must be functioning. To test the receptacle(s):

   a. Access the receptacle connections.

   b. Disconnect the return feed wire from the receptacle being tested.

   c. Install a jumper wire between the hot and return outlet terminals.

   d. Test for continuity across the feed terminals; no continuity indicates a faulty receptacle.

3. Start the generator, allowing it to reach operating temperature.

4. Test the AC voltage at one of the 120V receptacles; output should be 130-140 volts.

5. Stop the generator.

6. If the output measured in Step 4 was less than 130 volts but more than 6 volts, proceed to Step 8. If the output in Step 4 was less than 6 volts or zero voltage, attempt to restore residual rotor magnetism by using a 12V test battery to flash the fields.

   a. Check the wiring diagram to see if the unit has a "field boost" assembly in the circuit. If there is field boost, the field boost unit is faulty. Replace the field boost unit and repeat Steps 3-5. Flashing the fields is not required on generators equipped with field boost.

   b. Start the engine, bringing it up to rated no-load speed.

   c. Leaving the brush wires connected, connect the positive (+) test battery terminal to the positive (+) brush terminal (closest to the rotor bearing).

   d. For approximately five seconds, connect the negative (–) test battery terminal to the negative (–) brush terminal (closest to the rotor windings).

   e. Disconnect the test battery.

   f. Test AC voltage output.

   g. Stop, then restart, the engine.

   h. Retest AC voltage output.

7. If test Step 6 met specification, the rotor is good. If this test failed to meet specification, proceed with Step 8.

8. To isolate the system control board:

   a. Access the power regulator circuit board mounted inside the generator rear housing.

   b. Disconnect the white three-terminal snubber feedback cable connector from the board (Fig. GC104).

   c. Start the engine.

   d. Measure the AC output at the 120V receptacle; voltage should be 130-150 VAC.

   e. If AC voltage is within specification, stop the generator. Connect the two now-exposed outer circuit board terminals with a jumper lead (Fig. GC105). Start the generator. Measure the AC output at the 120V receptacle; voltage should be below 10 volts.

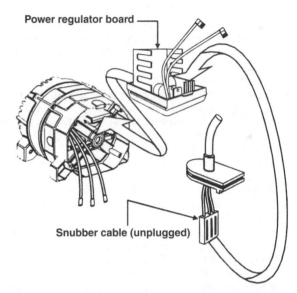

Fig. GC104 – View of the snubber cable which must be unplugged to test AC output. Refer to the text for the correct test procedure.

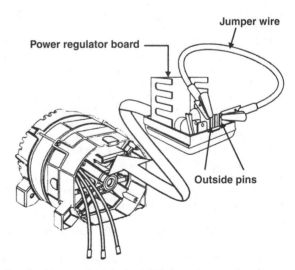

Fig. GC105 – View of the jumper connected to the power board to test AC output. Refer to the text for the correct test procedure.

- If voltage is below 10 volts, the system control board is faulty.
- If voltage still tests 130-150 volts, the power regulator board is faulty.

NOTE: The blue (2) wire must always connect to the outside power regulator board terminal (the terminal nearest to the rotor bearing), and the red (6) wire must always connect to the inner terminal.

9. To test the rotor:

The slip rings are an integral part of the rotor assembly and cannot be replaced.

Clean dirty or rough slip rings with a suitable solvent, then lightly sand with ScotchBrite or fine-grit sandpaper. Do not use steel wool, emery cloth or crocus cloth.

Replace the rotor if the slip rings are cracked, chipped, broken, or grooved.

a. Noting the wiring polarity, disconnect the brush wire connectors.

b. With the VOM set on the R × 1 scale, measure field coil winding resistance through the brushes. Compare the reading to the specification listed at the end of this section.

c. Remove the brush assembly.

d. Measure field coil winding resistance at the rotor slip rings. The resistance should be the same as Step B. On some units, limited access inside the rear bearing carrier housing may require removing the power cable grommet to test the slip rings.

e. Measure field coil winding resistance from each slip ring to a good ground on the rotor shaft or laminations. Resistance should be infinity/no-continuity.

f. If test substeps b, d, and e met specification, the rotor and brushes are in acceptable condition.

g. If test substep b failed specification but substeps d and e met specification, the rotor is good. Inspect the brushes and test each brush for continuity through its terminal. If it is faulty, it must be replaced.

h. If both test substeps b and d failed specification, or if substep e showed continuity, the rotor is faulty. Replace the rotor, then inspect and test the brushes.

j. The rotor can also be dynamically tested by performing a current draw test. To perform this test:

- With the VOM on the DCV scale, accurately note the voltage of a 12V test battery.
- With the VOM on the R × 1 scale, note the rotor winding resistance across the slip rings.
- Using Ohm's Law applicable to this test (I=V÷R), calculate the current flow (I) by dividing the rotor winding resistance (R) into the battery voltage (V), note the calculated amperage.
- Disconnect the brush feed wires. Isolate the brushes from the voltage regulator or bridge rectifier.
- Connect the positive (+) terminal of the test battery to the positive (+) side of the VOM/ammeter.
- Connect the ammeter negative (-) lead to the positive (+) brush terminal (wire #4).
- Connect a test wire from the negative (-) brush terminal (wire #1) to the negative (-) battery terminal.
- Note the field winding current draw on the ammeter.
- Start the engine. With the engine running at rated no-load RPM, note the amperage draw on the ammeter.
- Compare the ammeter reading with the engine running to both the ammeter reading with the engine stopped and the calculated amperage. If the running

amperage is significantly higher than the static or calculated amperage, the rotor has a running short. Replace the rotor. If there is NO amperage draw during the running test, the rotor has an open winding and must be replaced. If the running amperage is approximately the same as the static and the calculated amperage, the rotor is good.

- Stop the engine and disconnect the test equipment.

NOTE: When a rotor fails, the resulting momentary voltage surge can damage the voltage regulator system. If the rotor requires replacement, also test the power regulator and system control board.

10. To test the stator:

NOTE: Both the AC power winding and the excitation winding have blue and red wires. Do not confuse the power winding wires with the excitation winding wires. Compare all readings taken in the following tests to the specifications listed at the end of this section.

a. Disconnect and remove the power regulator board.

b. Using a 12V test battery, connect the battery positive terminal to the positive brush (closest to the rotor bearing), and connect battery negative terminal to the negative brush.

c. Start the generator.

d. Measure AC voltage output at the 120V receptacle and across the blue (6) and red (2) excitation winding wires. A minimum of 60VAC should be produced; a slightly higher output is acceptable.

e. Stop the generator. If all the tests in substep d met specification, the power regulator board is faulty. If the measurement in substep d tested low, proceed with substep f.

f. Disconnect the stator winding lead terminals, being careful not to allow the terminals to contact any part of the generator.

g. Restart the generator.

h. On single-voltage units, measure AC voltage output across the blue (11) to red (22) power winding terminals. The reading should be 60 volts or higher.

i. On dual-voltage units, measure AC voltage output across the blue (11) to red (22) and across the gray (44) to red (22) power winding terminals. The reading should be 60 volts or higher.

j. Stop the generator.

k. Set the VOM to ohms; note all readings in the following tests.

l. On single-voltage units, test resistance across the blue (11) to red (22) AC power winding terminals. Note the reading.

m. On dual-voltage units, test resistance across the blue (11) to red (22) and across the gray (44) to red (22) AC power winding terminals.

n. Individually test resistance from each blue (11), gray (44), and red (22) power winding terminal to any good, clean ground on the stator laminations. Resistance should read infinity/no-continuity.

o. Test resistance across the blue (2) to red (6) excitation winding terminals.

p. Individually test resistance from each blue (2) and red (6) excitation winding terminal to any good, clean ground on the stator laminations. Resistance should read infinity/no-continuity.

q. On single-voltage and dual-voltage units, test resistance across the black (55) to brown (66) battery charge winding terminals. Note the reading.

r. On dual-voltage units, also test resistance across the black (55) to brown (77) battery charge winding terminals.

s. Repeat substeps n and o with terminals (55) to (66A) and (55) to (77A).

t. Individually test resistance from each (55), (66), (66A), (77), and (77A) terminal to any good, clean ground on the stator laminations. Resistance should read infinity/no-continuity.

u. Replace the stator if it fails to meet test specifications.

INDIVIDUAL COMPONENT TESTING

Ground Fault Circuit Interrupter (GFCI)

To test the GFCI:

1. Start the generator and allow a brief no-load warm up period.
2. Press the TEST button. The RESET button should extend. If it does, proceed to Step 3. If the RESET button does not extend, the GFCI is faulty.
3. Press the RESET button. The RESET button should be flush with the TEST button. If the RESET button does not remain flush with the TEST button, the GFCI is faulty.
4. Stop the generator.

NOTE: If the GFCI seems to trip for no apparent reason, check to make sure the generator is not vibrating excessively due to faulty rubber mounts or debris between the generator and frame. Excessive vibration can trip the GFCI.

Diodes

To test each diode:

1. Set the VOM to the R × 1 scale.
2. Place one meter lead on each end of the diode and note the reading.
3. Reverse the meter leads and note the reading.
4. The meter should read continuity in one direction and infinity in the opposite direction. Any diode that reads continuity or infinity in both directions is faulty.

Idle Control System – XL & MC Generators, 3500-7500W

To Test the Switch

1. Access and disconnect the switch wires inside the control box.
2. The switch should test continuity with the switch on and no continuity with the switch off.

To Test System Input

1. Access the idle control wiring inside the control box.
2. With the idle switch OFF, start the generator.
3. Turn the idle switch ON.
4. Measure AC voltage between the blue/white switch output wire and the green wire on the Twistlok receptacle. 120 VAC should be indicated.
5. If 120 VAC is not present, check wiring integrity.

To Test the System Control Board Output

1. Access the control board inside the control panel.

2. Ensure that all wiring and connections are protected from grounding out.
3. With the idle switch off, start the generator.
4. With no load applied to the generator, turn the idle switch on.
5. Leaving the five-terminal connector plugged into the board, measure DC voltage across the two black wire terminals in the connector (Fig. GC106). Output should measure 120 VDC.

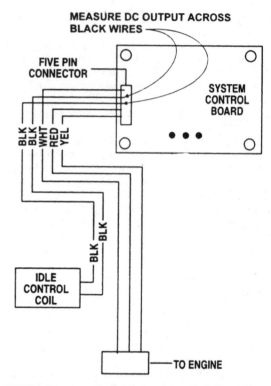

Fig. GC106 – Diagram view of the system control board showing the test terminals needed to measure DC output. Refer to the text for the correct test procedure.

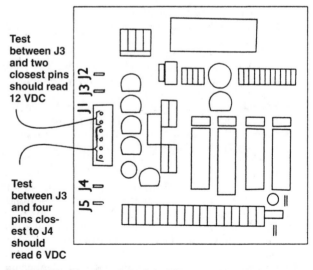

Fig. GC107 – Diagram view of the idle control board showing the board test points. Refer to the text for the correct test procedures.

Stepper Motor Harness Connector

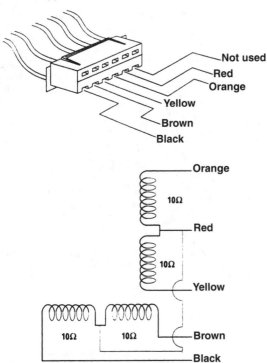

Fig. GC108 – Schematic and diagrammatic views of the stepper motor harness connector showing the color-coded wire circuits and placement. Refer to the text for the correct stepper motor test procedure.

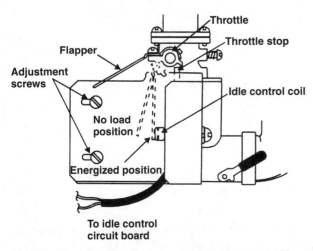

Fig. GC109 – Idle control coil bracket adjustment on older style one-cylinder side valve engines. Refer to the text for the correct idle speed adjustment procedure.

6. If 120 VDC is not present in Step 5, but 120 VAC does show in Step 4 of the previous blue/white to green test, the system control board is faulty. Stop the generator and replace the board.

   a. If the board is faulty, test the idle control solenoid (next test) prior to replacing the board.

   b. After replacing the board, adjust the voltage potentiometer prior to applying a load to the generator.

To Test the Idle Control Solenoid Coil

1. Remove the five-terminal connector from the system control board (Fig. GC106).
2. Test resistance between the two black wire terminals. Resistance should be approximately 1500 ohms.
3. Test resistance between each black wire terminal and ground. Resistance should be infinity/no-continuity.
4. Any resistance readings other than those specified in Steps 2 and 3 indicate a faulty solenoid which must be replaced.

Idle Control System – XL & MC Generators, 8000 & 10000W

To Test Idle Control Board Input

1. Access the idle control board inside the control box (Fig. GC107).
2. Board Pin J3 is the ground terminal for the board.
3. Set the VOM to the VDC scale.
4. Ensure that all wiring and connections are protected from grounding out.
5. With the idle switch OFF, start the generator.
6. Place the positive (+) meter probe on board terminal J2, and place the negative (-) meter probe on Terminal J3. The meter should indicate 12 VDC when the idle switch is moved to ON.
7. Stop the generator.

To Test the Idle Control Switch

1. Remove wires (83) and (84) from the control board.
2. Set the VOM to the R × 1 scale.
3. Connect the meter probes to wires (83) and (84).
   a. With the switch OFF, the meter should read infinity.
   b. With the switch ON, the meter should read less than 0.3 ohm.
4. Any readings other than those specified in step 3 indicates a faulty switch.

To Test the Idle Control Board Output

1. Set the VOM to the VDC scale.
2. Refer to Fig. GC107. Connect the negative (-) meter probe to board terminal J3.
3. With the idle switch OFF, start the generator.
4. While alternately turning the idle switch ON and OFF, individually test the two connector J1 pins closest to terminal J3. The meter will read approximately 12V DC only when the switch moves the stepper motor.
5. While alternately turning the idle switch ON and OFF, individually test the four connector J1 pins closest to terminal J4. The meter will read approximately 12V DC only when the switch moves the stepper motor.
6. Stop the generator.

To Test the Idle Stepper Motor

1. Unplug the five-wire stepper motor harness connector from the idle control board.
2. Set the VOM to the R × 1 scale.
3. Refer to Fig. GC108; connect one meter probe to the red harness terminal.
4. Alternately connect the other meter probe to the orange, yellow, brown, and black harness terminals. Each terminal should read approximately 10 ohms.
5. Test for continuity between each of the five connector terminals and the stepper motor housing (ground). All five tests should read infinity.

6. Any test results other than those specified in Steps 4 and 5 indicate a faulty stepper motor/harness assembly.

Idle Control Systems – Idle-speed Adjustments

Early 1-Cylinder Engines

1. Connect the tachometer/frequency meter to the engine/generator.
2. With the idle-control switch OFF, start the generator, allowing the engine to warm up to normal operating temperature.
3. Turn the idle control switch ON.
4. With the engine idle mixture properly set, the engine should slow to 2100-2400 RPM idle speed (35-40 Hz).
5. If necessary, loosen the three screws holding the control solenoid coil bracket to the engine (Fig. GC109). Then move the bracket right to decrease RPM/frequency or left to increase RPM/frequency. Tighten the screws when the correct idle is obtained.
6. Turn the carburetor idle mixture screw in slowly until the engine begins to run rough (lean); note the screw position.
7. Turn the carburetor idle mixture screw out slowly until the engine begins to run rough (rich); note the screw position.
8. Turn the idle mix screw in, midway between lean and rich.
9. Turn the idle control switch OFF. Note engine acceleration smoothness.
10. If the engine hesitates when accelerating, readjust the idle mixture slightly richer. Retest idle down and acceleration.

Early V-twins

1. Connect the tachometer/frequency meter to the engine/generator.
2. With the idle control switch OFF and all loads unplugged, start the engine and allow it to warm up to normal operating temperature. When warmed and under no load, the engine should be running 3720-3750 RPM (generator frequency: 62.0-62.5 Hz).
3. If necessary, adjust the speed adjust screw to obtain the correct RPM/frequency (Fig. GC110).
4. Load the generator to its rated capacity. If engine speed drops below 3420 RPM (57 Hz):
   a. Disconnect the load.
   b. Bend the throttle restrictor tang 1/16 in. (1.6 mm) toward DECREASE.
   c. Readjust the no-load speed adjust screw to obtain 3720-3750 RPM (62.0-62.5 Hz).
   d. Reapply the load and recheck RPM/Hz drop.
   e. Repeat the readjustment procedure until RPM/Hz remains above the specification in Step 4. Take care not to bend the tang more than 1/16 in. (1.6 mm) at a time.
5. Disconnect the generator load.
6. Manually hold the throttle against the carburetor idle speed adjust screw.
7. With the throttle held at idle, adjust the screw to bring the idle to 1440-1680 RPM (24-28 Hz).
8. Turn the carburetor idle mixture screw (Fig. GC111) clockwise until the RPM/frequency begins to decrease (lean). Note the screw position.
9. Turn the idle mixture screw counterclockwise until RPM/frequency increases and then begins to drop (rich). Note the screw position.

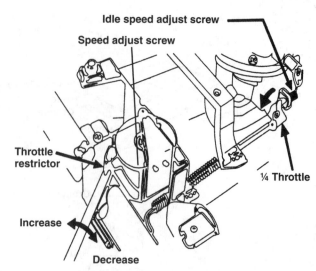

Fig. GC110 – *View showing the no-load frequency adjustment points on V-twin engines. Refer to the text for the correct adjustment procedure.*

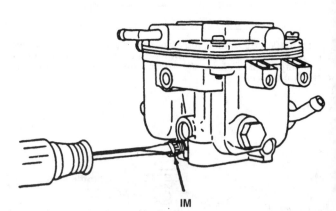

Fig. GC111 – *Drawing showing location of carburetor idle mixture screw (IM).*

10. Readjust the mixture screw to the halfway point between lean and rich.

11. Release the throttle.

12. Turn the idle control switch ON. The engine/generator speed should drop to 2400-2700 RPM (40-45 Hz).

13. If Step 12 does not meet the specification:

   a. Loosen the four screws holding the idle control solenoid bracket to the engine heat shield.

   b. Move the solenoid away from the engine to decrease RPM/frequency or toward the engine to increase RPM/frequency. With the generator unloaded and the idle control switch ON, the unit will auto-idle at the Step 12 specifications.

   c. Tighten the bracket screws.

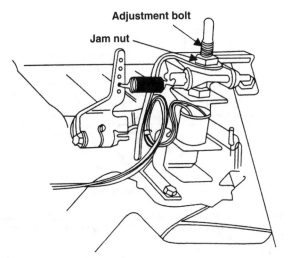

Fig. GC112 – View of idle solenoid adjustment on Generac engines. Refer to the text for the correct adjustment procedure.

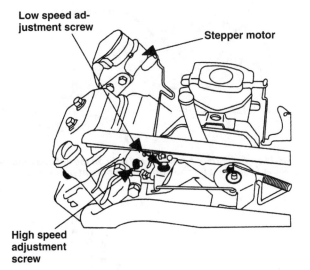

Fig. GC113 – View of the idle control adjustments on V-twin engines used on 8kW and 10kW EXL model generators. Refer to the text for the correct adjustment procedure.

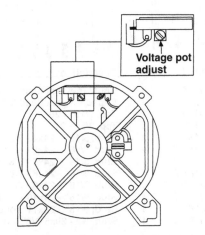

Fig. GC114 – Adjustment point for the endbell-mounted voltage regulator. Refer to the text for the correct adjustment procedure.

GN-190, -220, -320, -360, & -410 Engines

1. With the idle control switch OFF and all loads unplugged, connect the tachometer/frequency meter to the engine.
2. Start the engine and allow it to warm up to normal operating temperature.
3. The engine should be running at 3690-3810 RPM (61.5-63.5 Hz).
4. Turn the idle control switch on. The idle control solenoid should energize, pulling the throttle to 2400-2700 RPM idle speed (40-45 Hz).
5. If the idle speed does not meet the Step 4 specifications:
   a. Loosen the solenoid jam nut. (Fig. GC112)
   b. Turn the solenoid bolt clockwise to increase idle speed or counterclockwise to decrease idle speed.
   c. When the engine idle speed/Hz is within specification, tighten the jam nut.

EXL 8kW & 10kW (Late V-twins)

1. With the idle control switch OFF and all loads unplugged, connect the tachometer/frequency meter to the engine.
2. Start the engine and allow it to warm up to normal operating temperature.
3. Refer to Fig. GC113. Turn the idle switch ON.
4. Using a Phillips screwdriver:
   a. Adjust the low-speed adjustment screw to 2700 RPM (45 Hz).
   b. Turn the idle control switch OFF.
   c. Adjust the high-speed adjustment screw to 3720 RPM (62 Hz) no-load.
   d. Turn the idle control switch ON and OFF several times; verify RPM/Hz settings. Readjust as necessary.

Voltage Regulators (120/240 VAC Units With Regulator Mounted In Generator End Housing Or System Control Board Mounted In Control Panel Box)

To adjust the regulator or board:
1. Connect an AC voltmeter and a tachometer/frequency meter to the generator.
2. Start the engine and, with all loads unplugged, allow the generator to warm up to normal operating temperature.
3. At operating temperature, the unit should be running at 3720 RPM (62 Hz), with no-load output at 124 VAC (line-to-neutral) or 248 VAC (line-to-line).
4. If necessary, turn the voltage pot adjustment (Fig. GC114, GC115, or GC116) to meet the specifications in Step 3.
5. If the regulator has a gain adjustment potentiometer (Fig. GC116), connect a 60-watt light bulb to the generator after completing Steps 1-4. If the bulb flickers, adjust the gain pot until the flicker stops.
6. Stop the generator and disconnect all test equipment.

GENERATOR DISASSEMBLY (Brush-Style Units)

1. Disconnect the engine spark plug lead(s).
2. Remove the rear bearing carrier housing access cover.
3. Carefully remove the brush holder assembly from the rear housing. To prevent diode damage on the diode/brush units, leave the brush wires connected until the brush holder has been removed.

4. Unlock and remove the control panel harness connector from the plug at the rear of the panel, then unplug the stator connectors in the rear housing. If the rear housing or stator is being replaced, the locking tabs on the stator wire terminals need to be carefully bent back so the wires can be removed from the connectors and housing. Note the wiring orientation.

5. Remove the nuts holding the rear housing feet mounts to the cradle frame crossmember.

6. Remove the ground wire that connects the rear housing to the frame.

7. Carefully raise the generator, then install a temporary support, such as a wooden block, under the engine adapter so the rear housing mount studs will clear the crossmember.

8. Remove the four bolts holding the rear housing and stator to the engine adapter.

9. Carefully remove the rear housing from the rotor bearing and stator.

10. Carefully grip the stator and remove it from the generator, noting for reassembly, that the stator harness is in the nine o'clock position.

11. To remove the rotor (recommended method):

   a. Remove the rotor throughbolt.

   b. Inspect the inner diameter of the outer end of the rotor shaft. If the shaft is threaded, proceed to substep c; if it is not threaded :

    • Use rotor removal kit #41079 (Fig. GC117) or equivalent.

    • If the rotor throughbolt is ¼ in. (6.4 mm) diameter, tap the rotor shaft 3/8"-24, approximately 1 in. deep.

    • If the rotor throughbolt is 5/16 in. diameter, tap the rotor shaft 7/16 in.-20, approximately ¾-1 in. deep.

   c. Select a stud from the kit which, when threaded through the rotor shaft into the engine crankshaft, will be recessed into the rotor shaft approximately ½ in. (Fig. GC118).

   d. Insert the proper bolt into the threaded end of the rotor shaft, tightening the bolt against the stud.

   e. Carefully using a metal hammer with firm blows, use a tap-tighten-tap-tighten procedure to break the rotor free from the engine crankshaft taper.

   f. Remove the rotor shaft bolt and the long crankshaft stud.

12. To remove the rotor (alternate method):

   a. Remove the rotor throughbolt.

   b. Using a slide-hammer with the proper adapter threaded into the rotor shaft, tap the slide-hammer until the rotor breaks free from the engine crankshaft taper.

GENERATOR DISASSEMBLY (Brushless Units)

The rotor cannot be removed without first removing the stator housing assembly. Exercise caution when handling the stator and rotor so the windings and winding coatings are not damaged.

1. Disconnect the engine spark plug lead(s).

2. Remove the rear-bearing area housing cover.

3. Disconnect and remove the capacitor and the main harness.

4. Remove the nuts holding the rear housing feet mounts to the cradle frame crossmember.

5. Carefully raise the generator, then install a temporary support, such as a wooden block, under the engine adapter so the rear housing mount studs will clear the crossmember.

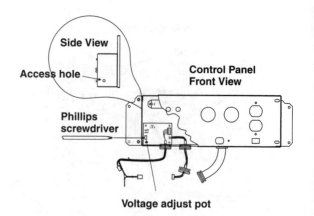

Fig. GC115–Adjustment point for the voltage regulator mounted on the control board. Refer to the text for the correct adjustment procedure..

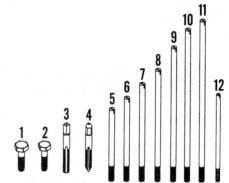

Fig. GC116 – Adjustment point for the gain on the voltage regulator. Refer to the text for the correct adjustment procedure.

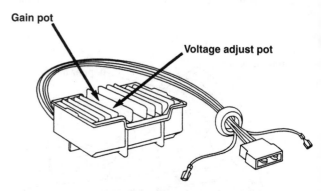

Item	Part #	Qty	Description
1.	63184	1	Capscrew, hex head 7/16" - 20x1½
2.	49472	1	Capscrew, hex head 3/8" - 24x1½
3.	63183	1	Tap, 7/16" - 20
4.	63182	1	Tap, 3/8" - 24
5.	63181	1	Stud, 5/16" - 24x4 - 7/8" long
6.	63181-F	1	Stud, 5/16" - 24x5 - ¼ " long
7.	63181-A	1	Stud, 5/16" - 24x5 - 7/8" long
8.	63181-B	1	Stud, 5/16" - 24x6 - 7/8" long
9.	63181-E	1	Stud, 5/16" - 24x8 - 7/8" long
10.	63181-C	1	Stud, 5/16" - 24x10 - ¼" long
11.	63181-D	1	Stud, 5/16" - 24x11" long
12.	63181-G	1	Stud, ¼" - 20x5-3/8" long

Fig. GC117 – View of the components of the rotor removal kit #41079. Refer to the text or the kit instructions for proper kit usage.

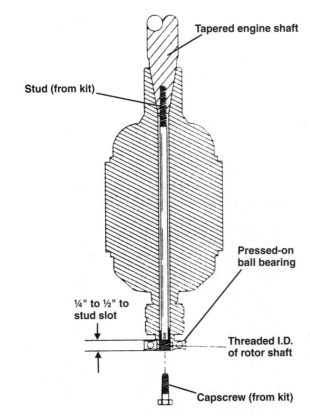

Fig. GC118 – Cross-section view of rotor mounted on crankshaft, showing usage of the removal kit components.

6. Remove the ground wire from the stator housing that connects to the frame.

7. Remove the right and left side engine adapter fan guard grilles.

8. Remove the four nuts holding the stator housing assembly to the engine adapter. If a stator housing stud comes out with a nut, that is acceptable.

9. With a helper holding the stator/end housing assembly, use a soft mallet and wooden block to tap outwardly and alternately on the end housing flanges so the assembly will break free from the engine adapter. Remove the assembly.

10. To remove the rotor, refer to Steps 11 and 12 in the Disassembly (Brush-Style Units) section.

GENERATOR REASSEMBLY (Brush-Style Units)

1. Install a temporary support, such as a wooden block, under the engine adapter so the rear housing mount studs will clear the frame crossmember.

2. Service the rotor bearing, if needed.

3. Ensure that the matching tapers on the engine crankshaft and inside the rotor shaft are clean and dry. Install the rotor onto the crankshaft. Install the rotor throughbolt and washer, then torque the bolt to 145-240 in.-lb. (12-20 ft.-lb., or 16.5-27.0 N•m).

4. Carefully holding the stator with the harness in the nine o'clock position, fit the stator over the rotor, seating the stator squarely against the engine adapter flange. With the harness in the nine o'clock position, the stator lamination alignment groove in the two o'clock position should align with the notch on the engine adapter bolt boss.

5. If the stator wires were removed from the rear housing grommet, reinstall them at this time.

6. Carefully fit the rear housing onto the rotor bearing and against the stator, insuring that the indexing roll pin in the housing's seven o'clock position (viewed from outside the housing) aligns with the stator-lamination groove in that position. Ensure that the stator fits snugly inside both the engine adapter and the end housing, with no gaps.

7. Install and finger-tighten the four bolts holding the rear housing and stator to the engine adapter. Torque the bolts to 25-70 in.-lb. (2.8-8.0 N•m), alternating in increments between the bolts.

8. Remove the engine spark plug(s), then rotate the rotor to check for interference between the rotor and stator. Correct any misalignment before proceeding. Reinstall the spark plug(s).

9. If the protective mesh tube was removed from the harness, reinstall it now.

10. If any wiring terminals were removed from their connectors during disassembly, refer to the proper wiring diagram and reinstall them into the connectors. The backs of the connectors have raised numbers corresponding to specific wire numbers.

11. Reconnect the control panel harness connector to the panel plug; lock as necessary.

12. Remove the temporary engine adapter support, setting the rear housing feet mount studs into the frame crossmember holes. Install the mount stud nuts, then torque the nuts 145-205 in.-lb. (12-17 ft.-lb., or 16.5-23.0 N•m).

13. Reconnect the ground wire between the rear housing and frame.

14. Connect the blue brush wire to the top terminal (J1) of the brush assembly, and connect the red brush wire to the bottom terminal (J2). with the brush-assembly index tab facing the engine (Fig. GC103), carefully mount the assembly into the rear housing, first finger-tightening the mount screws, then torquing the screws 20-70 in.-lb. (2.25-7.9 N•m).

15. Install the rear bearing housing cover, making sure that the louvers are positioned IN and UP, and that the top tab fits against the housing grommet. Torque the screws 20-70 in.-lb. (2.25-7.9 N•m).

16. Reconnect the engine spark plug leads.

GENERATOR REASSEMBLY (Brushless Units)

1. Install a temporary support, such as a wooden block, under the engine adapter so the rear housing mount studs will clear the frame crossmember.

2. Service the rotor bearing, if needed. Note that the rotor bearing is held in place with a snap ring.

3. Ensure that the matching tapers on the engine crankshaft and inside the rotor shaft are clean and dry. Install the rotor onto the crankshaft. Install the rotor throughbolt and washer, then torque the bolt 145-240 in.-lb.(16.5-27.0 N•m).

4. Carefully position the stator housing assembly over the rotor and onto the rear bearing while aligning the mounting studs or stud holes with the engine adapter holes.

5. Install the stator-to-adapter stud nuts. Torque the nuts 180 in.-lb.(20.0-20.5 N•m).

6. Remove the engine spark plug(s), then rotate the rotor to check for interference between the rotor and stator. Correct any misalignment before proceeding. Reinstall the spark plug(s).

7. Install the right and left side engine adapter fan-guard grilles.

8. Install the ground wire between the stator housing and frame.

9. Carefully raise the generator just enough to remove the temporary support, then lower the generator so the rear housing feet mount studs fit into the crossmember holes. Install the mount-stud nuts and tighten 145-205 in.-lb. (16.5-23.0 N•m).

10. Install and connect the main harness and capacitor.

11. Install the rear bearing area housing cover. Be sure the louvers are DOWN.

12. Reconnect the engine spark plug lead(s).

RESISTANCE SPECIFICATION TABLES

GENERAC

Model Number	Unit Designation	Rotor Winding	Main Ac Power	Excitation (Dpe) Ω	Winding
0413-0	PP5000	25.6-33.3	.23-.29/.24-.34	1.26-1.67	N/U
0415-0	SV-2400	20.9-27.1	.33-.42/ N/U	2.24-3.05	N/U
0416-0	SVT4200	21.0-27.2	.33-.39 ea.	1.07-1.42	N/U
0417-0	SVP5000	25.6-33.3	.23-.29/.24-.34	1.26-1.67	N/U
0421-0	SVP5000E	25.6-33.3	.23-.29/.24-.34	1.26-1.67	.07/.07
0435-0	SV4200	22.2	.33-.39/.33-.39	1.07-1.42	N/U
0455-0	PP5000	12.9-16.7	.38-.51/.42-.51	1.83-2.41	.14-.19/.16-.21
0633-0	4150XL	12.9-16.7	.38-.51/.42-.51	1.83-2.41	.14-.19/.16-.21
1006-0	MF6500	10.2-12.2	.22-.26/.22-.26	1.49-1.81	.07-.11/.07-.11
1010-0	PP5000	25.6-33.3	.23-.29/.24-.34	1.26-1.67	N/U
1011-0	PP5000	25.6-33.3	.23-.29/.24-.34	1.26-1.67	N/U
1013-0	7500XL	7.6-9.1	.20-.25/.20-.25	1.56-1.91	.05-.09/.06-.11
1016-0	MF2500	9.8-12.2	.62-.74/ N/U	5.82-7.05	.12-.16/.12-.16
1017-0	7500XL	7.6-9.1	.20-.25/.20-.25	1.56-1.91	.05-.09/.06-.11
1019-0	7500EXL	7.6-9.1	.20-.25/.20-.25	1.56-1.91	.05-.09/.06-.11
1021-0	6500MF	6.7-8.7	.22-.26/.22-.26	1.49-1.81	.07-.11/.07-.11
1140-0	PP5000T	25.6-33.3	.23-.29/.24-.34	1.26-1.67	N/U
1140-1	PP5000	2.04	.21	1.51	N/U
1140-2	PP5000	2.68	.33	1.46	N/U
1193-0	M2500	9.8-12.2	.62-.74/ N/U	5.82-7.05	.12-.16/.12-.16
1194.0	3000	18.5-22.4	.25-.31/ N/U	1.97-2.37	N/U
1263-0	ET2500	5.2	.7	2.3	N/U
1277-0	BBPS	7.6-9.1	.20-.25/.20-.25	1.56-1.91	.05-.09/.07-.11
1305-0	4200WATT	25.6-33.3	.23-.29/.24-.34	1.26-1.67	.07/.07
1306-0	5000CLI-GRD	25.6-33.3	.23-.29/.24-.34	1.26-1.67	.07/.07
1311-0	3000LT	18.5-22.4	.25-.31/ N/U	1.97-2.37	N/U
1312-0	SVP5000	25.6-33.3	.23-.29/.24-.34	1.26-1.67	.07/.07
1313-0	3500LW	12.9-16.7	.38-.51/.42-.51	1.83-2.41	.14-.19/.16-.21
1314-0	5500LW	9.3-12.0	.26-.32/.28-.35	2.26-2.8	.08-.12/.08-.12
1315-0	7500ELW	7.6-9.1	.20-.25/.20-.25	1.56-1.91	.05-.09/.07-.11
1329-0	SE5000	25.6-33.3	.23-.29/.24-.34	1.26-1.67	N/U
1338-0, -1	9kW	2.95	.30	1.1	0.082
1339-0	SE10000	2.95	.30	1.1	0.082
1340-0	5500LW	9.3-12.0	.26-.32/.28-.35	2.26-2.80	.08-.12/.08-.12
1356-0	SVP5000T	2.68	.33	1.46	N/U
1415-0	10000EXL	2.95	.30	1.1	0.082
5500-0	T24110	39.2	0.40	2.37	N/U
5501-0	T2411	39.2	0.40	2.27	N/U
5504-0	T4014	46.4	.41/.41	2.45	N/U
5505-0	T4015	46.4	.41/.41	2.45	N/U
5506-0	T4016	46.4	.41/.41	2.45	N/U
5507-0	T4017	46.4	.41/.41	2.45	N/U
5508-0	T4029	46.4	.41/.41	2.45	N/U
5509-0	S2410	24.0	0.34	2.46	0.26
5510-0	S2411	24.0	0.34	2.46	N/U
5511-0	S4012	23.7	.36/.36	1.53	N/U
5512-0	S4013	23.7	.36/.36	1.53	N/U
5513-0, -1	S4014	23.7	.36/.36	1.53	N/U
5514-0	S4015	23.7	.36/.36	1.53	N/U
5516-0	S4016	23.7	.36/.36	1.53	N/U
5517-0	S4017	23.7	.36/.36	1.53	N/U
5518-0	S4029	23.7	.36/.36	1.53	N/U
5521-0	S4014F	25.6-33.3	.26/.26	1.81	0.19
5522-0	S4015F	25.6-33.3	.26/.26	1.81	0.19

GENERAC (continued)

5523-0	S4016F	25.6-33.3	.26/.26	1.81	0.19
5524-0	S4017F	25.6-33.3	.26/.26	1.81	0.19
5525-0	S4029F	25.6-33.3	.26/.26	1.81	0.19
5526-0	S5019F	25.6-33.3	.26/.26	1.81	0.19
5527-0	S5020F	25.6-33.3	.26/.26	1.81	0.19
5528-0	S5021F	25.6-33.3	.26/.26	1.81	0.19
5529-0	C2410	24.0	0.34	2.46	0.26
5530-0	C2411	24.0	0.34	2.46	0.26
5531-0	C2412	24.0	0.34	2.46	0.26
5532-0	C2413	24.0	0.34	2.46	0.26
5541-0	C4018	25.6-33.3	.26-.26	1.81	0.19
5541-1	C4018	25.6-33.3	.26/.26	1.81	0.11/0.19
5542-0	C4019	25.6-33.3	.26/.26	1.81	0.19
5542-1	C4019	25.6-33.3	.26/.26	1.81	0.11/0.19
5543-0	C4020	25.6-33.3	.26/.26	1.81	0.19
5544-0	C4021	25.6-33.3	.26/.26	1.81	0.19
5546-0	C4022	25.6-33.3	.26/.26	1.81	0.19
5547-0	C4023	25.6-33.3	.26/.26	1.81	0.19
5548-0	C4024	25.6-33.3	.26/.26	1.81	0.19
5549-0	C4025	25.6-33.3	.26/.26	1.81	0.19
5552-0	C4028	25.6-33.3	.26/.26	1.81	0.19
5553-0	C4029	25.6-33.3	.26/.26	1.81	0.19
5556-0	C4032	25.6-33.3	.26/.26	1.81	0.19
5557-0	C4033	25.6-33.3	.26/.26	1.81	0.19
5558-0	C4034	25.6-33.3	.26/.26	1.81	0.19
5559-0	C5010	8.4	.28/.20	1.88	0.22
5559-1	C5000	25.6-33.3	.26/.26	1.80	0.1/0.1
5560-0	C5011	8.4	.28/.20	1.88	0.22
5561-0	C5012	8.4	.28/.20	1.88	0.14
5561-1	C5000	25.6-33.3	.26/.26	1.80	0.1/0.1
5563-0, -1, -2	C8010	12.6	.18/.18	1.13	0.4/0.6
5564-0, -2	C8011	12.6	.18/.18	1.13	0.4/0.6
5564-1	C8011	12.3	.15/.15	1.72	N/U
5565-0, -1, -2	C8012	12.6	.18/.18	1.13	0.4/0.6
5566-0, -1, -2	C8013	12.6	.18/.18	1.13	0.4/0.6
5567-0, -1	S4023	23.7	.36/.36	1.53	N/U
5568-0	S4018F	25.6-33.3	.26/.26	1.81	0.19
5569-0	T4023	46.4	.41/.41	2.45	N/U
5570-0	2.4kW	24.0	0.34	2.46	0.26
5571-0	4.0kW	23.7	.36/.36	1.53	N/U
5575-0	C5000	15.5	.26/.26	1.80	0.1/0.1
8479-0	G1450	25.8	0.47	1.50	0.10
8616-0	328220	46.0	.32/.32	1.95	N/U
8749-0	H1450	13.5	0.47	1.50	N/U
8750-0, -1, -2	H2450	37.5	0.40	2.50	N/U
8751-0, -1, -2	H4050	46.0	0.6/1.2	3.00	N/U
8752-0, -1, -2	G4050	46.0	0.6/1.2	3.00	N/U
8772-0	G1450	13.5	0.47	1.50	0.10
8773-0	G2450	37.5	0.40	2.50	N/U
8834-0	G1000	16.4	1.70	6.40	0.28
8835-0	G1600	22.7	0.52	3.20	0.19
8836-0	G2600	25.5	0.36	3.75	0.16
8837-0	G4000	19.4	0.22	3.70	0.10
8838-0	S1000	16.4	1.70	6.40	N/U
8839-0, -1	S1500	22.7	0.52	3.20	N/U
8840-0	S2400	25.5	0.30	2.90	N/U
8840-1	S2400	36.5	0.30	2.90	N/U
8865-0, -1, -2	L4000	7.4	.25-.56	2.40	N/U
8865-3, -4, -5	L4000	48.0	0.2/0.6	1.95	N/U
8865-6	L4000	25.1	0.29	1.50	N/U
8865-7	L4000	27.2	.16/.31	1.72	N/U
8865-8	L4000	25.6-33.3	.26/.26	1.81	0.19
8866-0, -1	L4000E	7.4	.26/.56	2.00	0.02
8866-2, -3	L4000E	48.0	.25/.56	2.00	0.02
8870-0	S4000	48.0	0.32	2.20	N/U
8870-1, -2	S4000	26.2	0.32	2.20	N/U

GENERAC (continued)

8871-0, -1	S4001	48.0	0.32	2.20	N/U
8872-0, -1	S5000	8.4	0.28	1.75	N/U
8876-0, -1	L5000	8.4	.20/.28	2.60	N/U
8877-0, -1	L5000E	8.4	.20/.28	2.60	0.20
8882-0	S3000T	23.1	0.2/0.6	0.80	0.02
8895-0, -1	R4000	7.4	.25/.56	2.40	N/U
8896-0	R5000	8.4	.20/.28	2.60	N/U
8905-0	S4002	48.0	.20/.60	2.20	N/U
8905-1	S4002	26.2	.20/.60	2.20	N/U
8906-0	L2400	36.5	0.40	2.40	N/U
8911-0, -1	L4001	7.4	.25/.56	2.40	N/U
8911-2, -3	L4001	15.5	.22/.43	1.74	N/U
8911-4	L4001	14.3	.22/.43	2.61	N/U
8915-0	L4002	48.0	.20/.60	1.95	N/U
8915-1	L4002	25.1	0.29	1.50	N/U
8915-2	L4002	27.2	0.16	1.72	N/U
8915-3	L4002	25.6-33.3	.26/.26	1.81	0.19
8953-0	H2250	21.8	0.40	2.50	0.12
8954-0	H4000	26.5	0.32	2.10	0.08
8967-0, -1	A1400	22.7	0.52	2.35	0.19
8967-2	A1400	20.17	0.76	2.50	0.52
8968-0, -1	A2400	25.5	0.30	2.14	0.14
8968-2	A2400	24.0	0.34	2.46	0.26
8969-0, -1	A4000	15.5	0.43	1.74	0.10
8969-2	A4000	14.3	0.23	2.61	0.11
8970-0, -1	S2400	25.5	0.30	2.14	N/U
8971-0, -1, -2	S4000	25.1	0.29	1.50	N/U
8971-3, -4	S4000	23.7	0.36	1.53	N/U
8972-0, -1, -2	S4001	25.1	0.29	1.50	N/U
8972-3, -4	S4001	23.7	0.36	1.53	N/U
8973-0, -1	S4002	27.2	.15/.32	2.32	N/U
8973-2, -3	S4002	25.6-33.3	.26/.26	1.81	0.19
8983-0, -1	S5002	27.3	0.21	2.32	N/U
8983-2, -3	S5002	27.3	0.21	1.91	N/U
8983-4, -5	S5002	25.6-33.3	.26/.26	1.81	0.19
8992-0	L7500E	14.5	0.16	0.88	0.15
9014-0	R4002D	15.5	0.22	1.76	N/U
9014-1	R4000	14.3	0.23	2.61	N/U
9023-0	L4001	15.0	.22/.43	1.74	N/U
9063-0	L4003	27.2	0.16	1.72	N/U
9063-1	L4003	25.6-33.3	.26/.26	1.81	0.19
9072-0	G3000	24.0	0.34	2.46	0.26
9073-0	G950	16.4	1.70	6.40	0.28
9074-0	G1700	22.7	0.52	3.20	0.19
9075-0	G6000	25.6-33.3	.26/.26	1.81	0.19
9076-0	G7000E	8.4	.20/.28	2.60	0.20
9078-0	4.0kW	48.0	.20/.60	0.80	0.02
9079-0	5.0kW	32.0	.20/.30	2.0	0.20
9091-0	S4002	27.2	.17/.35	1.26	N/U
9092-0, -1	HD3000	36.5	0.40	2.70	N/U
9093-0	HD6000	25.6-33.3	.26/.26	1.81	0.19
9094-0	S2400	24.0	0.34	2.46	0.26
9099-0, -1, -2	R6000	13.0	0.27	1.81	0.18
9100-0, -1, -2	R8000	13.0	0.27	1.81	0.18
9101-0, -1, -2	R8000	12.6	.18/.18	1.13	0.4/0.6
9101-3	R8000	12.6	.18/.18	1.13	0.4/0.6
9102-0, -1	6.0kW	13.0	.27/.27	1.80	0.18
9103-0, -1	8.0kW	14.5	.16/.16	1.30	0.15
9108-0	T2500	39.2	0.40	2.37	N/U
9109-0	T4000	46.4	.41/.41	2.45	N/U
9111-0	T4001	46.4	.41/.41	2.45	N/U
9157-0	L1400	19.2	0.76	2.50	N/U
9158-0	L2400	24.0	0.34	2.46	N/U
9159-0	L4004	25.8	0.26	1.81	N/U
9173-0	S4001	25.6-33.3	.26/.26	1.81	0.19
9184-0	4.0kW	7.4	.88/1.31 x 2	2.60	0.20

GENERAC (continued)

(3-phase stator)					
9185-0	T4000	46.4	.41/.41	2.45	N/U
9186-0	T4000	46.4	.41/.41	2.45	N/U
9214-0	4.0kW	7.4	.25/.56	2.00	0.02
9215-0	5.0kW	8.4	.20/.28	2.60	0.20
9216-1	5.0kW	15.5	.26/.29	1.76	0.09
9216-2, -3	5.0kW	12.4	.33/.33	1.40	0.15
9217-0	2.4kW	39.2	0.40	2.37	N/U
9219-0	EC4000D	15.5	.41/.52	2.95	.15/.15
9261-0, -1, -2	EC6401	12.6	.24/.24	1.59	1.0/1.2
9288-0, -1, -2	R8000	12.6	.18/.18	1.13	0.4/0.6
9288-3	R8000	12.6	.18/.18	1.70	N/U
9298-0	T4000	46.4	.41/.41	2.45	N/U
9325-0	3.0kW	25.6-33.3	.36/.36	1.53	N/U
9331-0	M2400	23.8	0.34	2.46	0.26
9332-0	M4000	23.7	.36/.36	1.53	N/U
9333-0	M4000	23.7	.36/.36	1.53	N/U
9334-0	M750	16.4	1.70	6.40	0.28
9335-0	4.0kW	25.6-33.3	.24/.26	1.80	0.19
9386-0, -1	C8011	12.3	.15/.15	1.72	N/U
9387-0	S3000	24.0	0.34	2.46	0.26
9397-0	C5032	25.6-33.3	.26/.26	1.81	.11/.19
9415-0, -1	S2435	24.0	0.34	2.46	0.26
9419-0	S2410	23.8	0.34	2.46	0.26
9420-0	S4014	23.7	.36/.36	1.53	N/U
9421-0	C4019	25.6-33.3	.26/.26	1.81	.11/.19
9423	C4019	25.6-33.3	.26/.26	1.81	.11/.19
9429-0	C5000	8.4	.20/.28	1.88	0.14
9429-1	C5000	15.5	.26/.26	1.76	.09/.09
9430-0	C5036	25.6-33.3	.26/.26	1.81	.11/.19
9430-1	C4019	25.6-33.3	.26/.26	1.81	.11/.19
9441-1, -4, -5	3500XL	12.9-16.7	.38-.51/.42-.51	1.83-2.41	.14-.19/.16-.21
9443-0	HD3000	24.0	0.34	2.46	0.26
9450-0	S5038	25.6-33.3	.24/.30	1.67	.12/.21
9456-0	S2410	25.5	.30	2.14	N/U
9458-0	S4014	25.8	0.43	1.74	.11/.19
9459-0	S4014	27.2	.26/.26	1.81	.11/.19
9460-0	S4014	23.49	0.29	1.50	N/U
9477-0	C5032	25.6-33.3	.26/.26	1.81	.11/.19
9478-0, -1	ET2100	23.8	0.8/0.8	2.74	N/U
9479-0, -1	HB2100	23.8	0.8/0.8	2.74	N/U
9482-0	G1000	16.4	1.70	6.40	0.28
9486-0	C5030	25.6-33.3	.26/.26	1.81	.11/.19
9487-0	C5031	25.6-33.3	.26/.26	1.81	.11/.19
9493-0	ET2100	23.8	0.8/0.8	2.74	N/U
9495-0	EC2100	23.8	0.8/0.8	2.74	N/U
9528-0, -1	EP5000	25.6-33.3	.26/.26	1.80	0.1/0.1
9530-0	2.4kW	24.0	0.34	2.46	0.26
9538-0, -1, -2	3000XL	12.9-16.7	.45-.57	3.66	.17/.17
9539-0, -1	3000XL	12.9-16.7	.48-.58/.55-.66	2.05-3.44	.11-.16/.11-.16
9547-0	2500XL	23.8	0.34	2.46	0.26
9551-0	2200XL	23.8	0.40	2.74	.28/.28
9663-0	3500XL	12.9-16.7	.38-.51/.42-.51	1.83-2.41	.14-.19/.16-.21
9672-0	2800XE	23.8	0.31	2.00	N/U
9676-0	S4023	23.7	.36/.36	1.53	N/U
9692-0	3500XE	12.9-16.7	0.31	2.00	N/U
9699-0	3500XLS	12.9-16.7	.38-.51/.42-.51	1.83-2.41	.14-.19/.16-.21
9700-0	EP4600	25.6-33.3	.27/.27	1.35	N/U
9704-0, -1, -2, -3	3500XL	12.9-16.7	.38-.51/.42-.51	1.83-2.41	.14-.19/.16-.21
9716-0	SV2400	24.0	0.34	2.46	0.26
9716-1, -2	SV2400	20.9-27.1	.33-.42	2.24-3.05	N/U
9717-2	SV4200	21.0-27.2	.33-.39/.33-.39	1.07-1.42	N/U
9718-0	SVT4200	23.7	.36/.36	1.53	N/U
9718-1, -2	SVT4200	21.0-27.2	.33-.39/.33-.39	1.07-1.42	N/U
9719-0	SVP5000	25.6-33.3	.26/.26	1.80	0.1/0.1
9719-1, -2	SVP5000	25.6-33.3	.23-.29/.24-.34	1.26/1.67	N/U

GENERAC (continued)

9720-0	SVL5000	25.6-33.3	.24/.30	1.67	.12/.20
9720-1, -2	SVT5000	25.6-33.3	.23-.29/.24-.34	1.26-1.53	N/U
9749-4	ET1500	3.9	2.1	4.9	N/U
9753-0	4000XL	12.9-16.7	.48-.58/.55-.66	2.05-3.44	.11-.16/.11-.16
9777-0, -1	4000XL	12.9-16.7	.38-.51/.42-.51	1.83-2.41	.14-.19/.16-.21
9778-0, -1, -2,-3, -4, -5, -6	5500XL	9.3-12.0	.26-.32/.28-.35	2.26-2.80	.08-.12/.08-.12
9779-0, -1, -2	6500XL	10.2-12.2	.22-.26/.22-.26	1.49-1.81	.07-.11/.07-.11
9780-0, -1	8000XL	8.6	.18/.18	2.36	.05/.05
9781-0, -1	10000XL	14.8	.13/.13	1.70	1.75
9794-0, -1, -2	3250XP	20.8	0.36	1.20	N/U
9795-0, -1	5500XEP	17.7	.15/.15	1.00	N/U
9797-0, -1, -2, -3, -4	5500EXL	9.3-12.0	.25-.32/.28-.35	1.93-2.37	.08-.12/.08-.12
9798-0, -1	6500EXL	10.2-12.2	.21-.26/.21-.26	1.49-1.81	.07-.11/.07-.11
9799-0, -1	SVP5000E	25.6-33.3	.23-.29/.24-.34	1.02-1.53	.05-.10/.05-.10
9800-0, -1	8000EXL	8.6	.20/.20	1.48	.04/.03
9801-0, -1, -2, -3, -4	10000EXL	14.8	.148/.148	1.70	.175
9801-5	10000EXL	2.95	.30	1.10	0.082
9802-0, -1	4000MC	12.9-16.7	.38-.51/.42-.51	1.83-2.41	.14-.19/.16-.21
9803-0, -1, -2	5500MC	9.3-12.0	.26-.32/.28-.35	2.26-2.80	.08-.12/.08-.12
9805-0, -1	6500MC	10.2-12.2	.22-.26/.22-.26	1.49-1.81	.07-.11/.07-.11
9807-0, -1, -2	8000MC	8.6	.18/.18	1.48	.05/.05
9809-0, -1	10000MC	14.8	.13/.13	1.70	1.75
9816-0,					
9817-0	4000XL	12.9-16.7	.48-.58/.55-.66	2.05-3.44	.11-.16/.11-.16
9827-0, -1	4.0kW	12.9-16.7	.38-.51/.42-.51	1.83-2.41	.14-.19/.16-.21
9829-0	2.4kW	23.83	0.33	2.59	0.27
9829-1	2.4kW	20.9-27.1	.33-.42	2.24-3.05	N/U
9830-0	4.2kW	22.15	0.36	1.53	N/U
9830-1	4.2kW	21.0-27.2	.33-.39/.33-.39	1.07-1.42	N/U
9831-0	5.0kW	25.6-33.3	.24/.30	1.67	N/U
9831-1	5.0kW	25.6-33.3	.23-.29/.24-.34	1.26-1.53	N/U
F09836-0	3600XP	2.36	.50/.50	2.34	N/U
9836-0, -1	3500XEP	22.1	.34/.34	1.60	.20/.20
9855-0, -1	1850W	19.2-24.8	.78-.93	2.09-3.54	.19-.40/.19-.40
9856-0	PP5500,				
9870-0	SVP5000	25.6-33.3	.23-.29/.24-.34	1.26-1.67	N/U
9878-0, -1	4000XL	12.9-16.7	.38-.51/.42-.51	1.83-2.41	.14-.19/.16-.21
9879-0	1.5kW	19.2-24.8	.78-.93	2.09-3.54	.19-.40/.19-.40
9880-0, -1, -2	3250XE	20.8	.36/.36	1.20	1.20
9882-0	SV2400	20.9-27.1	.32-.39	2.24-2.71	N/U
9883-0	SVP4200	21.0-27.2	.33-.39/.33-.39	1.07-1.32	N/U
9884-0	SVP5000	25.6-33.3	.23-.29/.24-.34	1.26-1.67	N/U
9885-0, -1, -2, -3	5500XL	9.3-12.0	.26-.32/.28-.35	2.26-2.80	.08-.12/.08-.12
9886-0, -1, -2	8000XL	8.6	.18/.18	1.89	.05/.05
9903-0, -1	6500XL	10.2-12.2	.22-.26/.22-.26	1.49-1.81	.07-.11/.07-.11
9904-0, -1	10000XL	14.8	.13/.13	1.70	1.75

CRAFTSMAN

Sears Model	Generac Model	Rotor Winding Ω	Main Ac Power Ω	Excitation (Dpe) Ω	Dc (Bcw) Winding Ω
N/U	9174-0	25.6-33.3	.26/.26	1.80	0.19
32670	0451-0	20.9-27.1	.33-.42	2.24-3.05	N/U
32672	0452-0	12.9-16.7	.33-.39/.33-.39	1.07-1.42	N/U
32674	0453-0	12.9-16.7	.38-.51/.42-.51	1.83-2.41	.14-.19/.16-.21
32676	9855-0, -1	19.2-24.8	.78-.93	2.09-3.54	.19-.40/.19-.40
32678: Either	9718-0	23.7	.36-.36	1.53	N/U
or	9718-1, -2	21.0-27.2	.33-.39/.33-.39	1.07-1.42	N/U
32682	9801-0, -1	14.8	1.48/1.48	1.70	.17/.15
32839	9785-0, -1	8.6	.20/.20	1.48	.03/.04
326970	9446-0,				

CRAFTSMAN (continued)

326971	9446-1	24.0	0.34	2.46	0.26
326980	9447-0	22.2	.34/.34	1.61	N/U
326990	1033-0	25.6-33.3	.23-.29/.24-.34	1.26-1.67	N/U
327020	9450-0	25.6-33.3	.24/.30	1.67	.12/.21
327030	9083-0,				
327031	9083-1	13.5	0.47	1.50	0.10
327040	9084-0	24.0	0.34	2.46	0.26
327050	9085-0,				
327051	9085-1	26.2	.20/.60	2.11	0.14
327052	9085-2,				
327053	9085-3	14.3	.23/.47	2.61	.11/.19
327054	9085-4	25.6-33.3	.26/.26	1.81	.11/.19
327060	9086-0	26.2	.32/.32	2.11	N/U
327070: Either	9162-0	12.6	.16/.16	1.33	0.15
or	9289-0	12.6	.18/.18	1.13	0.4/0.6
327071: Either	9162-1	12.6	.16/.16	1.33	0.15
or	9289-1	12.6	.18/.18	1.70	N/U
or	9784-0, -1	8.6	.18/.18	1.89	.05/.05
327072	9162-2	14.5	.16/.16	1.30	0.15
327073	9289-2,				
327075	9289-3	12.6	.18/.18	1.13	0.4/0.6
327100	9218-0	26.5	.32/.32	2.10	N/U
327101	9218-1	23.7	.36/.36	1.53	N/U
327112	1149-0	25.6-33.3	.23-.29/.24-.34	1.26-1.53	N/U
327120	1187-0	20.9-27.1	.32-.39	2.20-2.71	.18-.22/.18-.24
327130	1188-0	12.9-16.7	.33-.39/.33-.39	1.07-1.42	N/U
327140	1189-0	12.9-16.7	.38-.51/.42-.51	1.83-2.41	.14-.19/.16-.21
327150	9296-0	39.2	0.40	2.37	N/U
327151	9296-1	24.0	0.34	2.46	0.26
327152	1150-0	25.6-33.3	.24-.28/.24-.28	1.00-1.22	.07/.07
327160	1190-0	9.3-12.0	.26-.32/.28-.35	2.26-2.80	.08-.12/.08-.12
327161	9450-1	25.6-33.3	.24/.30	1.67	.12/.21
327181	1020-0	7.6-9.1	.20-.25/.20-.25	1.56-1.91	.05-.09/.07-.11
327183	9067-3	12.6	.16/.16	1.09	0.06
327190	10KW NG,				
327200	10KW NG	9.4	.14/.14	0.73	N/U
327201	1191-0	14.8	.148/.148	1.7	.17/.15
327230	20KW NG,				
327240	20KW NG	8.4	.05/.05	0.50	N/U
327250	9514-0,				
327251	9514-1,				
327252	9514-2,				
327253	9514-3	12.9-16.7	.38-.51/.42-.51	1.83-2.41	.14-.19/.16-.21
327270	0630-0	20.9-27.1	.33-.42	2.24-3.05	N/U
327280	9782-0,				
327281	9782-1,				
327282	9782-2,				
327283	9782-3,				
327284	9782-4	9.3-12.0	.26-.32/.28-.35	2.26-2.80	.08-.12/.08-.12
327290	9783-0,				
327291	9783-1	10.2-12.2	.22-.26/.22-.26	1.49-1.81	.07-.11/.07-.11
327700	0647-0	9.8-12.2	.62-.74	5.82-7.05	.12-.16/.12-.16
327750	1170-0	20.9-27.1	.32-.39	2.20-2.71	.18-.22/.18-.24
328160	8759-0	16.0	1.10	1.32	0.22
328170	8760-0	13.5	0.47	1.50	0.10
328171	8760-1,				
328172	8760-2,				
328180	8761-0,				
328181	8747-0,				
328182: Either	8746-0,		(N/U on 8746-0)		
or	8746-1,				
328183	8761-1	37.5	0.40	2.50	0.12
328190	8762-0,				
328191	8762-1,				
328192	8762-2	46.0	.32/.32	1.95	0.08
328210	8615-0	37.5	0.40	2.50	N/U

CRAFTSMAN (continued)

328220	8916-0	46.0	.32/.32	1.95	N/U
328231	8647-0, -1	16.3	.50/.50	1.80	0.08
328240	8648-0	17.9	.40/.40	1.30	0.10
328250	8649-0,				
328251	8649-1	21.4	.28/.28	1.20	0.08
328260	8650-0	19.4	.33/.33	1.20	0.07
328300	1335-0	2.95	.03	1.10	0.082
328310	8844-0	16.2	1.70	6.40	0.28
328320	8845-0	22.7	0.52	3.20	0.19
328321	8964-1	22.7	0.52	2.35	0.19
328330	8846-0	25.0	0.30	2.14	0.14
328331: Either	8965-0				
or 8965-1	25.5	0.30	2.14	0.14	
328332	8965-2	24.0	0.34	2.46	0.26
328340	8847-0	19.4	.22/.43	3.68	0.10
328341: Either	8966-0	15.0	.22/.43	1.76	0.13
or	8966-1	15.5	.22/.43	1.76	0.13
328342	8966-2	14.3	.23/.47	2.61	0.11
328350: Either	8892-0	8.4	.28/.20	1.88	0.20
or	9451-0	15.5	.26/.26	1.76	.09/.09
328351	8892-1	8.4	.20/.28	2.60	0.20
328360	8893-0	14.5	.16/.16	0.88	0.15
328390*	9374-0,				
328391*	9374-1,				
328392*	9374-2	12.6	.18/.18	1.13	.40/.60
328390 (1995)	9785-0,				
328391 (1995)	9785-1	8.6	.20/.20	1.48	.04/.03
328450	9024-0	27.3	.21/.21	1.91	N/U
328451	9024-1	27.2	.21/.21	1.91	N/U
328452	9024-2	25.6-33.3	.26/.26	1.81	0.19
328910	8495-0	17.9	.40/.40	1.30	0.10
328920	8496-0	4.6	.27/.27	1.87	N/U
328930	8497-0	4.6	.11/.11	1.00	N/U
329210	8205-0	44.8	.40/.40	1.30	0.10
329610	8259-0	15.4	0.47	2.60	0.10
329620	8260-0	17.9	.40/.40	1.30	0.10
329630	8261-0	17.9	.40/.40	1.30	0.10
329640	8262-0	17.9	.40/.40	1.30	0.10
329650	8263-0	21.4	.28/.28	1.20	0.08
675010	9114-0,				
675011	9114-1	13.5	0.47	1.50	0.10
675020	9115-0	24.0	0.34	2.46	0.26
675030	9116-0,				
675031	9116-1	26.2	0.20	2.11	0.14
676420	0566-0,				
676421	0566-1	22.15	.33-.39/.33-.39	1.07-1.32	N/U
676551	0567-0, -1	9.3-12.0	.26-.32/.28-.35	2.26-2.80	.08-.12/.08-.12
678200	1133-0	9.8-12.2	.62-.74	5.82-7.05	.12-.16/.12-.16
678241	1134-0	20.9-27.1	.32-.39	2.24-2.71	N/U
678420	1135-0	12.9-16.7	.38-.51/.42-.51	1.83-2.41	.14-.19/.16-.21

*except 1995 models

DAYTON

Dayton Model	Generac Model	Rotor Winding Ω	Main Ac Power Ω	Excitation (Dpe) Ω	Dc (Bcw) Winding Ω
1N166	9860-0, -1	9.3-12.0	.27/.27	2.34	.07/.07
1N167	9861-0, -1	10.2-12.2	.19/.19	1.54	.06/.05
1N168	9862-0, -1	8.6	.20/.20	1.48	.03/.03
1N169	9863-0	14.8	.14/.14	1.70	.15/.15
1N170	9864-0	19.2-24.8	.78-.93	2.09-3.54	.19-.40/.19-.40
1N171	9868-0	12.4	.33/.33	1.40	0.15
3W740	9439-0	22.7	.52	3.20	0.19
3W741: Either	9435-0	22.2	.34/.34	1.61	.14/.24
or	9580-0	23.7	.34/.34	1.60	.10/.10

DAYTON (continued)

3W741B	9580-1	21.0-27.2	.33-.39	1.41-1.68	.12-.16/.09-.13
3W742	9436-0	25.6-33.3	.26/.26	1.81	.11/.19
3W793	9431-0	20.2	0.76	2.50	0.52
3W794	9432-0	23.8	0.31	2.59	.27/.27
3W795: Either	9433-0	23.7	.34/.34	1.61	.14/.24
or	9578-0	23.7	.34/.34	1.60	.10/.10
3W795B	9578-1	21.0-27.2	.33-.39/.33-.39	1.07-1.32	.17-.21/.15-.19
3W796: Either	9434-0	25.6-33.3	.26/.26	1.81	.11/.19
or	9579-0	25.6-33.3	.26/.26	1.80	.10/.10
3W796B	9579-1	25.6-33.3	.24-.31/.24-.31	1.58-2.05	.08-.11/.10-.14
3W953	9540-0,				
3W953A	9540-1,				
3W953B	9540-2	12.9-16.7	.38-.51/.42-.51	1.83-2.41	.14-.19/.16-.21
3ZC11	9320-0, -1	24.0	0.34	2.46	0.26
3ZC11B: Either	9320-2,				
or	9520-2	20.9-27.1	.33-.42	2.24-3.05	N/U
3ZC12	9321-0,				
3ZC12A	9321-1	23.7	.36/.36	1.53	N/U
3ZC12D	9569-2	21.0-27.2	.33-.39/.33-.39	1.07-1.42	N/U
3ZC13	9322-0	25.6-33.3	.26/.26	1.81	0.19
3ZC13C	9570-2	25.6-33.3	.23-.29/.24-.34	1.26-1.53	N/U
3ZC14	9325-0	23.7	.36/.36	1.53	N/U
3ZC39	9396-0	25.6-33.3	.26/.26	1.81	.11/.19
3ZC40A	9404-1	12.6	.18/.18	1.13	0.4/0.6
4W108	8919-0,				
4W108A	8919-1	26.5	0.32	2.10	0.12
4W109	8920-0,				
4W109A	8920-1	8.4	0.28	1.75	0.14
4W110: Either	8922-0	7.4	.25/.56	2.00	0.20
or	9437-0	13.7	.23/.47	2.61	.11/.19
4W111	8925-0	7.4	0.50/1.30	2.60	0.20
4W112: Either	8923-0	8.4	.20/.28	21.60	0.20
or	9438-0	15.5	.26/.26	1.76	.09/.09
4W113	8924-0, -1	14.5	.16/.16	0.88	0.15
4W113A	9163-0, -1	12.6	.18/.18	1.13	.40/.60
4W113D: Either	9160-3	12.4	.33/.33	1.40	0.15
or	9163-2	12.6	.18/.18	1.13	.40/.60
4W114	8926-0	7.4	.25/.56	2.40	N/U
4W115	8928-0	15.0	0.43	1.74	0.10
4W115A: Either	8928-1	14.3	.23/.47	2.61	0.11
or	9160-0	15.5	.22/.43	1.76	0.09
4W115B	9160-1	15.5	.26/.26	1.76	.09/.09
4W115C	9160-2	12.4	.33/.33	1.40	0.15
4W116	8927-0	8.4	.20/.28	2.60	N/U
4W556	9175-0	25.6-33.3	.26/.26	1.81	0.19
5W260	8916-0	15.4	1.70	6.40	N/U
5W261	8917-0	19.5	0.46	1.55	0.26
5W262	8918-0	22.0	0.39	1.83	0.18
5W263	8921-0	36.5	0.40	2.40	N/U
5W963	9307-0,				
5W963A	9307-1	23.8	0.33	2.60	N/U
5W964	9308-0	22.1	.36/.36	1.53	N/U
5W964A	9308-1	22.1	.34/.34	1.35	N/U
5W965	9651-0,				
5W965A	9651-1	12.3	.18/.18	1.70	N/U

WIRING DIAGRAMS

Refer to the following cross-reference charts to locate the wiring diagram for the generator model being serviced.

Generac Model	Diagram Figure	Generac Model	Diagram Figure
0413-0	GC128	9325-0	GC119
0415-0	GC126	9397-0	GC128
0416-0	GC127	9441-1, -2, -3, -4, -5	GC120
0417-0	GC128	9459-0	GC119
0421-0	GC129	9460-0	GC119
0435-0	GC127	9486-0	GC128
0455-0	GC137	9528-0, -1	GC128
0633-0	GC120	9551-0	GC119
0810-0	GC129	9700-0	GC119
1006-0, -1	GC124	9704-0, -1, -2, -3	GC120
1010-0	GC128	9716-0, -1, -2	GC126
1011-0	GC128	9717-0	GC121
1013-0	GC139	9717-1, -2	GC127
1016-0	GC138	9718-0, -1	GC121
1017-0	GC142	9718-2	GC127
1019-0	GC141	9719-0, -1, -2, -3	GC128
1020-0	GC141	9720-1, -2	GC127
1021-0	GC140	9777-0, -1, -2	GC120
1140-0	GC128	9778-0, -1, -2, -3	GC123
1140-1, -2	GC150	9778-4, -5, -6, -7	GC123
1193-0	GC138	9779-0, -1, -2	GC123
1194-0	GC144	9780-0, -1	GC122
1198-0	GC120	9781-0, -1	GC122
1199-0	GC120	9785-0, -1, -2	GC126
1263-0	GC144	9794-0, -1, -2	GC127
1277-0	GC149	9797-0, -1, -2, -3, -4	GC134
1278-0	GC128	9798-0, -1, -2	GC134
1305-0	GC128	9799-0, -1	GC129
1306-0	GC128	9800-0, -1	GC132
1311-0	GC144	9801-0, -1, -2, -3	GC132
1312-0	GC128	9802-0, -1	GC119
1313-0	GC120	9803-0, -1, -2, -3	GC124
1314-0	GC123	9805-0, -1	GC124
1315-0	GC147	9807-0, -1, -2, -3	GC133
1329-0	GC128	9809-0, -1, -2	GC133
1339-0	GC148	9829-0, -1	GC126
1356-0	GC150	9831-0, -1	GC127
5513-0, -1	GC119	9855-0	GC130
5517-0	GC119	9856-0	GC128
5521-0	GC119	9870-0	GC128
5526-0	GC119	9878-0, -1	GC120
5527-0	GC119	9885-0, -1, -2, -3	GC135
5528-0	GC119	9886-0, -1, -2	GC136
5541-0, -1	GC128	9903-0, -1, -2	GC137
5567-0, -1	GC119	9904-0, -1	GC 136
8971-0, -1, -2, -3, -4	GC119		

Craftsman Model	Diagram Figure	Craftsman Model	Diagram Figure
32670	GC132	327200	GC151
326740	GC120	327250	GC120
326741	GC120	327251	GC120
326761	GC130	327252	GC120
326782	GC127	327253	GC120

326990	GC128	327270	GC126
327100	GC119	328390	GC132
327110	GC131	328391	GC132
327120	GC143	328392	GC132
327130	GC131	676350	GC120
327140	GC120	676351	GC120
327150	GC146	678200	GC138
327160	GC123	678420	GC120
327180	GC141	741700	GC120

Dayton Model	Diagram Figure	Dayton Model	Diagram Figure
1N170	GC130	3ZC12D	GC119
3ZC11	GC119	3ZC13	GC119
3ZC11A	GC119	3ZC13B	GC119
3ZC11B	GC119	3ZC13C	GC119
3ZC12A	GC119	3ZC14	GC119
3ZC12B	GC119	4W115A	GC119
3ZC12C	GC119		

WIRING DIAGRAM

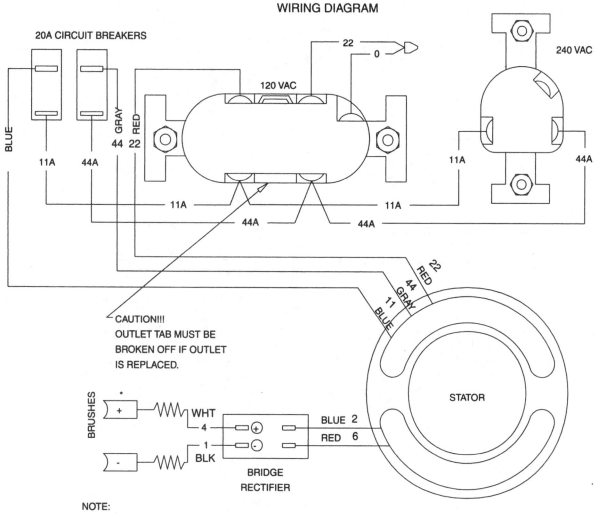

20A CIRCUIT BREAKERS

120 VAC

240 VAC

CAUTION!!!
OUTLET TAB MUST BE
BROKEN OFF IF OUTLET
IS REPLACED.

STATOR

BRUSHES

BRIDGE
RECTIFIER

NOTE:
*POSITIVE BRUSH CLOSEST TO BEARING

Fig. GC119

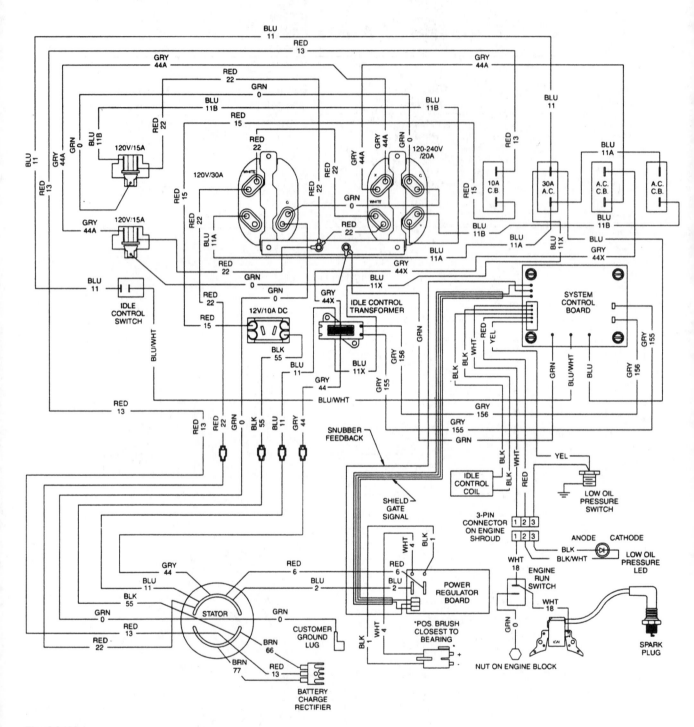

Fig. GC120

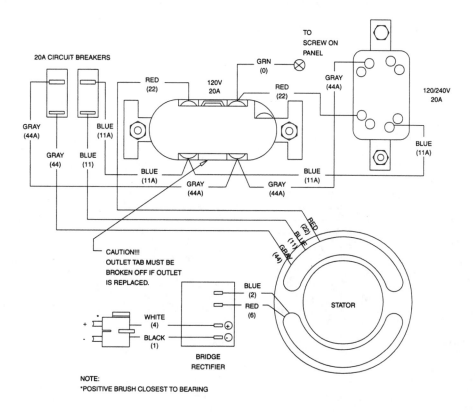

Fig. GC121

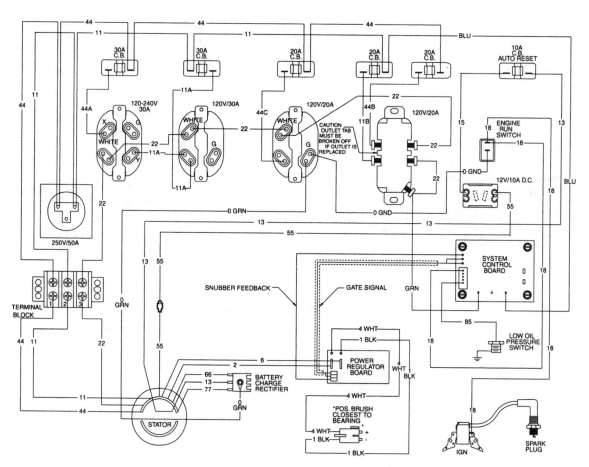

Fig. GC122

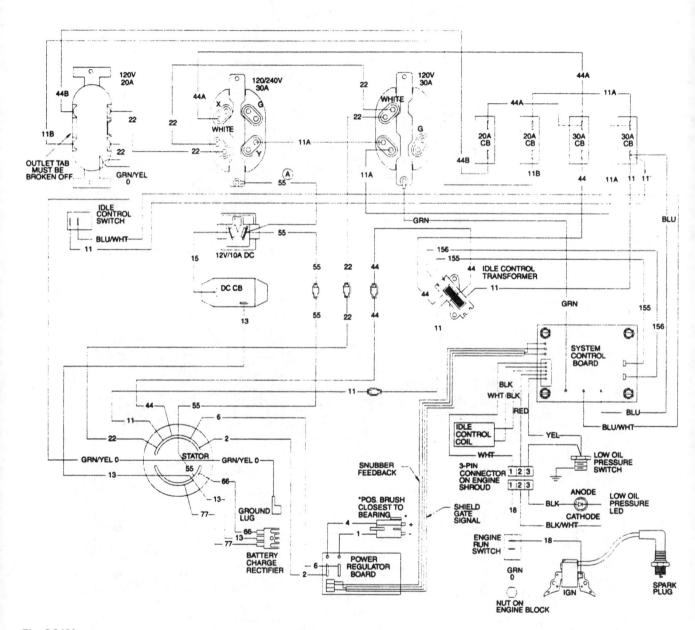

Fig. GC123

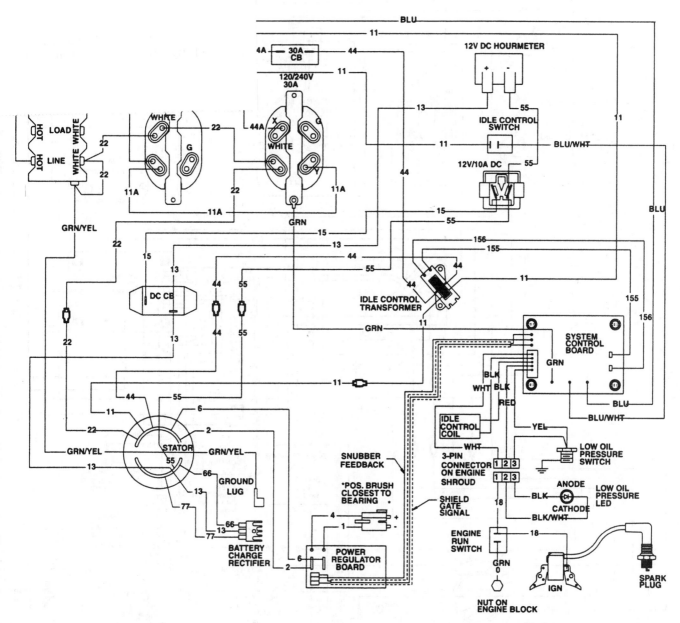

Fig. GC124

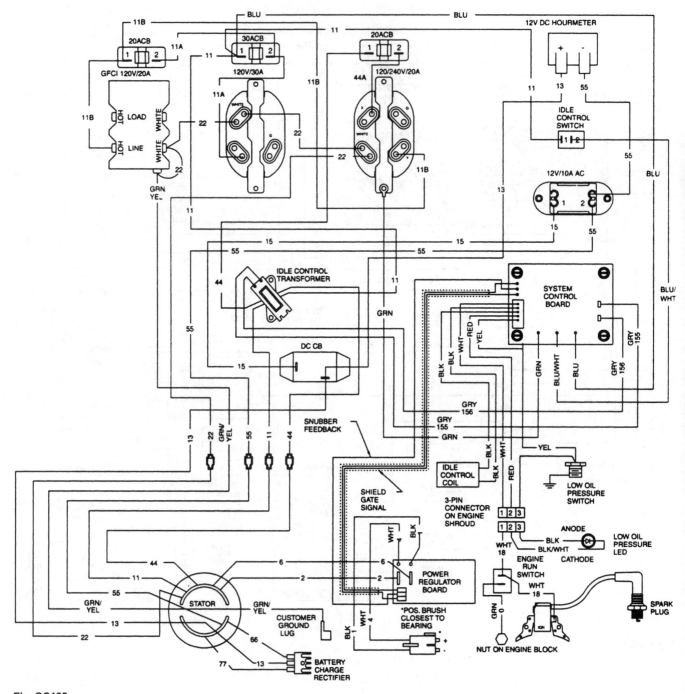

Fig. GC125

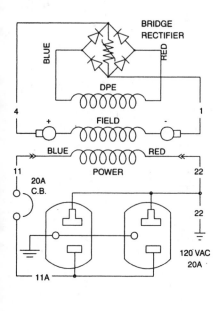

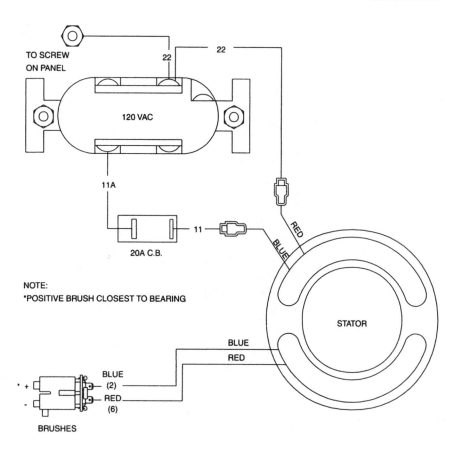

NOTE:
*POSITIVE BRUSH CLOSEST TO BEARING

Fig. GC126

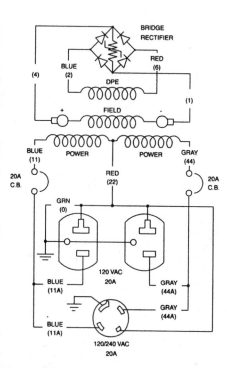

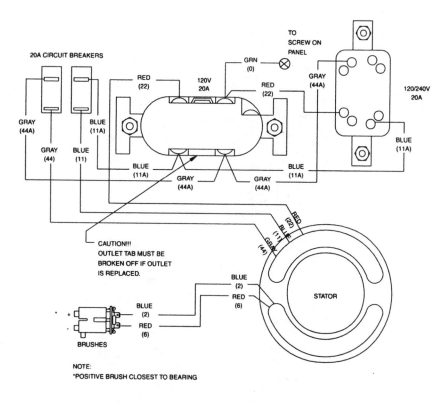

Fig. GC127

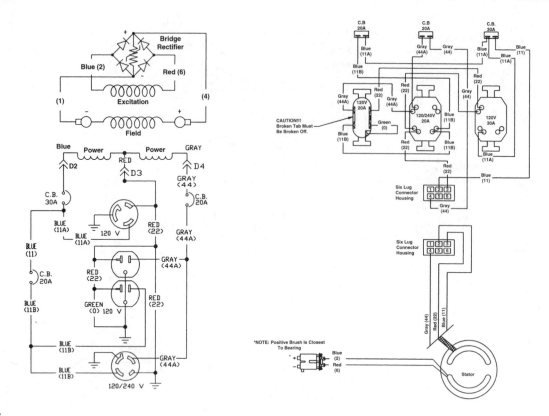

Fig. GC128

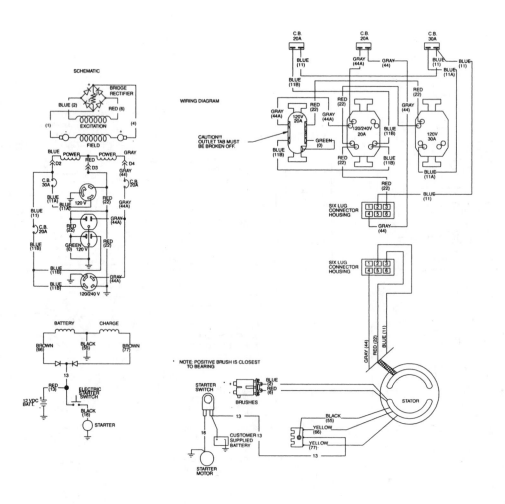

Fig. GC129

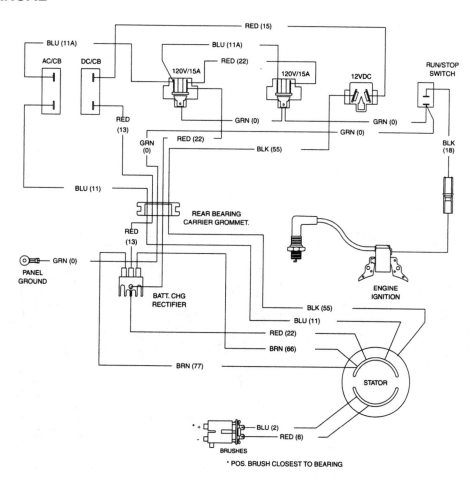

Fig. GC130

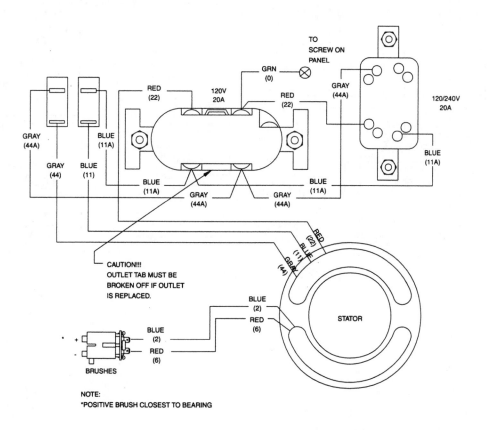

Fig. GC131

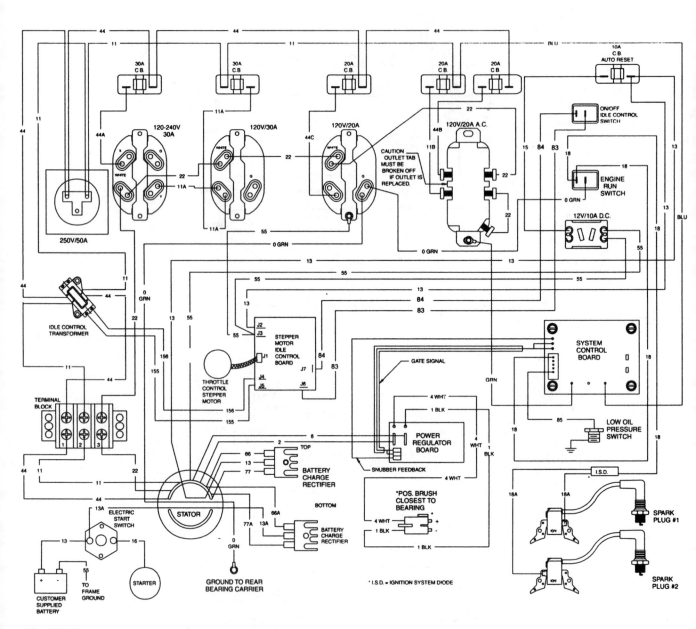

Fig. GC132

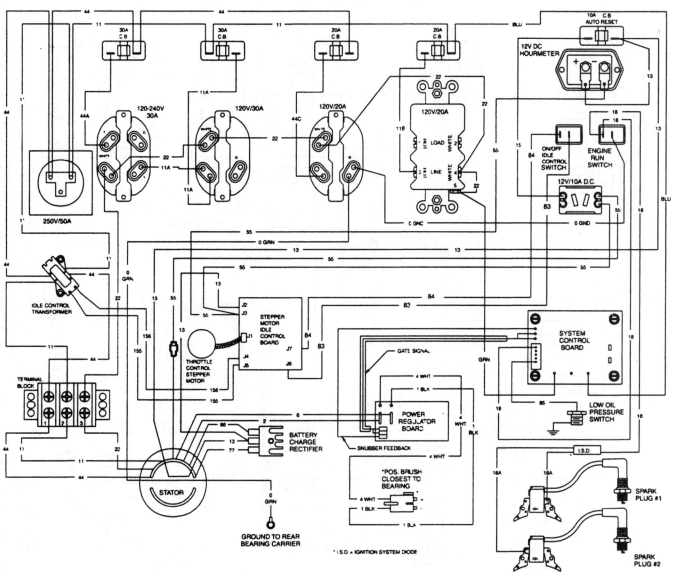

Fig. GC133

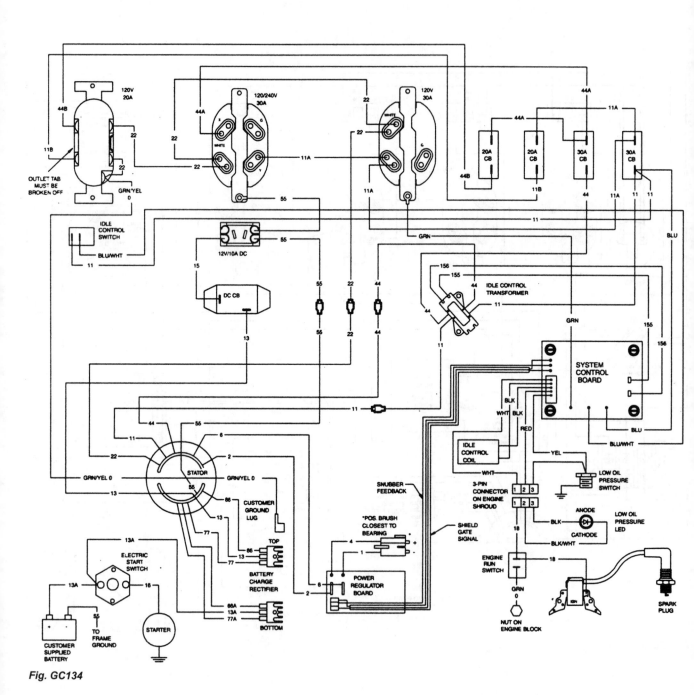

Fig. GC134

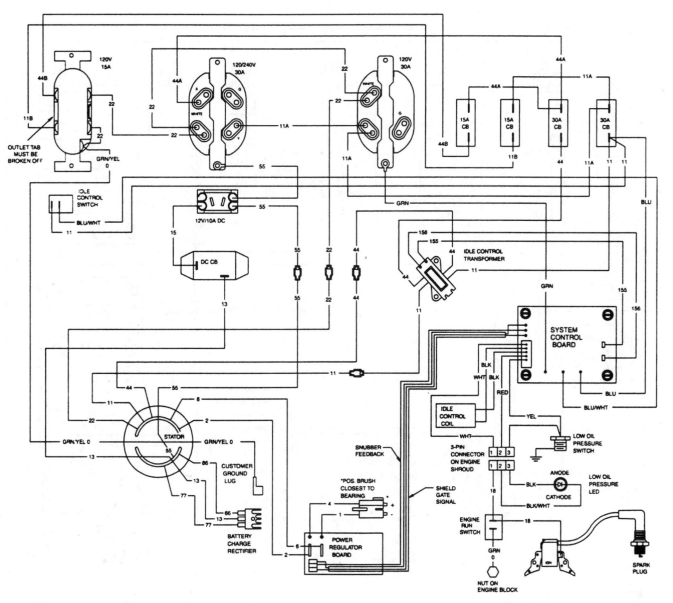

Fig. GC135

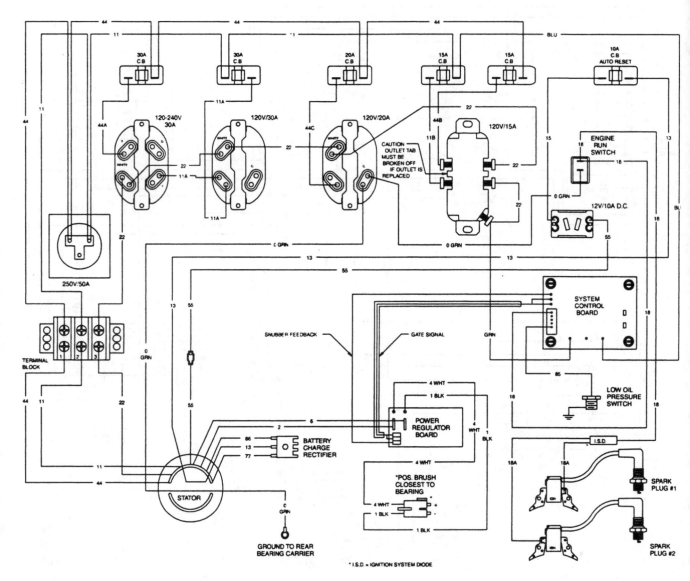

Fig. GC136

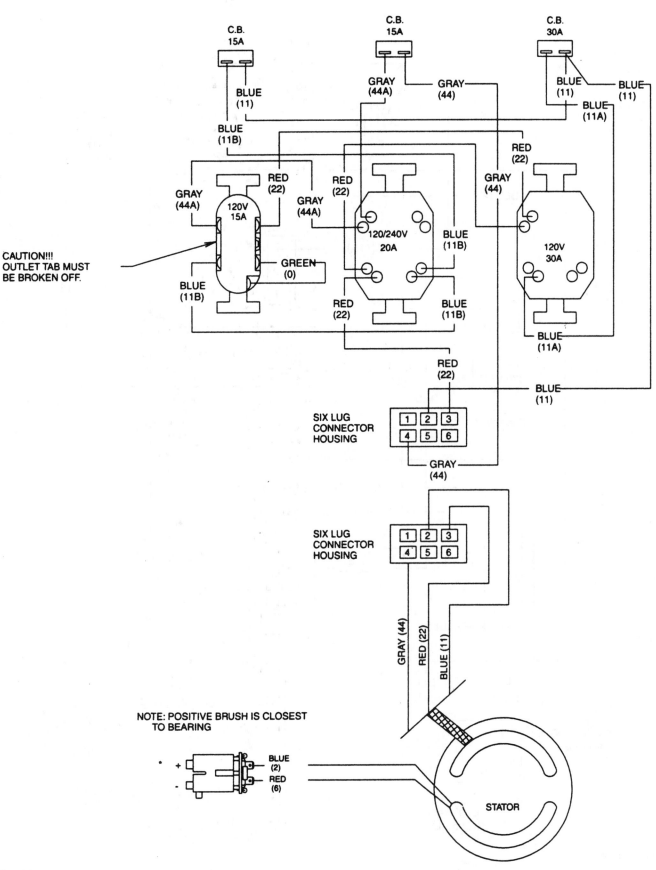

Fig. GC137

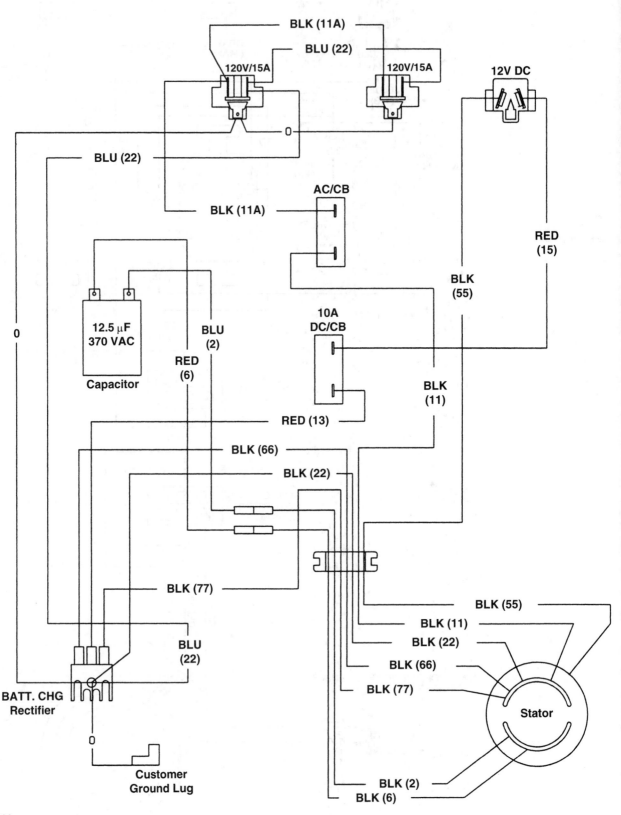

Fig. GC138

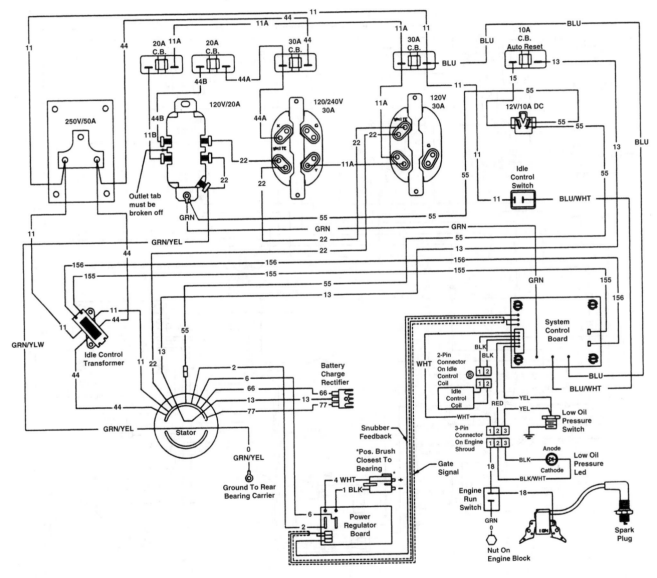

Fig. GC139

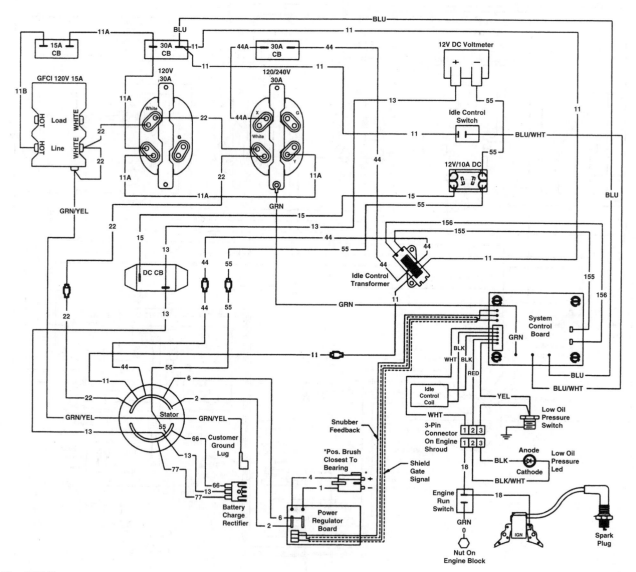

Fig. GC140

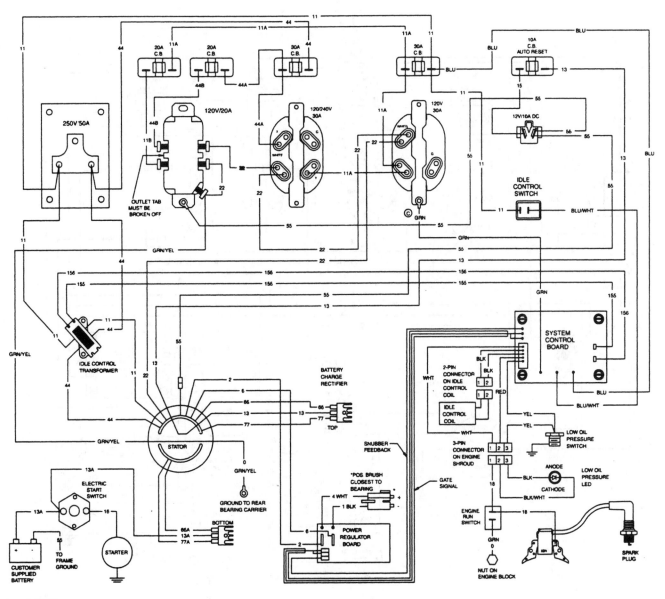

Fig. GC141

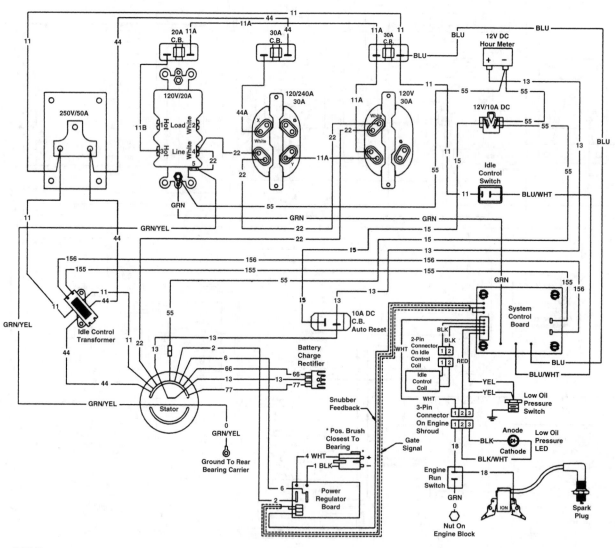

Fig. GC142

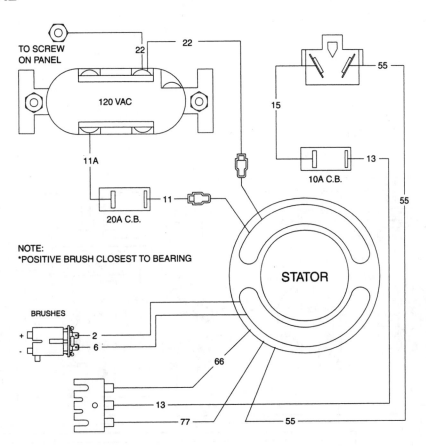

Fig. GC143

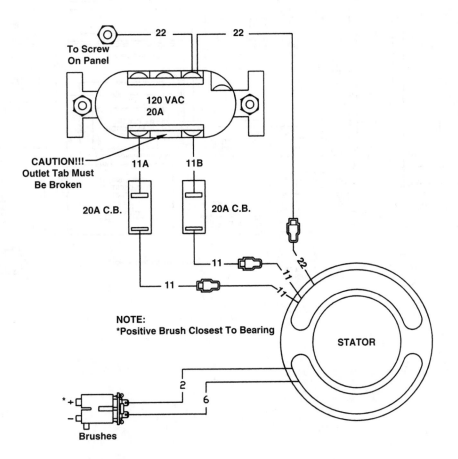

Fig. GC144

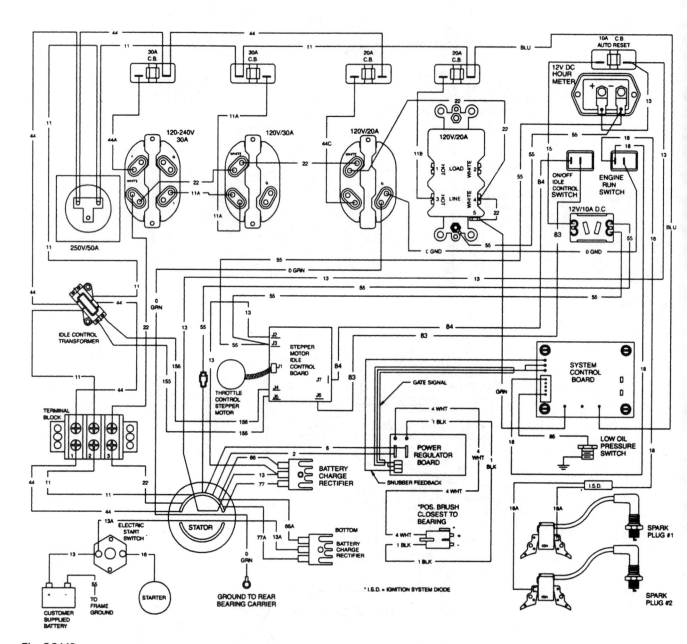

Fig. GC145

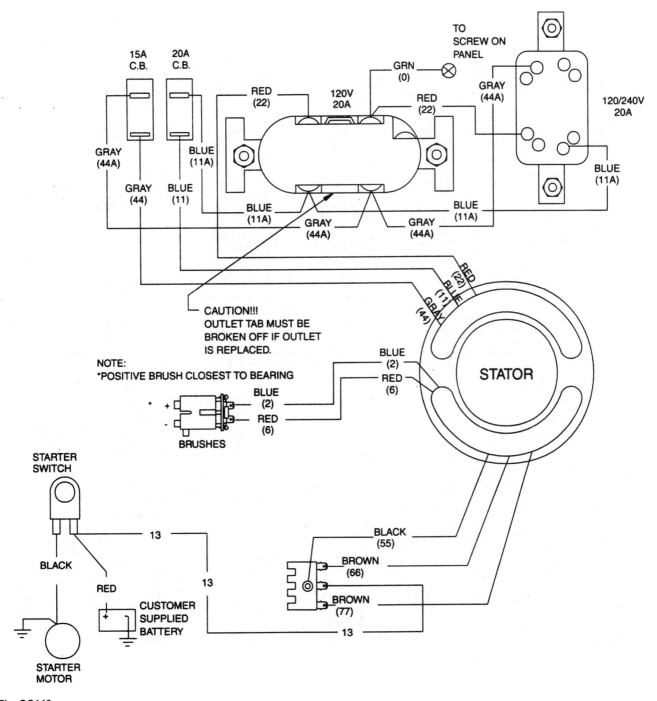

15A C.B.

20A C.B.

RED (22)

120V 20A

RED (22)

GRN (0)

TO SCREW ON PANEL

GRAY (44A)

120/240V 20A

GRAY (44A)

BLUE (11A)

GRAY (44)

BLUE (11A)

BLUE (11)

BLUE (11A)

BLUE (11A)

GRAY (44A)

GRAY (44A)

RED (22)

BLUE (11A)

GRAY (44)

CAUTION!!!
OUTLET TAB MUST BE BROKEN OFF IF OUTLET IS REPLACED.

NOTE:
*POSITIVE BRUSH CLOSEST TO BEARING

STATOR

BLUE (2)

RED (6)

+

-

BLUE (2)

RED (6)

BRUSHES

STARTER SWITCH

13

BLACK (55)

BLACK

BROWN (66)

RED

13

CUSTOMER SUPPLIED BATTERY

BROWN (77)

13

STARTER MOTOR

Fig. GC146

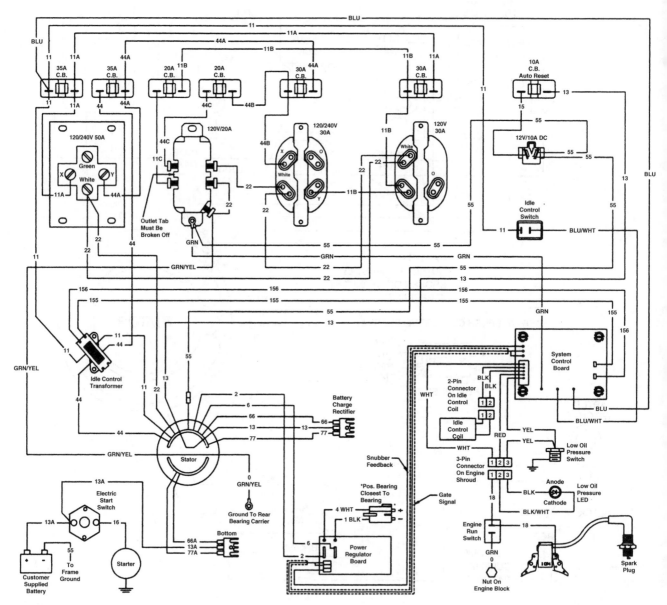

Fig. GC147

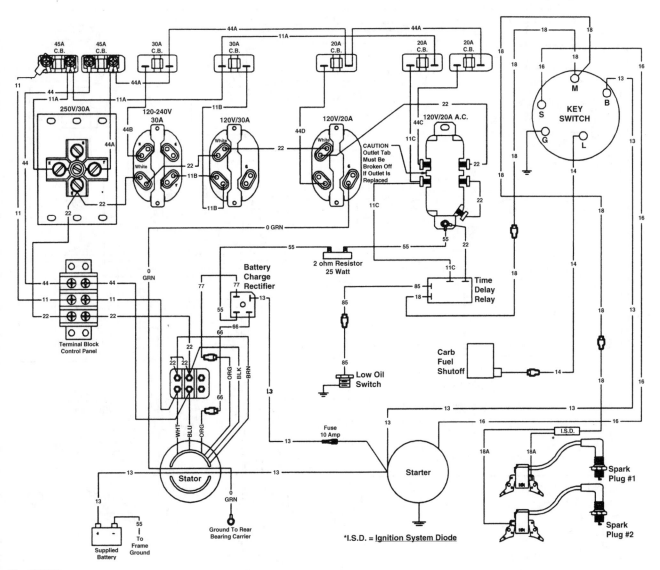

Fig. GC148

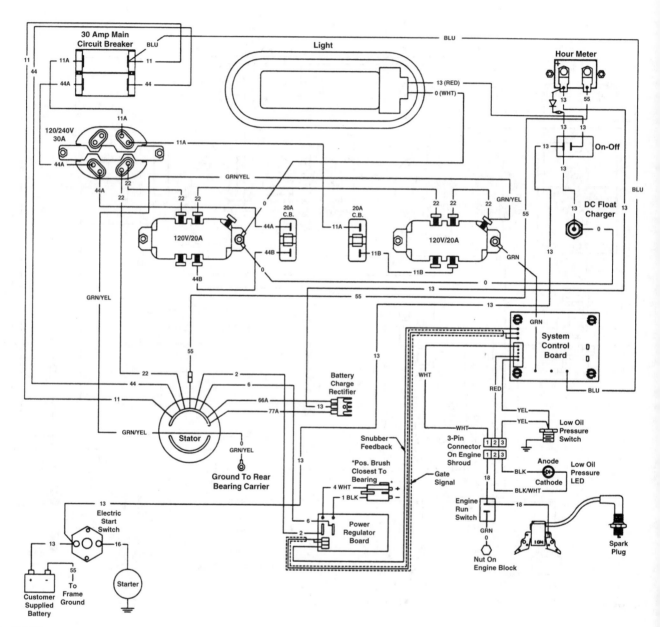

Fig. GC149

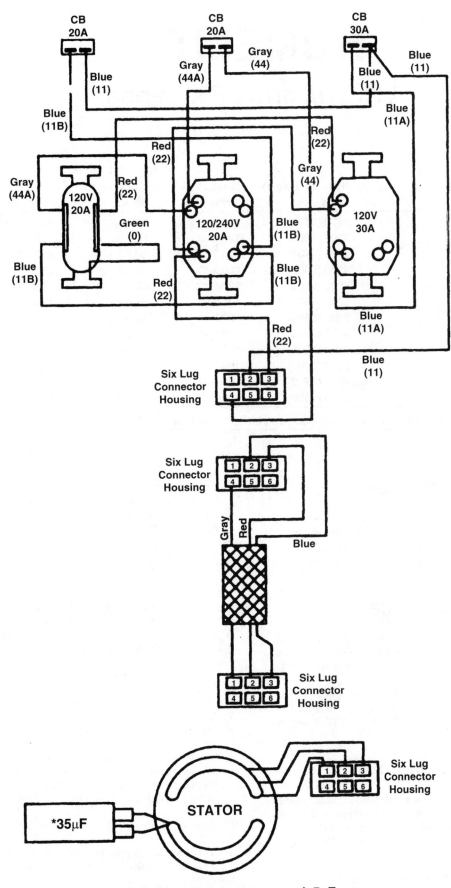

Fig. GC150

*Capacitor Value may vary +/- 5μF

HOMELITE
LR, LRI, and LRX Series

Generator Model	Rated (Surge) Watts	Output Volts	Engine Make	HP
LR2500	2300 (2500)	120	B&S	5.0
LR4300	3800 (4300)	120/240	Robin	7.5
LR(E)4400	4000 (4400)	120/240	B&S	8.0
LR5000T	4600 (5000)	120/240	Tecumseh	10.0
LR(E)5500	5000 (5500)	120/240	B&S	11.0
LR(E)5550	5000 (5550)	120/240	Robin	11.0
LRI2500	2300 (2500)	120	B&S	5.0
LRI(E)4400	4000 (4400)	120/240	B&S	8.0
LRI(E)5500	5000 (5500)	120/240	B&S	11.0
LRX3000	2300 (3000)	120	Robin	6.0
LRX(E)4500	4000 (4500)	120/240	Robin	8.5
LRX(E)5600	5000 (5600)	120/240	Robin	11.5

NOTE: An *E* in the model number prefix denotes electric start option.

IDENTIFICATION

These Homelite generators are brush style, single phase, 60 Hz units with integral electronic voltage regulation.

The LR series generators have endbell mounted receptacles. The LRI and LRX series generators have frame panel mounted receptacles.

MAINTENANCE

The endbell bearing and race should be inspected, cleaned, and relubricated any time the endbell is removed. If the bearing runs hot to the touch, replacement may be necessary.

Brushes must be at least 9/16 in. (14 mm) long; shorter brushes require replacement. Brushes can be accessed by removing the endbell. Refer to the DISASSEMBLY section in this chapter.

TROUBLESHOOTING

A Homelite generator analyzer tool #08371, field flasher tool #UP00457, and a VOM are recomended to troubleshoot generator malfunctions. Prior to any disassembly, initial testing should be performed with these two tools.

NOTE: Analyzer #08371 can only be used on generators with panel mounted receptacles. It will not work on generators with endbell mounted receptacles.

No load output voltage should be 113-140V on the 120-volt receptacles and 226-268V on the 240-volt receptacles. No load voltage with the idle control activated should be 80-95V. If little or no output is generated, the following items could be potential causes:

1. Engine RPM–The engine must maintain a minimum 3550 RPM under load. Normal no load speed is 3750-3800 RPM. No load speed with idle control activated should be 2600-2800 RPM on 2300-4000W models and 2400-2600 RPM on 4600-5000W models. Refer to the IDLE CONTROL section in this chapter for idle control service.
2. Loose component mounting fasteners (low or no voltage).
3. Circuit breakers or GFCI switch breaker (low or no voltage).
4. Faulty wiring connections (low or no voltage).
5. Faulty receptacle (low or no voltage).
6. Capacitor (low or no voltage).
7. Brushes (no voltage).
8. Loss of residual magnetism (no voltage).

NOTE: Residual magnetism can be restored without generator disassembly, and is the recommended first step if the generator has no output. Refer to the RESIDUAL MAGNETISM section.

9. Open or shorted windings (no or low voltage, respectively).
10. Dirty, broken, or disconnected slip rings (no voltage)
11. Stator windings (low voltage–shorted; no voltage–open)
12. Idle control malfunction (erratic or no idle; engine surging on LRI(E)4400 models).
13. Insufficient cooling system ventilation (overheating).

SIZE / PART NUMBER	RECEPTACLE	PLUG	CONFIGURATION
GENERATOR PLUGS AND RECEPTACLES			
VOLTAGE/AMPERAGE NEMA STANDARD HOMELITE P/N	120V 15AMP 5-15R 50991A	120V 15AMP 5-15P PURCHASE LOCALLY	
VOLTAGE/AMPERAGE NEMA STANDARD HOMELITE P/N	120V 15AMP 5-15R 04058-A	120V 15AMP 5-15P PURCHASE LOCALLY	
VOLTAGE/AMPERAGE NEMA STANDARD HOMELITE P/N	120V 20AMP 5-20R 51373	120V 20AMP 5-20P PURCHASE LOCALLY	
VOLTAGE/AMPERAGE NEMA STANDARD HOMELITE P/N	120V 20AMP TWIST LOCK L5-20R 48978	120V 20AMP TWIST LOCK L5-20P PURCHASE LOCALLY	
VOLTAGE/AMPERAGE NEMA STANDARD HOMELITE P/N	120V 30AMP TWIST LOCK L5-30R 42601	120V 30AMP TWIST LOCK L5-30P 43326	
VOLTAGE/AMPERAGE NEMA STANDARD HOMELITE P/N	240V 20AMP 6-20R 02863	240V 20AMP 6-20P 49709	
VOLTAGE/AMPERAGE NEMA STANDARD HOMELITE P/N	240V 20AMP TWIST LOCK L14-20R 46508	240V 20AMP TWIST LOCK L14-20P 47600	
VOLTAGE/AMPERAGE NEMA STANDARD HOMELITE P/N	240V 30AMP TWIST LOCK L14-30R 46718	240V 30AMP TWIST LOCK L14-30P 47601	

Fig. HM101 – Receptacles used on Homelite generators.

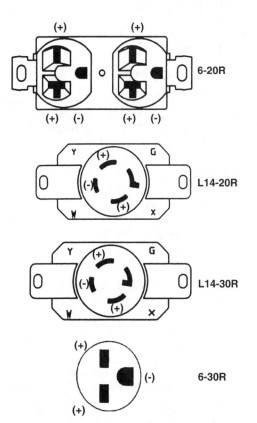

Fig. HM102 – Test connections for the 240V receptacles. Refer to text for correct test procedure.

GENERATOR REPAIR

PRELIMINARY TESTING AND RESIDUAL MAGNETISM

Receptacles

Fig. HM101 shows the various receptacles used on Homelite generators.

NOTE: Properly ground the generator prior to testing the receptacle output.

To test the receptacles:
1. Start the generator, verifying proper no-load speed of 3750-3800 RPM.
2. Ensure that the circuit breakers are set. If the circuit breakers fail to set, refer to the CIRCUIT BREAKER section in this chapter.
3. Set the VOM scale to AC volts.
4. For the 120V outlets, place the black VOM to ground and the red lead to the small flat leg of each duplex outlet. The reading should be 113-140 volts.
5. For the 240V receptacles, refer to the polarity markings of Fig. HM102.
   a. Positive (+) slots are Hot.
   b. Negative (-) slots are Neutral or Return;
   c. Lock slots, where used, are Frame Ground.
   Test between the positive (+) slots and both the negative (-) slots and the ground (G) slots; the reading should be 226-268 volts.
6. If the readings are not to specification:
   a. If only one receptacle shows no voltage, the receptacle may be faulty or may not be receiving output. Access and inspect the receptacle and wiring.
   b. A no voltage reading on all receptacles usually indicates a loss of residual magnetism: Refer to the following RESIDUAL MAGNETISM section of this chapter.
   c. A reading in the 1 to 4 volt range on all receptacles is a measurement of residual magnetism only. It indicates a fault in either the brushes, rotor, capacitor, bridge rectifier, or quad winding. Proceed with further testing.
   d. A half voltage reading (60V at the 120V receptacles; 120V at the 240V receptacles) usually indicates a faulty bridge rectifier or rectifier terminal connections. Proceed to BRIDGE RECTIFIER for further testing.
   e. A less than half voltage reading (30-50V) usually indicates a blown capacitor (capacitor bottom is bulged). A voltage reading (90V of 120V; 180V of 240V) usually indicates a faulty capacitor. Proceed to the CAPACITOR section for further testing.

NOTE: If the 120V receptacle is being replaced on 120/240V units, the connector tab on the hot side must be broken off: Failure to remove the tab will give an indication of overload upon generator start up. Torque terminal screws to 14 in.-lb. On duplexes, torque the ground screws to 10 in.-lb.

NOTE: When replacing receptacles, always make sure that the circuit breaker capacity matches the receptacle. If upgrading from an original 15-amp receptacle to a 20-amp receptacle, replace the breaker and/or GFCI with a matching unit.

Test Generator Output Using Both the #08371 Analyzer and the #UP00457 Field Flasher

NOTE: Analyzer #08371 can only be used on generators with panel mounted receptacles. It will not work on generators with endbell mounted receptacles.

1. Disconnect the large main harness connector and the small excitation harness connector from the back of the control panel (Fig. HM103).
2. Connect the main and excitation connectors to Analyzer #08371.
3. Start the generator and observe the analyzer.

NOTE: Do not unplug the analyzer while the generator is running.

a. If both neon lights are off:
   - Turn the Field Flasher (#UP00457) switch ON.
   - Plug the flasher into one of the analyzer receptacles.
   - When the flasher switch trips, the residual magnetism should be restored.
   - If both neon lights remain OFF, there is usually a problem in the excitation circuit.
   - If both lights light after flashing the fields, stop the generator, then restart it. If both lights are OFF again, the rotor will not hold residual magnetism. Refer to the ROTOR WINDING RESISTANCE section of this chapter and test the rotor. If the rotor is faulty, replace. If the rotor tests good, the circuit board is faulty and must be replaced.
b. If one neon light is ON (single voltage models) or if both neon lights are ON (dual voltage models), the rotor, stator, and brush head are good. Any problem will be with the voltage regulator board or the control box wiring.
c. If one green light is out, one output leg is open (3800-5000W units).
d. If one green light is dim, one output leg is partially shorted.

4. Continue to run the generator for five minutes. If there is a layer short, the windings should begin to smoke. Stop the generator. The stator is faulty and must be replaced.

Restore Residual Magnetism Using Only the #UP00457 Field Flasher

1. Start the generator and allow a brief warm up period.
2. Turn the Field Flasher switch ON.
3. Plug the flasher into one of the generator's 120V outlets.
4. When the flasher switch trips, residual magnetism should be restored.
5. Stop the generator, remove the flasher, then restart the generator.
6. Test the output at the 120V outlet. If there is no output, the rotor will not hold residual magnetism. Refer to the ROTOR WINDING RESISTANCE section of this chapter and test the rotor. If the rotor is faulty, replace it. If the rotor tests good, the circuit board is faulty and must be replaced.

Restore Residual Magnetism Without the #UP00457 Field Flasher

1. Obtain a 6 or 12-volt battery with two test leads attached to the battery.

Fig. HM103 – Typical rear view of the control panel showing the wiring connectors.

2. Start the generator.
3. Visually locate the brush holder through the slots on the left side of the endbell (left as viewed from the endbell end of the generator). There are two silver pins protruding from the brush holder–one pin is attached to the black (-) brush lead, and the other pin is attached to the red (+) brush lead. Note the polarity orientation.
4. Hold the negative (-) battery probe to the black (-) brush lead pin.
5. Touch the positive (+) battery probe to the red (+) brush lead pin for 1-2 seconds.
6. Stop and restart the generator.
7. Test generator output. If output tests within specification, the generator is good. If there is no output, the rotor will not hold residual magnetism. Refer to the ROTOR WINDING RESISTANCE section of this chapter and test the rotor.

IDLE CONTROL

NOTE: On LRI(E)4400 models, if the engine surges or hunts with a load of between 500 and 1000 watts, and the problem is not caused by the carburetor or governor, replace the paddle and electromagnet with kit #A08841. Follow installation instructions precisely.

Switch Test

To test the idle control switch:
1. Disconnect the spark plug wire.
2. Remove the Torx screws from the control panel faceplate, then carefully separate the faceplate from the rear cover.
3. Disconnect the wiring terminals from the back of the idle control switch.
4. Using a VOM set to the R × 1 scale, test for continuity between the switch terminals. The switch should have continuity in the IDLE position only. Any other test results indicate a faulty switch.

Paddle Setting

Refer to Figs. HM104 and HM105.

For proper paddle function on paddle design units:
1. Manually move the paddle against the electromagnet contact. Ensure that the flat face of the paddle (B) is parallel with the face of the electromagnet. Bend the paddle as necessary.

Fig. HM104 – View of idle control electromagnet and paddle on models LR2500 and LRI2500 generator.

Fig. HM105 – View of idle control electromagnet and paddle on generator models LR4400, LR5500, LRI4400 and LRI5500.

Fig. HM106 – With the idle control electromagnet energized, the carburetor throttle arm should not contact the slow idle speed adjusting screw. Refer to text.

2. Connect a tachometer to the engine. Start the generator and allow a brief warm up period.
3. Manually hold the paddle toward the electromagnet until the idle control linkage stop prevents further movement. Do not force. Observe the engine RPM. Idle control speed should be 2600-2800 RPM on models LRI2500 and LRI(E)4400, and 2400-2600 RPM on Model LRI(E)5500. To

adjust idle control speed to this range, refer to the ELECTROMAGNET TEST and ADJUSTMENT.
4. Release the paddle.
5. Move the idle control switch to the IDLE position: The paddle should pull to the electromagnet. If the paddle does not pull to the electromagnet, proceed with the ELECTROMAGNET TEST and ADJUSTMENT.
6. Turn the idle control switch OFF and stop the generator.

Electromagnet Test and Adjustment

To test the electromagnet:
1. With the generator stopped and the idle control switch OFF, disconnect the two black wires from the idle control connector terminals at the rear of the control panel (Fig. HM103).
2. Using the VOM, test for continuity between the two black wires. Continuity should test 240-285 ohms.
3. Test for continuity between each black wire and the electromagnet body. There should be NO continuity.
4. Failure of the continuity tests in either Step 2 or Step 3 indicates a faulty electromagnet. Replace the electromagnet.
5. If the electromagnet tests good, and the idle control switch passes the switch test, but the electromagnet still does not energize when the switch is moved to the IDLE position, check the continuity of the wiring between the idle control switch and the circuit board. Refer to the Disassembly Control Panel section.
6. If the wiring continuity tests good, reassemble, then test the voltage to the electromagnet.
7. Start the generator and, with the VOM set to read DC voltage, connect the VOM leads to the idle control connector terminals exposed in Step 1. VOM should read 35-45V DC. Insufficient DC voltage indicates a faulty circuit board; replace the board.

To adjust the electromagnet:
1. Connect a tachometer to the engine, then start the generator and allow a brief warm up period.
2. Move the idle control switch to IDLE.
3. Observe the tachometer for the proper idle control speed. 2600-2800 RPM on 2300-4000W models, and 2400-2600 RPM on 4600-5000W models.
4. On paddle design units (Figs. HM104 and HM105), insure that the electromagnet is positioned close enough to the paddle so that the paddle is held to the electromagnet. If necessary, loosen the electromagnet locknuts (A) and move the electromagnet for a stronger pull. The paddle needs to contact the electromagnet so that the carburetor idle speed adjusting screw (Fig. HM106) does **not** contact the throttle arm (B).
5. On spring-loaded lever design units, adjust the solenoid to put a light tension on the spring.

GROUND FAULT CIRCUIT INTERRUPTER (GFCI)

To test the GFCI:
1. Start the generator and allow a brief warm up period.
2. Depress the TEST button. The RESET button should extend; if it does, proceed to Step 3. If the RESET button does not extend, the GFCI is faulty.
3. With the RESET button extended, plug a test light into the GFCI receptacle. If the light lights with the RESET button extended, the GFCI line and load terminals have been wired backwards. To correct, stop the generator, then:
  a. Refer to Fig. HM107.

b. Access the GFCI connections (refer to the control panel disassembly section in this chapter).

c. Determine whether the GFCI receptacle is a Bryant or an Arrow Hart.

d. Make sure that the input power wires connect to the proper line screws, and, when applicable, any protected receptacles downstream of the GFCI connect to the proper load screws. When replacing a GFCI receptacle, always note the line and load connection orientation.

e. Make sure that the white (neutral) wires connect to the correct **silver colored** screws, and the black or red (hot) wires connect to the correct **brass** screws. The green (ground) wire always connects to the green screw.

4. To restore power, restart the generator and depress the RESET button firmly into the GFCI until a definite click is heard. If properly reset, the RESET button is flush with the surface of the TEST button. When the RESET button stays in, the GFCI is ON.

5. Stop the generator.

NOTE: If the GFCI seems to trip for no apparent reason, check to make sure the packing material was removed from under the engine when the generator was initially prepared for service. If the block was not removed, the vibration isolation mounts will not properly prevent engine vibration from reaching the control panel, and the vibration will trip the GFCI.

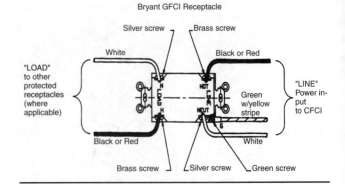

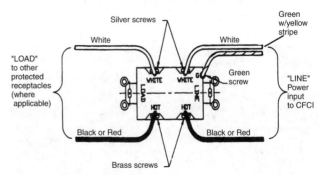

Fig. HM107 – Wire side views of Bryant and Arrow Hart GFCI receptacles showing wiring connection differences.

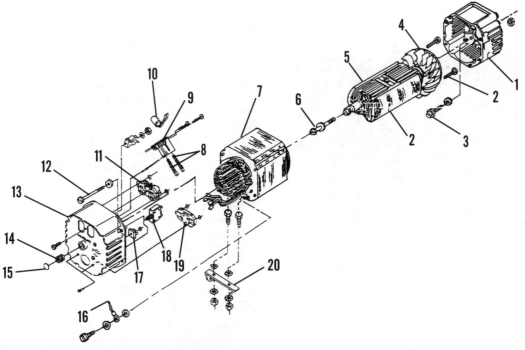

Fig. HM108 – Typical view of Homelite LR series generator components. LR series uses endbell mounted receptacles. Single voltage units do not have twistlok receptacle.

1. Engine adpter	8. Brushes	15. Plug
2. Screw	9. Brush holder	16. Ground wire
3. Bolt	10. Capacitor	17. Rectifier
4. Fan	11. Receptacle	18. Circuit breaker
5. Rotor	12. Bolt	19. Locking receptacle
6. Rotor bolt	13. Endbell	20. Support
7. Stator assy.	14. Needle bearing	

CIRCUIT BREAKERS

NOTE: When replacing receptacles, always make sure that the circuit breaker capacity matches the receptacle. If upgrading from an original 15-amp receptacle to a 20-amp receptacle, replace the breaker and/or GFCI with a matching unit.

To test each circuit breaker:
1. Make sure the breaker is set. If tripped, reset.
2. Disassemble as necessary to access the breaker (refer to DISASSEMBLY, CONTROL PANEL), then remove the wires from the rear of the breaker.
3. Using the R × 1 scale on the VOM, test for continuity across the breaker terminals.
4. If no continuity reading is noted, or if the breaker fails to reset, the breaker is faulty. Replace the breaker.

DISASSEMBLY

NOTE: Before beginning any disassembly, allow the generator to cool.

Refer to Figs. HM108, HM109, and HM110 for component identification.

Brushes

The brushes can be accessed after removing the endbell. Proceed with disassembly.

Generator Head

1. On electric start models, disconnect the battery cables and remove the battery.
2. Shut off the fuel valve, disconnect the fuel hose, and remove the fuel tank.

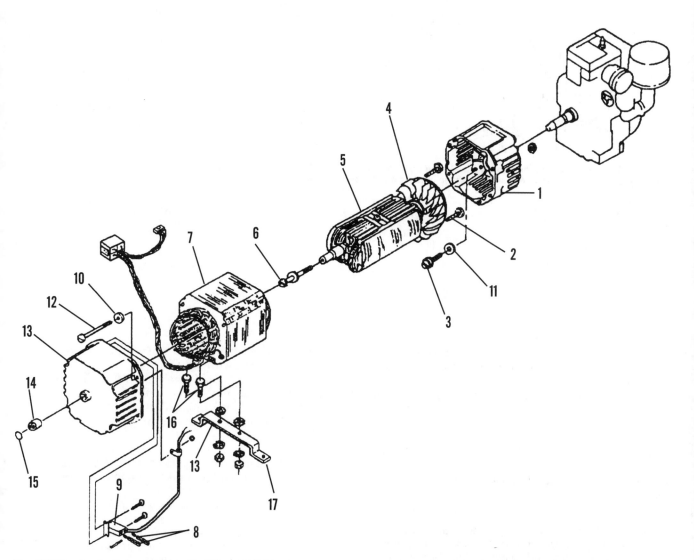

Fig. HM109 – Typical view of Homelite LRI and LRX-series generator head components with panel mounted receptacles. The support bracket will vary depending on the Model.

1. Engine adapter	7. Stator assy.	13. Endbell
2. Screw	8. Brushes	14. Needle bearing
3. Bolt	9. Brush holder	15. Plug
4. Fan	10. Washer	16. Bolts
5. Rotor	11. Washer	17. Support
6. Rotor bolt	12. Bolt	

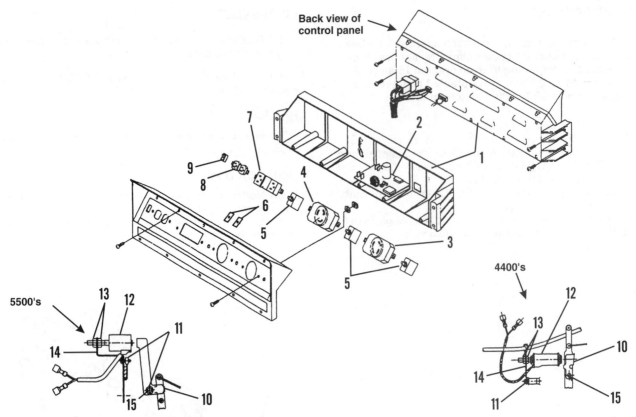

Fig. HM110 – *Typical view of the control panel used on Homelite LRI and LRX-series generators. This view is the panel used on the dual voltage models. The panel on the single voltage models has fewer receptacles and breakers.*

1. Control panel
2. Control board
3. 240V receptacle
4. 120V receptacle
5. Circuit breakers
6. Rubber grommets
7. GFCI receptacle
8. Receptacle
9. Switch
10. Paddle
11. Nuts
12. Electromagnet
13. Nuts
14. Bracket
15. Screw

3. On models with panel mounted receptacles, disconnect the wiring connectors from the rear of the control panel (Fig. HM103).

4. Remove the hardware which fastens the generator support bracket to the mounts on the generator cradle frame.

5. Place a temporary support block under the engine adapter, if necessary.

6. Remove the four stator throughbolts from the endbell flanges.

7. Gently pull the stator/endbell assembly straight out from the generator.

NOTE: After the endbell is partially pulled out, the brushes will spring out of their holder. Do not lose the brushes. Use caution when setting the stator on the workbench so as not to scrape or damage the stator winding varnish.

8. On units with panel mounted receptacles, remove the endbell from the stator by sliding the stator harness and its protective rubber grommet from the lower slot in the endbell. On units with endbell mounted receptacles, sepa-

rate the endbell from the stator, then disconnect the stator to endbell wiring.

9. To inspect and service the endbell bearing, remove the bearing cap plug from the endbell.

10. Remove the rotor throughbolt and lock washer.

11. Remove the rotor using one of the following two methods:

NOTE: To prevent rotor damage, do not let the rotor fall from the engine crankshaft upon removal.

a. The recommended method is to use a slide hammer with a 3/8-16 end threaded into the rotor shaft. Screw the slide hammer into the end of the rotor shaft. Tap the slide hammer until the rotor is loose from the engine crankshaft.

b. An alternate method is to make a push tool to fit inside the rotor shaft (Fig. HM111):
• Obtain a length of ¼ in. all thread rod or a Homelite #22272 Rotor Rod.
• Thread the rod through the rotor shaft and into the engine crankshaft until it bottoms (do not force), then unscrew it one turn.

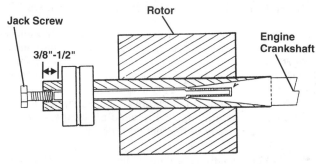

Fig. HM111 – Alternate push rod tool method of rotor removal. Tool can be constructed in shop. Refer to text.

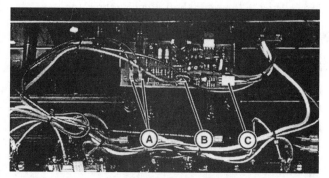

Fig. HM112 – View showing the control panel faceplate separated from the back cover, as well as most of the components mounted inside the control panel assembly.

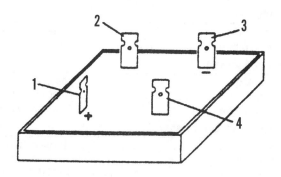

Analog VOM Testing

	Pos. No.	Black Test Lead (-)			
		1 (+)	2 (AC)	3 (-)	4 (AC)
Red Test Lead (+)	1 (+)		Conducting	Conducting	Conducting
	2 (AC)	Non Conducting		Conducting	Non Conducting
	3 (-)	Non Conducting	Non Conducting		Non Conducting
	4 (AC)	Non Conducting	Non Conducting	Conducting	
		Resistance Readings			

Fig. HM113 – Bridge rectifier diode test chart. Refer to text for correct test procedure.

- Mark the rod flush with the end of the rotor shaft.
- Remove the rod and note the depth of the internal rotor shaft threads from the end of the shaft. Older generators had threads flush with the end of the shaft; newer generators have recessed threads.
- Cut the in rotor length of the rod 3/8 in.-1/2 in. (10-13 mm) shorter than the mark on older generators, and shorter than the start of the threads on newer generators.
- Cut a screwdriver slot in the outer end of the rod.
- Lightly screw the rod back through the rotor shaft and into the engine crankshaft, then back it out one turn, as before.
- Using a 3/8-16 bolt as a jack screw, tighten the bolt into the end of the rotor shaft and against the rod. With a metal hammer, alternately tighten and tap until the rotor breaks loose.

Control Panel

1. If not previously done, disconnect the wiring connectors from the rear of the control panel (Fig. HM103).
2. Remove the four Torx head screws holding the rear cover of the control panel to the generator cradle frame (two on each side).
3. Carefully remove the control panel assembly.
4. Unscrew the Torx screws holding the control panel faceplate to the rear cover, then carefully separate the faceplate assembly from the rear cover.

Control Board

Refer to Fig. HM112.

NOTE: If plastic tie wraps must be cut and removed during disassembly, note their positions and install new tie wraps during reassembly.

1. Disassemble control panel following the instructions in Control Panel.
2. Disconnect the two black wire spade connectors (A) from the circuit board.
3. Disconnect the red and black wires which feed through the circuit board coil (B) from the circuit breakers.
4. Disconnect the red white and black wire four-pin connector (C) from the side of the circuit board.
5. Remove the circuit board by sliding it from the control panel slots.

CAPACITOR

NOTE: Even with the generator stopped, the capacitor may still retain a charge. Do not touch the capacitor terminals, as electric shock may result. Always discharge the capacitor by shorting across the terminals with a screwdriver or similar insulated tool.

To test the capacitor:
1. Place the VOM on the highest ohm scale.
2. Disconnect the capacitor leads, noting the polarity orientation.
3. Connect the VOM leads to the capacitor terminals. The VOM needle should rise quickly to continuity, then move slowly back toward infinity until capacitor resistance is indicated. If the capacitor has its value marked on its side, the meter reading can be compared to this value.

NOTE: Digital meters will advance towards infinity until the capacitor reaches full charge, then show no continuity/overscale.

4. Reverse the VOM leads. The meter should move rapidly toward zero ohms, followed by the same slow rise toward capacitor value as noted in Step 3.

    a. If the VOM reads straight continuity, the capacitor is shorted. Replace the capacitor.

    b. If the VOM fluctuates between continuity and infinity, the capacitor is leaking; replace the capacitor.

NOTE: When reconnecting the capacitor leads, always observe proper polarity orientation.

BRIDGE RECTIFIER (REGULATOR)

Rectifiers can be either square or round in shape, with four male spade terminals. Refer to Fig. HM113 for testing.

1. Disconnect the rectifier from the circuit, noting the wiring orientation.
2. Set the VOM to the R × 1 scale.
3. Connect the meter leads to any pair of adjacent or diagonal terminals (1-2, 1-4, 3-2, 3-4, 1-3, or 2-4); note the reading.
4. Reverse the meter leads and again note the reading.
5. Repeat Steps 3 and 4 with the other terminal pairs.
6. Compare the readings with the Fig. HM113 chart.

    a. If one test set shows continuity both directions, the rectifier has a short circuit and must be replaced. A shorted rectifier may have fed AC current to the rotor. After replacing the rectifier, flash the fields to re-establish residual magnetism.

    b. If one test set shows no continuity in both directions (except the 2-4 set, which should show no continuity both directions), the rectifier has an open circuit. Replace the rectifier.

    c. If one or more reading is significantly higher or lower than the others, the rectifier is faulty and must be replaced.

ROTOR

Rotor Slip Rings

The slip rings (Fig. HM114) are an integral part of the rotor assembly and cannot be replaced.

Clean dirty or rough slip rings with a suitable solvent, then lightly sand with ScotchBrite or fine grit sandpaper. Do not use steel wool, emery cloth, or crocus cloth.

Cracked, chipped, broken, or grooved slip rings cannot be repaired. In these cases, the rotor must be replaced.

Rotor Winding Resistance

The rotor can be tested without removing it from the generator.

To test resistance:

1. Using a VOM set on the R × 1 scale, hold the VOM leads to the rotor slip rings.
2. If no reading is observed upon initial testing, inspect the winding to slip ring connection and solder joint. If the solder joint is faulty, and if there is sufficient wire length, resolder the joint and retest.

3. Rotor resistance specifications at 77° F (25° C) are as follows:

LRX3000	40.7 ohms
2300W units, except LRX3000	46.7 ohms
4000W units	67.0 ohms
4600W and 5000W units	76.0 ohms

NOTE: All resistance readings are approximate and may vary depending on the winding temperature and the accuracy of the test equipment.

4. If the VOM readings are lower than specified, the rotor has shorted windings. If the VOM reads infinity, the rotor has open windings. Either way, replace the rotor.
5. Test for continuity between each slip ring and the rotor shaft. If either test shows continuity, the rotor windings are shorted to the rotor shaft. Replace the rotor.

STATOR

Winding Resistance

To test the stator windings:
1. Set the VOM to the R × 1 scale.
2. Connect the meter leads to the following color coded wiring pairs.
3. Measure the winding resistances and compare to the following specifications:

All Units:

Green (ground) to stator body	< 0.5 ohm
Black to stator body, White to stator body, Red to stator body, and Yellow to stator body	Infinite resistance/no continuity

LR2500:

Black to White	0.366-0.412 ohm
Yellow to Yellow	1.110-1.250 ohms

LR4300, LR4400:

Black to White	0.261-0.294 ohm
Brown to Red	0.261-0.294 ohm
Yellow to Yellow	0.939-1.059 ohms

LR5000, LR5500, LR5550:

Black to White	0.196-0.220 ohm
Brown to Red	0.196-0.220 ohm
Yellow to Yellow	0.915-1.028 ohms

LRI2500:

Black to White	0.449-0.506 ohm
Yellow to Yellow	1.564-1.764 ohms

LRI4400, LRX4500:

Black to White	0.350-0.394 ohm
Red to White	0.350-0.394 ohm
Yellow to Yellow	1.342-1.514 ohm

LRI5500, LRX5600:

Black to White	0.280-0.316 ohm
Red to White	0.280-0.316 ohm
Yellow to Yellow	1.306-1.472 ohms

LRX3000:

Black to White	0.443-0.500 ohm
Yellow to Yellow	0.994-1.064 ohms

4. If test results do not meet specification, replace the stator assembly.

REASSEMBLY

Control Board

NOTE: Replace any plastic cable ties that were cut during disassembly.

1. Clean all contacts on the board prior to installation.
2. When installing a new control board into single voltage generators, remove the two jumpers (A) from the board (Fig. HM115). Jumpers **must** be left in position on the board for all other installations, or the generator will have no 240V output.
3. Assemble the circuit board to the back cover of the control panel by sliding it into its slots.
4. Connect the red white and black wire four-pin connector (C—Fig. HN112) to the side of the circuit board.
5. Reconnect the red and black wires which feed through the circuit board coil (B) to the circuit breakers.
6. Connect the two black spade connectors (A) to their proper board terminals.

Control Panel

1. Carefully mate the rear cover to the control panel faceplate, insuring that no wires are pinched. Insert and

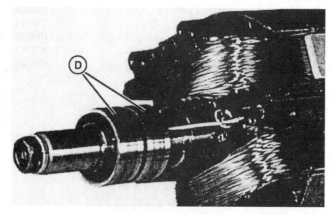

Fig. HM114 – View showing the brush slip rings (D) on the rotor.

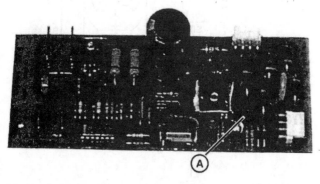

Fig. HM115 – View of a replacement control board (circuit board) showing the 120/240V jumper connector (A). Refer to text for jumper application.

tighten, but do not force, the Torx head screws which hold the faceplate to the rear cover.
2. Fit the control panel to the generator cradle frame, then fasten the panel with the four Torx head screws at the back of the rear cover (two on each side).
3. Reconnect the wiring connectors from the generator head to the rear of the control panel (Fig. HM103).

Generator Head

1. If the engine adapter was removed during disassembly, reinstall at this time. The unslotted side is up. Torque the adapter bolts as follows:

2300W units–120-150 in.-lb. (13.6-16.9 N•m)

4000 and 5000W units–240-250 in.-lb. (27.1-28.2 N•m)

2. Place a temporary 2 × 4 support block under the engine adapter, if necessary.

3. Insure that the matching tapers on the engine crankshaft and inside the rotor shaft are clean and dry. Install the rotor onto the crankshaft. Install the rotor bolt and lockwasher. Tighten the rotor bolt **finger tight** at this time.

4. Service the rotor bearing in the endbell, if needed.

5. Install the brushes in the brush holder. Hold the brushes in place against their springs by inserting a straightened paper clip or similar tool through the endbell hole next to the bearing. Insert the tool from the outside in, making certain that it passes through the brush holder and over both brushes.

6A. On units with panel mounted receptacles, assemble the endbell to the stator by feeding the stator harness through the lower slot in the endbell, making sure that the protective grommet is properly installed in the slot.

6B. On units with endbell mounted receptacles, reconnect the stator to endbell wiring, then fit the endbell to the stator.

7. Carefully install the stator/endbell assembly over the rotor, correctly aligning the rotor endbell bearing and the stator-to-engine adapter locator pins.

8. Slide the two lower stator throughbolts into position, tightening them finger tight.

9. Slide the two upper stator throughbolts into position, then torque all four throughbolts to 60-80 in.-lb. (6.8-9.0 N•m).

10. Make sure that the rotor turns freely inside the stator, then torque the rotor bolt to 100-140 in.-lb. (11.3-15.8 N•m).

11. Remove the brush holder tool, and install the rotor bearing plug cap.

12. Install the generator support bracket bolts. Torque the bolts to 145-155 in.-lb. (16.4.-17.5 N•m).

13. Reconnect the wiring connectors to the rear of the control panel (Fig. HM103).

14. Install the fuel tank and reconnect the fuel hose.

15. On electric start models, install the battery and reconnect the battery cables.

FASTENER TORQUE CHART

Fasteners are listed in the approximate order of reassembly and service.

Application	Torque Value
Engine adapter to engine:	
5/16-24 bolts	120-150 in.-lb. (13.6-17.0 N•m)
3/8-16 bolts	240-250 in.-lb. (27.1-28.3 N•m)
Fan to rotor	12-16 in.-lb. (1.4-1.8 N•m)
Rotor throughbolt	100-140 in.-lb. (11.3-15.8 N•m)
Brush holder (Plastite screws)	12-16 in.-lb. (1.4-1.8 N•m)
Brush holder	18-22 in.-lb. (2.0-2.5 N•m)
Brush head cover	18-22 in.-lb. (2.0-2.5 N•m)
Rectifier (8-32 × .50 screws)	18-22 in.-lb. (2.0-2.5 N•m)
Endbell/Stator througholts:	
Except 1/4–20 × 10.5 in. bolts	60-80 in.-lb. (6.8-9.0 N•m)
With 1/4-20 × 10.5 in. bolts	65-75 in.-lb. (7.3-8.5 N•m)
Receptacles:	
Mounting screws	9-13 in.-lb. (1.0-1.5 N•m)
Terminal screws	14 in.-lb. (1.6 N•m)
Ground screws	10 in.-lb. (1.1 N•m)
Circuit board to control panel	15-20 in.-lb. (1.8-3.4 N•m)
Control panel front to back cover	15-20 in.-lb. (1.8-2.4 N•m)
Idle bracket (clamp)	14-18 in.-lb. (1.6-2.0 N•m)
Idle paddle clamp to governor arm	14-18 in.-lb. (1.6-2.0 N•m)
Idle bracket to engine	60-70 in.-lb. (6.8-7.9 N•m)
Throttle arm to engine	35-45 in.-lb. (3.4-5.1 N•m)
Stator to support bracket	150-155 in.-lb. (17.0-17.5 N•m)
Stator bracket to isolator	145-155 in.-lb. (16.4-17.5 N•m)
Isolator to frame	145-155 in.-lb. (16.4-17.5 N•m)
Engine support to isolator	145-155 in.-lb. (16.4-17.5 N•m)
Engine to engine support	145-155 in.-lb. (16.4-17.5 N•m)
Ground strap fastener	145-155 in.-lb. (16.4-17.5 N•m)
Ground wire to stator	45-55 in.-lb. (5.1-6.2 N•m)
Fuel tank support to frame	25-35 in.-lb. (2.8-3.9 N•m)
Heat shield to tank support	8-12 in.-lb. (0.9-1.4 N•m)
Start switch to battery plate	70-80 in.-lb. (7.9-9.0 N•m)
Battery strap to battery plate	12-16 in.-lb. (1.4-1.8 N•m)
Battery cables to battery	40-50 in.-lb. (4.5-5.6 N•m)
Battery cables to starter switch	50-60 in.-lb. (5.6-6.8 N•m)
Battery cable to starter	30-40 in.-lb. (3.4-4.5 N•m)
Generator to frame crossmember	220-250 in.-lb. (24.9-28.2 N•m)

WIRING DIAGRAMS

Generator	Figure
LR2500	HM116
LR4300	
LR4400	
LR5000T,	
LR5500, and	
LR5550	HM120
LRI2500	
Pre-1999 U.S.	HM117
Pre-1999 Canadian	HM118
1999 and newer	HM119
LRI4400	
Pre-1999 U.S.	HM121
Pre-1999 Canadian	HM122
1999 and newer	HM125
LRI5500	
Pre-1999 U.S.	HM123
Pre-1999 Canadian	HM124
1999 and newer	HM125
LRX3000	HM119
LRX4500 and LRX5600	HM125

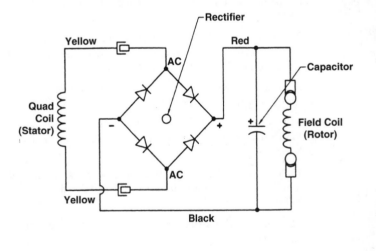

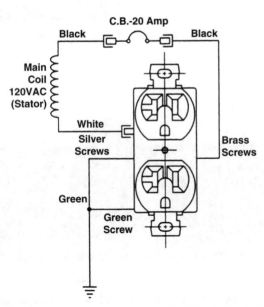

Fig. HM116 – Wiring diagram for model LR2500 generators. Pre 1999 Canadian generators are identical except for the use of 15-amp circuit breakers. Always make sure the receptacle capacity matches the breaker capacity.

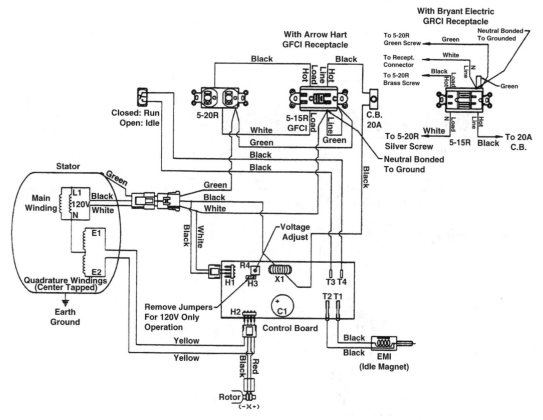

Fig. HM117 – Wiring diagram for pre 1999 U.S. model LRI2500 generators.

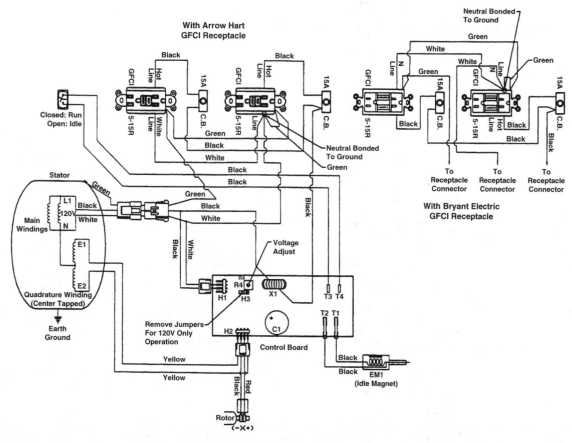

Fig. HM118 – Wiring diagram for pre 1999 Canadian model LRI2500 generators.

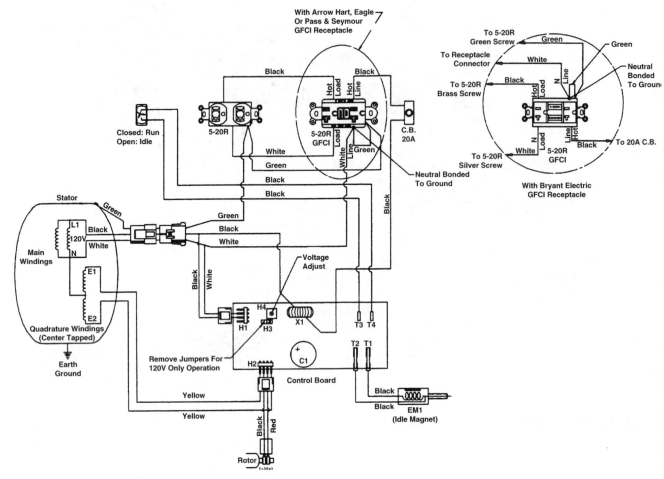

Fig. HM119 – Wiring diagram for LRX3000 and 1999 and newer model LRI2500 generators.

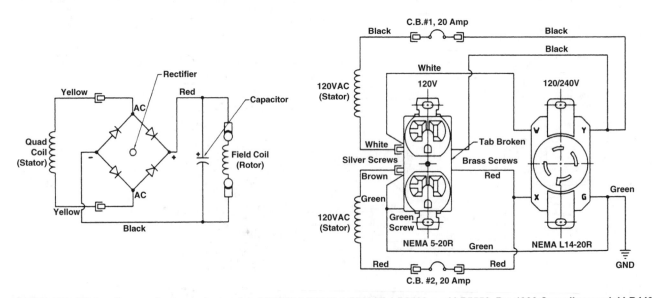

Fig. HM120 – Wiring diagram for generator models LR4300, LR4400, LR5000T, LR5500, and LR5550. Pre 1999 Canadian model LR4400 is identical except for the use of 15-amp circuit breakers. Some pre 1999 U.S. model LR5500 generators used 25-amp circuit breakers. Always make sure the receptacle capacity matches the breaker capacity.

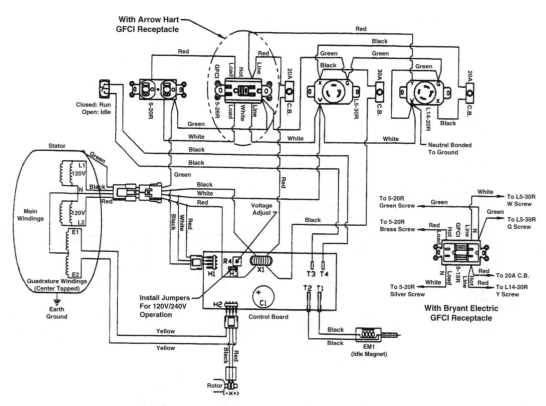

Fig. HM121 – Wiring diagram for pre 1999 U.S. model LRI4400 generators.

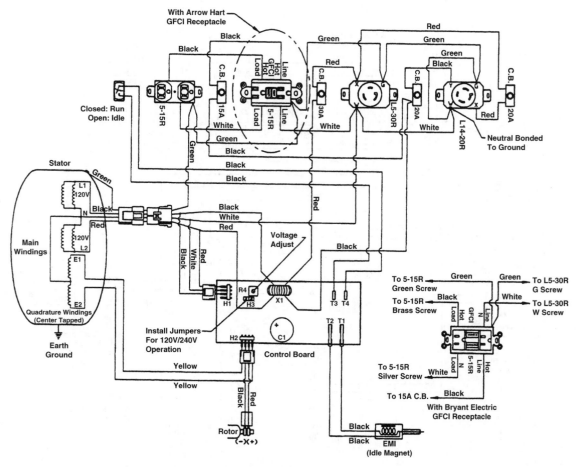

Fig. HM122 – Wiring diagram for pre 1999 Canadian model LRI4400 generators.

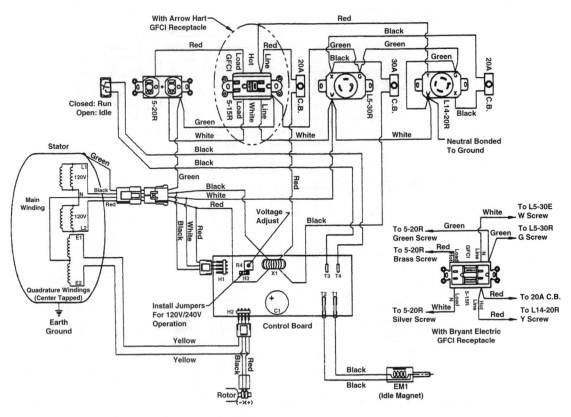

Fig. HM123 – Wiring diagram for pre 1999 U.S. model LRI5500 generators.

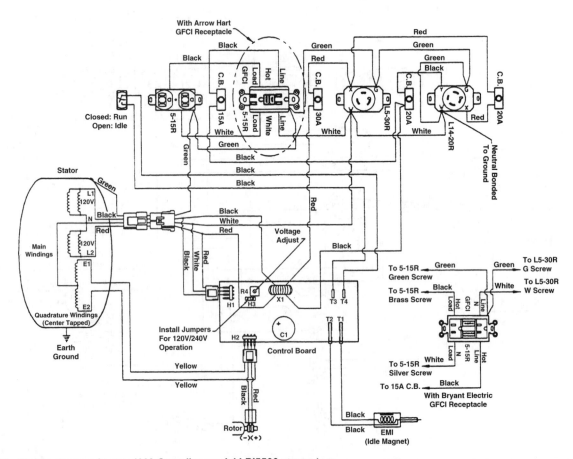

Fig. HM124 – Wiring diagram for pre 1999 Canadian model LRI5500 generators.

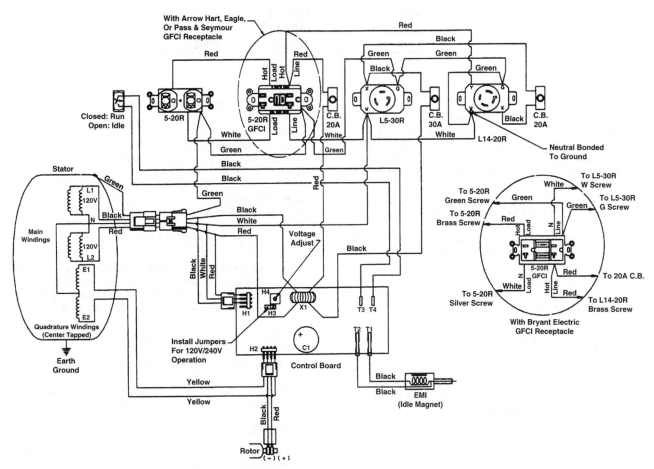

Fig. HM125 – Wiring diagram for models LRX4500 and LRX5600 and 1999 and newer modelsLRI4400 and LRI5500 generators showing voltmeter and elapsed time (hour) meter used on some models.

HOMELITE

EH/HL Series

Generator model	Rated (Surge)	Output Volts	Engine Make	Watts HP
EH2500HD	2300 (2500)	120	B&S	5
EH4400HD	4000 (4400)	120/240	B&S	8
EH5000HD-CSA	4800 (5000)	120/240	B&S	11
EH5500HD	5000 (5500)	120/240	B&S	11
HL2500	2300 (2500)	120	B&S	5
HL4400	4000 (4400)	120/240	B&S	8
HL5000CSA	4800 (5000)	120/240	B&S	10
HL5500	5000 (5500)	120/240	B&S	10

NOTE: Full output on these units can only be achieved by dividing the load between two or more receptacles. The surge, or intermittent, rating should not be used more than 10 minutes per hour of operation or the generator will be damaged.

IDENTIFICATION

Homelite EH and HL series generators are 60Hz units with integral electronic voltage regulation. They can be either brushless or brush style. Brushless generators have a round endbell. brush style units can have either a round or a square endbell. Figures HM201-HM204 show the various component designs.

MAINTENANCE

On round endbell units, the endbell bearing is a prelubricated, sealed ball bearing. Inspect for smooth operation. A faulty bearing can cause abnormal drag loading on the engine, or rotor to stator rubbing.

On square endbell units, the endbell bearing and race should be inspected, cleaned, and relubricated any time the endbell is removed. If the bearing runs hot to the touch, replacement may be necessary.

On brush style units, brushes must be at least 9/16 in. (14 mm) long. shorter brushes require replacement. Brushes can be accessed by removing the endbell. Refer to DISASSEMBLY in this chapter. When checking the brushes, also inspect the brush leads for continuity and for proper contact with the brush holder.

TROUBLESHOOTING

A Homelite field flasher tool #UP00457 and a VOM may be used to troubleshoot generator malfunctions.

Output voltage should be 113-127 V on the 120-volt receptacles and 226-254 V on the 240-volt receptacles. If little or no output is generated, the following items could be potential causes:

1. Engine RPM–The engine must maintain a minimum 3550 RPM under load. Normal no load speed is 3750-3800 RPM.
2. Loose component mounting fasteners (low or no voltage).
3. Circuit breakers or GFCI switch breaker (low or no voltage).
4. Loose wiring connections (low or no voltage).
5. Faulty receptacle (low or no voltage).
6. Capacitor (low or no voltage).
7. Brushes (no voltage).
8A. Rotor– Loss of residual magnetism (no voltage).

NOTE: Residual magnetism can be restored without generator disassembly, and is the recommended first step if the generator has no output. Refer to RESIDUAL MAGNETISM Section.

8B. Rotor–Open or shorted windings (no or low voltage, respectively).
8C. Rotor–Dirty, broken, or disconnected slip rings (no voltage).
9. Stator windings (low voltage–shorted, no voltage–open).
10. Insufficient cooling system ventilation (overheating).

RECEPTACLES

Fig. HM205 shows the various receptacles used on Homelite generators.

NOTE: The generator should be grounded before testing the receptacle output.

To test the receptacles:
1. Start the generator.
2. Make sure that the circuit breakers are set.
3. Set the VOM scale to AC volts.
4. For the 120V outlets, place the black VOM to ground and the red lead to the small flat leg of each duplex outlet. The reading should be 113-127 volts.
5. For the 240V receptacles, refer to the polarity markings of Fig. HM206:
   a. Positive (+) slots are HOT.
   b. Negative (-) slots are NEUTRAL or RETURN.
   c. Lock-slots, where used, are FRAME GROUND.

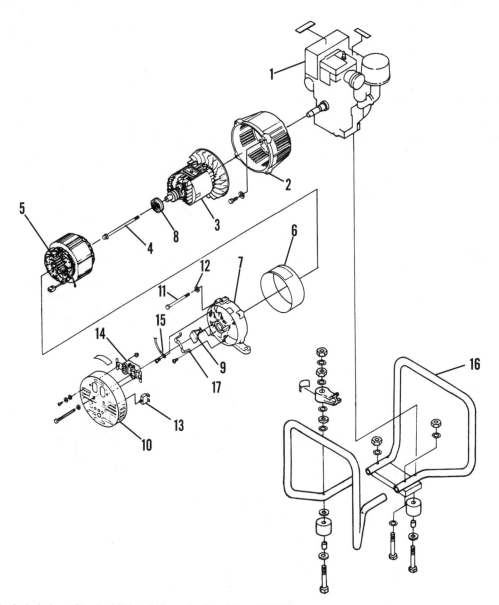

Fig. HM201 – Exploded view of typical EH and HL series brushless 2300W generator components.

1. Engine
2. Engine adapter/front housing
3. Rotor assembly with fan
4. Rotor bolt and washer
5. Stator assembly

6. Stator cover
7. Rear housing/end housing
8. Rotor bearing
9. Capacitor
10. End cover

11. Endbell/stator bolt (4)
12. Lock washer (4)
13. AC circuit breaker
14. 120V AC receptacle
15. Ground wire
16. Frame
17. Capacitor holder

Test between the positive (+) slots and both the negative (–) slots and the ground (G) slots. the reading should be 226-254 volts.

6. If the readings are not to specification:

a. If only one receptacle shows no voltage, the receptacle may not be receiving output. Access and inspect the receptacle wiring.

b. A no voltage reading on all receptacles usually indicates a loss of residual magnetism. Refer to the following RESIDUAL MAGNETISM section of this chapter.

c. A reading in the 1 to 4 volt range on all receptacles is a measurement of residual magnetism only. It indicates a fault in either the brushes, rotor, capacitor, bridge rectifier, or quad winding. Proceed with further testing.

d. A half voltage reading (60 volts at the 120V receptacles. 120 volts at the 240V receptacles) usually indicates a faulty bridge rectifier or rectifier terminal connections. Proceed to BRIDGE RECTIFIER further testing.

e. A less than half voltage reading (30-50V) usually indicates a blown capacitor (capacitor bottom isbulged). A ¾-voltage reading (90V of 120V, 180V of 240V) usually indicates a faulty capacitor. Proceed to the CAPACITOR section for further testing.

NOTE: If the 120V receptacle is being replaced on 120/240V units, the connector tab on the hot side must be broken off. Failure to remove the tab will give an indication of overload upon generator start up. Always torque terminal screws to 14 in.-lb. On duplexes, torque the ground screws to 10 in.-lb.

GENERATOR REPAIR

RESIDUAL MAGNETISM

To restore residual magnetism using the Homelite #UP00457 field flasher:
1. Start the generator and allow a brief warm up period.
2. Turn the field flasher switch on.
3. Plug the flasher into one of the generator's 120V outlets.
4. When the flasher switch trips, residual magnetism should be restored.
5. Stop the generator, remove the flasher, then restart the generator.

6. Test the output at the 120V outlet. If the output tests within specifications, the generator is good. If there is no output, the rotor will not hold residual magnetism. Refer to the ROTOR WINDING RESISTANCE section of this chapter and test the rotor.

To restore residual magnetism without the Homelite field flasher on *brush-style* units:
1. Obtain a 6 or 12-volt battery with two test leads attached to the battery.
2. Start the generator.
3. There are two silver pins protruding from the brush holder—one pin is attached to the black (-) brush lead, and the other pin is attached to the red (+) brush lead. Note the polarity orientation.
4. Hold the negative (-) battery probe to the black (-) brush-lead pin.
5. Touch the positive (+) battery probe to the red (+) brush lead pin for 1-2 seconds.
6. Stop the generator and restart it.
7. Test the generator output. If the output tests within specification, the generator is good. If there is no output, the rotor will not hold residual magnetism. Refer to the ROTOR WINDING RESISTANCE in this chapter and test the rotor.

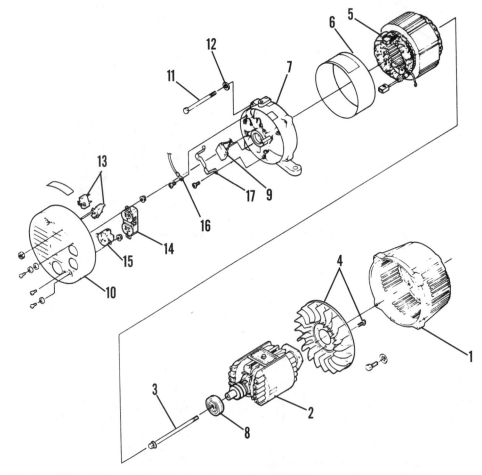

Fig. HM202 – Exploded view of typical EH and HL series brushless 4000 and 5000W generator components.

1. Engine adapter/front housing
2. Rotor assembly
3. Rotor bolt and washer
4. Fan with four screws
5. Stator assembly
6. Stator cover

7. Endbell/rear housing
8. Rotor bearing
9. Capacitor
10. End cover
11. Endbell/stator bolt (4)
12. Lock washer (4)

13. AC circuit breakers
14. 120V AC receptacle
15. 240V AC receptacle
16. Ground wire
17. Capacitor holder

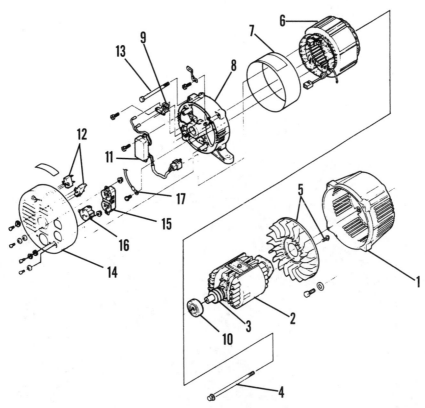

Fig. HM203 – Exploded view of typical round-endbell EH and HL series brush style 4000 and 5000W generator components.

1. Engine adapter/front housing
2. Rotor assembly
3. Slip rings
4. Rotor bolt
5. Fan w/5 screws

6. Stator assembly
7. Stator cover
8. Endbell/rear housing
9. Brush holder with brushes
10. Bearing

11. Regulator
12. AC circuit breakers
13. Endbell/stator through bolts (4)
14. End cover
15. 120V AC receptacle
16. 240V AC receptacle
17. Ground wire

To restore residual magnetism without the Homelite field flasher on *brushless* units:

1. Construct a jumper tool (Fig. HM207) using a male two-prong 120V household plug and two 16-gauge (minimum) wire leads.

NOTE: Mark the jumper wires so that the wire coming from the narrow spade on the 120V plug connects to the positive (+) battery terminal. The wire from the wide spade must connect to the negative (-) battery terminal. An alligator clip on the negative (-) jumper end would be helpful.

2. Start the generator.

3. Plug the jumper tool into one of the generator's 120V receptacle outlets, then connect the negative (-) jumper lead to the negative (-) terminal of a 12 volt battery.

4. Touch the positive (+) jumper lead to the positive (+) battery terminal for 1-2 seconds.

5. Completely disconnect the jumper tool.

6. Stop the generator and restart it.

7. Test generator output. If output tests within specification, the generator is OK. If there is no output, the rotor will not hold residual magnetism. Refer to the ROTOR WINDING RESISTANCE in this chapter and test the rotor.

DISASSEMBLY

Refer to Figs. HM201-HM204 for component identification. Always allow the generator to cool down before disassembly. On some models, it may be helpful to remove the fuel tank for easier access to the generator head.

1. On round endbell units, unscrew the fasteners holding the end cover to the endbell, then unscrew the fasteners holding the receptacle(s) and circuit breaker(s) to the end cover.

2. On EH models with a stator cover (wrapper), unbend the two tabs holding the wrapper to the stator, then remove the wrapper.

3. Place a temporary support, such as a 2 × 4, under the engine adapter.

4. On round endbell units, remove the bolts holding the endbell feet to the generator cradle frame.

5. On square endbell units, remove the fasteners holding the bottom of the stator to the stator support bracket.

6. Remove the four stator throughbolts.

7. Pull the endbell/stator assembly from the generator.

NOTE: On brush style units, as the endbell pulls away from the rotor, the spring loaded brushes will pop out from the brush holder. Do not lose the brushes. Use caution when setting the stator on the workbench so as not to scrape or damage the stator winding varnish.

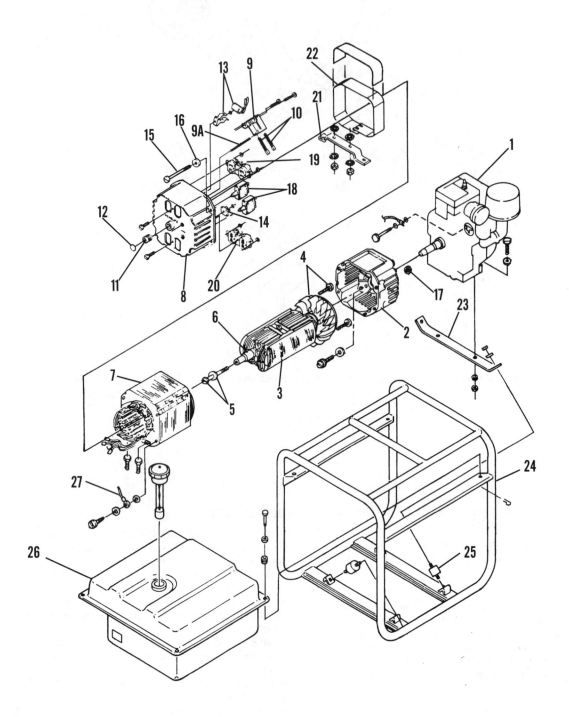

Fig. HM204 – Exploded view of typical square endbell EH and HL series brush style 4000 and 5000W generator components.

1. Engine
2. Engine adapter/front housing
3. Rotor assembly
4. Fan with four screws
5. Rotor through bolt and washer
6. Slip rings
7. Stator assembly
8. Endbell/rear housing
9. Brush holder

9A. Brush retainer pin
10. Brushes
11. Rotor bearing
12. Rotor bearing cap plug
13. Capacitor with mount bracket
14. Diode rectifier
15. Endbell/stator throughbolt (4)
16. Flat washer (4)
17. Hex nut (4)

18. AC circuit breakers
19. 120V AC receptacle
20. 240V AC receptacle
21. Stator mount bracket
22. Stator cover
23. Support
24. Frame
25. Vibration isolator
26. Fuel tank
27. Ground wire

GENERATOR PLUGS AND RECEPTACLES

SIZE / PART NUMBER	RECEPTACLE	PLUG	CONFIGURATION
VOLTAGE/AMPERAGE NEMA STANDARD HOMELITE P/N	120V 15AMP 5-15R 50991A	120V 15AMP 5-15P	
VOLTAGE/AMPERAGE NEMA STANDARD HOMELITE P/N	120V 15AMP 5-15R 04058-A	120V 15AMP 5-15P	
VOLTAGE/AMPERAGE NEMA STANDARD HOMELITE P/N	120V 20AMP 5-20R 51373	120V 20AMP 5-20P	
VOLTAGE/AMPERAGE NEMA STANDARD HOMELITE P/N	120V 20AMP TWIST LOCK L5-20R 48978	120V 20AMP TWIST LOCK L5-20P	
VOLTAGE/AMPERAGE NEMA STANDARD HOMELITE P/N	120V 30AMP TWIST LOCK L5-30R 42601	120V 30AMP TWIST LOCK L5-30P 43326	
VOLTAGE/AMPERAGE NEMA STANDARD HOMELITE P/N	240V 20AMP 6-20R 02863	240V 20AMP 6-20P 49709	
VOLTAGE/AMPERAGE NEMA STANDARD HOMELITE P/N	240V 20AMP TWIST LOCK L14-20R 46508	240V 20AMP TWIST LOCK L14-20P 47600	
VOLTAGE/AMPERAGE NEMA STANDARD HOMELITE P/N	240V 30AMP TWIST LOCK L14-30R 46718	240V 30AMP TWIST LOCK L14-30P 47601	

Fig. HM205 – Receptacles used on Homelite generators.

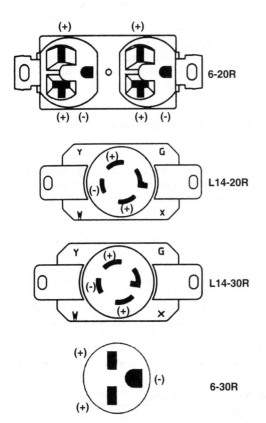

Fig. HM206 – Test connections for the 240V receptacles. Refer to text for correct test procedure.

8. Disconnect the stator to endbell wiring connectors and separate the endbell from the stator.

9. Use a ½ in. wrench to remove the rotor throughbolt.

10. Apply penetrating oil through the rotor shaft onto the engine crankshaft taper, then carefully tap the rotor laminations with a mallet to separate the rotor from the engine.

NOTE: Do not allow the rotor to drop once it breaks free. On some models, the rotor fan is part of the rotor and cannot be replaced separately. On replaceable fan models, the fan can be removed at this time.

BRUSHES

Refer to MAINTENANCE in this Chapter.

CAPACITOR

NOTE: Even with the generator stopped, the capacitor could have a retained charge. Do not touch the capacitor terminals, as electric shock may result. Always discharge the capacitor by shorting across the terminals with a screwdriver or similar tool with an insulated handle.

To test the capacitor:

1. Place the VOM on the highest ohm scale.

2. Disconnect the capacitor leads, noting the polarity orientation.

3. Connect the VOM leads to the capacitor terminals. The VOM needle should rise quickly to continuity, then move more slowly back toward infinity until the capacitor resistance is indicated. If the capacitor has its value marked on its side, the meter reading can be compared to this value.

NOTE: Digital meters will advance towards infinity until the capacitor reaches full charge, then show no continuity/overscale.

4. Reverse the VOM leads. The meter should move rapidly toward zero ohms, followed by the same slow rise toward capacitor value as noted in Step 3.

   a. If the VOM reads straight continuity, the capacitor is shorted and must be replaced.

   b. If the VOM fluctuates between continuity and infinity, the capacitor is leaking. Replace the capacitor.

NOTE: When reconnecting the capacitor leads, always observe proper polarity orientation.

DIODES

To test the diodes on diode equipped units:

1. Remove at least one end of the diode from the circuit. Use caution not to break the diode lead.

2. Set the VOM to read ohms.

3. Connect the meter leads to the diode terminals and observe the reading.

4. Reverse the meter leads and observe the reading again.

5. A good diode will show a reading in only one direction of Steps 3 and 4.

6. If a reading is observed in both directions, the diode is shorted. If no reading is observed in either direction, the diode is open. Replace the diode.

BRIDGE RECTIFIER (REGULATOR)

NOTE: If, after testing, the rectifier shows shorted, it may have fed AC to the rotor, destroying residual magnetism. If the generator shows no output after replacing the rectifier, flash the fields. Refer to the preceding RESIDUAL MAGNETISM section.

On rectifiers with integral wiring leads, usually round endbell units, refer to Fig. HM203 for identification:
1. Disconnect the wires from the circuit, noting the orientation of the red and black wires to the brush holder terminals (on some models the brush wires are orange and blue).
2. Set the VOM to the R × 1 scale.
3. Connect one meter lead to either yellow wire, and connect the other meter lead to the orange or red wire. note the reading.
4. Reverse the meter leads and again note the reading.
5. One of the Step 3 and 4 tests should show continuity. one should show no continuity.
6. Repeat Steps 3 and 4 with the other Yellow wire, again noting the readings.
7. Now repeat Steps 3, 4, and 6 using the Yellow wires and the blue or black wire.
8. After all the wires have been forward and reverse tested, compare the readings.
   a. If one test set shows continuity in both directions, the rectifier has a short circuit and must be replaced. A shorted rectifier may have fed AC current to the rotor. after replacing the rectifier, flash the fields to re establish residual magnetism.
   b. If one test set shows no continuity both directions, the rectifier has an open circuit. replace the rectifier.
   c. If one or more reading is significantly higher or lower than the others, the rectifier is faulty. replace the rectifier.

On square rectifiers with four male spade terminals, usually square endbell units. refer to Fig. HM204 for identification and Fig. HM208 for testing:
1. Disconnect the rectifier from the circuit, noting the wiring orientation.
2. Set the VOM to the R × 1 scale.
3. Connect the meter leads to any pair of adjacent or diagonal terminals (1-2, 1-4, 3-2, 3-4, 1-3, or 2-4). note the reading.
4. Reverse the meter leads and again note the reading.
5. Repeat Steps 3 and 4 with the other five terminal pairs.
6. Compare the readings with the Fig. HM208 chart.
   a. If one test set shows continuity both directions, the rectifier has a short circuit. replace the rectifier. A shorted rectifier may have fed AC current to the rotor. After replacing the rectifier, flash the fields to re establish residual magnetism.
   b. If one test set shows no continuity in both directions (except the 2-4 set, which should show no continuity in both directions), the rectifier has an open circuit. Replace the rectifier.
   c . If one or more reading is significantly higher or lower than the others, the rectifier is faulty. Replace the rectifier.

CIRCUIT BREAKERS

To test each circuit breaker:
1. Make sure the breaker is set. If tripped, reset.
2. If not already done, disassemble as necessary to access the breaker (refer to the preceding DISASSEMBLY section

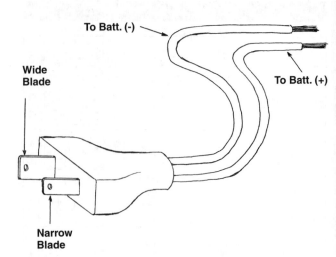

Fig. HM207 – Homemade jumper tool for remagnetizing the rotor (flashing the field).

in this chapter), then remove the wires from the rear of the breaker.
3. Using the R × 1 scale on the VOM, test for continuity across the breaker terminals.
4. If no continuity reading is noted, or if the breaker fails to reset, the breaker is faulty. Replace the breaker.

ROTOR

Rotor Slip rings

On brush style units, the slip rings (Fig. HM203, typical) are an integral part of the rotor assembly. They are not replaceable.

Clean dirty or rough slip rings with a suitable solvent, then lightly sand with ScotchBrite or fine grit sandpaper. Do not use emery cloth or crocus cloth.

Replace the rotor if the slip rings are cracked, chipped, broken, or grooved

Rotor Winding Resistance

NOTE: All resistance readings are approximate and may vary depending on the winding temperature and the accuracy of the test equipment.

The rotor can be tested without removing it from the generator.

To test resistance:
1. Use a VOM set on the R × 1 scale. Hold one VOM lead to each rotor slip ring.
2. If no reading is observed upon initial testing, inspect the winding to slip ring connection and solder joint. If the solder joint is faulty, and if there is sufficient wire length, resolder the joint and retest.
3. Rotor resistance specifications at 77E F (25E C) are as follows:

Model	Ohms	Model	Ohms
EH2500	43.9 – 49.5	HL2500	41.2 – 46.4
EH4400	63.0 – 71.0	HL4400	57.4 – 64.8
EH5500	71.4 – 80.6	HL5500	67.5 – 76.1

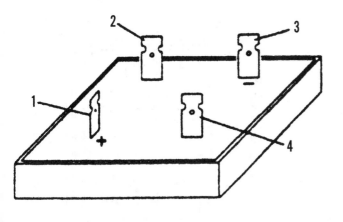

Digital VOM OHM Reading

Pos lead on #1 terminal Neg lead on #2 terminal	1.675-1.731
Pos lead on #1 terminal Neg lead on #3 terminal	3.215-3.301
Pos lead on #1 terminal Neg lead on #4 terminal	1.588-1.612
Pos lead on #2 terminal Neg lead on #3 terminal	1.612-1.616
Pos lead on #2 terminal Neg lead on #1 or #4 terminal	No Reading
Pos lead on #3 terminal Neg lead on #1, #2, or #4 terminal	No Reading
Pos lead on #4 terminal Neg lead on #1 or #2 terminal	No Reading
Pos lead on #4 terminal Neg lead on #3 terminal	1.863-1.871

Analog VOM Testing

Pos. No.	Black Test Lead (-)			
	1 (+)	2 (AC)	3 (-)	4 (AC)
1 (+)		Conducting	Conducting	Conducting
2 (AC)	Non Conducting		Conducting	Non Conducting
3 (-)	Non Conducting	Non Conducting		Non Conducting
4 (AC)	Non Conducting	Non Conducting	Conducting	
	Resistance Readings			

Red Test Lead (+)

Fig. HM208 – Bridge rectifier diode test chart. Refer to text for correct test procedure

4. If the VOM readings are lower than specified, the rotor has shorted windings. If the VOM reads infinity, the rotor has open windings. Either way, replace the rotor.
5. Connect one VOM lead to one slip ring, and place the other lead on the rotor shaft.
6. Repeat Step 5 with the other slip ring.
7. There should be **no** continuity in the tests in Steps 5 and 6. If continuity is observed on either test, the windings are shorted to the core. Replace the rotor.

STATOR

Winding Resistance

> NOTE: All resistance readings are approximate and may vary depending on the winding temperature and the test equipment accuracy.

To test the stator windings:
1. Set the VOM to the R × 1 scale.
2. With the stator leads unplugged, connect the VOM leads to the following color coded wiring pairs, and compare the readings to the specifications listed:

White – Black	EH2500	0.366-0.412	HL2500	0.431-0.485
Yellow – Yellow	EH2500	1.11-1.25	HL2500	1.02-1.15
White – Black	EH4400	0.261-0.294	HL4400	0.295-0.331
Red – Brown	EH4400	0.261-0.294	HL4400	0.295-0.331
Yellow – Yellow	EH4400	0.94-1.06	HL4400	0.83-0.94
White – Black	EH5500	0.196-0.220	HL5500	0.215-0.241
Red – Brown	EH5500	0.196-0.220	HL5500	0.215-0.241
Yellow – Yellow	EH5500	0.91-1.03	HL5500	0.79-0.89

3. If either wire pair test shows no continuity, or if the test shows substantially lower readings than the specification, the stator is faulty.
4. Testing one stator wire at a time, connect one VOM lead to each stator wire, then connect the other VOM lead to a clean, dry spot on the stator laminations. There should be no continuity on any test. If continuity is observed, the respective winding is shorted and the stator must be replaced.

REASSEMBLY

Refer to Figs. HM201-HM204 for component identification
1. If the engine adapter was removed, install the adapter. On 2500W units, torque the bolts 120-150 in.-lb. (13.6-17.0 N.m). On 4-5KW units, torque the bolts 240-250 in.-lb. (27.1-28.3 N•m).
2. Place a temporary support, such as a 2 × 4, under the engine adapter.
3. If the fan is replaceable, snap the fan into position on the rotor, then install and torque the fan screws 12-16 in.-lb. (1.4-1.8 N•m).
4. Make sure that the matching tapers on the engine crankshaft and inside the rotor shaft are clean and dry. install the rotor onto the crankshaft. Install the rotor bolt and lockwasher, then torque the rotor bolt 100-140 in.-lb. (11.3-15.8 N•m).
5. Service the rotor bearing in the endbell, if needed.
6. Slide the stator into position over the rotor and against the engine adapter. Ensure that the stator fits snugly against the adapter.
7. On brush style units, torque the brush holder to the endbell 12-16 in.-lb. (1.4-1.8 N•m), then install the brushes into the brush holder. Hold the brushes in place against

their springs by inserting a straightened paper clip or similar tool (Homelite #02857) through the endbell hole next to the bearing. Insert the tool from the outside in, making certain that it passes through the brush holder and over both brushes.

> **NOTE: If using the old brushes, ensure the curve on each brush face correctly aligns with the slip ring diameter.**

8. Connect the stator to endbell wiring, then position the endbell onto the stator, insuring that the wires are not pinched or kinked.

9. Install the four stator throughbolts. using the sequence shown in Fig. HM209. Torque the bolts 60-80 in.-lb. (6.8-9.0 N•m).

10. On square endbell units, install the stator to support bracket fasteners. Torque the fasteners 150-155 in.-lb. (17.0-17.5 N•m).

11. On round endbell units, install the bolts holding the endbell feet to the generator cradle frame. Torque the bolts 145-155 in.-lb. (16.4-17.5 N•m).

12. Manually rotate the engine crankshaft to make sure that the rotor turns freely and does not drag against the stator or endbell. Check and realign as necessary.

13. On EH models with a stator cover, assemble the cover over the stator. The edge of the cover must be snug against the endbell. Fit the cover end tabs into their respective slots, pull the cover tight, then bend the tabs back snugly over the cover.

14. On round endbell units, mount the receptacle(s) and circuit breaker(s) to the endbell cover, then mount the end cover to the endbell. Make certain that the wires are not pinched or kinked.

15. If a ground strap was removed from one of the endbell or stator mount positions during disassembly, reinstall it now and torque the fastener 145-155 in.-lb. (16.4-17.5 N•m).

TORQUE CHART

Application	Torque Value
Engine adapter to engine:	
2500W Units	120-150 in.-lb. (13.6-17.0 N•m)
4000-5000W Units	240-250 in.-lb. (27.1-28.3 N•m)
Fan to rotor	12-16 in.-lb. (1.4-1.8 N•m)
Rotor throughbolt	100-140 in.-lb. (11.3-15.8 N•m)
Brush holder to endbell	12-16 in.-lb. (1.4-1.8 N•m)
Endbell/stator throughbolts	60-80 in.-lb. (6.8-9.0 N•m)
Receptacles:	
Mounting screws	9-13 in.-lb. (1.0-1.5 N•m)

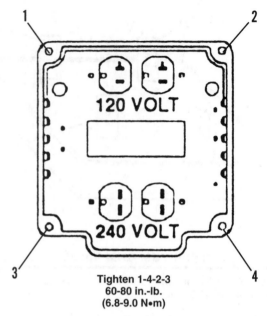

Tighten 1-4-2-3
60-80 in.-lb.
(6.8-9.0 N•m)

Fig. HM209 – Typical torque sequence and value for endbell reassembly.

Terminal screws	14 in.-lb. (1.6 N•m)
Ground screws	10 in.-lb. (1.1 N•m)
Stator to support bracket	150-155 in.-lb. (17.0-17.5 N•m)
Stator bracket to isolator	145-155 in.-lb. (16.4-17.5 N•m)
Isolator to frame	145-155 in.-lb. (16.4-17.5 N•m)
Engine support to isolator	145-155 in.-lb. (16.4-17.5 N•m)
Engine to wngine support	145-155 in.-lb. (16.4-17.5 N•m)
Ground strap fastener	145-155 in.-lb. (16.4-17.5 N•m)
Ground wire to stator	45-55 in.-lb. (5.1-6.2 N•m)
Fuel tank support to frame	25-35 in.-lb. (2.8-3.9 N•m)
Heat shield to tank support	8-12 in.-lb. (0.9-1.4 N•m)
Start switch to battery plate	70-80 in.-lb. (7.9-9.0 N•m)
Battery strap to battery plate	12-16 in.-lb. (1.4-1.8 N•m)
Battery cables to battery	40-50 in.-lb. (4.5-5.6 N•m)
Battery cables to starter switch	50-60 in.-lb. (5.6-6.8 N•m)
Battery cable to starter	30-40 in.-lb. (3.4-4.5 N•m)

WIRING DIAGRAMS

Refer to Fig. HM210 for wiring diagram for models EH2500 and HL2500.

Refer to Fig. HM211 for wiring diagram for models EH4400 and HL4400.

Refer to Fig. HM212 for wiring diagram for models EH5000, EH5500, HL5000, and HL5500.

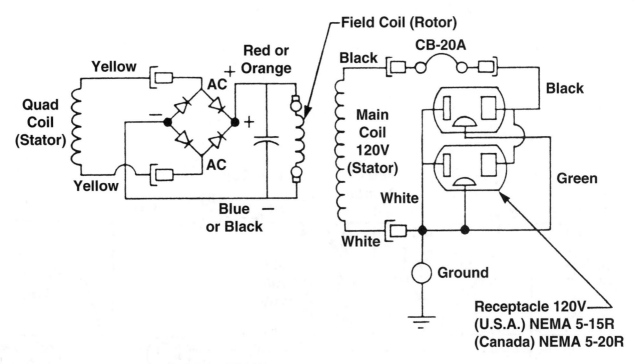

Fig. HM210 – Wiring diagram for models EH2500 and HL2500.

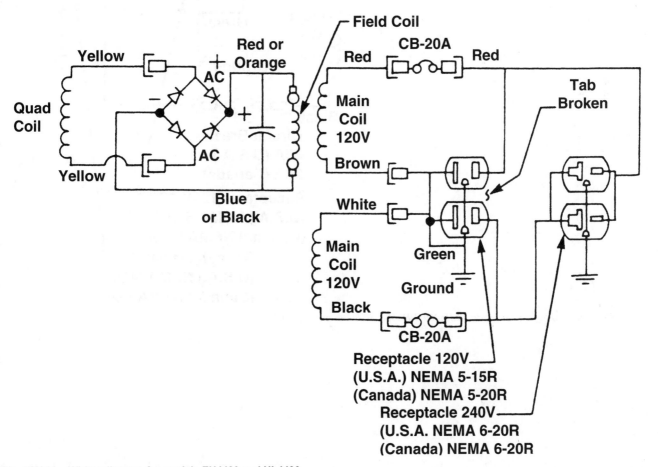

Fig. HM211 – Wiring diagram for models EH4400 and HL4400.

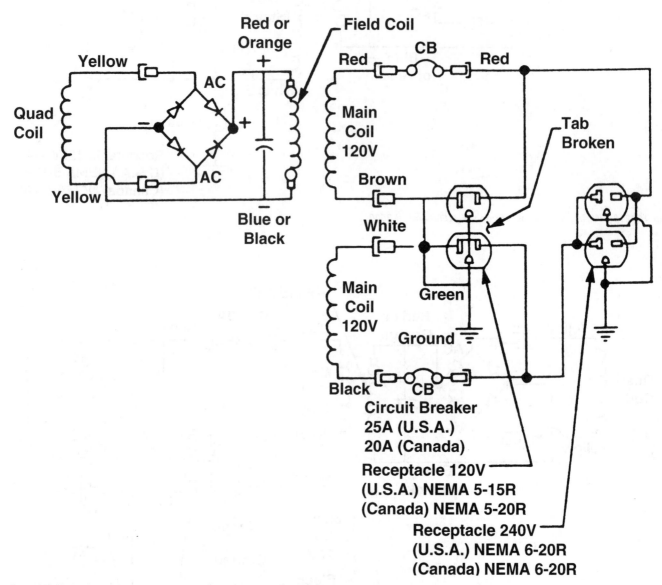

Fig. HM212 – Wiring diagram for models EH5000, EH5500, HL5000, and HL5500.

HOMELITE
CG ("CONTRACTOR") Series

Generator Model	Rated (Surge) Watts	Output Volts	Engine Make	HP
CG4400	4000 (4400)	120/240	Honda	8.0
CG4800	4250 (4800)	120/240	Honda	8.0
CG5200	4500 (5200)	120/240	Honda	9.0
CG(E)5800	5200 (5800)	120/240	Honda	11.0
CG(E)6300	5500 (6300)	120/240	Honda	11.0

NOTE: Models with *E* in the model number prefix denotes electric-start option.

IDENTIFICATION

These Homelite generators are brush-style, single-phase, 60 Hz units with integral electronic voltage regulation. The receptacles are located on the side-mounted control panel.

Refer to Fig. HM301 for generator-component identification.

MAINTENANCE

The endbell bearing and race should be inspected, cleaned, and relubricated any time the endbell is removed. If the bearing runs hot to the touch, replacement may be necessary.

Brushes must be at least 3/8 in. (9.5 mm) long. Shorter brushes require replacement. Brushes can only be accessed by removing the endbell/brush head cover and fan. Refer to the BRUSHES section in this chapter.

TROUBLESHOOTING

A Homelite generator field flasher tool #UP00457 and a VOM may be used to troubleshoot generator malfunctions. Prior to any disassembly, initial testing should be performed with these two tools.

No-load output voltage should be 117-130V on the 120-volt receptacles and 234-260V on the 240-volt receptacles. No-load voltage with the idle control activated should be 80-95V on the 120V receptacle. If little or no output is generated, the following items could be potential causes:
1. Engine RPM–The engine must maintain a minimum 3550 RPM under load. Normal no-load speed is 3750-3800 RPM. No-load speed with idle control activated should be 2650-2800 RPM. Refer to the appropriate ENGINE section for engine service. Refer to the idle control section in this chapter for idle control service.
2. Loose component mounting fasteners (low or no voltage).
3. Circuit breakers or GFCI switch breaker (low or no voltage).
4. Faulty wiring connections, especially loose spade terminals (low or no voltage).

5. Faulty receptacle (low or no voltage).
6. Capacitor (low or no voltage).
7. Brushes (no voltage).
8. Electronic voltage regulator board (over-voltage or no output).
9. Rotor:
   a. Loss of residual magnetism (no voltage).

NOTE: Residual magnetism can be restored without generator disassembly if the Homelite field flasher tool #UP00457 is used, or with only minimal disassembly without using the field flasher tool, and is the recommended first step if the generator has no output. Refer to the RESIDUAL MAGNETISM section.

   b. Open or shorted windings (no or low voltage, respectively).
   c. Dirty, broken or disconnected sliprings (no voltage).
10. Stator windings (low voltage – shorted; no voltage – open).
11. Idle control malfunction (erratic or no idle).
12. Insufficient cooling system ventilation (overheating).

GENERATOR REPAIR

PRELIMINARY TESTING AND RESIDUAL MAGNETISM

Receptacles

Fig. HM302 shows the various receptacles used on Homelite generators.

NOTE: The generator should be properly grounded to earth prior to testing the receptacle output.

To test the receptacles:
1. Start the generator, verifying proper no load speed of 3750-3800 RPM.
2. Make sure that the circuit breakers are set. If the circuit breakers fail to set, refer to the CIRCUIT BREAKER section in this chapter.
3. Set the VOM scale to AC volts.
4. For the 120V outlets, place the black VOM to neutral/ground and the red lead to the small flat leg of each duplex outlet. The reading should be 117-130 volts.

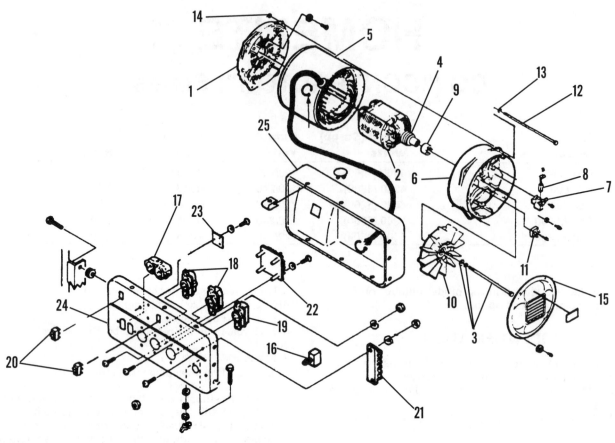

Fig. HM301 – View of typical CG series generator components.

1. Engine adapter
2. Rotor assembly
3. Rotor bolt and washers
4. Slip rings
5. Stator assembly
6. Endbell/brush head
7. Brush holder
8. Brush (2)
9. Rotor bearing

10. Fan
11. Diode rectifier
12. Endbell/stator throughbolt (4)
13. Lockwasher (4)
14. Hex nut (4)
15. End cover
16. AC circuit breaker
17. 120V AC receptacle
18. 120V AC Twistlok receptacle

19. 240V AC Twistlok receptacle
20. Rocker switches
21. Terminal block
22. Voltage regulator
23. Idle-control unit
24. Control panel
25. Panel cover

5. For the 240V receptacles, refer to the polarity markings of Fig. HM303.

   a. Positive (+) slots are hot.

   b. Negative (-) slots are neutral or return.

   c. Lock slots, where used, are frame ground.

6. Test between the positive (+) slots and both the negative (-) slots and the ground (G) slots. The reading should be 234-260 volts.

7. If the readings are not to specification:

   a. If only one receptacle shows no voltage, the receptacle may be faulty or may not be receiving output. Access and inspect the receptacle and wiring.

   b. A no voltage reading on all receptacles usually indicates a loss of residual magnetism. Refer to the following residual magnetism section of this chapter.

   c. A reading in the 1 to 3 volt range on all receptacles is a measurement of residual magnetism only. It indicates a fault in either the brushes, rotor, voltage reg-

ulator, bridge rectifier, or quad winding. Proceed with further testing.

   d. A half voltage reading (60V at the 120V receptacles; 120V at the 240V receptacles) usually indicates a faulty bridge rectifier or rectifier terminal connections. Proceed to BRIDGE RECTIFIER section of this chapter for further testing.

   e. A less than half voltage reading (30-50V) usually indicates a blown capacitor (capacitor bottom is bulged). A 3/4-voltage reading (90V of 120V; 180V of 240V) usually indicates a faulty capacitor. The capacitor is part of the voltage regulator assembly and is not serviced separately. Proceed with further testing.

   f. An over voltage reading (150-160V at the 120V receptacles; 300V at the 240V receptacles) indicates a faulty EVR board, if engine RPM is correct. Proceed to the ELECTRONIC VOLTAGE REGULATOR section of this chapter for further testing.

SIZE / PART NUMBER	RECEPTACLE	PLUG	CONFIGURATION
GENERATOR PLUGS AND RECEPTACLES			
VOLTAGE/AMPERAGE NEMA STANDARD HOMELITE P/N	120V 15AMP 5-15R 50991A	120V 15AMP 5-15P PURCHASE LOCALLY	
VOLTAGE/AMPERAGE NEMA STANDARD HOMELITE P/N	120V 15AMP 5-15R 04058-A	120V 15AMP 5-15P PURCHASE LOCALLY	
VOLTAGE/AMPERAGE NEMA STANDARD HOMELITE P/N	120V 20AMP 5-20R 51373	120V 20AMP 5-20P PURCHASE LOCALLY	
VOLTAGE/AMPERAGE NEMA STANDARD HOMELITE P/N	120V 20AMP TWIST LOCK L5-20R 48978	120V 20AMP TWIST LOCK L5-20P PURCHASE LOCALLY	
VOLTAGE/AMPERAGE NEMA STANDARD HOMELITE P/N	120V 30AMP TWIST LOCK L5-30R 42601	120V 30AMP TWIST LOCK L5-30P 43326	
VOLTAGE/AMPERAGE NEMA STANDARD HOMELITE P/N	240V 20AMP 6-20R 02863	240V 20AMP 6-20P 49709	
VOLTAGE/AMPERAGE NEMA STANDARD HOMELITE P/N	240V 20AMP TWIST LOCK L14-20R 46508	240V 20AMP TWIST LOCK L14-20P 47600	
VOLTAGE/AMPERAGE NEMA STANDARD HOMELITE P/N	240V 30AMP TWIST LOCK L14-30R 46718	240V 30AMP TWIST LOCK L14-30P 47601	

Fig. HM302 – Receptacles used on Homelite generators.

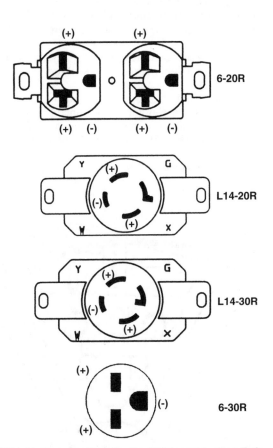

Fig. HM303 – Test connections for the 240V receptacles. Refer to text for correct test procedure.

NOTE: If replacing the 120V receptacle on 120/240V units, the connector tab on the hot side must be broken off. Failure to remove the tab will give an indication of overload upon generator start up. Torque the terminal screws to 14 in.-lb. On duplexes, torque the ground screws to 10 in.-lb.

NOTE: When replacing receptacles, always make sure that the circuit breaker capacity matches the receptacle. If upgrading from an original 15-amp receptacle or GFCI to a 20-amp receptacle or GFCI, replace the breaker with a matching unit.

To Test Residual Magnetism Voltage

1. Disassemble the control panel as necessary to access the electronic voltage regulator ("EVR" – wiring diagram item VR1).

2. Remove and insulate the black wire.

3. Start the generator, verifying proper no-load engine speed of 3750-3800 RPM.

4. Measure DC voltage between the red (+) and white (-) wire terminals on the EVR.

   a. Six volts (6VDC) or higher is sufficient to activate the EVR and produce rated output.

   b. Voltage below 6VDC requires flashing the fields.

To Restore Residual Magnetism Using the #UP00457 Field Flasher

1. Start the generator and allow a brief warm up period.

2. Turn the field flasher switch on.

3. Plug the flasher into one of the generator's 120 V outlets.

4. When the flasher switch trips, residual magnetism should be restored.

5. Stop the generator, remove the flasher, then restart the generator.

6. Test the output at the 120V outlet. If there is no output, the rotor will not hold residual magnetism. Refer to the rotor winding resistance section of this chapter and test the rotor. If the rotor is faulty, replace the rotor.

To Restore Residual Magnetism Without the #UP00457 Field Flasher

1. Obtain a 6- or 12-volt battery with two probe-tipped test leads attached to the battery.

2. Remove the endbell/brush head cover, rotor bolt, and fan.

3. Reinstall and torque the rotor bolt 100-140 in.-lb. (11.3-15.8 N•m).

4. Start the generator. Do not run the generator without the rotor bolt properly torqued.

5. Hold the negative battery probe to the innermost brush terminal with the black wire (B – Fig. HM304).

6. Touch the positive battery probe to the outer brush terminal with the red wire (R – Fig. HM304) for 1-2 seconds.

7. Stop, then restart, the generator.

8. Test generator output. If output tests within specification, the generator is OK. If there is no output, the rotor will not hold residual magnetism. Proceed with further testing.

GROUND FAULT CIRCUIT INTERRUPTER (GFCI)

To test the GFCI:

1. Start the generator and allow a brief warm-up period.
2. Depress the TEST button. The RESET button should extend. if it does, proceed to Step 3. If the RESET button does not extend, the GFCI is faulty.
3. With the RESET button extended, plug a test light into the GFCI receptacle. If the light lights with the RESET button extended, the GFCI line and load terminals have been wired backwards. To correct, stop the generator, then:

    a. Refer to Fig. HM305.
    b. Access the GFCI connections (refer to control panel disassembly in this chapter).
    c. Determine whether the GFCI receptacle is a Bryant or an Arrow Hart.
    d. Make sure that the input power wires connect to the proper *line* screws, and, when applicable, any protected receptacles downstream of the GFCI connect to the proper *load* screws. When replacing a GFCI receptacle, always note the line and load connection orientation.
    e. Ensure that the white (neutral) wires connect to the correct **silver colored** screws, and the black or red (hot) wires connect to the correct **brass** screws. The green (ground) wire always connects to the green screw.

4. To restore power, restart the generator and depress the RESET button firmly into the GFCI until a definite click is heard. If properly reset, the RESET button is flush with the surface of the TEST button. When the RESET button stays in, the GFCI is on.
5. Stop the generator.

CIRCUIT BREAKERS

> NOTE: When replacing receptacles, always make sure that the circuit breaker capacity matches the receptacle. If upgrading from an original 15-amp receptacle or GFCI to a 20-amp receptacle or GFCI, replace the breaker with a matching unit.

To test each circuit breaker:

1. Insure the breaker is set. If tripped, reset.
2. Disassemble as necessary to access the breaker, then remove the wires from the rear of the breaker.
3. Using the R × 1 scale on the VOM, test for continuity across the breaker terminals.
4. If no continuity reading is noted, or if the breaker fails to reset, the breaker is faulty. Replace the breaker.

MAXIMUM (FULL) POWER SWITCH

Some CG models are equipped with a MAX POWER switch.

With the switch in the 120/240V position, output is generated with the main stator windings connected in series (Fig. HM306) so that, to utilize full generator capacity, the maximum 120V load capable of being generated is only half the rated output. A full rated 240V load could also be supported if no 120V load was applied.

With the switch in the 120V position, the main stator windings are connected in parallel (Fig. HM307). This allows the full rated load to be drawn at 120V, with no 240V output.

To test the max-power switch:

1. Disassemble the control panel as necessary to access the switch.
2. **Carefully note** the wiring color code orientation, then disconnect the switch wires.
3. With the VOM set on the R × 1 scale, place the meter leads on the following switch wire pairs, and compare the readings to the specifications listed:

120V Position

Black to Brown	Continuity
Blue to White	Continuity
Black to White	No continuity/Infinity
Blue to Brown	No continuity/Infinity

120/240V Position

Blue to Red	Continuity
White to Brown	Continuity
Red to White	No continuity/Infinity
Blue to Brown	No continuity/Infinity

4. Any reading other than those specified indicates a faulty switch. replace the switch.

IDLE CONTROL

System Test

Place the idle control auto/start switch in the *start* position. Start the generator, allowing a brief warm up period. Move the switch to the AUTO position. The engine should idle down within 3-5 seconds. Apply a 50-watt minimum load. The engine should immediately return to full-throttle. Disconnect the load and stop the generator.

If the engine did not idle down with the switch in the AUTO position, perform the following tests:

1. Set the VOM to the R × 1 scale.
2. Test the fuse. If the fuse is blown:
    a. Inspect the electromagnet lead wires for chafing or shorts.
    b. Replace the fuse with a 0.5-amp fuse **only** (Homelite #49318). Higher rated fuses will not protect the control board.
3. Test the switch:
    a. Disassemble the generator control panel as necessary to access the internal components.
    b. Note which switch terminals are connected to the yellow wires.
    c. Disconnect the yellow wires.

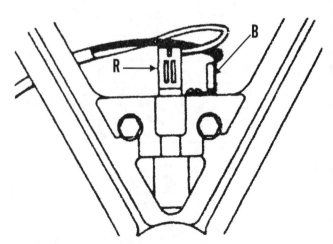

Fig. HM304 – View showing location of red and black brush wires and brush-holder screws.

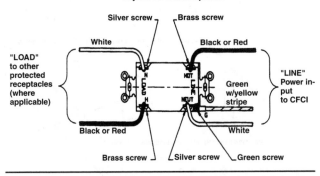

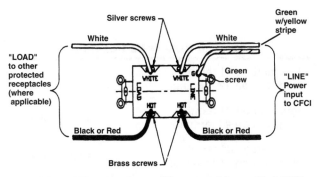

Fig. HM305 – Wire side views of Bryant and Arrow-Hart GFCI receptacles showing wiring-connection differences.

Conventional Wiring

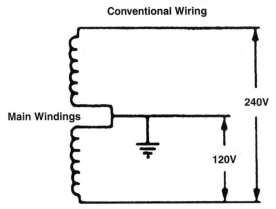

Fig. HM306 – View of series connected main stator windings with max power switch in 120/240V position. This position will produce full rated 240 volts output but only half of rated output.

Maximum Power Switch Parallels 120V Windings

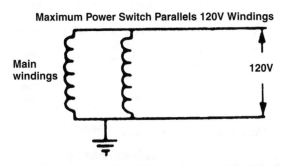

Fig. HM307 – View of parallel connected main stator windings with the max power switch in the 120V position. This position will produce full rated output at 120 volts but not 240-volt output.

d. Test the yellow wire terminals for continuity in both switch positions: There should be continuity in the AUTO position only. Any other readings indicate a faulty switch.

4. Test the electromagnet wiring:
   a. Disassemble the generator control panel as necessary to access the internal components.
   b. Remove the yellow and red electromagnet wires from the control board.
   c. Connect the VOM leads to the yellow and red wires. The reading should be 240-260 ohms.
   d. Connect one VOM lead to the yellow wire. Connect the other VOM lead to the outer casing of the electromagnet. There should be **no** continuity.
   e. Repeat the preceding substep d using the red wire and the magnet casing.
   f. **Any** reading lower than that specified in substep c or any continuity readings in substeps d or e indicate a faulty electromagnet.

5. Test the electromagnet position strength:
   a. Connect a tachometer to the engine.
   b. Start the generator.
   c. Manually hold the governor arm paddle against the electromagnet. Note the RPM. Idle control engine speed should be 2650-2800 RPM.

NOTE: Make sure the carburetor throttle lever does not contact the idle speed adjusting screw with the paddle in the idle position. Make sure the paddle is parallel to the electromagnet in the *idle* position for proper magnetic pull.

   d. Stop the generator.
   e. If necessary, adjust the holding nuts to move the electromagnet in or out.

6. Inspect the control box wiring. Faulty wiring or poor connections in any of the following wires will cause improper idle control operation:
   a. The yellow wire between the idle control B=board (IC1 – wiring diagram) and the *auto/start* switch (SW1 – wiring diagram).
   b. The yellow wire between the *auto/start* switch and the terminal block (TB1, wiring diagram).
   c. The white wire between the idle control board and the terminal block.

If the generator idles with the max power switch in the 120/240V position only, inspect the control box wiring routing. Either the #1 or #4 terminal board wire is routed through the transformer incorrectly. To correct:

Disconnect one end of either the #1 or #4 lead (wire numbers are marked on the board). Remove the disconnected wire from the transformer bobbin. Reroute the wire back through the transformer from the opposite direction. Reconnect the disconnected end.

Start the generator and check the idle control operation.

Control Board

If the engine did not return to full no-load throttle (*System Test*, Step 4), and all other tests show no faults, the control board is faulty and must be replaced.

NOTE: Prior to restarting the generator after replacing the control board, make sure that the electromagnet has been properly continuity tested: An electromagnet with a winding short can allow the idle control to operate for 1-3 cycles, then ruin the board.

Tests For Generator RPM Surging

If the generator repeatedly cycles from full speed to idle, then back to full speed, check for the following:

1. Idle speed too low. Idle speed should be 2650-2800 RPM. Idle speeds below 2650 RPM do not sufficiently produce consistent electromagnet voltage to hold the paddle steady. Adjust idle speed as necessary. Refer to Step 5 under the preceding **System Test**.

2. Bent governor arm paddle. The face of the paddle on the governor arm must be parallel to the face of the electromagnet for proper pull strength. Bend as necessary to align.

3. Paddle too far from electromagnet. Adjust the holding nuts to move the electromagnet closer to the paddle.

BRUSHES

The brushes should be replaced every 1000 hours or whenever the brushes reach a length of 3/8 in. (9.5 mm) or less (Fig. HM308).

To service the brushes:

1. Remove the endbell/brush head cover.
2. Remove the rotor throughbolt and the rotor fan.

> **NOTE: Do not run the generator with the rotor bolt removed.**

3. Test brush continuity as follows:
   a. Noting the wiring orientation, remove the red wire from the bridge rectifier. remove the black wire from the brush holder.
   b. Set the VOM to the R × 1 scale.
   c. Connect one meter lead to the red rectifier wire terminal and the other lead to the rotor slip ring nearest the bearing. Continuity should be observed.
   d. Repeat substep c with the black wire terminal and the slipring nearest the rotor windings. Continuity should again be observed.
   e. Failure to note continuity in either substep c or d indicates faulty wires, terminals, or brushes. Disconnect the questionable brush lead and test the lead for continuity, and/or proceed with further testing.

4. Noting the wiring orientation, remove the brush leads from the brushes, then remove the brush holder from the endbell (Fig. HM304).

5. Examine the brushes for length, spring tension, and integrity, and make sure that the brushes slide freely in the brush holder. Replace questionable brushes.

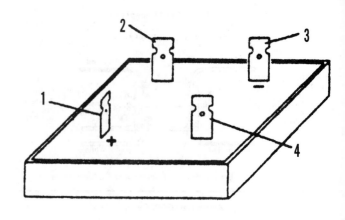

Digital VOM OHM Reading

Pos lead on #1 terminal Neg lead on #2 terminal	1.675-1.731
Pos lead on #1 terminal Neg lead on #3 terminal	3.215-3.301
Pos lead on #1 terminal Neg lead on #4 terminal	1.588-1.612
Pos lead on #2 terminal Neg lead on #3 terminal	1.612-1.616
Pos lead on #2 terminal Neg lead on #1 or #4 terminal	No Reading
Pos lead on #3 terminal Neg lead on #1, #2, or #4 terminal	No Reading
Pos lead on #4 terminal Neg lead on #1 or #2 terminal	No Reading
Pos lead on #4 terminal Neg lead on #3 terminal	1.863-1.871

Analog VOM Testing

	Pos. No.	Black Test Lead (-)			
		1 (+)	2 (AC)	3 (-)	4 (AC)
Red Test Lead (+)	1 (+)		Conducting	Conducting	Conducting
	2 (AC)	Non Conducting		Conducting	Non Conducting
	3 (-)	Non Conducting	Non Conducting		Non Conducting
	4 (AC)	Non Conducting	Non Conducting	Conducting	
	Resistance Readings				

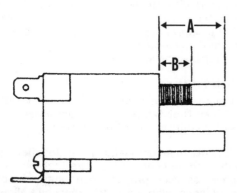

Fig. HM308 – Worn brush length (B) should not be less than 3/8 in. (9.5 mm) when measured as shown. Brush length (A) of a new brush is 3/4 inch (19 mm).

Fig. HM309 – Bridge rectifier diode test chart. Refer to text for correct test procedure.

NOTE: When replacing brushes, note that the red brush lead always goes from the positive (+) bridge rectifier terminal to the brush nearest the bearing.

6. Reinstall the rotor fan and throughbolt. Torque the bolt 100-140 in.-lb. (11.3-15.8 N•m).
7. Replace the endbell cover.

BRIDGE RECTIFIER

The rectifier is square in shape, with four male spade terminals. It is located on the left side of the endbell/brush head, behind the rotor fan. Refer to Fig. HM301 for identification and Fig. HM309 for testing.
1. Remove the endbell/brush head cover.
2. Remove the rotor throughbolt and the rotor fan.

NOTE: Do not run the generator with the rotor bolt removed.

3. Disconnect the rectifier from the circuit, noting the wiring orientation.
4. Set the VOM to the R × 1 scale.
5. Connect the meter leads to any pair of adjacent or diagonal terminals (1-2, 1-4, 3-2, 3-4, 1-3, or 2-4). Note the reading.
6. Reverse the meter leads and again note the reading.
7. Repeat Steps 5 and 6 with the other five terminal pairs.
8. Compare the readings with the Fig. HM309 chart.
   a. If one test set shows continuity both directions, the rectifier has a short circuit. Replace the rectifier. A shorted rectifier may have fed AC current to the rotor. After replacing the rectifier, flash the fields to re-establish residual magnetism.
   b. If one test set shows no continuity in both directions (except the 2-4 set, which *should* show no continuity both directions), the rectifier has an open circuit and must be replaced.
   c. If one or more reading is significantly higher or lower than the others, the rectifier is faulty. Replace the rectifier.
9. When reassembling, torque the rotor throughbolt 100-140 in.-lb. (11.3-15.8 N•m).

ROTOR

Rotor Slip Rings

The slip rings are an integral part of the rotor assembly. They are not replaceable.
Clean dirty or rough slip rings with a suitable solvent, then lightly sand with Scotch Brite or sandpaper. Do not use steel wool, emery cloth or crocus cloth.
Cracked, chipped, broken, or grooved sliprings are not repairable. In these cases, replace the rotor.

Rotor Winding Resistance

The rotor can be tested without removing it from the generator.
To test resistance:
1. Use a VOM set on the R × 1 scale. hold the VOM leads to the rotor slip rings.
2. If no reading is observed upon initial testing, inspect the winding-to-slip ring connection and solder joint. If the solder joint is faulty, and if there is sufficient wire length, resolder the joint and retest.

3. Rotor resistance specifications at 77° F (25° C) are as follows:

CG4400, CG4800, and
  CG5200 . . . . 42.5 ohms
CG5800 and
  CG6300 . . . . 50.0 ohms

NOTE: All resistance readings are approximate and may vary depending on the winding temperature and the accuracy of the test equipment.

4. If the VOM readings are lower than specified, the rotor has shorted windings. If the VOM reads infinity, the rotor has open windings. Either way, replace the rotor.
5. Test for continuity between each slip ring and the rotor shaft. If either test shows continuity, the rotor windings are shorted to the rotor shaft. Replace the rotor.

STATOR WINDING RESISTANCE

To test the stator windings:
1. Set the VOM set to the R × 1 scale.
2. Noting the wiring orientation, disconnect the stator windings from their respective terminals.

NOTE: It will be necessary to disassemble the generator control panel to access the one yellow excitation winding wire which connects to the circuit breaker.

3. Connect the meter leads to the following color coded wiring pairs.
4. Measure the winding resistances and compare to the following specifications:

All Units
Blue to stator laminations,
  Brown to stator laminations,
  White to stator laminations,
  Black to stator laminations, and
  Yellow (each) to stator laminations . . . . . No continuity/infinity

CG4400, CG4800, CG5200

Blue to Brown	0.338-0.378 ohm
White to Black	0.338-0.778 ohm
Yellow to Yellow	1.110-1.250 ohms

CG5800, CG6300

Blue to Brown	0.211-0.239 ohm
White to Black	0.211-0.239 ohm
Yellow to Yellow	1.034-1.166 ohms

5. If test results do not meet specification, replace the stator assembly.

ELECTRONIC VOLTAGE REGULATOR ("EVR")

If the generator is producing over voltage (150-160V at the 120V receptacles, or 300+V at the 240V receptacle), and the engine is running at the correct RPM, the EVR board is faulty. Replace the board.

If there is no output after testing all the other generator components, suspect the EVR board. The EVR system can be tested as follows.
1. Disassemble the generator control panel as necessary to access the internal components.
2. There are two wires – one yellow and one white – between the EVR board (VR1 wiring diagram) and the termi-

nal block (TB1 – wiring diagram). Disconnect these yellow and white wires at the terminal block end.

3. Disconnect the red wire from the EVR board. Insulate the red wire terminal with electrical tape.

4. Disconnect the small black wire from the EVR board terminal strip Position #1, then connect it to Position #3 with the two small white leads.

5. Start the generator.

6. With the VOM set to a proper AC-Volt scale, measure the output at the 120V receptacles. A meter reading of 150-160V of unregulated voltage indicates a faulty EVR board. Stop the generator and replace the board, carefully noting wiring orientation.

FASTENER-TORQUE CHART

Fasteners are listed in the approximate order of reassembly and service.

Application	Torque Value
Engine adapter to engine:	
5/16-24 bolts	120-150 in.-lb. (13.6-17.0 N•m)
3/8-16 bolts :	240-250 in.-lb. (27.1-28.3 N•m)
Rotor throughbolt	100-140 in.-lb. (11.3-15.8 N•m)
Brush holder (Plastite screws)	12-16 in.-lb. (1.4-1.8 N•m)
Brush holder	18-22 in.-lb. (2.0-2.5 N•m)
Brush head cover	18-22 in.-lb. (2.0-2.5 N•m)
Rectifier (8-32 × .50 screws)	18-22 in.-lb. (2.0-2.5 N•m)
Endbell/stator throughbolts:	
Except 1/4-20x10.5 in. Bolts	60-80 in.-lb. (6.8-9.0 N.m)

With 1/4-20x10.5 in. bolts	65-75 in.-lb. (7.3-8.5 N•m)
Receptacles:	
Mounting screws	9-13 in.-lb. (1.0-1.5 N•m)
Terminal screws	14 in.-lb. (1.6 N•m)
Ground screws	10 in.-lb. (1.1 N•m)
Circuit board to control panel	15-20 in.-lb. (1.8-2.4 N•m)
Control panel front to back cover	15-20 in.-lb. (1.8-2.4 N•m)
Idle bracket (clamp)	15-20 in.-lb. (1.8-2.4 N•m)
Idle paddle clamp to governor arm	14-18 in.-lb. (1.6-2.0 N•m)
Idle bracket to engine	60-70 in.-lb. (6.8-7.9 N•m)
Throttle arm to engine	35-45 in.-lb. (3.4-5.1 N•m)
Stator to support bracket	150-155 in.-lb. (17.0-17.5 N•m)
Stator bracket to isolator	145-155 in.-lb. (16.4-17.5 N•m)
Isolator to frame	145-155 in.-lb. (16.4-17.5 N•m)
Engine support to isolator	145-155 in.-lb. (16.4-17.5 N•m)
Engine to engine support	145-155 in.-lb. (16.4-17.5 N•m)
Ground strap fastener	145-155 in.-lb. (16.4-17.5 N•m)
Ground wire to stator	45-55 in.-lb. (5.1-6.2 N•m)
Fuel tank support to frame	25-35 in.-lb. (2.8-3.9 N•m)
Heat shield to tank support	8-12 in.-lb. (0.9-1.4 N•m)
Start switch to battery plate	70-80 in.-lb. (7.9-9.0 N•m)
Battery strap to battery plate	12-16 in.-lb. (1.4-1.8 N•m)
Battery cables to battery	40-50 in.-lb. (4.5-5.6 N•m)
Battery cables to starter switch	50-60 in.-lb. (5.6-6.8 N•m)
Battery cable to starter	30-40 in.-lb. (3.4-4.5 N•m)
Generator to frame	
crossmember	220-250 in.-lb. (24.9-28.2 N•m)

WIRING DIAGRAM

Refer to Fig. HM310 for the CG series wiring diagram.

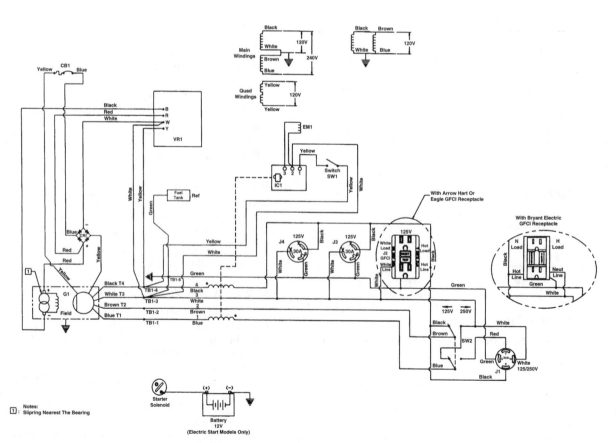

Fig. HM310 – Wiring diagram for CG series generators.

HONDA
EX700c Generator

Generator Model	Rated (Surge) Watts	Output Volts	DC Amps	Engine Model
EX700c	600 (700)	120	6	GXH50-GCAL

IDENTIFICATION

These generators are direct drive, brushless, 60 Hz, single phase units with both 120VAC and 12VDC output available from panelmounted receptacles.

The generator serial number is located on the outside of the side cover, just below the engine switch knob and the recoil starter handle grip.

Refer to Figs. HA101, HA102 and HA 103 to view the essential generator components.

WIRE COLOR CODES

The following color code abbreviations are used to identify the generator wiring in this series:

Bl – Black	Bu – Blue	W – White
R – Red	G – Green	Lb – Light blue
Y – Yellow	Br – Brown	Lg – Light green
Gr – Gray	P – Pink	O – Orange

A wire with a slash separated ID is a 2-color wire; e.g. R/W is a red wire with a white tracer stripe.

GENERATOR REPAIR

TROUBLESHOOTING

Before testing any electrical components, check the integrity of the terminals and connectors. While testing individual components, also check the wiring integrity.

Resistance readings may vary with temperature and test equipment accuracy. For access to the components listed in this **Troubleshooting** section, refer to Fig. HA101 and the later **Disassembly** section. All operational testing must be performed with the engine running at its governed noload speed of 4550-4850 RPM, unless specific instructions state otherwise. Always load-test the generator after repairing defects discovered during troubleshooting.

To troubleshoot the generator:
1. Start the engine and verify the following:
   a. AC pilot light:
      If the pilot light is ON, stop the generator, perform Step 2 and proceed to Step 3.
      If the pilot light is OFF, stop the generator, perform Step 2 and proceed to Step 4.
   b. AC circuit breaker button IN (AC circuit breaker will be covered in Step 4).
   c. DC circuit breaker button IN (DC circuit breaker will be covered in Step 5).
2. Remove the four front panel cover screws, then remove the cover and panel.
3. To test the AC receptacle:

a. Disconnect either red feed wire by inserting a small flat blade screwdriver into the slot next to the wire, then pulling the wire terminal out (Fig. HA103).
b. Install a jumper wire across one pair of terminals.
c. The receptacle should now test continuity across the feed terminals. Failure of the receptacle to test continuity indicates a faulty receptacle.
d. The receptacle should also test continuity between its mounting flange and the ground terminal. Failure to test ground continuity indicates a faulty receptacle.
4. To test for AC output:
   a. Access and test the AC circuit breaker: With the button IN, there should be continuity between the breaker terminals. If the button will not remain IN, there is no continuity with the button IN, or there is continuity with the button OUT the breaker is faulty. Replace the breaker.
   b. Disconnect the 2-pin (orange and gray) connector and the single black wire connector from the harness, and disconnect the red, white, and blue wire terminals from the cyclo-converter (Fig. HA102).
   c. Use electrical tape to insulate the red, white, blue, and black wire terminals.
   d. Remove the engine spark plug.
   e. Using caution not to let the meter leads contact each other, test the AC voltage output at the orange and gray terminals while rapidly pulling the starter rope; output should be 0.5-2.0 volts.
   f. Test resistance between the orange and gray terminals. Resistance should be 0.28-0.58 ohms. If test Steps 4e and 4f meet specification, the cyclo-converter is faulty; replace it. If there was no voltage output or resistance was beyond specification, the stator sub coil is faulty. Replace the stator. If the voltage output was below specification, the rotor magnets are weak. Replace the rotor.
   g. Test continuity between the orange terminal and any good, clean engine ground.
   h. Repeat Step 4g with the gray terminal.
   i. If either Step 4g or 4f showed continuity, the stator sub coil is shorted. Replace the stator.
   j. Untape the black wire terminal, and alternately untape the red, white, and blue terminals for the next test.
   k. Again using caution not to let the terminals and meter leads short against anything, rapidly pull the starter rope and test AC voltage output between black-red, black-white, and black-blue. Each test should produce 3-30 volts.
   l. Test resistance between the black terminal and each red, white, and blue terminal; resistance should be

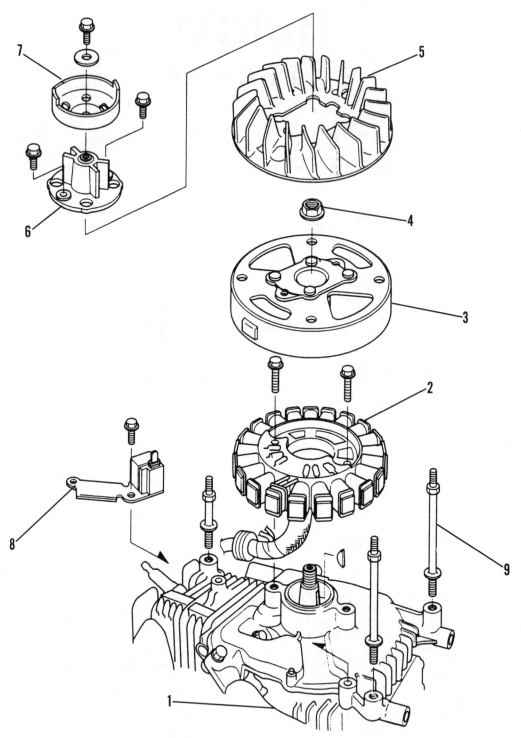

Fig. HA101 – Exploded view of the electrical generation components.

1. Engine
2. Stator assembly
3. Rotor assembly
4. Rotor nut
5. Fan
6. Fan boss
7. Starter pulley
8. Ignition pulse generator
9. Fan housing stud (3)

0.49-0.64 ohm for each test. If test Steps 4k and 4l meet specification, the cyclo-converter is faulty and must be replaced. If there was no voltage output or if resistance tested beyond specification on at least one test, the stator main winding is faulty. Replace the stator. If the voltage output was below specification, the rotor magnets are weak. Replace the rotor.

m. Test continuity between the black terminal and any good, clean engine ground.

n. Repeat Step 4k with the red, white, and blue terminals.

o. If any test in Steps 4k or 4l showed continuity, the stator main winding is shorted; replace the stator.

p. When reconnecting the blue, white, and red wires to the cyclo-converter, the wires must be mounted in the positions and angle shown if Fig. HA102.

5. To test for DC output, if none is present at the DC receptacle:

  a. Access the back of the DC circuit breaker/receptacle and unplug the harness connector, then depress the reset button. If the button does not stay in, the receptacle is faulty. If the button stays in:

    • Place a jumper wire between the receptacle output terminals (Fig. HA104).

    • Test for continuity between the input terminals. Continuity indicates a good receptacle.

    • Test for continuity between the input terminal fed by the white/red wire and the vertical output terminal. Continuity indicates a good breaker.

  b. Access the diode-rectifier assembly at the back of the cyclo-converter unit (Fig. HA105) and disconnect the harness using a VOM set to the R × 1 scale, test continuity between any one pair of adjacent or diagonal terminals (1-2, 1-3, 2-4, 3-4, 1-4, or 2-3). Note the reading. Reverse the meter leads and again note the reading. Compare the readings with the chart. Any variation indicates a faulty rectifier.

NOTE: Depending on meter polarity, continuity/infinity readings may be exactly opposite those shown in the chart. If so, reverse the (+) and (−) probe notations in the chart.

  c. Access the 4-pin diode-rectifier harness connector: Resistance test the two brown wire terminals in the rectifier connector; resistance should test 0.06-0.26 ohm. Measure the resistance between each brown-wire terminal and any ground position on the engine block. There should be infinity/no-continuity. If the test results in Steps 5a and 5b met specification, any fault is in the wiring. Repair or replace as necessary. If the test results in Steps 5a or 5b did not meet specification and the wiring integrity is secure, the DC stator winding is faulty. Replace the stator.

6. To test the stator exciter coil winding:

  a. Access the 3-wire (black/blue, green, and blue) connector behind the control panel.

  b. Continuity test the black/blue-to-green wire terminals. Resistance should be 0.55-0.85 ohm.

  c. If Step 6b test results do not meet specification, proceed to the **Disassembly** section.

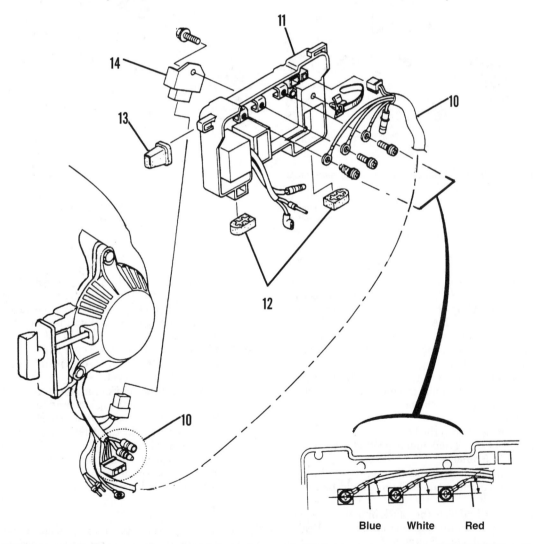

Blue   **White**   **Red**

Fig. HA102 – AC stator coil winding connections should always be mounted to the cyclo-converter in a left-to-right blue-white-red order at 30° from horizontal.

10. Stator harness   12. Rubber converter mount (2)   14. Rectifier
11. Cyclo-converter   13. Rubber converter mount

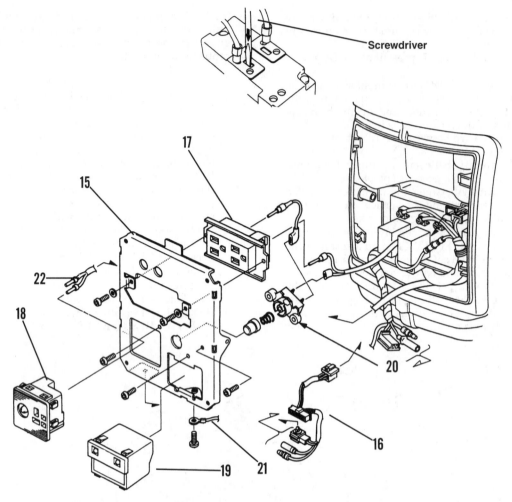

Fig. HA103 – Removing the AC connections to test the receptacle.

15. Control panel
16. Control panel harness
17. AC receptacle

18. DC receptacle
19. Ignition control module
20. AC circuit breaker

21. Ground wire
22. DC output terminal

7. To test the ignition control module:
   a. Leave the control panel harness connected to the panel, but disconnect the 10-pin module connector from the module (Fig. HA106).
   b. Test resistance between the black wire terminal (coil primary circuit) and engine ground. Resistance should be 0.7-1.1 ohms.
   c. Test for continuity between the yellow-wire terminal (oil-level switch) and engine ground. There should be infinity/no-continuity with the crankcase oil at the proper level.
   d. Test resistance between the blue wire terminal (ignition pulse generator) and engine ground. Resistance should be 25-39 ohms.
   e. Test for continuity between the green-wire terminal (ground) and engine ground. There should be continuity.
   f. Test resistance between the black/blue-wire terminal (stator exciter coil) and engine ground. Resistance should be 0.55-0.85 ohm.
   g. If all the test results are within specification, the ignition control module is faulty. Replace the module.

   h. If any individual test fails to test within specification, proceed with further testing.
   (The ignition pulse generator will be tested in **Disassembly**, and the ignition coil will be tested in **ENGINE, Ignition System**.)
8. To initially test the oil level switch:
   a. Insure that the crankcase oil is full and that the generator is sitting level.
   b. Disconnect the spark plug lead. Connect an approved spark tester to the plug lead and the engine ground.
   c. Pull the starter rope. If no spark is visible at the tester, proceed to the **ENGINE, Ignition System** section of this chapter.
   d. If there is spark, drain the crankcase oil and pull the starter rope again. If there is still spark, proceed to the ENGINE, Ignition System section of this chapter.

Disassembly

Refer to Fig. HA101 to view the essential generator components. When disassembling, note the position of all wiring cable ties. Ties removed during disassembly must be replaced when reassembling.

To disassemble the generator:

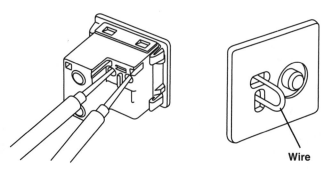

Fig. HA104 – Testing the DC breaker/receptacle for continuity by installing a jumper wire across the outlet terminals. Refer to text for correct test procedure.

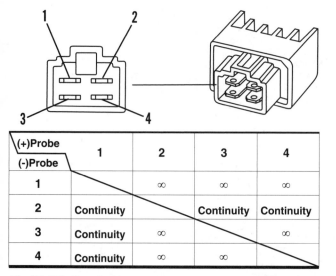

(+)Probe (–)Probe	1	2	3	4
1		∞	∞	∞
2	Continuity		Continuity	Continuity
3	Continuity	∞		∞
4	Continuity	∞	∞	

Fig. HA105 – Terminal test positions for the diode rectifier. Refer to text for the correct test procedure

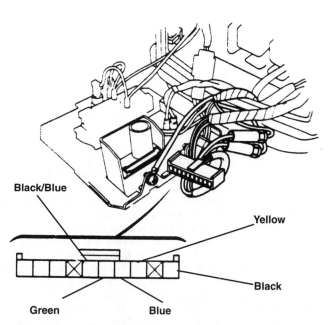

Fig. HA106 – View showing the 10 pin connector removed to test the ignition control module. Refer to text for correct test procedure.

1. Remove the muffler protective cover, control panel front cover, and side panel maintenance cover.
2. Remove and disconnect the control panel assembly.
3. Carefully drain the fuel tank and the crankcase oil.

NOTE: The oil should be drained hot, but generator disassembly should be done with the unit cool.

4. Unscrew the carburetor bowl drain screw to drain the bowl. Disconnect the bowl drain hose, vent hose, air intake hose, and throttle linkage. Remove the carburetor.
5. Remove the side covers, noting the positions of the internal rubber mounts and insulator collars. Disconnect the fuel valve hose from the bottom of the fuel tank.
6. Remove the fuel tank, noting the position and condition of the tank mounts. The left and right rubber mounts are not interchangeable.
7. Disconnect and remove the cyclo-converter unit, noting the position and condition of the three converter mounts.
8. Remove the engine bed mount plate assembly (Fig. HA107); note the position and condition of the bed mounts and collars, and the position of the wiring harness ground wire.
9. Remove the air filter assembly and tube.
10. Remove the fan cover assembly, taking care not to lose the three collars that fit between the cover and the mount studs. The recoil starter and the ignition coil are mounted to the fan cover. They do not have to be removed unless they are being repaired or replaced.
11. Remove the fan shroud halves, noting the position of the top insulator and collar assembly and the harness clamp.
12. Remove the ignition pulse generator unit. The pulse unit can be tested at this time by measuring resistance between terminals A and B of Fig. HA108. The resistance should be 25-39 ohms.
13. Holding the rotor with a strap wrench, remove the starter pulley, cooling fan boss, cooling fan, and rotor nut.

NOTE: Do not strike the rotor; striking the rotor will damage the rotor.

14. Using a 2-slot puller, remove the rotor from the engine crankshaft.
15. With the rotor removed, the stator exciter coil can be isolation-tested:
   a. Continuity test between the black/blue wire terminal and the stator lamination core. Resistance should be 0.55-0.85 ohm.
   b. If test result is not as specified, the stator is faulty. Replace the stator.
   c. If the winding passes the continuity test but failed the **GENERATOR REPAIR; Troubleshooting** Step 6 test, the stator is OK but the stator to control panel wiring is faulty. Repair or replace as necessary.

Reassembly

Refer to Figs. HA101, HA 102 and HA 103 to view the essential generator components.

To reassemble the generator:
1. Ensure that:
   a. The matching tapers on the engine crankshaft and inside the rotor hub are clean and dry, and the

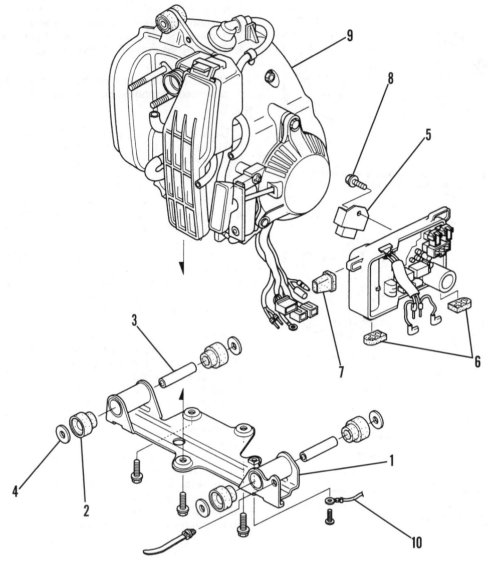

Fig. HA107 – View showing proper placement of engine mounts.

1. Engine mount
2. Inuslatur
3. Collar
4. Washer

5. Rectifier
6. Rubber mounts (A)
7. Rubber mount (B)

8. Screw
9. Engine
10. Ground wire

matching engine block shoulder and stator hole are clean.

b. There is no metallic foreign material stuck to the rotor magnets.

c. The crankshaft Woodruff key is properly positioned in its slot.

2. If the stator was removed, install it now and torque the stator mount bolts 5.9 N•m (52 in.-lb.). Make sure that the stator harness is routed correctly; it must be held in place by the ignition pulse generator bracket.

3. Install the rotor; holding the rotor with a strap wrench, torque the rotor nut 27 N•m (240 in.-lb.). Rotate the rotor by hand to make sure that the rotor does not rub against the stator. Correct any misalignment or interference.

4. Install the cooling fan and fan boss.

5. Position the starter pulley onto the fan boss with the lock tab in one of the V-grooves of the fan boss. Holding the rotor

with a strap wrench, install the pulley flat washer and bolt; torque the bolt 9.8 N•m (85-90 in.-lb.).

6. Install the ignition pulse generator unit, making sure that the pulse unit mounting arm secures the stator harness. The air gap clearance between the rotor projection and terminal A of the pulse unit (Fig. HA108) should be set at 0.4-0.6 mm (0.016-0.024 in.). Torque the pulse-unit bolts 5.9 N•m (52 in.-lb.). Reconnect the blue pulse unit wire.

7. Be sure the rubber cylinder fin inserts are correctly positioned (Fig. HA109) and the shroud seals are correctly fastened to the shrouds. Apply a 1.0 mm (3/64 in.) bead of Loctite 515 liquid gasket to the crankcase breather tube prior to inserting it into the crankcase (Fig. HA 110). Mount the side shrouds to the engine assembly. Note the position of the top collar, insulators, and washers.

8. Mount the three positioning collars onto the engine studs, then mount the fan cover assembly onto the engine. Tighten the cover nuts. When installing the ignition coil

ground wire terminal onto the lower left stud, insure that the washer is installed *between* the fan cover and the terminal.

9. Install the air filter assembly and tube.

10. Install the engine bed mount plate assembly, insuring proper placement of the insulators, collars, and washers (Fig. HA107).

11. Verify the correct routing of the hoses and tubes on the carburetor side of the engine assembly (Fig. HA111). When installing the tube protector over the carburetor hose, do not allow the long end to cover the 90° bend in the hose (long end goes toward pump).

12. Reconnect the cyclo-converter unit, then reconnect the wiring harness ground wire to the weld nut on the bottom of the bed mount.

13. Reinstall the fuel tank, verifying the correct positioning of the non-interchangeable rubber tank mounts. The mount with the longer, channel-style foot fits on the tank extension next to the fuel hose fitting.

14. Verify the correct routing of the hoses and wiring inside the left-side generator cover (Fig. HA112).

15. Connect the generator harness to the 2-pin engine switch plug in the left-side cover (Fig. HA112). Secure the wire with the cable tie (Fig. HA113).

16. Laying the left-side cover down, fit the engine and inverter unit into the mount positions in the cover, verifying the proper placement of the rubber mounts, washers, and collars. Make all hose and wiring connections as necessary.

17. Verify the integrity of the cover seals in the right-side cover, then place the right-side cover onto the left-side cover. Install and tighten the cover fasteners.

18. Stand the generator upright. Install the top snap ring and the gas tank neck seal.

19. Using new gaskets and O-ring, install the carburetor, reconnecting the air filter, fuel, and vent hoses and the throttle linkage. Make sure that the straight side of the outer carburetor manifold insulator is UP

20. Connect and install the control panel assembly.

21. Install the side panel maintenance cover, control panel cover, and muffler protective cover.

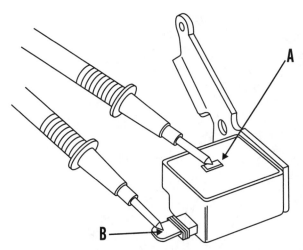

Fig. HA108 – Resistance between Terminals A and B of the ignition pulse generator should be 25-39 ohms.

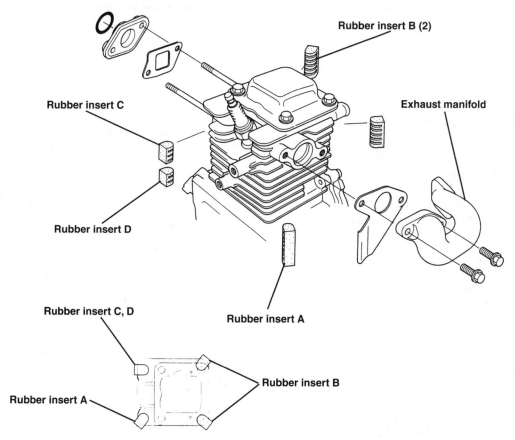

Fig. HA109 – View showing the proper placement of the rubber cylinder fin inserts, inner carburetor insulator assembly, and exhaust manifold.

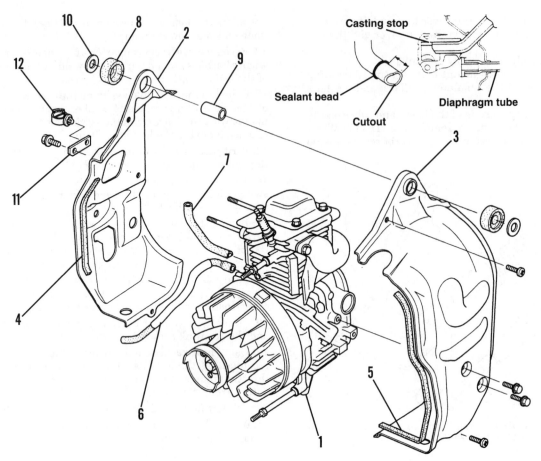

Fig. HA110 – View showing air shroud placement over engine assembly. Note correct placement and installation of crankcase breather hose and fuel pump pulse hose.

1. Engine
2. Left shroud
3. Right shroud
4. Seal
5. Seal
6. Diaphragm tube
7. Breather tube
8. Insulator
9. Collar
10. Washer
11. Harness clip bracket
12. Harness clip

WIRING DIAGRAM

Refer to Fig. HA125 to view the wiring diagram.

ENGINE

The model GXH50 engine is a 4-stroke design with push rod-operated overhead valves. The engine serial number is stamped on a flat boss on the lower side of the crankcase, to the left of the oil fill plug.

Refer to Fig. HA114 for an exploded view of the mechanical engine components.

To access the engine, refer to the **GENERATOR REPAIR, Disassembly** section of this chapter.

MAINTENANCE

Spark Plug

Recommended spark plug is NGK CR5HSB, Denso U16FSR-UB, or equivalent. Recommended gap is 0.6-0.7 mm (.024-.028 in.). Torque the plug 12 N•m (105 in.-lb.).

CAUTION: Abrasive blasting to clean spark plugs is not recommended, as this may introduce abrasive material into the engine and cause extensive damage.

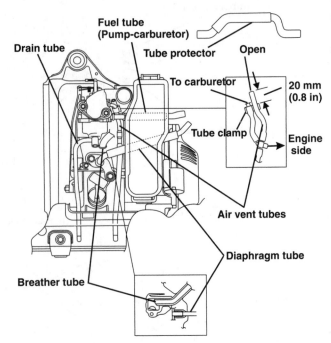

Fig. HA111 – View showing proper routing of engine hoses and tubes.

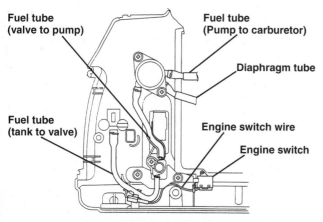

Fig. HA112 – View showing fuel system hose routing and engine switch wire routing inside the left-side generator cover.

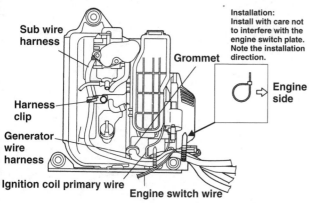

Fig. HA113 – View showing correct cable fastening for proper wiring security inside the generator cover.

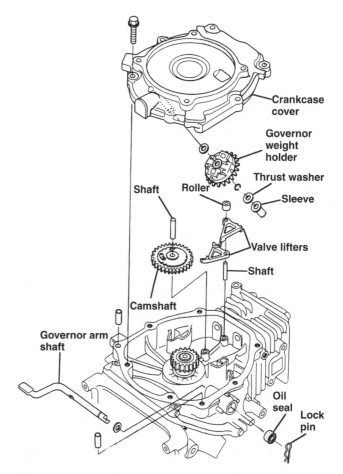

Fig. HA114 – Exploded view of mechanical engine components.

Lubrication

The engine is splash-lubricated by a dipper on the connecting rod cap.

Oil capacity is 0.25 litre (0.26 US qt). The crankcase is full when the oil level is at the top of the oil fill hole with the generator sitting level.

Use oil with an API service classification of SH, SJ, or better. Use SAE 30 oil if temperatures are consistently above 10° C (50° F); SAE 10W30 in any temperature; or SAE 5W30 if temperatures are consistently below freezing. DO NOT USE 10W40 oil.

Drain the old oil when the oil is hot to insure the removal of all impurities. To drain the oil:

Remove the maintenance cover panel, unscrew the oil-fill cap, tilt the generator.

Whenever the crankcase oil has been drained, dry-test the operation of the low oil shutdown switch:

1. Make sure that the generator is sitting level.
2. Remove the spark plug.
3. Connect an approved spark tester to the plug lead and cylinder block.
4. Pull the starter rope and check for spark. There should be none.
5. If the tester shows spark with the crankcase empty of oil, the shutdown switch is faulty. Refer to the subsequent **Ignition System** section for testing and replacement.

Air Filter

1. To access the filter, remove the maintenance cover panel and the snap fit filter cover.
2. Gently clean the sponge filter element in a warm detergent/water solution. Rinse thoroughly, then allow to dry completely.
3. Pour clean engine oil onto the element, squeezing out any excess oil.
4. Reinstall the element.
5. Inspect the snap fit filter cover, insuring that the channel mounted cover seal fits tightly in the cover all the way to the channel ends. Replace if necessary.
6. Reinstall the filter cover, insuring a proper, tight fit. Install the maintenance cover panel.

Carburetor

To access the carburetor, refer to the **GENERATOR REPAIR, Disassembly** section, Steps 1-4.

1. Disassemble the carburetor using the exploded view drawing of Fig. HA115. Clean and renew components as necessary.
2. Inspect the float valve and seat, float pivot arm, float pin, and carb body float pin stanchions for wear. Renew as necessary.
3. The throttle shaft and choke shaft are not serviced separately. If wear exceeds reasonable limits, the carburetor must be replaced.

4. The pilot screw limiter cap prevents removal of the pilot screw without breaking the screw. If carburetor repair requires pilot screw removal and replacement:

   a. Install the spring onto the screw.

   b. Lightly seat the screw all the way into the carb.

   c. Back the screw out 2-5/8 turns.

   d. Apply Loctite 638 or equivalent inside the limiter cap, then fit the cap to the pilot screw so the cap stop prevents the screw from rotating counterclockwise. DO NOT turn the screw while installing the cap.

5. The pilot jet must be installed prior to installing the throttle stop screw. Lightly oil the jet O-ring prior to installation.

6. To check float height:

   a. Set the carburetor vertically on the workbench, manifold flange DOWN and float pin UP, so the float is hanging (Fig. HA116).

   b. Pressing lightly against the float to cause the float valve to contact its seat, measure the distance from the bowl-mount flange to the bottom edge of the float on the side of the float opposite the valve. It should be 12.0 mm (0.47 in.). Any other dimension necessitates replacing the float.

   c. Insure that the float operates freely.

7. Install the float bowl so that the bowl drain screw fitting is on the choke side of the carburetor.

Governor Linkage

To view the governor linkage, refer to Fig. HA117. To set or adjust the governor:

1. Make sure that the governor link fits inside the link spring. The link ends **must** be installed in the outermost holes in the throttle shaft and the governor arm. The shorter straight arm of the spring must go toward the carburetor.

2. Make sure that the governor spring is installed between the control base arm and the lower hole of the governor arm. The longer straight arm of the spring must go toward the governor arm.

3. Loosen the governor arm nut.

4. Manually hold the throttle wide open.

5. Using small pliers, turn the governor arm shaft to its maximum clockwise position. Do not force.

6. While holding the throttle and governor shaft, torque the governor arm nut 7 N•m (62 in.-lb.).

7. Check the governor linkage for free movement.

8. Start the engine and allow it to warm up, then check no-load speed. The engine should be running at 4550-4850 RPM. Use the adjusting screw as necessary.

Fuel Pump

The fuel pump cannot be repaired. If it does not pump, it must be replaced as a unit. To troubleshoot the pump system:

1. Check the integrity of the four preformed fuel system hoses and the fuel filter in the tank outlet fitting. If any hoses require removal or replacement, refer to Fig. HA112 and insure correct orientation.

2. Remove the pulse hose from the crankcase fitting and, while pulling the starter rope, insure that crankcase pressure/vacuum pulsations pass the fitting.

3. Make sure that the carburetor is not blocked, preventing pump operation. Loosen the carb bowl drain screw. While pulling the starter rope, observe fuel flow from the bowl drain hose.

Valve Clearance

Checking and adjusting valve clearance must be done with the engine cold.

1. Remove the valve cover screws.

2. Carefully and alternately pry up each corner of the valve cover to loosen and remove the cover. Excessively forcing one corner more than the others will deform the cover; a deformed cover must be replaced.

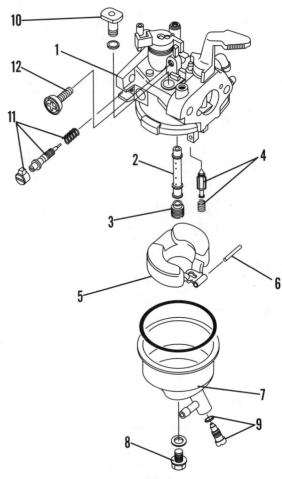

Fig. HA115 – Exploded view of carburetor components.

1. Carburetor body	7. Float bowl
2. Main nozzle	8. Bolt
3. Main jet	9. Drain screw
4. Fuel inlet valve	10. Pilot jet
5. Float	11. Pilot screw
6. Pivot pin	12. Throttle stop screw

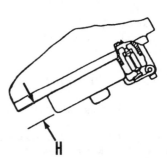

Fig. HA116 – View of proper carburetor position to check float level setting.

3. Place the piston at TDC compression stroke by removing the spark plug, rotating the flywheel, and observing the piston head through the spark plug hole.

4. Using a feeler gauge, check valve clearance:

Intake . . . . **0.06-0.10 mm (0.003-0.004 in.)**
Exhaust . . . . **0.09-0.13 mm (0.0035-0.005 in.)**

5. To adjust valve clearance, loosen the rocker arm adjusting screw lock nut, then turn the adjusting screw in to decrease clearance or out to increase clearance. Torque lock nut 5.4 N•m (48 in.-lb.), then recheck clearance and readjust, if necessary.

6. Remove old valve cover sealer from the cover and block surfaces.

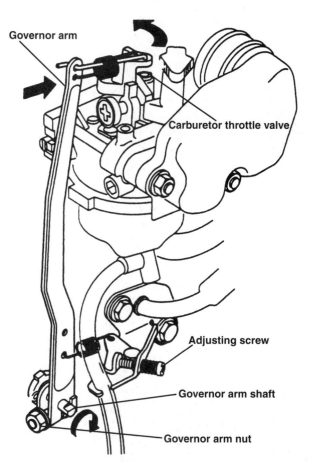

Governor arm

Carburetor throttle valve

Adjusting screw

Governor arm shaft

Governor arm nut

Fig. HA117 – View of governor linkage on the engine. Refer to text for proper adjustment and installation.

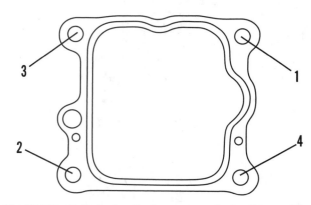

3

1

2

4

Fig. HA118 – Cylinder head/valve cover sealant pattern and fastener tightening sequence. Refer to text.

7. Apply a 1.0 mm (3/64 or .040 in.) diameter bead of Loctite 515 liquid gasket to the cylinder block valve-cover surface as shown in Fig. HA118. Install the cover. Incrementally torque the valve cover bolts 5.9 N•m (52 in.-lb.) using the sequence shown in the figure.

Ignition System

To Test the Ignition Coil

1. Test the resistance of the primary circuit by placing one VOM lead on the black primary winding wire terminal. Place the other VOM lead on any clean lamination core ground. Resistance should be 0.7-1.1 ohms.

2. Test the resistance of the secondary circuit by placing one VOM lead on the spark plug terminal inside the boot. Place the other VOM lead on any clean lamination core ground. Resistance should be 1200-2100 ohms.

3. Any readings other than those specified in Steps 1 and 2 indicate a faulty coil.

4. If replacing the coil, apply a light coating of Loctite 242 or equivalent to the self-tapping coil screw threads, then torque the screws 1.8 N•m (16 in.-lb.).

To Test the Engine Stop (Ignition) Switch

1. Access and disconnect the black and green 2-pin engine switch connector from behind the control panel.

2. With the switch mounted inside the left side cover, continuity test the switch wire terminals while turning the switch knob on the outside of the cover. There should be continuity in the OFF position only.

3. Continuity in the ON position or no continuity in the OFF position indicates a faulty switch.

To Test the Low-Oil Shutdown Switch

1. Unplug the yellow feed wire from the generator harness. Unbolt the green ground wire terminal and the other two switch mount bolts from the engine.

2. Test for continuity between the yellow and the green terminals.

    a. There should be continuity with the switch in its normal upright position.

    b. There should be no continuity with the switch turned upside down.

3. Test the switch float by immersing the switch in a container of oil.

    a. There should be no continuity with the switch below oil level.

    b. Continuity should occur as the switch is lifted from the oil.

4. Any test results not consistent with Steps 2 or 3 indicate a faulty oil switch.

Rewind Starter

To view the starter components, refer to Fig. HA119. To service the starter:

1. Disassemble, clean, and inspect as necessary.

2. Renew all worn or damaged components.

3. When installing a new rope, wind the rope onto the reel in a counterclockwise direction when viewing the engine side of the reel.

4. Pretension the spring approximately three turns.

5. When assembling the swing arm and collar assembly, insure that the tips of the swing arms are fitted into the lower flat parts of the reel-ramp slots.

Muffler and Spark Arrester (Fig. HA120)

1. Service the muffler and spark arrester *only* with the engine cold.
2. Carefully remove carbon deposits from the spark arrester screen with a wire brush.

3. Remove carbon deposits from the muffler by lightly tapping the seam flange.

4. Always install a new muffler gasket. Apply a light coating of Loctite 271 or equivalent to the muffler bolt threads. Torque the muffler bolts 7.8 N•m (70 in.-lb.).

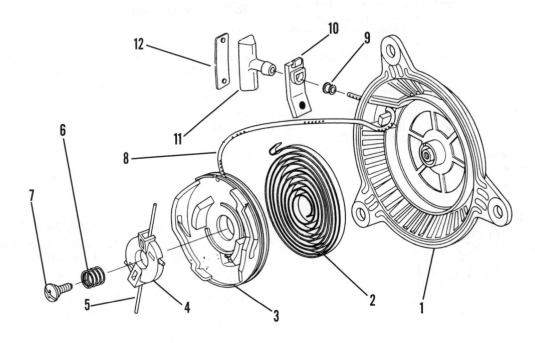

Fig. HA119 – Exploded view of recoil starter assembly.

1. Starter housing	5. Swing arm	9. Grommet
2. Rewind spring	6. Friction spring	10. Rope guide
3. Starter reel	7. Screw	11. Rope handle
4. Swing arm collar	8. Starter rope	12. Cover

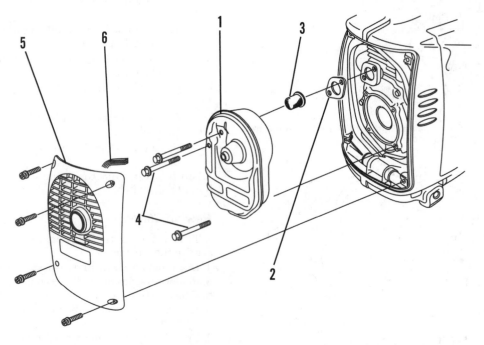

Fig. HA120 – Exploded view of muffler components and muffler cover.

1. Muffler	3. Spark arrester	5. Muffler cover
2. Gasket	4. Bolts	6. Seal

REPAIRS

Specification Chart

Component	Part	Std. Dimension	Wear Limit
Cylinder/crankcase and crankcase side cover	Cylinder bore	41.800-41.815 mm (1.6457-1.6463 in.)	41.900 mm (1.6496 in.)
	Valve guide I.D. (IN and EX)	4.000-4.018 mm (0.1575-0.1582 in.)	4.060 mm (0.1598 in.)
	Camshaft bearing	5.005-5.029 mm (0.1970-0.1978 in.)	
	shaft bore I.D.	5.005-5.023 mm (0.1970-0.1978 in.)	5.050 mm (0.1988 in.)
	Valve lifter bore I.D.		
Rocker arm shaft bore I.D.		4.000-4.018 mm (0.1575-0.1582 in.)	4.050 mm (0.1594 in.)
Valve lifter shaft and camshaft O.D.		4.990-5.000 mm (0.1959-0.1965 in.)	4.950 mm (0.1949 in.)
Rocker arm shaft O.D.		3.990-4.000 mm (0.1571-0.1575 in.)	3.950 mm (0.1555 in.)
Rocker arm journal I.D.		4.005-4.025 mm (0.1577-0.1585 in.)	4.050 mm (0.1594 in.)
Camshaft	Lobe height	27.972 mm (1.1013 in.)	26.972 mm (1.0619 in.)
	Journal I.D.	5.020-5.050 mm (0.1976-0.1988 in.)	5.100 mm (0.2008 in.)
Valve lifter journal I.D.		5.005-5.025 mm (0.1970-0.1978 in.)	5.050 mm (0.1988 in.)
Valve stem O.D.	Intake	3.970-3.985 mm (0.1563-0.1569 in.)	3.900 mm (0.1535 in.)
	Exhaust	3.935-3.950 mm (0.1549-0.1555 in.)	3.880 mm (0.1528 in.)
Valve clearance	Intake		0.06-0.10 mm (.003-.004 in.)
	Exhaust	0.09-0.13 mm (.0035-.005 in.)	
Valve springs, IN and EX	Free length	22.8-23.7 mm (0.90-0.93 in.)	
Crankshaft	Rod journal O.D.	14.973-14.984 mm (0.5895-0.5899 in.)	14.940 mm (0.5882 in.)
Connecting rod	Wrist-pin bore	10.006-10.017 mm (0.3939-0.3944 in.)	10.050 mm (0.3957 in.)
Crankshaft journal bore		15.000-15.011 mm (0.5906-0.5910 in.)	15.040 mm (0.5921 in.)
Crankshaft journal oil clearance		0.016-0.038 mm (0.0006-0.0015 in.)	0.100 mm (0.0039 in.)
Crankshaft journal side clearance		0.1-0.6 mm (0.04-0.24 in.)	0.8 mm (0.031 in.)
Piston	Skirt O.D.	41.770-41.790 mm (1.6445-1.6453 in.)	41.700 mm (1.6417 in.)

Piston-to-cylinder Clearance	0.010-0.045 mm (0.004-0.00187 in.)	0.120 mm (0.0047 in.)
Wrist pin bore	10.002-10.008 mm (0.3938-0.3940 in.)	10.050 mm (0.3957 in.)
Wrist pin O.D.	9.994-10.000 mm (0.3935-0.3937 in.)	9.950 mm (0.3917 in.)
Wrist pin-to-bore Clearance	0.002-0.014 mm (0.0001-0.0006 in.)	0.100 mm (0.0039 in.)
Ring thickness: Top	0.77-0.79 mm (0.030-0.031 in.)	0.720 mm (0.0283 in.)
Second	0.97-0.99 mm (0.038-0.039 in.)	0.920 mm (0.0362 in.)
Ring side clearance, Top and Second	0.015-0.050 mm (0.0006-0.0020 in.)	0.120 mm (0.0047 in.)
Ring end gap, Top and Second	0.150-0.300 mm (0.0059-0.0118 in.)	0.600 mm (0.0236 in.)

Cylinder Compression

Cylinder compression should test at 415-445 kPa (60-65 PSI).

Disassembly

NOTE: The cylinder head cannot be removed. It is one piece with the cylinder block. The engine must be disassembled to service the valves and valve seats.

1. Remove the valve cover screws.
2. Carefully and alternately pry up each corner of the valve cover to loosen and remove the cover. Excessively forcing one corner more than the others will deform the cover; a deformed cover must be replaced.
3. Place the piston at TDC compression stroke by removing the spark plug, rotating the crankshaft, and observing the piston head through the spark plug hole. This must be done to facilitate rocker arm removal and to prevent damage to the oil level switch.
4. Note the rocker arm orientation. Loosen the rocker arm lock nuts and adjusting screws. Remove the rocker arm shaft by pulling it out from the exhaust side of the head.
5. Note the push rod placement. Remove the push rods.
6. Place the engine on the workbench with the exposed crankshaft down. Support the crankcase so as not to damage the crankshaft threads.
7. Remove the crankcase side cover bolts. Utilizing the pry bosses between the cover and crankcase, carefully separate the cover from the crankcase.
8. Remove the governor slider spool, spool thrust washer, and governor gear/weight retainer clip from the side cover shaft. Remove the gear/weight and thrust washer.

9. Note valve lifter and camshaft lobe orientation, then remove lifter collar, lifters, lifter shaft, camshaft, and camshaft roller shaft.
10. Flip the engine over.
11. Remove the oil level switch bolts. Remove the switch.
12. Remove the oil case bolts. Utilizing the pry bosses between the oil case and the cylinder block, carefully separate the case from the block.
13. Remove and discard the crankshaft oil seal.
14. Noting the position of the connecting rod cap alignment marks, remove the cap bolts, oil dipper, and cap.
15. Rotate the crankshaft to clear the connecting rod, then remove the crankshaft.
16. Pull the connecting rod/piston assembly from the cylinder.
17. Line the cylinder with protective paper or equivalent to prevent damage to the cylinder bore while doing valve work.
18. While pushing down on the valve spring retainer, slide the retainer sideways to allow the valve stem to fall through the keyhole slot in the retainer. Remove the retainers and springs, then remove the valves through the cylinder.
19. Remove the crankcase breather screw, valve, and limit plate from the cylinder block.

Valve Service

NOTE: The cylinder head is not removable; it is one piece with the cylinder block. The valves and valve seats must be serviced through the cylinder bore.

1. Line the cylinder with protective paper or equivalent to prevent damage to the cylinder bore while doing valve work.

2. Using an electric drill and a rotary wire brush, carefully clean the combustion deposits from the combustion chamber.

3. Check valve seats and faces for pitting or scoring; renew as necessary.

4. Check valve stems for bending or excessive wear. Check the push rods for bending.

5. Measure all valve system components using the dimensions in the specification chart. Renew as necessary. If the valve guides are worn beyond specification, the cylinder assembly will need to be replaced.

Camshaft and Crankcase Cover Service

Use the dimensions in the specification chart for these measurements.

1. Measure all shaft bore diameters in both the cover and the cylinder assembly.

2. Measure the shaft bore diameters in the camshaft and the valve lifters.

3. Measure the camshaft and valve lifter shafts.

4. Inspect the timing gears and valve lifter lobe ramps and push rod sockets for wear.

5. Measure the camshaft lobe.

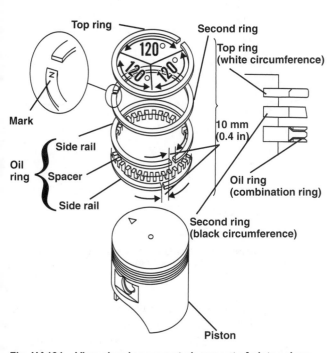

Fig. HA121 – View showing correct placement of piston rings.

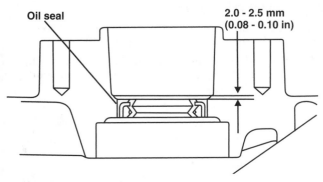

Fig. HA122 – View showing correct depth of crankshaft oil seal inside oil case.

Cylinder and Oil Case Service

Use the dimensions in the specification chart for needed measurements.

1. Inspect and measure the cylinder bore: For accurate dimensioning, the bore should be measured at the top, middle, and bottom of the piston travel area, and on both the X axis and Y axis (parallel to the piston pin and perpendicular to the piston pin, respectively).

2. Inspect the crank shaftbearing journals.

3. Inspect the breather valve seat.

Piston, Rings, and Connecting Rod

Use the dimensions in the specification chart for needed measurements.

1. Measure the piston 90° from the wrist pin bore and 10 mm (0.4 in.) from the bottom of the piston skirt.

2. Measure piston ring thicknesses and ring-to-land side clearances.

3. If the cylinder bore is to specification, insert the rings into the bore one at a time (push them in with the piston so they are perpendicular to the bore), and measure the ring end gaps.

4. Measure the wrist pin O.D. and the wrist pin bore I.D. in the piston and the connecting rod.

5. Install and torque the crankshaft journal rod cap 5.9 N•m (52 in.-lb.), then measure the journal I.D.

Crankshaft

Use the dimensions in the specification chart for needed measurements.

1. Measure the connecting rod journal in at least two positions 90° apart.

2. Using Plastigage, measure the crank journal to connecting rod clearance. Follow the Plastigage instructions exactly.

3. Inspect the crankshaft ball bearings, cam drive gear, rotor taper, taper keyway, and threads.

Reassembly

Install the crankcase breather valve, valve stop plate, and screw into the cylinder block. Insure that the chamfer on the valve and the chamfer on the stop plate match the chamfer in the block.

1. Install the intake and exhaust valves, valve springs, and spring retainers. Compress the valve spring lightly while sliding the retainer keyhole slot onto the valve stem.

2. Assemble the piston to the connecting rod:

   a. Insure that the "Δ" mark on the piston head points toward the intake side and the connecting rod alignment marks face the open side of the cylinder block.

   b. Insure that the gap of the wrist pin retaining ring **does not** align with the notch in the wrist pin bore face.

3. Assemble the piston rings to the piston:

   a. Install the rings into the respective grooves as shown in Fig. HA121.

   b. Insure that the gaps in the ends of the oil ring side rail rings are positioned 10 mm (0.4 in.) to either side of the spacer ring gap.

   c. Stagger the top, second, and oil spacer ring end gaps 120° apart, with no gap in line with the wrist pin.

4. Lubricate the piston and rings with engine oil, then carefully install the piston assembly into the cylinder. Insure that the rod cap alignment mark is visible. Slide the assembly all the way into the cylinder.

5. Install the crankshaft into the cam gear side of the cylinder block.

6. Rotate the crankshaft so the connecting rod journal is at BDC, oil the journal, then slide the connecting rod down onto the crankshaft journal. With the rod cap marks aligned, install the oil dipper so that the dipper is on the flywheel side of the rod. Torque the rod cap bolts 5.9 N•m (52 in.-lb.).

7. Install the oil seal into the oil case so that the outer face of the seal is even with the outer edge of the seal mount bore, 2.0-2.5 mm (0.08-010 in.) below the outer edge of the seal bore lead taper (Fig. HA122).

8. Insuring that the oil case and cylinder block mating surfaces are degreased, clean and dry:

    a. Apply a 1.5-2.0 mm diameter (1/16-5/64 in.) bead of Loctite 515 liquid gasket or equivalent to the cylinder block surface as shown in Fig. HA123.

    b. Insure that the two locator dowel pins are properly positioned in the cylinder block.

    c. Using a seal protector to prevent oil seal damage, install the oil case onto the cylinder block. Torque the eight case bolts incrementally in the sequence shown in Fig. HA123 to a final torque of 7.4 N•m (65 in.-lb.).

9. Rotate the crankshaft to place the piston at TDC, then install the oil level switch O-ring and switch. Snugly tighten the switch bolts.

10. Flip the engine over.

11. Install the camshaft shaft, valve lifter shaft, and two cover locator dowels into their proper block positions.

12. Lubricate and install the valve lifters, lifter collar, and camshaft. Ensure that the "Δ" camshaft gear timing mark aligns with the "•" crankshaftgear timing mark.

13. Install the governor gear thrust washer, gear/weight assembly, gear retainer clip, slider spool thrust washer, and slider spool onto the side cover shaft.

14. Make sure that the side cover and cylinder block mating surfaces are degreased, clean, and dry:

    a. Apply a 1.5-2.0 mm diameter (1/16-5/64 in.) bead of Loctite 515 liquid gasket or equivalent to the cylinder block surface as shown in Fig. HA124.

    b. Insure that the two locator dowel pins are properly positioned in the cylinder block.

    c. Carefully aligning the governor gear/crank gear teeth, install the side cover onto the cylinder block. Torque the cover bolts incrementally in the sequence shown in Fig. HA124 to a final torque of 7.4 N•m (65 in.-lb.).

15. Noting the removal orientation, oil and install the rocker arm push rods, rocker arms, and rocker arm shaft.

16. Place the piston at TDC compression stroke by rotating the crankshaft and observing the piston through the spark plug hole. (If the crankshaft was not rotated after aligning the crank gear/cam gear timing marks during previous assembly, the piston is already at TDC compression.)

17. Insure that the push rod ball ends are properly positioned in their valve lifter and rocker arm sockets.

18. Using a feeler gauge, and adjusting the rocker arm screws as necessary, set the valve clearance:

Intake . . . . **0.06-0.10 mm (.003-.004 in.)**
Exhaust . . . . **0.09-0.13 mm (.0035-.005 in.)**

19. Torque the rocker arm locknuts 5.4 N•m (48 in.-lb.), then recheck clearance and readjust, if necessary.

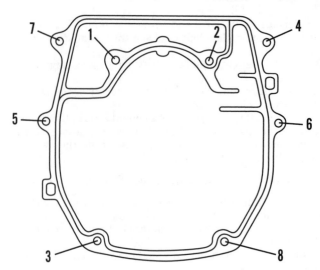

Fig. HA123 – Cylinder block/oil case sealant pattern and fastener tightening sequence. Refer to text.

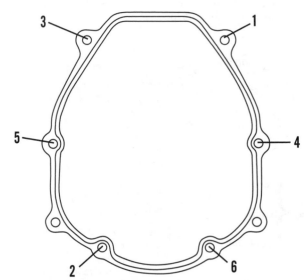

Fig HA124 – Cylinder block/side cover sealant pattern and fastener tightening sequence. Refer to text.

20. Insuring that the valve cover and cylinder head mating surfaces are degreased, clean, and dry:

21. Apply a 1.0 mm diameter (3/64 or .040 in.) bead of Loctite 515 liquid gasket to the cylinder block valve cover surface as shown in Fig. HA118.

22. Install the valve cover.

23. Incrementally torque the valve cover bolts 5.9 N•m (52 in.-lb.) in the sequence shown in Fig. HA118.

24. If the carburetor insulator and/or exhaust manifold were removed, always use new gaskets when reinstalling.

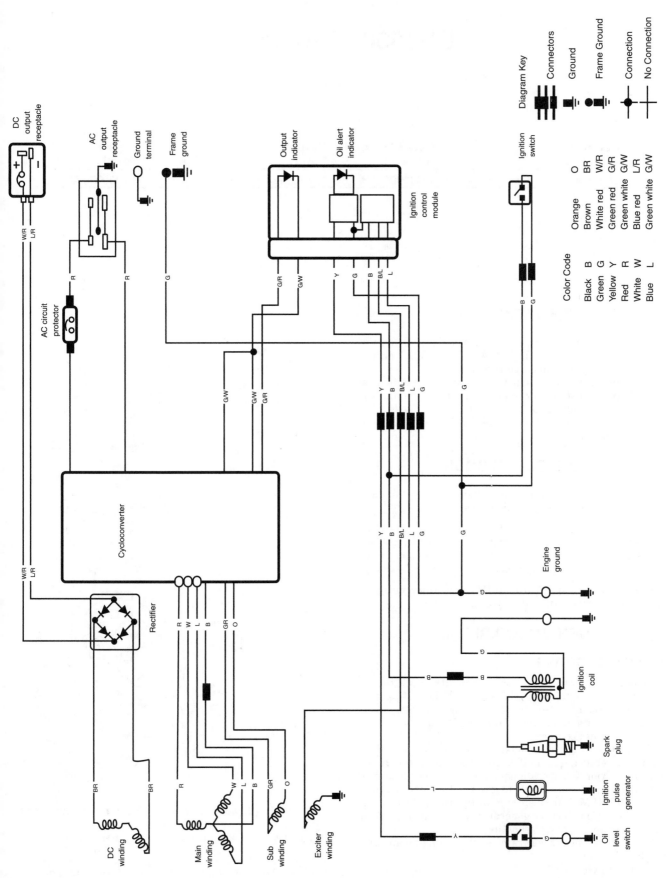

Fig. HA125 – Wiring diagram for EX700c generator.

HONDA
EU1000i Generator

Generator Model	Rated (Surge) Watts	Output Volts	DC Amps	Engine Model
EU1000i	900 (1000)	120 AC/12 DC	8	GXH50-GCAL

IDENTIFICATION

These generators are direct drive, brushless, 60 Hz, single-phase units with both 120VAC and 12VDC output available from panel-mounted receptacles.

The generator serial number is located on the outside of the side cover, just below the engine switch knob and the recoil starter handle grip.

Refer to Figs. HA151, HA152 and HA153 to view the generator components.

WIRE COLOR CODES

The following color-code abbreviations are used to identify the generator wiring in this series:

Bl – Black	Bu – Blue	W – White
R – Red	G – Green	Lb – Light blue
Y – Yellow	Br – Brown	Lg – Light green
Gr – Gray	P – Pink	O – Orange

A wire with a slash-separated ID is a two-color wire; (e.g. R/W is a red wire with a white tracer stripe).

GENERATOR REPAIR

TROUBLESHOOTING

Before testing any electrical components, check the integrity of the terminals and connectors. While testing individual components, also check the wiring integrity.

Resistance readings may vary with temperature and test equipment accuracy.

For access to the components listed in this **Troubleshooting** section, refer to Fig. HA151 and the later **Disassembly** section. All operational testing must be performed with the engine running at its governed no-load speed of 5400-5600 RPM, unless specific instructions state otherwise. Always load test the generator after repairing defects discovered during troubleshooting.

1. To troubleshoot the generator, start the engine and verify the following:

   a. ECO switch OFF,

   b. Overload Indicator Light OFF,

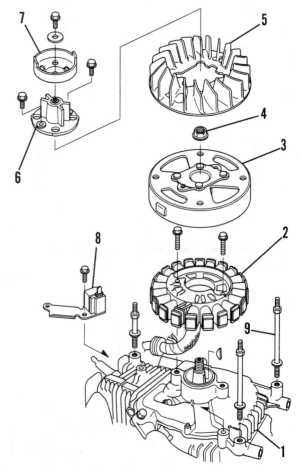

Fig. HA151 – Exploded view of the electrical generation components.

1. Engine	6. Fan boss
2. Stator assembly	7. Starter pulley
3. Rotor assembly	8. Ignition pulse generator
4. Rotor flange nut	9. Fan housing stud (3)
5. Cooling fan	

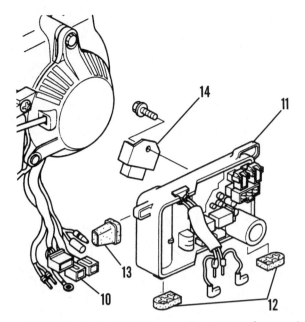

Fig. HA152 – Removing the AC connections to test the receptacle.

10. Starter harness
11. Inverter unit
12. Rubber mounts

13. Rubber mount
14. Rectifier

c. If the AC pilot light is ON, stop the generator, perform Step 2 then proceed to Step 3. If the pilot light is OFF, stop the generator, perform Step 2 then proceed to Step 4.

d. DC circuit breaker button IN (circuit breaker will be covered in Step 5).

2. Remove the four front panel cover screws, then remove the cover and panel.

3. To test the AC receptacle:

a. Disconnect either red feed wire by inserting a small flat blade screwdriver into the slot next to the wire, then pulling the wire terminal out (Fig. HA153).

b. Install a jumper wire across one pair of terminals.

c. The receptacle should now test continuity across the feed terminals. Failure of the receptacle to test continuity indicates a faulty receptacle.

d. The receptacle should also test continuity between its mounting flange and the ground terminal. Failure to test ground continuity indicates a faulty receptacle.

e. The AC *Composite Sockets*, or *Parallel Operation Terminals*, mounted above the AC receptacle on the control panel, can be continuity tested by placing one meter lead inside the socket and one lead on the back terminal. Continuity should be present. Failure to read continuity on either socket indicates a faulty socket.

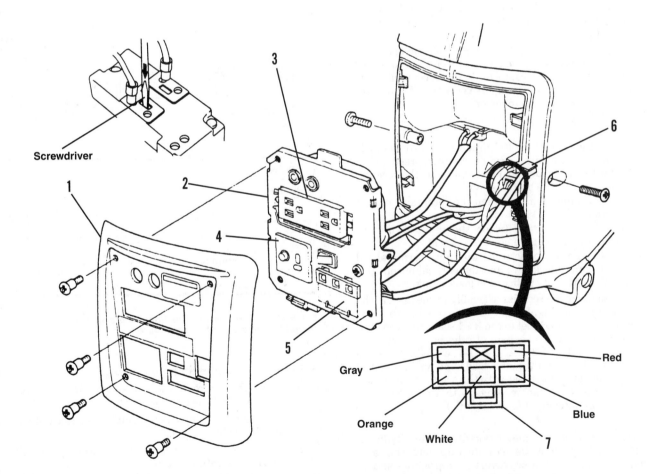

Fig. HA153 – Accessing the six-pin connector to test AC output. Refer to text for the correct test procedure.

1. Front cover
2. Control panel
3. 120V receptacle

4. 12V receptacle
5. Ignition control module

6. Control panel harness
7. 6-Pin connector

4. To test for AC output:

a. Access and identify the six-pin (red, white, blue, orange, gray, and black) inverter-harness connector (Fig. HA153), but do not unplug the connector.

b. Start the generator.

c. Test the main coil AC voltage between the red-blue, blue-white, and white-red paired terminals (1-2, 2-3, and 3-1). Voltage should be 227-263 volts at each pair.

d. Test the sub coil AC voltage between the gray-orange wire terminal pair (4-5). Voltage should be 11-15 volts.

e. Stop the generator, then unplug the 6-pin connector.

f. Using an ohmmeter and the same three terminal pairs as Step 4c, test the winding resistance. Each winding should test 2.100-3.300 ohms.

g. Using an ohmmeter and the same gray-orange terminal pair as Step 4d, test the winding resistance. The winding should test 300-500 ohms.

h. Still using an ohmmeter, test for continuity between any good, clean ground position on the engine block and each of the five wire terminals. There should be infinity/no-continuity.

i. If one or more of the results of test Steps 4c or 4d failed to meet specification, or if any of the results of test Steps 4f-4h failed to meet specification, the stator windings are faulty. Replace the stator.

j. If all of the results of test Steps 4f and 4g tested equally low, the rotor magnetism is weak. Replace the rotor.

k. If all of the preceding test results met the specification but there is still no AC output, return to Step 3 and test the AC receptacle. If the receptacle tests good, and the wiring integrity is correct, the inverter is faulty. Replace the inverter.

5. To test for DC output, if none is present at the DC receptacle:

a. Access the back of the DC circuit breaker/receptacle and unplug the harness connector, then press the reset button. If the button does not stay in, the receptacle is faulty. If the button stays in:

• Place a jumper wire between the receptacle output terminals (Fig. HA154).

• Test for continuity between the input terminals. Continuity indicates a good receptacle.

• Test for continuity between the input terminal fed by the white/black wire and the vertical output terminal. Continuity indicates a good breaker.

b. Access the diode rectifier assembly at the back of the inverter unit (Fig. HA155) and disconnect the harness. Using a VOM set to the R × 1 scale, test continuity between any one pair of adjacent or diagonal terminals (1-2, 1-3, 2-4, 3-4, 1-4, or 2-3). Note the reading. Reverse the meter leads and again note the reading. Repeat the test with the other five terminal pairs. Compare the readings with the chart. Any variation indicates a faulty rectifier.

NOTE: Depending on meter polarity, continuity/infinity readings may be exactly opposite those shown in the chart. If so, reverse the positive and negative probe notations in the chart.

c. Access the four-pin diode rectifier harness connector:

Test the resistance of the two brown-wire terminals in the rectifier connector. Resistance should test 0.1-0.2 ohm. Measure the resistance between each brown wire terminal and any good, clean ground position on the engine block. There should be infinity/no continuity. If the test results met specification, any fault is in the wiring; repair or replace as necessary. If the test results did not meet specification and the wiring integrity is secure, the DC stator winding is faulty. Replace the stator.

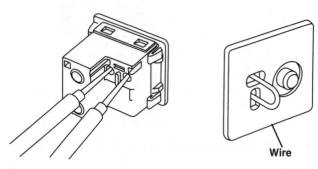

Fig. HA154 – Testing the DC breaker/receptacle for continuity by installing a jumper wire across the outlet terminals. Refer to text for the correct test procedure.

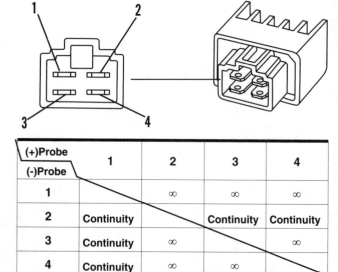

(+)Probe / (−)Probe	1	2	3	4
1		∞	∞	∞
2	Continuity		Continuity	Continuity
3	Continuity	∞		∞
4	Continuity	∞	∞	

Fig. HA155 – Terminal test positions for the diode rectifier. Refer to text for the correct test procedure.

6. To test the stator exciter-coil winding:

a. Access the 2-wire (black/blue and yellow/green) connector behind the control panel.

b. Continuity test the wire terminals; resistance should be 0.5-0.9 ohm.

c. If test results do not meet specification, proceed to the **Disassembly** section.

7. To test the ECO throttle control system:

a. Access the ECO switch and continuity test the terminals. There should be continuity in the ON position only.

NOTE: The ECO throttle switch should always be installed in the control panel with the ON and OFF markings up.

b. Access the throttle control motor mounted on top of the carburetor and test the resistances between terminals 1-3 and terminals 2-4 (Fig. HA156). Resistance should be 50-70 ohms at each noted terminal pair. Resistance readings beyond this range indicate a faulty motor.

c. Move the throttle lever and check the ECO motor for smooth operation. If binding is noted, disconnect the throttle linkage to determine the source of the binding. Correct as necessary.

d. The ECO switch is faulty if it fails to function as follows:
  • The no-load engine speed should increase with the ECO switch OFF.
  • The no-load engine speed should decrease with the ECO switch ON.
  • The engine speed should increase when a load is applied to the generator with the ECO switch ON.

e. To operationally check the ECO motor after replacing the inverter:
  • While observing the throttle, start the engine, allow it to run briefly, then stop the engine.
  • At start up, the throttle should move from wide-open to normal run.
  • At shut down, the throttle should return to the wide-open position.

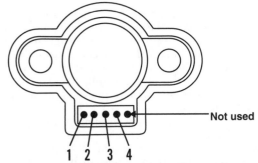

Fig. HA156 – Terminal test positions for the ECO throttle control motor. Refer to text for the correct test procedure.

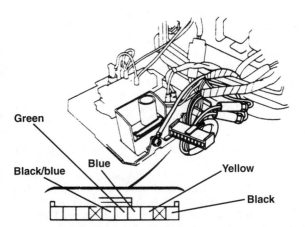

Fig. HA157 – View showing the 10-pin connector removed to test the ignition control module. Refer to text for the correct test procedure.

  • If the ECO motor does not perform properly, replace the motor.

8. To test the ignition control module:
  a. Leave the control panel harness connected to the panel, but disconnect the 10-pin module connector from the module (Fig. HA157).
  b. Test resistance between the black wire terminal (coil primary circuit) and engine ground. Resistance should be 0.7-1.1 ohms.
  c. Test for continuity between the yellow wire terminal (oil-level switch) and engine ground. There should be infinity/no continuity with the crankcase oil at the proper level.
  d. Test resistance between the blue wire terminal (ignition pulse generator) and engine ground. Resistance should be 25-39 ohms.
  e. Test for continuity between the green wire terminal (ground) and engine ground. There should be continuity.
  f. Test resistance between the black/blue wire terminal (stator exciter coil) and engine ground. Resistance should be 0.5-0.9 ohm.
  g. If all the above tests conclude within specification, the ignition control module is faulty. Replace the module.

9. To initially test the oil level switch:
  a. Insure that the crankcase oil is full and that the generator is sitting level.
  b. Disconnect the spark plug lead. Connect an approved spark tester to the plug lead and the engine ground.
  c. Pull the starter rope. If no spark is visible at the tester, proceed to the **ENGINE, Ignition System** section of this Chapter.
  d. If there is spark, drain the crankcase oil and pull the starter rope again. If there is spark, proceed to the **ENGINE, Ignition System** section of this chapter.

Disassembly

To disassemble the generator:
1. Remove the muffler protective cover, control panel front cover, and side panel maintenance cover.
2. Remove and disconnect the control panel assembly.
3. Carefully drain the fuel tank and the crankcase oil.

NOTE: The oil should be drained hot, but generator disassembly should be performed with the unit cool.

4. Unscrew the carburetor bowl drain screw to drain the bowl. Disconnect the bowl drain hose, vent hose, air-intake hose, and throttle control motor harness. Remove the carburetor.
5. Remove the side covers, noting the positions of the internal rubber mounts and insulator collars. Disconnect the fuel valve hose from the bottom of the fuel tank.
6. Remove the fuel tank, noting the position and condition of the tank mounts. The left and right rubber mounts are not interchangeable.
7. Disconnect and remove the inverter unit, noting the position and condition of the inverter mounts.
8. Remove the engine bed mount plate assembly; note the position and condition of the bed mounts and collars, and the position of the wiring harness ground wire.
9. Remove the air filter assembly and tube.

10. Remove the fan cover assembly, taking care not to lose the collars which fit between the cover and the mount studs. The recoil starter and the ignition coil are mounted to the fan cover. They should not have to be removed unless they are being repaired or replaced.

11. Remove the fan shroud halves, noting the position of the top insulator and collar assembly and the harness clamp.

12. Remove the ignition pulse generator unit. The pulse unit can be tested at this time by measuring resistance between Terminals 'A' and 'B' of Fig. HA158. The resistance should be 25-39 ohms.

13. Holding the rotor with a strap wrench, remove the starter pulley, cooling fan boss, cooling fan, and rotor nut.

> **NOTE: Do not strike the rotor. Striking the rotor will damage it.**

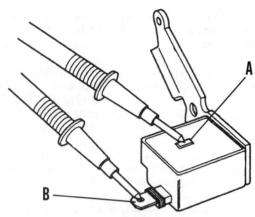

Fig. HA158 – Resistance between Terminals A and B of the ignition pulse generator should be 25-39 ohms.

14. Using a 2-slot puller, remove the rotor from the engine crankshaft.

15. With the rotor removed, perform an islation test on the stator exciter coil.

   a. Test for continuity between the black/blue wire terminal and the stator lamination core. Resistance should be 0.5-0.9 ohm.

   b. Any test result other than that specified indicates a faulty stator. Replace the stator.

   c. If the winding passes the preceding Step A test but failed the **troubleshooting** continuity test, the stator is working but the stator to control panel wiring is faulty. Repair or replace as necessary.

Reassembly

To reassemble the generator:

1. Insure that:

   a. The matching tapers on the engine crankshaft and inside the rotor hub are clean and dry, and the matching engine block shoulder and stator hole are clean.

   b. There is no metallic foreign material stuck to the rotor magnets.

   c. The crankshaft woodruff key is properly positioned in its slot.

2. If the stator was removed, install it now and torque the stator mount bolts 5.9 N•m (52 in.-lb.). Make sure that the stator harness is routed correctly. It must be held in place by the ignition pulse generator bracket.

3. Install the rotor. Holding the rotor with a strap wrench, torque the rotor nut 27 N•m (240 in.-lb.). Rotate the rotor by hand to make sure that the rotor does not rub against the stator. Correct any misalignment or interference.

4. Install the cooling fan and fan boss.

5. Position the starter pulley onto the fan boss with the lock tab in one of the V-grooves of the fan boss. Holding the rotor with a strap wrench, install the pulley flat washer and bolt. Torque the bolt 9.8 N•m (85-90 in.-lb.).

6. Install the ignition pulse generator unit, making sure that the pulse unit mounting arm secures the stator harness. The air gap clearance between the rotor projection and Terminal A of the pulse unit (Fig. HA158) should be set at 0.4-0.6 mm (.016-.024 in.). Torque the pulse unit bolts 5.9 N•m (52 in.-lb.). Reconnect the blue pulse unit wire.

7. Verify that the rubber cylinder fin inserts are correctly positioned (Fig. HA159) and the shroud seals are correctly fastened to the shrouds. Apply a 1.0 mm diameter (3/64 in.) bead of Loctite 515 liquid gasket or equivalent to the

crankcase breather tube prior to inserting it into the crankcase (Fig. HA160). Mount the side shrouds to the engine assembly. Note the position of the top collar, insulators, and washers.

8. Mount the positioning collars onto the engine studs, then mount the fan cover assembly onto the engine. Tighten the cover nuts. When installing the ignition coil ground wire terminal onto the lower left stud, make sure that the washer is installed between the fan cover and the terminal.

9. Install the air filter assembly and tube.

10. Install the engine bed mount plate assembly, ensuring proper placement of the insulators, collars, and washers (Fig. HA161).

11. Verify the correct routing of the hoses and tubes on the carburetor side of the engine assembly (Fig. HA162). When installing the tube protector over the fuel pump to carburetor hose, do not allow the long end to cover the 90° bend in the hose (the long end goes toward the pump).

12. Reconnect the inverter unit, then reconnect the wiring harness ground wire to the weldnut on the bottom of the bed mount.

13. Reinstall the fuel tank, verifying the correct positioning of the non-interchangeable rubber tank mounts. The mount with the longer, channel style foot fits on the tank extension next to the fuel hose fitting.

14. Verify the integrity and correct routing of the hoses and wiring inside the left side generator cover (Fig. HA163).

15. Connect the generator harness to the two-pin engine switch plug in the left side cover (Fig. HA163). Secure the wire with the cable tie (Fig. HA164).

16. Laying the left cover down, fit the engine/inverter unit into the mount positions in the cover, verifying the proper placement of the rubber mounts, washers, and collars. Make all hose and wiring connections as necessary.

17. Verify the integrity of the cover seals in the right cover, then place the right cover onto the left cover. Install and tighten the cover fasteners.

18. Stand the generator upright. Install the top snap ring and the gas tank neck seal.

19. Using new gaskets and O-ring, install the carburetor, reconnecting the air filter, fuel, and vent hoses and the throttle control motor harness. Make sure that the straight side of the outer carburetor manifold insulator is UP

20. Connect and install the control panel assembly.

21. Install the side panel maintenance cover, control panel cover, and muffler protective cover.

ENGINE

The model GXH50 engine is a four-stroke design with push-rod operated overhead valves. The engine serial number is stamped on a flat boss on the lower side of the crankcase, to the left of the oil fill plug. Exploded views of the engine are shown in Figs. HA165A-HA165D.

To access the engine, refer to the **GENERATOR REPAIR, Disassembly** section of this chapter.

MAINTENANCE

Spark Plug

The recommended spark plug is NGK CR5HSB, Denso U16FSR-UB, or equivalent. Recommended gap is 0.6-0.7 mm (.024-.028 in.). Torque plug 12 N•m (105 in.-lb.).

CAUTION: Do not use abrasive blasting to clean spark plugs as this may introduce abrasive material into the engine and cause extensive damage.

Lubrication

The engine is splash lubricated by a dipper on the connecting rod cap.

Oil capacity is 0.25 litre (0.26 US qt). The crankcase is full when no more oil can be added to the oil fill hole with the generator sitting level.

Use oil with an API service classification of SH, SJ, or better. Use SAE 30 oil if temperatures are consistently above 10° C (50° F), SAE 10W30 in any temperature, or SAE 5W30 if temperatures are consistently below freezing. DO NOT USE 10W40 oil.

Drain the old oil when it is hot to insure the removal of all impurities. To drain the oil:
1. Remove the maintenance cover panel.
2. Unscrew the oil fill cap.
3. Tilt the generator.

Whenever the crankcase oil has been drained, dry test the operation of the low oil shutdown switch:
1. Insure that the generator is sitting level.
2. Remove the spark plug.

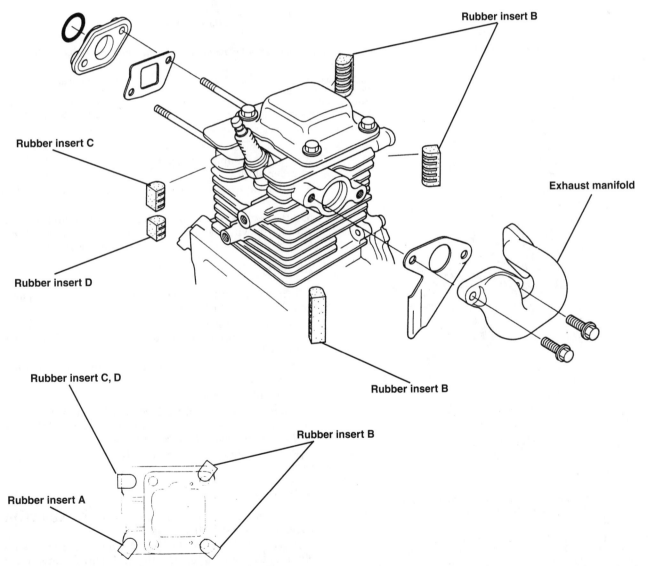

Fig. HA159 – View showing the proper placement of the rubber cylinder fin inserts, inner carburetor insulator assembly, and exhaust manifold.

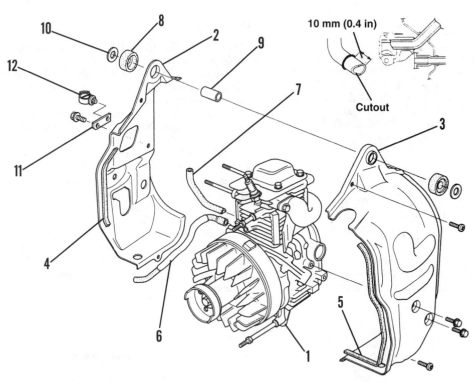

Fig. HA160 – View showing air shroud placement over engine assembly. Note correct placement and installation of crankcase breather hose and fuel pump pulse hose.

1. Engine assembly
2. Left shroud
3. Right shroud
4. Left shroud seal

5. Right shroud seal
6. Pump diaphragm tube
7. Breather tube
8. Shroud insulator (2)

9. Insulator collar
10. Insulator washer (2)
11. Harness clip bracket
12. Harness clip

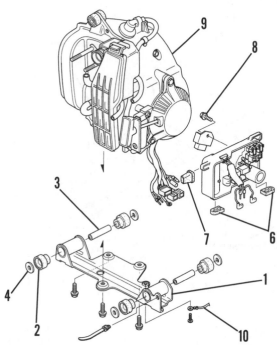

Fig. HA161 – View showing proper placement of engine mounts, inverter unit, and diode rectifier case breather hose and fuel pump pulse hose.

1. Engine bed frame
2. Vibration isolators (4)
3. Isolator collars (2)
4. Isolator washers (4)
5. Inverter unit

6. Bottom inverter mounts (2)
7. Side inverter mount
8. Rectifier
9. Engine/generator unit
10. Ground wire

3. Connect an approved spark tester to the plug lead and cylinder block.

4. Pull the starter rope and check for spark; there should be none.

5. If the tester shows spark with no oil in the crankcase, the shutdown switch is faulty. Refer to the subsequent **Ignition System** section for testing and replacement.

Air Filter

1. To access the filter, remove the maintenance cover panel and the snap fit filter cover.

2. Gently clean the sponge filter element in a warm detergent/water solution. Rinse thoroughly, then allow to dry completely.

3. Pour clean engine oil onto the element, squeezing out any excess oil.

4. Reinstall the element.

5. Inspect the snap fit filter cover, insuring that the channel mounted cover seal fits tightly in the cover all the way to the channel ends. Replace if necessary.

6. Reinstall the filter cover, and the maintenance cover panel.

Carburetor

To access the carburetor, refer to the **GENERATOR REPAIR, Disassembly** section.

1. If necessary, remove the throttle control motor assembly (Fig. HA166). To test the throttle control motor, refer to the **GENERATOR REPAIR, Troubleshooting** section.

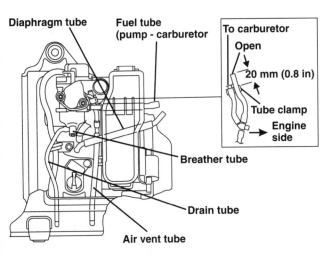

Fig. HA162 – View showing proper routing of engine hoses and tubes.

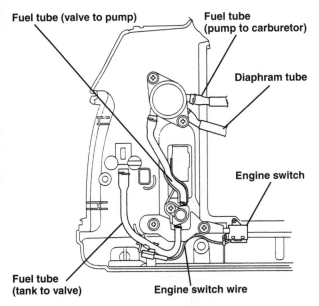

Fig. HA163 – View showing fuel system hose routing and engine switch wire routing inside the left side generator cover.

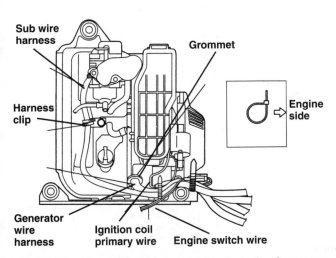

Fig. HA164 – View showing correct cable fastening for proper wiring security inside the generator cover.

2. Disassemble the carburetor referring to the exploded drawing of Fig. HA167. Clean and renew components as necessary.

3. Inspect the float valve and seat, float pivot arm, float pin, and carb body float pin stanchions for wear. Renew as necessary.

4. The throttle shaft and choke shaft are not serviced separately. If wear exceeds reasonable limits, the carburetor must be replaced.

5. The pilot screw limiter cap prevents removal of the pilot screw without breaking the screw. If carburetor repair requires pilot screw removal and replacement:

   a. Install the spring onto the screw.

   b. Lightly seat the screw all the way into the carb.

   c. Back the screw out 2-5/8 turns.

   d. Apply Loctite 638 or equivalent inside the limiter cap, then fit the cap to the pilot screw so the cap stop prevents the screw from being rotated counterclockwise. DO NOT turn the screw while installing the cap.

6. The pilot jet must be installed prior to installing the throttle stop screw. Lightly oil the jet O-ring prior to installation.

7. To check float height:

   a. Set the carburetor vertically on the workbench, manifold flange *down* and float pin *up*, so the float is *hanging* (Fig. HA168).

   b. Pressing lightly against the float to cause the float valve to contact its seat, measure the distance from the bowl mount flange to the bottom edge of the float on the side of the float opposite the valve. It should be 12.0 mm (0.47 in.). Any other dimension necessitates replacing the float.

   c. Insure that the float operates freely.

8. Install the float bowl so that the bowl drain screw fitting is on the choke side of the carburetor.

9. Prior to reinstalling the throttle control motor assembly, fit the link lever spring onto the link lever pin. Then position the link lever onto the throttle lever with the spring in its recess. Ensure that the throttle control motor drive pin fits into the link lever slot correctly.

Fuel Pump

The fuel pump cannot be repaired. If it does not pump, it must be replaced as a unit. To troubleshoot the pump system:

1. Check the integrity of the pre-formed fuel system hoses and the fuel filter in the tank outlet fitting. If any hoses require removal or replacement, refer to Fig. HA162 and verify correct orientation.

2. Remove the pulse hose from the crankcase fitting. While pulling the starter rope, make sure that crankcase pressure/vacuum pulsations pass the fitting.

3. Ensure that the carburetor is not blocked, or otherwise preventing pump operation. Loosen the carb bowl drain screw; while pulling the starter rope, observe fuel flow from the bowl drain hose.

Valve Clearance

Checking and adjusting valve clearance must be done while the engine is cold.

1. Remove the valve cover screws.

2. Carefully pry up each corner of the valve cover to loosen and remove the cover. Alternate between the corners while prying. Excessively forcing one corner more than the others will deform the cover; a deformed cover must be replaced.

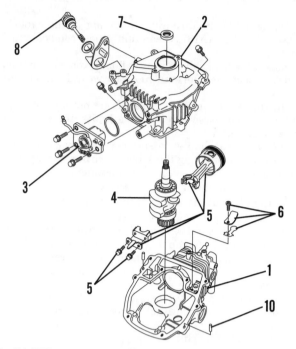

Fig. HA165A

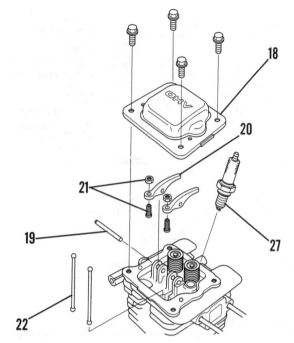

Fig. HA165B

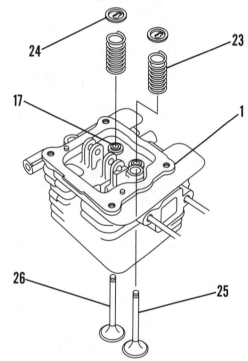

Fig. HA165C

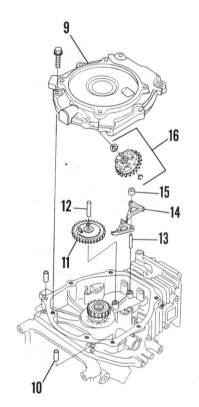

Fig. HA165D

Fig. HA165A-D – Exploded view of mechanical engine components.

1. Cylinder block
2. Oil case
3. Oil-level switch and O-ring
4. Crankshaft with cam-drive gear
5. Piston/rings/connecting rod assembly
6. Breather assembly
7. PTO seal
8. Oil cap and gauge
9. Crankcase cover

10. Locator dowel (2 each side)
11. Camshaft
12. Camshaft roller shaft
13. Valve lifter roller shaft
14. Valve lifter (2)
15. Valve lifter collar
16. Governor assembly
17. Valve guide (2)
18. Valve cover

19. Rocker arm shaft
20. Rocker arm (2)
21. Adjuster screw and locknut (2 each)
22. Push rod (2)
23. Valve spring (2)
24. Valve spring retainer (2)
25. Intake valve
26. Exhaust valve
27. Spark plug

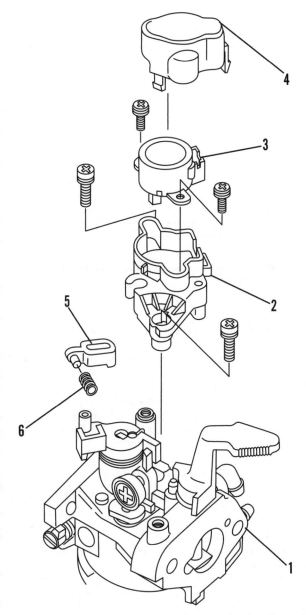

Fig. HA166 – View showing the mounting of the throttle control motor atop the carburetor. Refer to text for proper lever installation.

1. Carburetor
2. Throttle control motor base
3. Throttle control motor
4. Throttle control motor cover
5. Throttle link lever
6. Throttle link lever spring

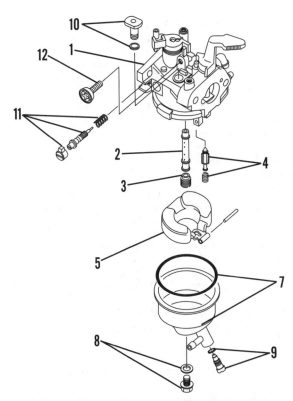

Fig. HA167 – Exploded view of carburetor components.

1. Carburetor body
2. Main nozzle
3. Main jet
4. Float valve assembly
5. Float
6. Float hinge pin
7. Float bowl and gasket
8. Float bowl bolt and washer
9. Float bowl drain screw and O-ring
10. Pilot jet
11. Idle-mix screw assembly
12. Throttle stop screw

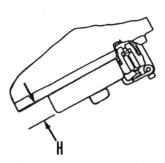

Fig. HA168 – View of proper carburetor position to check float level setting.

3. Place the piston at TDC by removing the spark plug, rotating the flywheel, and observing the piston head through the spark plug hole.

4. Using a feeler gauge, check valve clearance:

Intake . . . . 0.06-0.10 mm (.003-.004 in.)
Exhaust . . . . 0.09-0.13 mm (.0035-.005 in.)

5. To adjust valve clearance, loosen the rocker arm adjusting screw locknut, then turn the adjusting screw in to decrease clearance or out to increase clearance. Torque locknut 5.4 N•m (48 in.-lb.), then recheck clearance and readjust, if necessary.

6. Remove old valve cover sealer from the cover and block surfaces.

7. Apply a 1.0 mm (3/64 or .040 in.) diameter bead of Loctite 515 liquid gasket to the cylinder block valve cover surface as shown in Fig. HA169. Install the cover and incrementally torque the valve cover bolts 5.9 N•m (52 in.-lb.) using the sequence shown in Fig. HA169.

Ignition System

To Test the Ignition Coil

1. Test the resistance of the primary circuit by placing one VOM lead on the black primary winding wire terminal. Place the other VOM lead on any clean lamination core ground. Resistance should be 0.7-1.1 ohms.

2. Test the resistance of the secondary circuit by placing one VOM lead on the spark plug terminal inside the boot. Place the other VOM lead on any clean lamination core ground. Resistance should be 1200-2100 ohms.

3. Any readings other than those specified in Steps 1 and 2 indicate a faulty coil.

4. If replacing the coil, apply a light coating of Loctite 242 or equivalent to the 6 mm screw threads. Torque the screws 1.8 N•m (16 in.-lb.).

To Test the Engine Stop (Ignition) Switch

1. Access and disconnect the black and green two-pin engine switch connector from behind the control panel.

2. With the switch mounted inside the left side cover, test the switch wire terminals while turning the switch knob on the outside of the cover. There should be continuity in the OFF position only.

3. Continuity in the ON position or no continuity in the OFF position indicates a faulty switch.

To Test the Low Oil Shutdown Switch

1. Unplug the yellow feed wire from the generator harness; unbolt the green ground wire terminal and the other two switch mount bolts from the engine.

2. Test for continuity between the yellow and the green terminals.

    a. There should be continuity with the switch in its normal upright position.

    b. There should be no continuity with the switch turned upside down.

3. Test the switch float by immersing the switch in a container of oil.

    a. There should be no continuity with the switch below oil level.

    b. When switch is lifted from the oil, there should be continuity.

4. Any test results not consistent with Steps 2 or 3 indicate a faulty oil switch.

Rewind Starter

To view the starter components, refer to Fig. HA170. To service the starter:

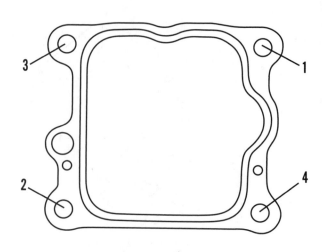

Fig. HA169 – Cylinder head/valve cover sealant pattern and fastener tightening sequence.

1. Disassemble, clean, and inspect as necessary.

2. Renew all worn or damaged components.

3. When installing a new rope, wind the rope onto the reel in a counterclockwise direction when viewing the engine side of the reel.

4. Pretension the spring approximately three turns.

5. When assembling the swing arm and collar assembly, insure that the tips of the swing arms are fitted into the lower flat parts of the reel ramp slots.

Muffler and Spark Arrester

1. Service the muffler and spark arrester (HA 171) *only* with the engine cold.

2. Carefully remove carbon deposits from the spark arrester screen with a wire brush.

3. Remove carbon deposits from the muffler by lightly tapping the seam flange.

4. Always install a new muffler gasket; apply a light coating of Loctite 271 or equivalent to the threads. Torque the muffler bolts 7.8 N•m (70 in.-lb.).

REPAIRS

Specification Chart

Component	Part	Std. Dimension	Wear Limit
Cylinder/crankcase and crankcase side cover	Cylinder bore	41.800-41.815 mm (1.6457-1.6463 in.)	41.900 mm (1.6496 in.)
	Valve guide I.D. (IN and EX)	4.000-4.018 mm (0.1575-0.1582 in.)	4.060 mm (0.1598 in.)
	Camshaft		
	Shaft bore I.D. and	5.005-5.023 mm	5.050 mm
	Valve lifter	(0.1970-0.1978 in.)	(0.1988 in.)
	Shaft bore I.D.		
	Rocker arm		
	Shaft bore I.D.	4.000-4.018 mm (0.1575-0.1582 in.)	4.050 mm (0.1594 in.)
Valve lifter shaft and camshaft shaft O.D.		4.990-5.000 mm (0.1959-0.1965 in.)	4.950 mm (0.1949 in.)

Rocker arm shaft O.D.		3.990-4.000 mm (0.1571-0.1575 in.)	3.950 mm (0.1555 in.)
Rocker arm journal I.D.		4.005-4.025 mm (0.1577-0.1585 in.)	4.050 mm (0.1594 in.)
Camshaft	Lobe height	27.972 mm (1.1013 in.)	26.972 mm (1.0619 in.)
	Journal I.D.	5.020-5.050 mm (0.1976-0.1988 in.)	5.100 mm (0.2008 in.)
Valve lifter journal I.D.		5.005-5.025 mm (0.1970-0.1978 in.)	5.050 mm (0.1988 in.)
Valve stem O.D.	Intake	3.970-3.985 mm (0.1563-0.1569 in.)	3.900 mm (0.1535 in.)
	Exhaust	3.935-3.950 mm (0.1549-0.1555 in.)	3.880 mm (0.1528 in.)
Valve clearance	Intake	0.06-0.10 mm (.003-.004 in.)	
	Exhaust	0.09-0.13 mm (.0035-.005 in.)	
Valve springs, IN and EX	Free length	22.8-23.7 mm (0.90-0.93 in.)	
Crankshaft	Rod journal O.D.	14.973-14.984 mm (0.5895-0.5899 in.)	14.940 mm (0.5882 in.)
Connecting rod	Wrist pin bore	10.006-10.017 mm (0.3939-0.3944 in.)	10.050 mm (0.3957 in.)
	Crankshaft journal bore	15.000-15.011 mm (0.5906-0.5910 in.)	15.040 mm (0.5921 in.)
	Crankshaft journal oil clearance	0.016-0.038 mm (0.0006-0.0015 in.)	0.100 mm (0.0039 in.)
	Crankshaft journal side clearance	0.1-0.6 mm (0.04-0.24 in.)	0.8 mm (0.031 in.)
Piston	Skirt O.D	41.770-41.790 mm (1.6445-1.6453 in.)	41.700 mm (1.6417 in.)
	Piston-to-cylinder clearance	0.010-0.045 mm (0.004-0.00187 in.)	0.120 mm (0.0047 in.)
	Wrist pin bore	10.002-10.008 mm (0.3938-0.3940 in.)	10.050 mm (0.3957 in.)
	Wrist pin O.D.	9.994-10.000 mm (0.3935-0.3937 in.)	9.950 mm (0.3917 in.)
	Wrist pin-to-bore clearance	0.002-0.014 mm (0.0001-0.0006 in.)	0.100 mm (0.0039 in.)
	Ring thickness: Top	0.77-0.79 mm (0.030-0.031 in.)	0.720 mm (0.0283 in.)
	Second	0.97-0.99 mm (0.038-0.039 in.)	0.920 mm (0.0362 in.)
	Ring side clearance, Top and Second	0.015-0.050 mm (0.0006-0.0020 in.)	0.120 mm (0.0047 in.)
	Ring end gap, Top and Second	0.150-0.300 mm (0.0059-0.0118 in.)	0.600 mm (0.0236 in.)

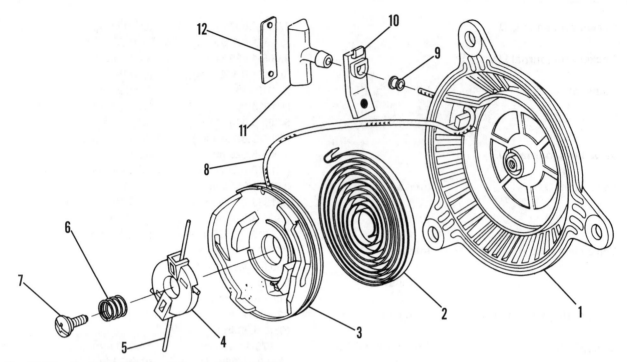

Fig. HA170 – Exploded view of recoil starter assembly.

1. Starter case
2. Spring
3. Reel
4. Swing-arm collar

5. Swing arm (2)
6. Friction spring
7. Retainer screw
8. Rope

9. Rope grommet
10. Rope guide
11. Grip
12. Grip cover

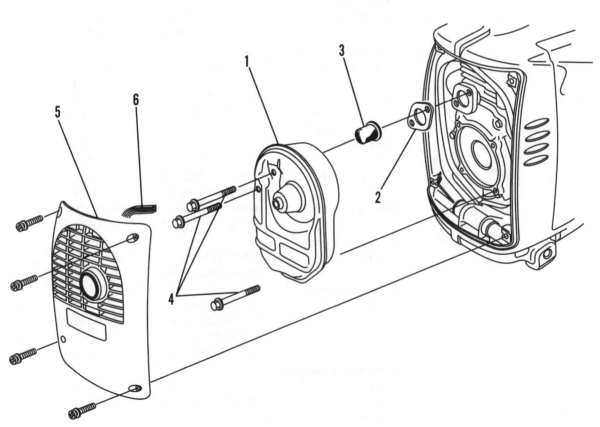

Fig. HA171 – Exploded view of muffler components and muffler cover.

1. Muffler assembly
2. Muffler gasket

3. Spark-arrester screen
4. Muffler bolts (3)

5. Muffler cover
6. Muffler-cover seal

Cylinder Compression

Cylinder compression should test at 415-445 kPa (60-65 PSI).

Disassembly Sequence

NOTE: The cylinder head cannot be removed. It is part of the cylinder block. The engine must be disassembled to service the valves and valve seats.

1. Remove the four valve cover screws.
2. Carefully pry up each corner of the valve cover to loosen and remove the cover and alternate between corners. Excessively forcing one corner more than the others will deform the cover; a deformed cover must be replaced.
3. Place the piston at TDC compression stroke by removing the spark plug, rotating the crankshaft, and observing the piston head through the spark plug hole. This must be done to facilitate rocker arm removal and to prevent damaging the oil level switch.
4. Note the rocker arm orientation, loosen the rocker arm locknuts and adjusting screws, remove the rocker arm shaft by pulling it out from the exhaust side of the head.
5. Note the push rod placement. Remove the push rods.
6. Place the engine on the workbench with the exposed crankshaft down. Support the crankcase so as not to damage the crankshaft threads.
7. Remove the crankcase side cover bolts. Utilizing the pry bosses between the cover and crankcase, carefully separate the cover from the crankcase.
8. Remove the governor gear retainer clip from the side cover shaft. Remove the gear and thrust washer.
9. Note the valve lifter and camshaft lobe orientation, then remove lifter collar, lifters, lifter shaft, camshaft, and camshaft roller shaft.
10. Flip the engine over.
11. Remove the oil level switch bolts, then remove the switch.
12. Remove the oil case bolts. Utilizing the pry bosses between the oil case and the cylinder block, carefully separate the case from the block.
13. Remove and discard the crankshaft oil seal.
14. Noting the position of the connecting rod cap alignment marks, remove the cap bolts, oil dipper, and cap.
15. Rotate the crankshaft to clear the connecting rod, then remove the crankshaft.
16. Pull the connecting rod/piston assembly from the cylinder.
17. Line the cylinder with protective paper to prevent damaging the cylinder bore while doing valve work.
18. While pushing down on the valve spring retainer, slide the retainer sideways to allow the valve stem to fall through the keyhole slot in the retainer. Remove the retainers and springs, then remove the valves through the cylinder.
19. Remove the crankcase breather screw, valve, and limit plate from the cylinder block.

Valve Service

NOTE: The cylinder head cannot be removed. It is part of the cylinder block. The valves and valve seats must be serviced through the cylinder bore.

1. Line the cylinder with protective paper to prevent damaging the cylinder bore while doing valve work.

2. Using an electric drill and a long shank rotary wire brush, carefully remove combustion deposits from the combustion chamber.
3. Check valve seats and faces for pitting or scoring. Renew as necessary.
4. Check valve stems for bending or excessive wear. Check push rods for bending.
5. Measure all valve system components using the dimensions in the specification chart. Renew as necessary. If the valve guides are worn beyond specification, the cylinder assembly must be replaced.

Camshaft and Crankcase Cover Service

Use the dimensions in the specification chart for these measurements.
1. Measure all shaft bore diameters in both the cover and the cylinder assembly.
2. Measure the shaft bore diameters in the camshaft and the valve lifters.
3. Measure the camshaft and valve lifter shafts.
4. Inspect the timing gears and valve lifter lobe ramps and push rod sockets for wear.
5. Measure the camshaft lobe.

Cylinder and Oil Case Service

Use the dimensions in the specification chart for needed measurements.
1. Inspect and measure the cylinder bore. Measure the bore at the top, middle, and bottom of the piston travel area, both parallel to the piston pin and perpendicular to the piston pin.
2. Inspect the crankshaft bearing journals.
3. Inspect the breather valve seat.

Piston, Rings, and Connecting Rod

Use the dimensions in the specification chart for needed measurements.
1. Measure the piston 90° from the wrist pin bore and 10 mm (0.4 in) from the bottom of the piston skirt.
2. Measure piston ring thicknesses and ring-to-land side clearances.
3. If the cylinder bore is to specification, insert the rings into the bore one at a time Push them in with the piston so they are perpendicular to the bore. Measure the ring end gaps.
4. Measure the wrist pin outer diameter and the wrist pin bore inner diameter in the piston and the connecting rod.
5. Install and torque the crankshaft journal rod cap 5.9 N•m (52 in.-lb.), then measure the journal inner diameter.

Crankshaft

Use the dimensions in the specification chart for required measurements.
1. Measure the connecting rod journal in at least two positions 90° apart.
2. Using Plastigage, measure the crank journal to connecting rod clearance. Follow the Plastigage instructions exactly.
3. Inspect the crankshaft ball bearings, cam drive gear, rotor taper, taper keyway, and threads.

Reassembly

Use Loctite 242 or equivalent on all metal fastener threads except the connecting rod bolts.
1. Install the crankcase breather valve, valve stop plate, and screw into the cylinder block. Insure that the chamfer

on the valve and the chamfer on the stop plate match the chamfer in the block.

2. Install the intake and exhaust valves, valve springs, and spring retainers. Compress the valve spring lightly while sliding the retainer keyhole slot onto the valve stem.

3. Assemble the piston to the connecting rod:

   a. Make sure that the "V" mark on the piston head points toward the Intake side and the connecting rod alignment marks face the open side of the cylinder block.

   b. Make sure that the gap of the wrist pin retaining ring **does not** align with the notch in the wrist pin bore face.

4. Assemble the piston rings to the piston:

   a. Install the rings into the respective grooves as shown in Fig. HA172.

   b. Make sure that the gaps in the ends of the oil ring side rail rings are positioned 10 mm (0.4 in.) to either side of the spacer ring gap.

   c. Stagger the top, second, and oil spacer ring end gaps 120° apart, with no gap in line with the wrist pin.

5. Lubricate the piston and rings with engine oil, then carefully install the piston assembly into the cylinder. Make sure that the rod cap alignment mark is visible. Slide the assembly all the way into the cylinder.

6. Install the crankshaft into the cam gear side of the cylinder block.

7. Rotate the crankshaft so the connecting rod journal is at BDC, oil the journal, then slide the connecting rod down onto the crankshaft journal. With the rod cap marks aligned, install the oil dipper so that the dipper is on the flywheel side of the rod. Torque the rod cap bolts 5.9 N•m (52 in.-lb.).

8. Install the oil seal into the oil case so that the outer face of the seal is even with the outer edge of the seal mount bore, 2.0-2.5 mm (0.08-0.10 in.) below the outer edge of the seal bore lead taper (Fig. HA173).

9. Clean and degrease the oil case and cylinder block mating surfaces.

   a. Apply a 1.5-2.0 mm (1/16-5/64 in.) diameter bead of Loctite 515 liquid gasket or equivalent to the cylinder block surface as shown in Fig. HA174.

   b. Make sure that the locator dowel pins are properly positioned in the cylinder block.

   c. Using a seal protector to prevent oil seal damage, install the oil case onto the cylinder block. Torque the case bolts incrementally in the sequence shown in Fig. HA174 to a final torque of 7.4 N•m (65 in.-lb.).

10. Rotate the crankshaft to place the piston at TDC, then install the oil level switch O-ring and switch. Snug tighten the switch bolts.

11. Flip the engine over.

12. Install the camshaft shaft, valve lifter shaft, and cover locator dowels into their proper block positions.

13. Lubricate, then install the valve lifters, lifter collar, and camshaft. Make sure that the "L" camshaft gear timing mark aligns with the "•" crankshaft gear timing mark.

14. Install the governor gear thrust washer, gear, and gear retainer clip onto the side cover shaft.

15. Make sure that the side cover and cylinder block mating surfaces are degreased, clean, and dry.

   a. Apply a 1.5-2.0 mm diameter (1/16-5/64 in.) bead of Loctite 515 liquid gasket or equivalent to the cylinder block surface as shown in Fig. HA175.

   b. Make sure that the locator dowel pins are properly positioned in the cylinder block.

   c. Carefully align the governor gear/crank gear teeth, then install the side cover onto the cylinder block.

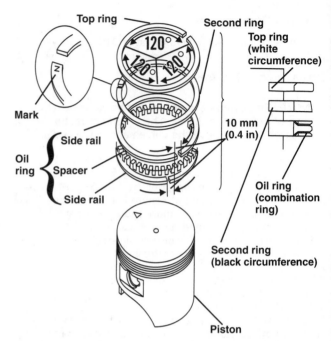

Fig. HA172 – Cylinder block/oil case sealant pattern and fastener tightening sequence.

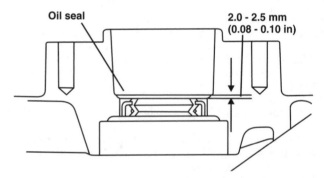

Fig. HA173 – View showing correct depth of crankshaft oil seal inside oil case.

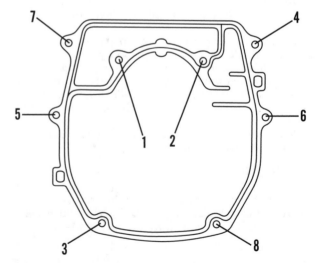

Fig. HA174 – Cylinder block/oil case sealant pattern and fastener tightening sequence.

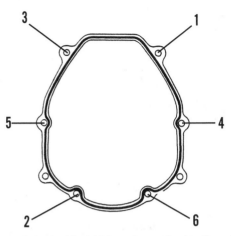

Fig. HA175 – Cylinder block/side cover sealant pattern and fastener tightening sequence.

Torque the cover bolts incrementally in the sequence shown in Fig. HA175 to a final torque of 7.4 N•m (65 in.-lb.).

16. Noting the removal orientation, lubricate and install the rocker arm push rods, rocker arms, and rocker arm shaft.

17. Place the piston at TDC compression stroke by rotating the crankshaft and observing the piston through the spark plug hole. (If the crankshaft was not rotated after aligning the crank gear/cam gear timing marks during previous assembly, the piston is already at TDC compression.)

18. Make sure that the push rod ball ends are properly positioned in their valve lifter and rocker arm sockets.

19. Adjust the rocker arm screws as necessary and use a feeler arm gauge to set the valve clearance.

Intake . . . . 0.06-0.10 mm (.003-.004 in.)
Exhaust . . . . 0.09-0.13 mm (0.0035-0.005 in.)

Torque the rocker-arm locknuts 5.4 N•m (48 in.-lb.), then recheck clearance and readjust, if necessary.

20. Ensure that the valve cover and cylinder head mating surfaces are degreased, clean, and dry.

    a. Apply a 1.0 mm (3/64 or .040 in.) diameter bead of Loctite 515 liquid gasket to the cylinder block valve cover surface as shown in Fig. HA169.

    b. Install the valve cover.

    c. Incrementally torque the valve cover bolts 5.9 N•m (52 in.-lb.) in the sequence shown in Fig. HA169.

21. If the carburetor insulator and/or exhaust manifold were removed, always use new gaskets when reinstalling.

WIRING DIAGRAM

Refer to Fig. HA176 to view the wiring diagram.

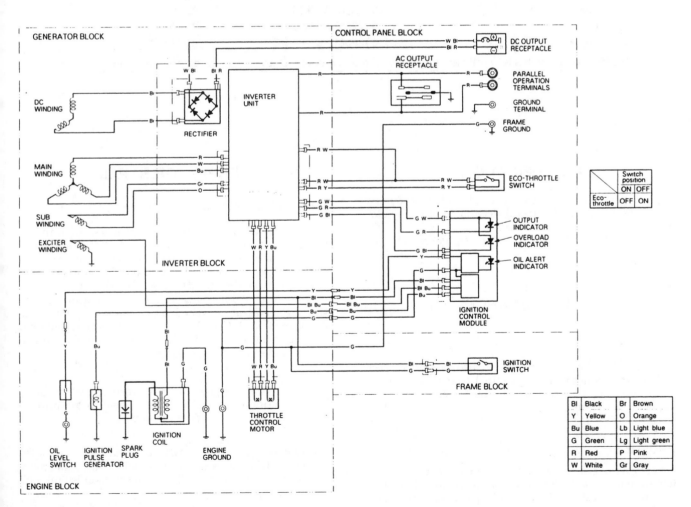

Fig. HA176 – Wiring diagram for EU1000i generator.

HONDA
EB, EG, and EM Series 120V
Brush-type Generators

Generator Model	Rated (Surge) Watts	Output Volts
EB2200X	2000 (2200)	120
EB2500X K1	2300 (2500)	120
EG1000Z	900 (1000) AC/100 DC	120 AC/12 DC
EG1400X, K1	1200 (1400) AC/100 DC	120 AC/12 DC
EG1400Z	1200 (1400) AC/100 DC	120 AC/12 DC
EG1800Z	1500 (1800) AC/100 DC	120 AC/12 DC
EG2200X	2000 (2200) AC/100 DC	120 AC/12 DC
EG2200Z	1800 (2200) AC/100 DC	120 AC/12 DC
EG2500X K1	2300 (2500) AC/100 DC	120 AC/12 DC
EM1600X	1400 (1600) AC/100 DC	120 AC/12 DC
EM1800X, K1	1500 (1800) AC/100 DC	120 AC/12 DC
EM2200X	2000 (2200) AC/100 DC	120 AC/12 DC
EM2500X K1	2300 (2500) AC/100 DC	120 AC/12 DC

NOTE: models with a K1 suffix are upgrades of basic models. Output ratings remain the same.

IDENTIFICATION

These generators are direct drive, brush-style, 60 Hz, single-phase units with integral electronic voltage regulation and panel mounted receptacles.

Refer to Figs. HA201-HA203 to view the generator components.

MAINTENANCE

ROTOR BEARING

The brush-end rotor bearing is a sealed bearing. Inspect and replace as necessary. If a seized bearing is encountered, also inspect the rotor journal and the end cover bearing cavity for scoring.

BRUSHES

Replace the brushes when the brush length reaches 0.5 in. (13 mm) with the brush out of the brush holder (Fig. HA204) or 0.3125 in. (8 mm) with the brush protruding from the holder.

GENERATOR REPAIR

TROUBLESHOOTING

Before testing any electrical components, check the integrity of the terminals and connectors. While testing individual components, also check the wiring integrity. For access to the components listed in this **Troubleshooting** section, refer to the later **Disassembly** section.

Resistance readings may vary with temperature. The resistance specifications listed in this chapter are for a temperature of 68° F (20° C). Always load test the generator after repairing defects discovered during troubleshooting.

To troubleshoot the generator:

NOTE: The positive (+) brush terminal on these generators is always the one closest to the rotor bearing, and has a LG/W tracer wire.

1. Circuit breakers must be ON. To test either the AC or DC (if equipped) breakers, access the breaker wiring and continuity test the breaker terminals. The breaker is faulty if:

   a. It will not stay set.

   b. It shows continuity in the OFF position.

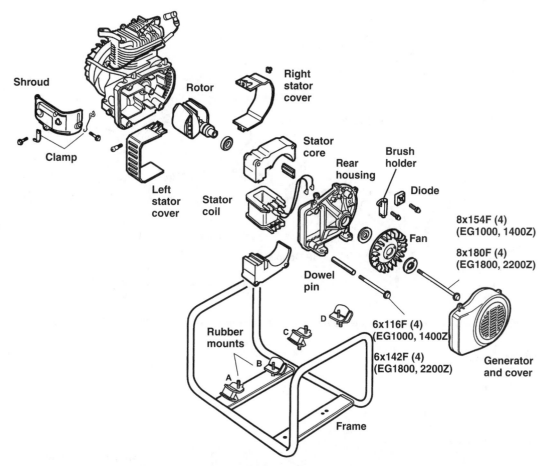

Fig. HA201 – Exploded view of typical 1.0-2.5 KW EB, EG, and EM series generator

c. It does not show continuity in the ON position.

2. Remove the generator end cover, rotor throughbolt, fan, and two fan plates (Fig. HA201).

3. Install steel washers on the rotor throughbolt equal in thickness to the fan and plates removed in Step 2. Reinstall the rotor throughbolt. Torque the bolt 240-275 in.-lb. (28-32 N•m).

NOTE: The generator can be run with the fan removed for testing purposes only. Do not apply a load to the generator with the fan removed.

4. Start the engine and verify proper no-load speed of 3750 RPM.

5. Test the AC winding.

a. Test the 120V receptacle output (Figs. HA202 and HA203).

b. Test the output at the white and red stator wires (1 and 2 respectively – Fig. HA205).

6. AC output should be 115-125 volts on EG-Z models, 102-138 volts on all other units. If there is output at the white and red wires but none at the receptacles:

a. Insert a jumper wire between the receptacle plugs (Fig. HA206).

b. Disconnect either the white or red wire from the rear of the receptacle.

c. Test for ohms/resistance across the white and red receptacle terminals. No continuity indicates a faulty

receptacle. Continuity indicates faulty wiring between the stator and the receptacle.

7. If the unit is equipped with a GFCI receptacle, refer to the later Individual Component Testing section in this chapter for the GFCI test.

8. Test the DC windings:

a. Test the 12V terminals for output.

b. Test the output at the white/red (+) and black/red (-) connections (10 and 4–Fig. HA205)

c. DC output should be 10-14 volts. If there is output from the white/red (+) and black/red connections, but none at the 12-volt terminals, inspect the control box wiring.

9. If the DC output from the white/red and black/red connections is not to specification, test the DC diode assembly (Fig. HA207).

a. Set the VOM to the R × 1 scale.

b. Stop the generator.

c. Disconnect the two gray wires and the one white/red wire from the diode assembly (Fig. HA205).

d. Place one meter lead on the positive (+) diode terminal and the other meter lead on either AC diode terminal; note the reading.

NOTE: On original old style 4-terminal diodes, the negative (-) diode terminal is not used. Newer style 3-terminal diodes and replacement diodes for older style 3-terminal diodes have a white/red wire

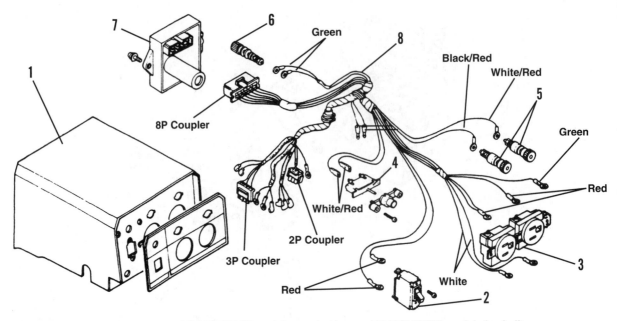

Fig. HA202 – Exploded view of control box on EG1400-2200X models; control box on EG1000-2200Z models is similar.

1. Control box
2. AC circuit breaker
3. AC receptacle

4. DC circuit breaker
5. DC terminals
6. Ground terminal

7. AVR
8. Control box harness

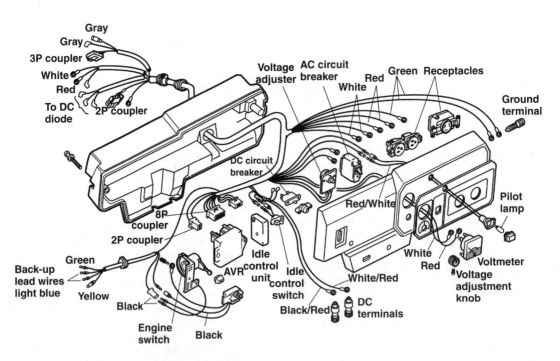

Fig. HA203 – Typical exploded view of control box on EB2200-2500X and EM1600-2500X models. Components vary by model.

attached to the positive (+) diode terminal to prevent connection errors. Refer to Fig. HA208 for diode identification.

e. Reverse the meter leads and repeat Step 9d. Note the reading.

f. The meter should show continuity in one direction and infinity/no continuity in the opposite direction.

g. Repeat Steps 9 d-e with the other AC diode terminal. If either test shows continuity in both directions or

infinity in both directions, the diode assembly is faulty.

h. Repeat Steps 9 d-e except use both AC diode terminals. There should be no continuity in either direction. A continuity reading between the AC terminals indicates a faulty diode assembly.

i. If the diode assembly tests good, proceed to the Stator, DC Winding Test.

NOTE: If replacing the diode assembly, carefully note the wiring orientation. Improper wiring connection will damage the stator coil. Always install the diode assembly with the beveled corner to the lower right.

10. Test the rotor winding slip ring brush DC voltage.
    a. Briefly stop the generator, disconnect the brush holder harness connector, then restart the generator.
    b. Light green/white wire is positive (+) brush. Light green/black wire is negative (-) brush (8 and 9 – Fig.

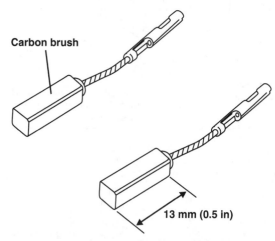

Carbon brush

13 mm (0.5 in)

Fig. HA204 – Minimum brush length should be 0.5 in. (13 mm) out of holder; 0.3125 in. (8 mm) with the brush protruding from the holder.

HA209 – EB series; 11 and 12 – Fig. HA205 – EG and EM series).
    c. Brush output should be 10-14 volts.
    The slip rings are an integral part of the rotor assembly; they are not replaceable.
11. Inspect the brushes for minimum length of 0.5 in. (13 mm) with the brush out of the brush holder (Fig. HA204) or 5/16 in. (8 mm) with the brush protruding from the holder. Check continuity between wire terminals and brush tip. Check for freedom of movement, unusual wear. Replace if necessary. When replacing brushes, always fit the brushes into the brush holder so the concave end aligns with the slip ring diameter.
    On EG-Z series, brush replacement is made easier with the special brush removal and installation tool #07999-ZA40000. Follow the instructions included with the tool.
12. If the brush voltage in Step 10 tests low, and the brushes meet specifications in Step 11, test the back up coil for AC voltage:
    a. Stop the engine.
    b. Disconnect the back up coil wires from the engine. These will be two yellow wires on EG-Z models, two blue wires on all other models (Fig. HA210).
    c. Start the engine and test the coil output.
    AC output should be 34-40 volts on EG-Z models, 25-35 volts on all other models.
13. If the back up coil voltage tests low, stop the engine and test the coil resistance (Fig. HA210): Coil should test 2.1-2.7 ohms.
14. Test the rotor winding resistance.
    a. Disconnect the brush wires and remove the brush holder.
    b. Connect the VOM leads to the slip rings and test resistance:

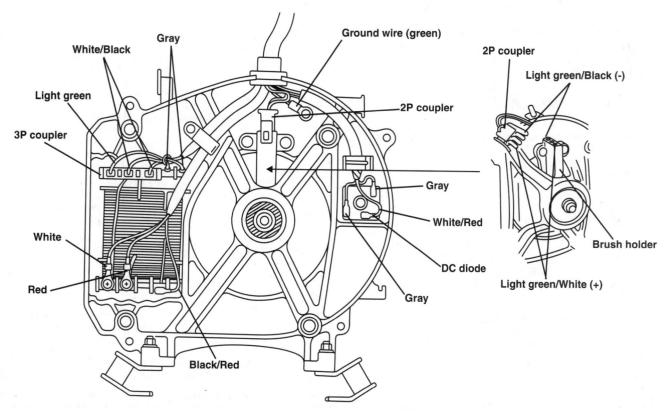

Fig. HA205 – Troubleshooting test connections on generators with both AC and DC output, except for some tests on the EG-Z series.

Model	Ohms
EB2200X	14-19
EB2500X	16-20
EG1000Z	12-15
EG1400X	12-16
EG1400Z	14-16
EG1800Z	14.5-17
EG2200X	14-19
EG2200Z	16.5-19
EG2500X	14-19
EM1600X	13-17
EM1800X	14-19
EM1800X K1	13-17
EM2200X	14-19
EM2500X	16-20

A rotor that does not test within specifications is faulty. Replace the rotor.

15. To test the stator:

a. Obtain a fully charged 12-volt battery and a pair of probe end test leads.

b. Disconnect the two-prong brush wire connector.

c. Completely disconnect the AVR from the generator and set it aside.

d. Connect the positive (+) battery post to the light green/white (+) brush holder terminal. Connect the negative (-) battery post to the light green/black (-) brush holder terminal.

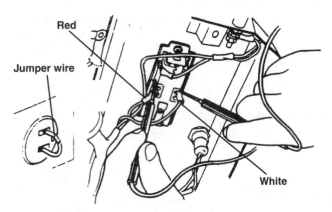

Fig. HA206 – Troubleshooting-test connections on generators with AC output only.

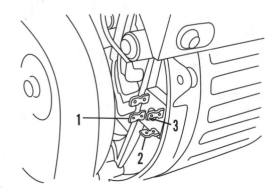

Fig. HA207 – To test DC diodes, check continuity between terminals.

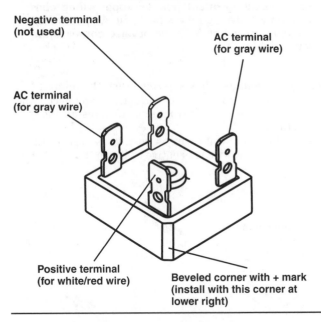

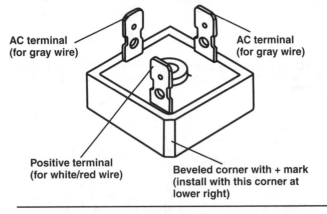

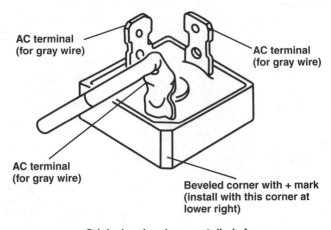

Fig. HA208 – View of the various diode assemblies used on generators with DC output. Refer to text for usage differences and correct test procedures.

NOTE: Improper battery to brush connections will damage the generator.

  e. Start the generator, verifying the correct 3750 RPM no load speed.

  f. Referring to Figs. HA205 and HA209, test the stator output voltages using the following color coded test connections. Note the readings.

NOTE: EB series units do not have DCV windings.

1. Main Output Winding Test – red to white

Model	AC Voltage
All EBs, EG1400X, EG2200X, EG2500X, and all EMs	102-138
EG1000Z, EG1400Z, EG1800Z, and EG2200Z	115-125

2. Exciter winding test – light green to white. Refer to Fig. HA211.

Model	AC Voltage
All EBs, EG1400X, EG2200X, EG2500X, and all EMs	112-148
EG1000Z, EG1400Z, EG1800Z, and EG2200Z	38-48

3. Signal winding test – white/black to white/black.

Model	AC Voltage
All EBs, All EMs	13-17
EG1400X, EG2200X, and EG2500X	14-20
EG-Z series	Not Applicable

4. DC Winding Test – gray to gray.

Model	AC Voltage
All EBs	Not applicable
All EG-Zs	None specified; resistance-test only.
EG1400X, EG2200X, and EG2500X	19-29
All EMs	17-23

5. DC Output Test – (+) white/red to (-) black/red.

Model	DC Voltage
All EBs	not applicable
All EGs and All EMs	10-14

  g. Some generators also specify stator winding resistances. If any running tests in the preceding steps. did not meet specifications, stop the generator, disconnect the applicable color coded wire terminals, and test the questionable winding using the following specifications.

1. Main output winding – red to white.

Model	Ohms
EG1400X	0.2-0.4
EB2500X, EG2500X, EM1800X, and EM2500X	0.16
EG1000Z	0.6
EG1400Z	0.4
EG1800Z	0.3
EG2200Z	0.2

2. Exciter winding – light green to white. Refer to Fig. HA207.

Model	Ohms
EG1400X	0.3-0.5
EB2500X, EG2500X, EM1800X, and EM2500X	1.23
EG1000Z	1.5
EG1400Z	1.2
EG1800Z	1.0
EG2200Z	0.9

3. Signal Winding – white/black to white/black.

Model	Ohms
EG1400X	12-16
EB2500X, EG2500X, EM1800X, and EM2500X	18.15
EG-Z series	Not applicable

4. DC Winding – gray to gray. Refer to Fig. HA212.

Model	Ohms
EG1400X	0.2-0.4
EB2500X, EG2500X, EM1800X, and EM2500X	0.15
EG-Z series	0.2

  h. If one or more stator tests do not meet specification, the stator is faulty. If *all* the windings test low, retest the rotor resistance.

16. On EM models with manual voltage adjustment, disconnect the two blue wires from the voltage adjuster and test the resistance: The adjuster should test zero ohms in the Maximum position and 1000 ohms in the Minimum position. If faulty, replace.

17. Disconnect the 12-volt test battery, properly reconnect all generator component wiring, and connect a DC voltmeter to the brush leads and a tachometer to the engine.

18. Start the generator and promptly verify 3750 RPM no load engine speed and 10-14 VDC brush output. If brush voltage is good, proceed to Step 19. If brush voltage is beyond specification (low or high), **immediately** stop the generator and test the AVR (Fig. HA213) by measuring resistance between the following terminal pairs:

| Meter Lead | | Ohm |
Positive	Negative	Specification
5		
1		
6		
2		
4	8	
8	4	
2	3	
2		
3	7 and	
6	7	∞ (infinity)
7	3	±19
7	6	±150

An AVR which does not meet specification is faulty. Replace the regulator.

19. If replacing the regulator, repeat Step 19 after replacement. If brush output is within specification.

  a. Stop the generator.

  b. Remove the rotor throughbolt, then remove the washers added in Step 3.

  c. Position the outer fan plate, fan, and inner fan plate on the rotor throughbolt, then install the bolt (Fig. HA214).

d. Make sure that the holes in the outer fan plate properly align with the fan lugs, then torque the rotor throughbolt 240-275 in.-lb. (28-32 N•m).

e. Install the generator end cover.

f. Start the generator and apply a full test load.

g. Verify load test, then stop the generator.

Individual Component Testing

Ground Fault Circuit Interrupter (GFCI)

To test the GFCI:

1. Start the generator and allow a brief no load warm up period.

2. Depress the TEST button; the RESET button should extend. If it does, proceed to Step 3. If the RESET button does not extend, the GFCI is faulty.

3. Depress the RESET button. The RESET button should then be flush with the TEST button. If the RESET button does not remain flush with the TEST button, the GFCI is faulty.

4. Stop the generator.

NOTE: If the GFCI seems to trip for no apparent reason, check to make sure the generator is not vibrating excessively due to faulty rubber mounts or debris between the generator and frame. Excessive vibration can trip the GFCI.

Idle Control (Auto-Throttle)

NOTE: Auto-Throttle (Fig. HA215) activates when the generator detects loads of less than 1 amp (120 watts). Do not operate an auto throttle equipped generator with less than 120-watt load or the generator will idle down, causing insufficient voltage for the load and possibly damaging the load and the generator.

The auto throttle switch should show continuity in the ON position and no continuity in the OFF position.

If the auto throttle switch is OK and the idle control solenoid moves freely, but the engine will not idle down when in a no load condition:

1. Check to make sure the red wire is properly routed through the idle control unit (Fig. HA216) and the wiring is

secure between the idle control unit and the throttle solenoid.

2. Disconnect the throttle solenoid wires and test for continuity. The solenoid should test approximately 50-450

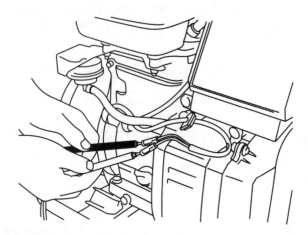

Fig. HA210 – Testing the back up coil.

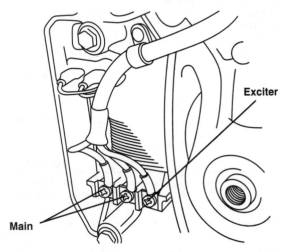

Fig. HA211- Troubleshooting test connections on EG-Z series generators for the main output winding and the exciter winding.

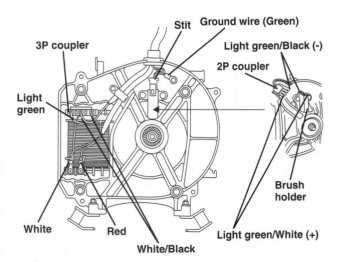

Fig. HA209 – Troubleshooting test connections on EG and EM Series generators

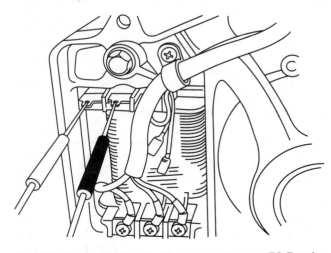

Fig. HA212 – Troubleshooting test connections on EG-Z series generators for the DC winding.

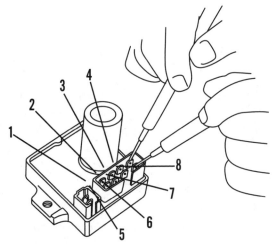

Fig. HA213 – Test connections for the automatic voltage regulator. Refer to text for specifications.

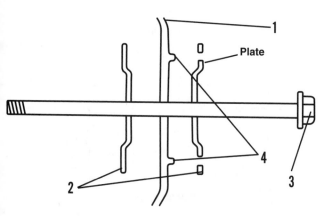

Fig. HA214 – The holes in the outer fan plate must correctly align with the fan lugs; the inner fan plate must mount with the small inner diameter flat against the back side of the fan.

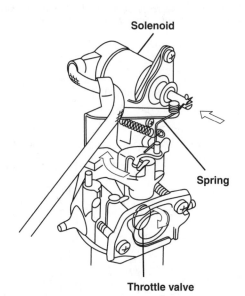

Fig. HA215– View showing the construction and operation of the auto throttle idle control system. When the control unit detects that the load has dropped below 1 amp (120 watts), the control activates the solenoid (arrow), moving the carburetor throttle to the idle position.

ohms. No continuity or excessively high continuity indicates a faulty solenoid.

3. If the solenoid tests OK, the idle control unit is faulty.

Voltmeter, if Equipped

To test the voltmeter, disconnect either meter wire, then test across the meter for continuity: Resistance should be 100 ohms or more. A lower resistance reading denotes a faulty voltmeter.

Pilot Lamp, if Equipped

To test the lamp:

1. Disconnect the red/white and black wires from the rear of the lamp.

2. Test across the lamp leads; there should be continuity.

3. If the lamp shows no continuity:

   a. Pull the lens off the outside of the lamp.

   b. Remove the bulb from the socket.

   c. Test the bulb; it should have continuity. If not, replace.

   d. If the bulb tests OK for continuity, test the socket wires for continuity; if the wires show no continuity, the socket is faulty.

Engine Oil Alert Lamp, if Equipped

To test the lamp:

1. Disconnect the black and yellow lamp wires from the generator harness.

2. Obtain a fully charged 6-volt lantern battery.

3. Connect the positive (+) battery terminal to the black lamp wire and the negative (-) battery terminal to the yellow wire.

4. If the lamp does not light, it is faulty.

Engine Switch

To test the engine switch, check for continuity between the black switch lead and the switch body ground: There should be continuity with the switch in the OFF position only.

Disassembly

To disassemble the generator:

1. Remove the upper frame and the fuel tank.

2. Remove the generator end cover mounts, end cover, rotor throughbolt, fan, and fan plates.

3. Disconnect the generator to control box wiring connectors. Loosen the control box fasteners and remove the control box. On two-piece control boxes, carefully separate the front panel from the rear cover to access the control box components.

4. Remove the muffler assembly. On EB-M and EB-X models, note the location of the inner heat shield bracket, where it fits into the end cover support, for ease of later reassembly.

5. Remove the generator end cover rubber mounts as necessary.

6. Remove the brush holder.

7. Remove the four rear bearing housing bolts, then remove the housing, followed by the stator core and coil assembly.

NOTE: Do not disassemble the core/coil assembly.

8. Use a slide hammer with a 10 × 1.25 metric thread adapter to remove the rotor. Use caution to prevent dropping the rotor when it breaks free from the engine crankshaft.

Reassembly

1. If the rotor bearing is being replaced:
   a. Press the new bearing onto the rotor shaft with the bearing slot towards the rotor windings and the outer bearing face 0.197 in. (5.00 mm) in from the end of the rotor shaft (Fig. HA217).
   b. Install the retaining ring in the bearing slot so the raised part of the ring is centered over the deepest part of the slot, with the ring gap at the shallow part of the slot (Fig. HA218).
2. Make sure that the matching tapers inside the rotor shaft and on the engine crankshaft are clean and dry. Verify the integrity of the locator pin inside the rotor taper. On EG-Z series, instead of an inside locator pin, the rotor has a raised boss on the outside of the tapered rotor shaft.
3. On all but the EG-Z series units, align the rotor locator pin with the crankshaft keyway. Fit the rotor onto the crankshaft and place both fan plates (without the fan) onto the rotor throughbolt. Then hand tighten the bolt at this time.
4. On the EG-Z series units, align the rotor shaft boss with the punch mark on the end of the engine crankshaft. Place both fan plates (without the fan) onto the rotor throughbolt, then hand tighten the bolt at this time. The following Step 7 does not apply to the EG-Z series.
5. Fit the stator core and coil assembly over the rotor and onto the crankcase cover adapter.
6. Install the rear housing and finger tighten the bolts.
7. With the rear housing in position, verify the rotor pin and keyway alignment (Fig. HA219): With the rotor insulator in approximately the 9:30 position and dimensioned as noted, the "V" mark on the engine starter pulley must align with the 12:00 hole in the blower housing.

NOTE: On units with the Oil Alert System, if the rotor is positioned incorrectly, the oil alert may shut the engine down even with sufficient crankcase oil.

8. Torque the rotor throughbolt 240-275 in.-lb. (28.0-32.0 N•m) to seat the rotor.
9. Torque the rear housing bolts 70-105 in.-lb. (8.0-12.0 N•m).
10. With the engine spark plug removed, rotate the crankshaft and inspect for rotor to stator core rubbing or binding. A feeler gauge should be used to verify that .020 in. (0.5 mm) clearance exists around the entire rotor inside the stator. If clearance is insufficient on one side, or if rubbing occurs, check for debris on the crankshaft/rotor taper. Realign as necessary.
11. Install the end cover rubber mounts. Refer to Fig. HA220 for proper mount orientation.
12. Install the brush holder.
13. Install the muffler assembly. On EB-X and EM-X models, insure that the inner heat shield bracket is properly positioned on the end cover support.
14. On two-piece control boxes, insure that the harness wiring is properly routed through the grommeted hole in the rear cover. Fasten the rear cover to the front panel,

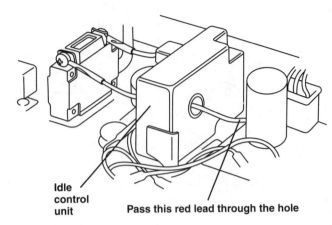

Fig. HA216 – Control box view of the idle control unit with the red wire correctly routed through the unit.

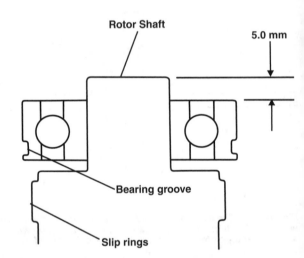

Fig. HA217 – Install a new rotor bearing with the slot toward the windings and with the outer face 0.197in. (5.00 mm) in from the shaft end.

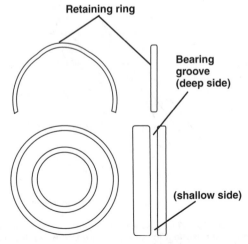

Fig. HA218 – Install the rotor bearing ring with the raised part centered over the deep side of the slot and with the gap centered over the shallow side of the slot.

making sure no wires are pinched or kinked. Install any cable ties removed during disassembly.

15. Remove the rotor throughbolt; mount the fan and fan plates onto the bolt as shown in Fig. HA214. Retorque the throughbolt 240-275 in.-lb. (28.0-32.0 N•m).

16. Install the generator end cover, fuel tank, and upper frame.

WIRING DIAGRAMS

Refer to the following list for the respective wiring diagram:

Generator Model	Wiring Diagram Figure
EB2200X	HA221
EB2500X K1	HA222
EG1000Z	HA223
EG1400X	HA224
EG1400X K1	HA225
EG1400Z	HA223
EG1800Z	HA223
EG2200X	HA224
EG2200Z	HA223
EG2500X K1	HA225
EM1600X	HA226
EM1800X	HA227
EM1800X K1	HA228
EM2200X	HA226
EM2500X K1	HA228

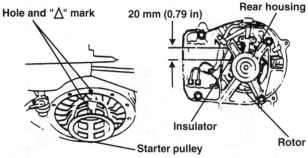

Fig. HA219 – For correct rotor alignment, refer to text.

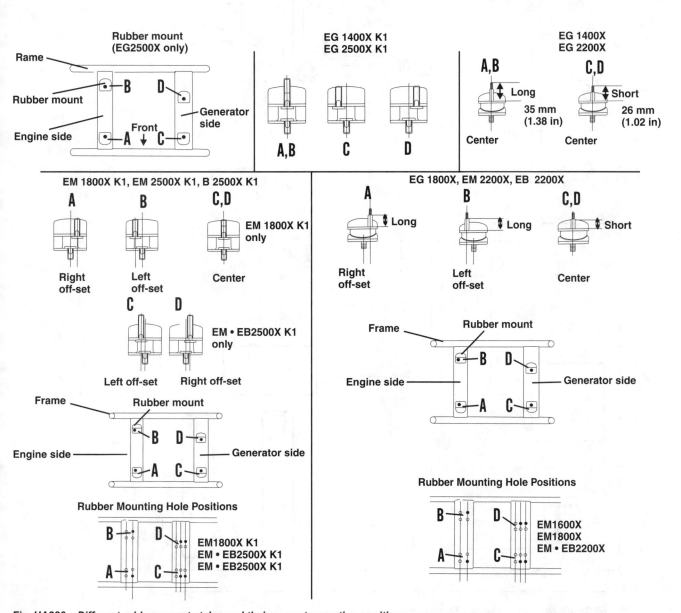

Fig. HA220 – Different rubber mount styles and their correct mounting positions.

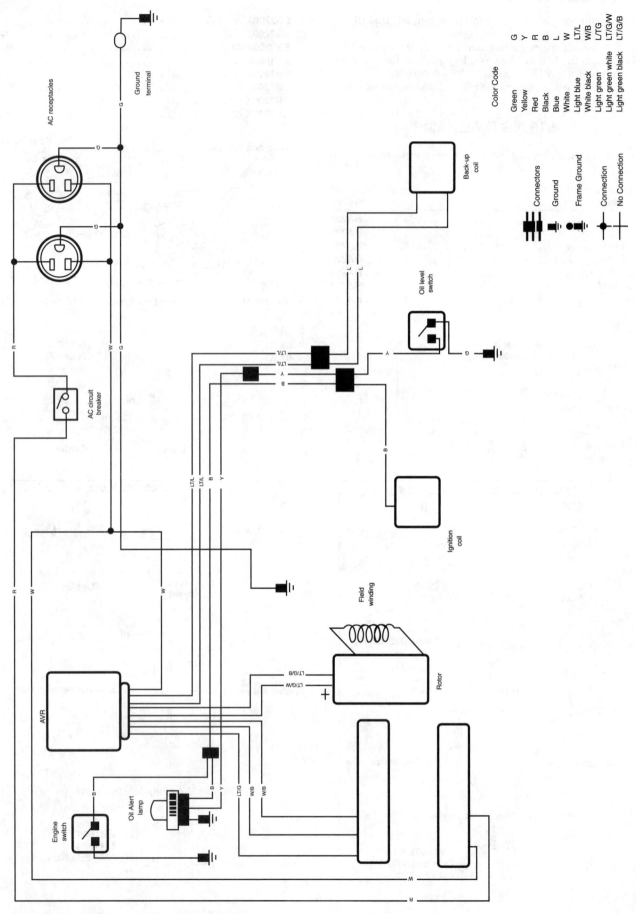

Fig. HA221 – Wiring diagram for model EB2200X generator.

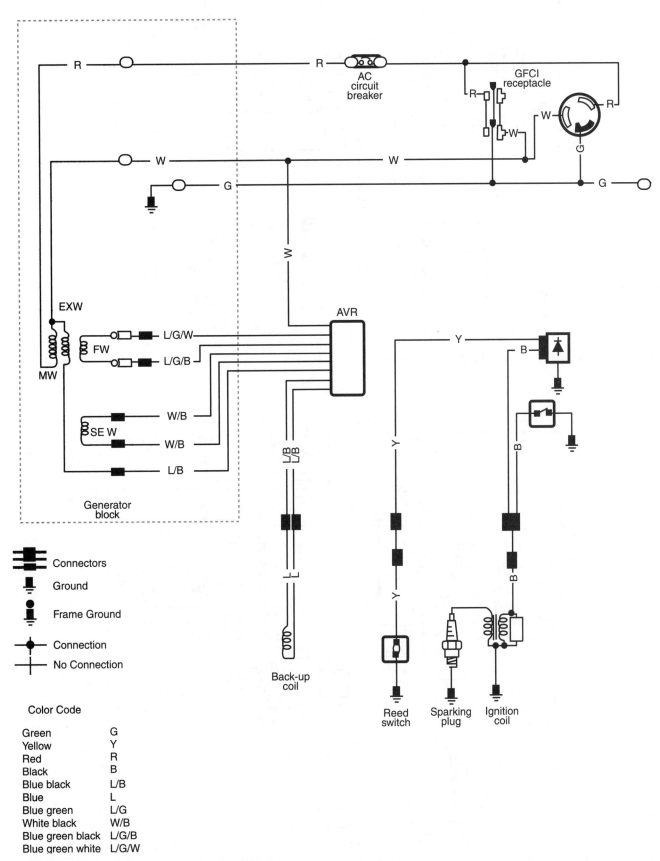

Fig. HA222 – Wiring diagram for model EB2500X K1 generator.

Connectors
Ground
Frame Ground
Connection
No Connection

AC circuit breaker

GFCI receptacle

EXW
FW
MW
SE W

Generator block

AVR

Back-up coil

Reed switch

Sparking plug

Ignition coil

Color Code

Green	G
Yellow	Y
Red	R
Black	B
Blue black	L/B
Blue	L
Blue green	L/G
White black	W/B
Blue green black	L/G/B
Blue green white	L/G/W

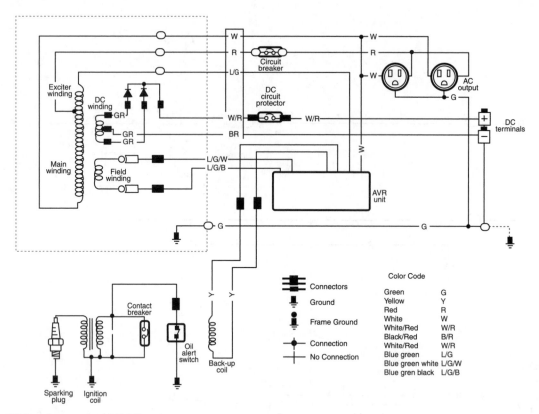

Fig. HA223 – Wiring diagram for all EG-Z series generators.

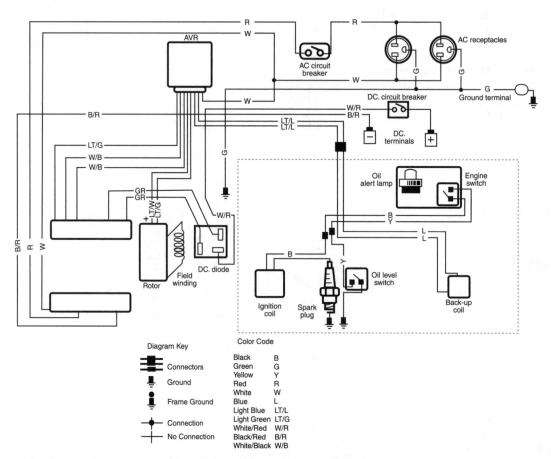

Fig. HA224 – Wiring diagram for generator models EG1400X (except K1) and EG2200X.

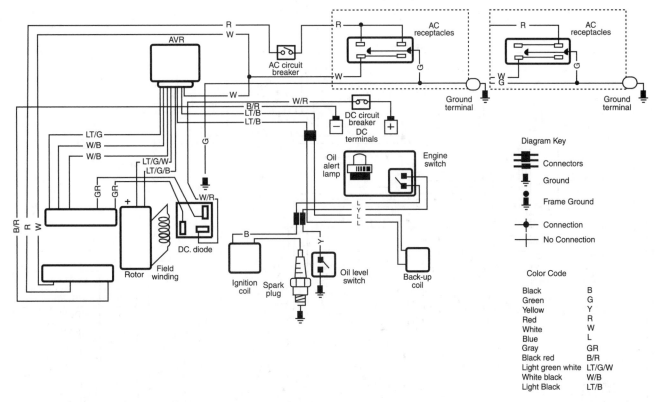

Fig. HA225 – Wiring diagram for generator models EG1400X K1 and EG2500X K1.

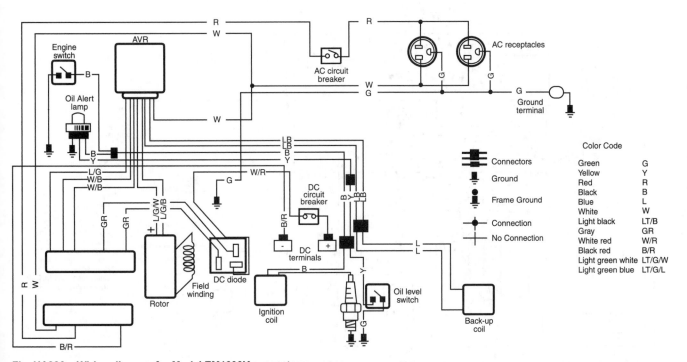

Fig. HA226 – Wiring diagram for Model EM1600X generator.

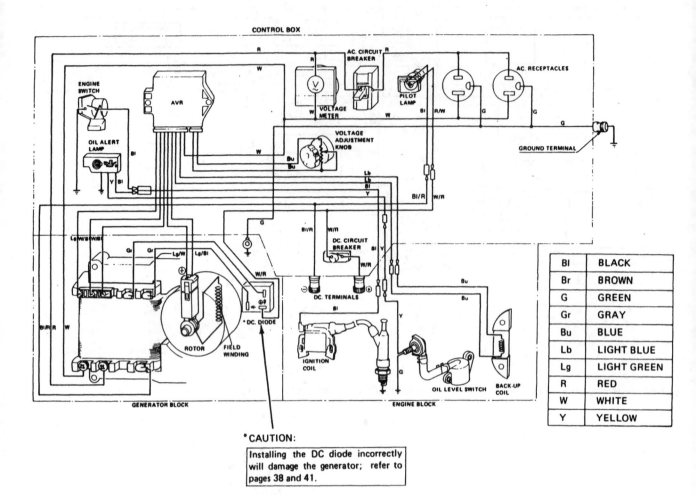

CAUTION:

Installing the DC diode incorrectly will damage the generator; refer to pages 38 and 41.

Fig. HA227 – Wiring diagram for generator models EM1800X and EM2200X.

Bl	BLACK
Br	BROWN
G	GREEN
Gr	GRAY
Bu	BLUE
Lb	LIGHT BLUE
Lg	LIGHT GREEN
R	RED
W	WHITE
Y	YELLOW

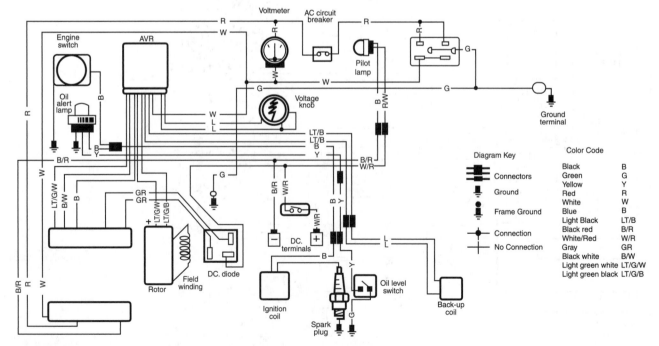

Diagram Key

Connectors
Ground
Frame Ground
Connection
No Connection

Color Code

Black	B
Green	G
Yellow	Y
Red	R
White	W
Blue	B
Light Black	LT/B
Black red	B/R
White/Red	W/R
Gray	GR
Black white	B/W
Light green white	LT/G/W
Light green black	LT/G/B

Fig. HA228 – Wiring diagram for generator models EM1800X K1 and EM2500X K1.

HONDA

EN Series 120V Brushless Generators

Generator Model	Rated (Surge) Watts	Output Volts
EN2000	1800 (2000)	120
EN2500	2300 (2500)	120

IDENTIFICATION

These generators are direct drive, brushless, 60 Hz, single phase units with the receptacles mounted in the rear stator housing end cover.

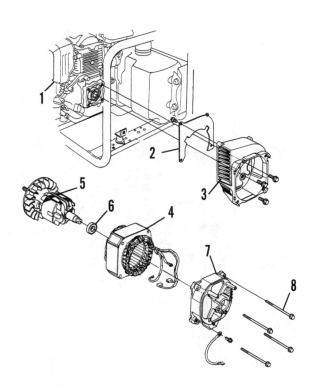

Fig. HA301A – Exploded view of generator components.

1. Engine
2. Housing plate
3. Engine adapter/front housing
4. Stator assembly
5. Rotor assembly
6. Bearing
7. Stator housing
8. Bolt

Refer to Figs. HA301A and 301B to view the generator components.

MAINTENANCE

The rear housing rotor bearing is a sealed bearing. Inspect and replace as necessary. If a seized bearing is encountered, inspect the rotor journal and the rear housing bearing cavity for scoring.

GENERATOR REPAIR

Troubleshooting

Before testing any electrical components, check the integrity of the terminals and connectors. While testing individual components, also check the wiring integrity. Resistance readings may vary with temperature and test equipment accuracy. All operational testing must be performed with the engine running at its governed no load speed of 3800-3900 RPM, unless specific instructions state otherwise. Always load test the generator after repairing defects discovered during troubleshooting.

To troubleshoot the generator:

1. Remove the generator end cover.

2. Circuit breaker must be ON. To test the breaker, disconnect the breaker wiring and continuity test the breaker terminals. If the breaker will not stay set, shows continuity in the Off position, or does not show continuity in the ON position, the breaker is faulty.

3. Start the engine and test for AC output at the receptacles.

   a. If there is low or no output, test for AC output at the red and white receptacle feed wires (Fig. HA302). Rated output at the feed wires indicates a faulty receptacle. Proceed to Step 10.

   b. If there is low or no output at the feed wires, manually move the throttle/governor arm to momentarily produce 4000+ RPM. If there is rated output now, with the engine at high RPM, stop and restart the engine.

   c. Continued rated output at high RPM indicates a faulty rotor. Proceed with further testing. No or low output indicates a faulty condenser. Proceed with further testing.

4. Test AC voltage output at the two light green condenser winding terminals (Fig. HA302). Output should be:

EN2000 . . . . **180V**
EN2500 . . . . **200V**

Readings beyond these specifications may indicate a faulty stator. Stop the engine and proceed with further testing.

5. To test the condenser stator winding, refer to Fig. HA302. Disconnect the two light green wires from the condenser and measure resistance between light green terminals. Specified resistance values are:

Light green to light green, model EN2000: . . . . **3.8-4.6 ohms**
Light green to light green, model EN2500: . . . . **3.5-4.3 ohms**

If resistance reading does not meet specification, the stator is faulty.

6. To test the condenser:
   a. Using a screwdriver or similar tool with an insulated handle, short across the condenser terminals to discharge the condenser.
   b. Using a VOM set on the highest ohm scale, test continuity across the terminals
   c. The meter needle should initially show continuity, momentarily returning to infinity.
   d. A reading other than that described in Step C, especially continuous continuity or continuous infinity, indicates a faulty condenser.

If replacing the condenser, insure that the new condenser is positioned inside the five alignment casting pins on the rear housing, with the terminals UP (Fig. HA303).

7. To test the main stator winding, refer to Fig. HA302; disconnect the red circuit breaker wire and the white receptacle feed wire and measure resistance betwen red and white terminals. Specified resistance values are:

Red to white, model EN2000: . . . . 0.8-1.1 ohms
Red to white, model EN2500: . . . . 0.6-0.8 ohms
Red to stator laminations continuity/infinity
White to stator laminations continuity/infinity

If readings do not meet specification, the stator is faulty.

8. To test for winding shorting, make the following color coded wire resistance tests:

Red to each light green, continuity/infinity
White to each light green, continuity/infinity

Any reading other than those specified indicates a shorted stator.

9. To test the rotor:
   a. Connect one lead of a 12V/15W test light to one terminal of a 12-volt test battery.
   b. Connect the test light assembly probes to the rotor terminals.
   c. Reverse the test light connections.
   d. The test light should light brightly one direction and dim the other direction.

If the light does not light either direction or lights brightly both directions, the rotor has a faulty coil. Replace the rotor.

10. To test the receptacle:
   a. Disconnect either the red or the white feed wire.
   b. Install a jumper wire across one pair of terminals (Fig. HA304).
   c. The receptacle should now test continuity across the feed terminals. Failure of the receptacle to test continuity indicates a faulty receptacle.

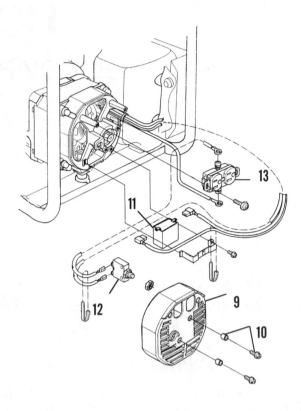

Fig. HA301B - Exploded view of generator components.

  9. End bell/rear housing
10. Bolt & spacer
11. Condenser

12. Circuit breaker
13. AC receptacle

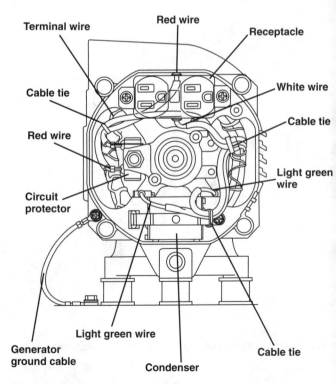

Fig. HA302 – Generator end view (end cover removed) of components and wiring connections.

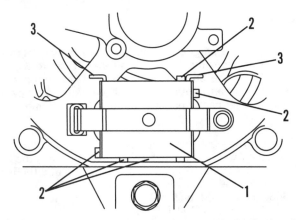

Fig. HA303 – Condenser must be mounted inside the five locator pins, with the terminals UP.

1. Capacitor/condenser
2. Locator pins (5)
3. Terminals (2)

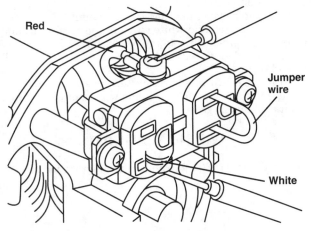

Fig. HA304 – To test the receptacle, install a jumper wire across one pair of terminals, and disconnect at least one feed wire.

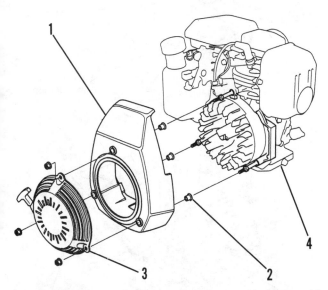

Fig. HA305 – Recoil starter and fan cover must be removed to lock the flywheel for removal of the generator rotor. Refer to text for correct procedure.

1. Fan cover housing
2. Fan cover housing collar (3)
3. Recoil starter assembly
4. Engine

Disassembly

Refer to Figs. HA301A and HA301B for component identification.

To disassemble the generator:

1. Place a temporary support, under the generator front housing.

2. Remove the generator end cover and disconnect the two red circuit breaker wires.

3. Disconnect the red and white receptacle wires and the two light green condenser wires.

4. Remove the rear housing/stator throughbolts and the rubber housing mount.

5. Remove the rear housing assembly from the stator, noting the stator wire routing.

6. Carefully remove the stator assembly. Do not allow the stator windings to be damaged while handling the stator.

7. To remove the rotor:

    a. Place some protective padding, such as a shop rag, onto the generator frame crossmember, below the rotor.

    b. Remove the recoil starter assembly and fan cover from the engine (Fig. HA305).

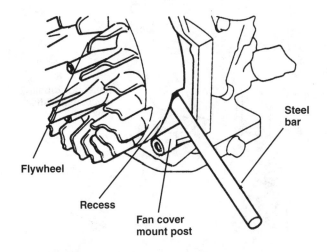

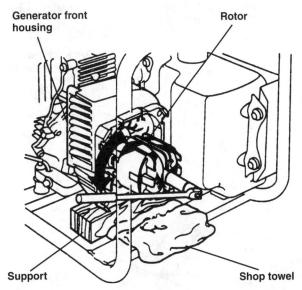

Fig. HA306 – View showing flywheel locked for rotor removal. Refer to text for correct procedure.

c. Lock the flywheel by placing a steel bar between the flywheel back side recess opposite the magnets and the top side of the lower right fan cover mount post (Fig. HA306).

d. While holding the flywheel bar, place a wrench on the end of the rotor shaft and, applying force in a counter clockwise direction, break the threaded rotor shaft loose from the engine crankshaft. Do not strike the rotor.

e. Remove the rotor being careful not to damage the rotor winding or fan.

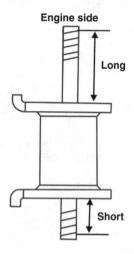

Reassembly

NOTE: If the engine and rubber mounts were also removed from the frame for service, refer to Fig. HA307 for the correct positioning of the mounts. A-type has an engine mounted gas tank. AL-type has a larger capacity tank mounted to the frame.

1. Support the generator front housing.
2. Lock the flywheel by placing a steel bar between the flywheel back side recess opposite the magnets and the bottom side of the lower right fan cover mount post (Fig. HA308). Thread the motor shaft onto the crankshaft and tighten to 22.0 N•m (190-195 in.-lb.).
3. Fit the stator assembly over the rotor and against the front housing.
4. Feed the stator wires through the proper opening in the rear housing, then fit the housing to the stator.
5. Install the rear housing/stator throughbolts and tighten to 11.0 N•m (95-100 in.-lb.).
6. Install the housing to frame rubber mount.
7. Connect the red and white wires to the receptacle, and the two light green wires to the condenser.
8. Reconnect the two red circuit breaker wires, then install the end cover.
9. Install the fan cover and recoil starter assembly onto the engine, making sure the spark plug wire is properly fitted into its alignment groove.

WIRING DIAGRAM

Refer to Fig. HA309 for the wiring diagram for this series.

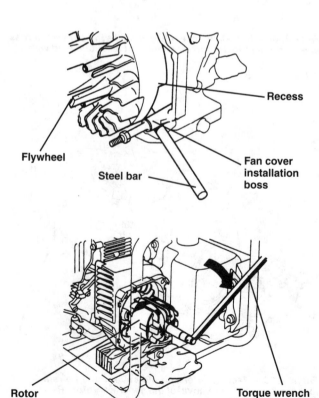

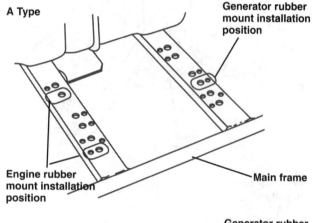

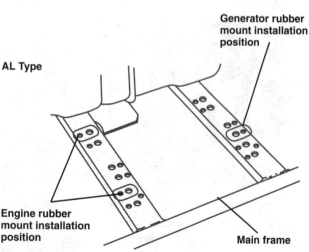

Fig. HA307 – View of correct installation of rubber frame mounts.

Fig. HA308 – View showing flywheel locked for rotor installation. Refer to text for correct procedure.

GENERATOR BLOCK

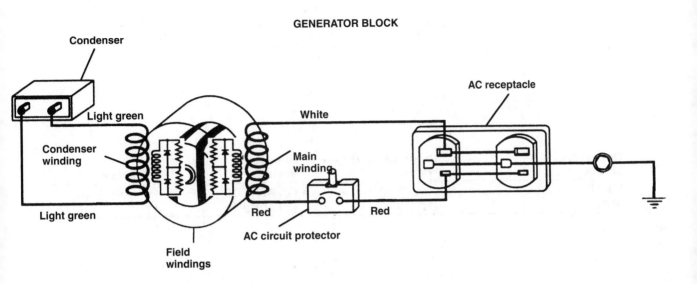

ENGINE BLOCK

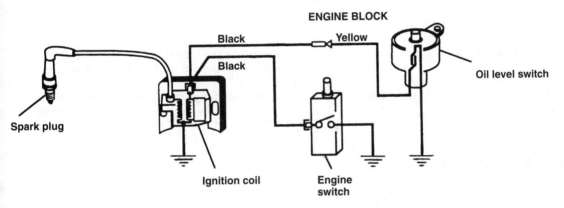

Fig. HA309– Wiring diagram.

HONDA

EU2600i and 3000is Brushless Generators

Generator Model	Rated (Surge) Watts	Output Volts	DC Amps
EU2600i	2400 (2600)	120 AC/12 DC	10
EU3000is	2800 (3000)	120 AC/12 DC	12

IDENTIFICATION

These generators are direct drive, brushless, 60 Hz, single phase units with both 120VAC and 12VDC output available from panel mounted receptacles.

Refer to Fig. HA351A and HA351B to view the generator components.

WIRE COLOR CODES

Honda wiring schematics use the following color code abbreviations to identify the generator wiring in this series:

Bl – Black	Bu – Blue	W – White
R – Red	G – Green	Lb – Light blue
Y – Yellow	Br – Brown	Lg – Light green
Gr – Gray	P – Pink	O – Orange

GENERATOR REPAIR

TROUBLESHOOTING

Before testing any electrical components, check the integrity of the terminals, connectors, and fuse. While testing individual components, also check the wiring integrity. Resistance readings may vary with temperature and test equipment accuracy. All operational testing must be performed with the engine running at its governed no load speed of 3500 RPM, unless specific instructions state otherwise. Always load test the generator after repairing defects discovered during troubleshooting.

To troubleshoot the generator:

1. Start the engine and verify the following:
   a. ECO switch is OFF,
   b. Overload indicator light is OFF,
   c. DC circuit breaker IN (circuit breaker will be covered in Step 5).
2. To test the AC receptacle:

a. Disconnect either the red or the white feed wire.
b. Install a jumper wire across one pair of terminals (Fig. HA352).
c. The receptacle should now test continuity across the feed terminals. Failure of the receptacle to test continuity indicates a faulty receptacle.

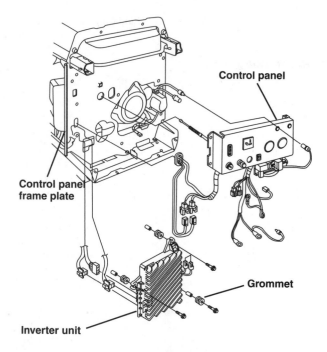

Fig. HA351A – View of generator components. Electric start components are not used on EU2600 series.

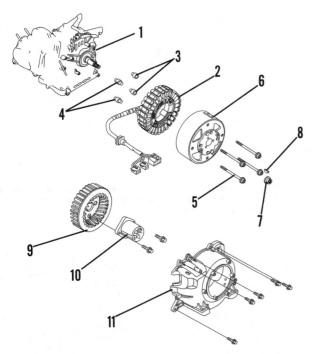

Fig. HA351B – Exploded view of generator components.

1. Engine
2. Stator
3. Stator collar
4. Stator collar
5. Throughbolt
6. Rotor
7. Nut
8. Woodruff key
9. Fan
10. Starter clutch
11. Fan cover

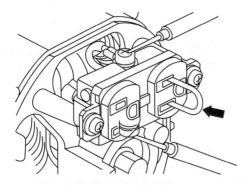

Fig. HA352 – To test the receptacle, install a jumper wire across one pair of terminals, and disconnect at least one feed wire. There should then be continuity across the feed terminals.

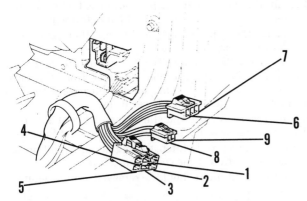

Fig. HA353 – View of stator connectors unplugged for winding tests. Refer to text for terminal identification and the correct test procedure.

3. Remove the air filter cover and filter, then open the maintenance panel to access the harness connectors.
4. To test for AC output:
   a. Remove the 6-pin connector from the harness.
   b. Start the generator.
   c. Refer to Fig. HA353.
   d. Test the main coil AC voltage between the red wire paired terminals 1-2, 2-3, and 3-1. Voltage should be 182-218 volts at each pair.
   e. Test the sub coil AC voltage between the gray wire terminal pair 4-5: Voltage should be 13-17 volts.
   f. Stop the generator.
   g. Using the same three red wire terminal pairs as Step d, pull the starter rope and note the voltage readings. Output should read 18VAC minimum on each pair.
   h. Using the same gray wire terminal pair as Step 4e, pullthe starter rope and check the voltage readings. Output should read 1VAC minimum.
   i. Using an ohmmeter and the same three red wire terminal pairs as Steps 4d and 4g, test the winding resistance. EU2600 series should test 1.4 ohms at each pair. EU3000 series should test 1.1 ohms at each pair.
   j. Using an ohmmeter and the same gray wire terminal pair as Steps 4e and 4h, test the winding resistance. EU2600 series should test 0.08 ohm. EU3000 series should test 0.07 ohm.
   k. Test for continuity between a ground position on the engine block and each of the five previously noted red and gray wire terminals. There should be infinity/no continuity.
   l. If one or more, but not all, of the results of test Steps 4d, e, g, or h failed to meet specification, or if any of the results of test Steps i, j, or k failed to meet specification, the stator windings are faulty. Replace the stator.
   m. If all of the results of test Steps 4d, e, g, and h tested equally low, the rotor magnetism is weak. Replace the rotor.
   n. If all of the preceding test results met specification but there is still no AC output the inverter is faulty. Replace the inverter.
5. To test for DC output:
   a. Access the back of the DC circuit breaker/receptacle and unplug the harness connector, then depress the reset button. If the button does not stay in, the receptacle is faulty. If the button stays in:
   • Place a jumper wire between the receptacle output terminals (Fig. HA354).
   • Test for continuity between the input terminals. Continuity indicates a good receptacle.
   • Test for continuity between the input terminal fed by the White/red wire and the vertical output terminal. Continuity indicates a good breaker.
   b. Access the diode rectifier assembly (Fig. HA355) and disconnect the harness. Using a VOM set to the R × 1 scale, test continuity between any one pair of adjacent or diagonal terminals (1-2, 1-3, 2-4, 3-4, 1-4, or 2-3). Note the reading. Reverse the meter leads and again note the reading. Repeat the test with the other terminal pairs. Compare the readings with the chart: Any variation indicates a faulty rectifier. The 1 (+) to 2 (-) continuity reading should be approximately 90 ohms. All other continuity readings should be approximately 18 ohms.

NOTE: Depending on meter polarity, continuity/infinity readings may be exactly opposite those shown in the chart. If that is the case, reverse the (+) and (-) probes in the chart.

c. Remove the 3-pin (2-black/yellow, 1-blank) connector from the wiring harness (Fig. HA353 – terminals 6 and 7). Using a VOM set to the R × 1 scale, measure the resistance between the two black/yellow wire terminals. Reading should be:

EU2600	0.1 ohm
EU3000	0.09 ohm

d. Measure the resistance between each black/yellow wire and a ground position on the engine block: There should be infinity/no continuity. Set the VOM to the lowest AC voltage scale and, while pulling the engine starter rope, test the output of the two black/yellow wires. The minimum reading should be 1 volt.

e. If the test results met specification, any fault is in the wiring. Repair or replace as necessary. If the test results do not meet specification and the wiring integrity is secure, the DC stator winding is faulty. Replace the stator.

6. To test for battery charge output (EU3000 series):

a. Access and disconnect the 2-pin stator connector (Fig. HA353 – Terminals 8 and 9 – both brown wires).

b. Set the VOM to the lowest AC voltage scale and, while pulling the engine starter rope, test the output of the two brown wires. The minimum reading should be 1 volt.

c. Set the VOM on the R × 1 scale, then test the resistance of the two brown wires. The reading should be 0.09 ohm.

d. With the VOM still set on the R × 1 scale, measure the resistance between each brown wire and a ground position on the engine block. There should be infinity/no continuity.

e. If any of test steps did not meet specification, the stator charge winding is faulty. Replace the stator.

7. To test the battery charge regulator/rectifier, refer to Fig. HA356. Test the resistance of the numbered terminals, then compare the readings with the chart. Any variation indicates a faulty rectifier.

8. To test the ECO throttle (idle control) system:

a. Continuity test the ECO switch terminals. There should be continuity in the ON position only.

NOTE: The ECO throttle switch should always be installed in the control panel with the ON position up.

b. Access the throttle control motor mounted on top of the carburetor, and test resistances between terminals 1-3 and terminals 2-4 (Fig. HA357). Resistance should be 50-70 ohms at each noted terminal pair. Resistance readings beyond this range indicate a faulty motor. After testing or replacing the throttle motor, make certain that the motor harness is routed correctly (Fig. HA358).

c. Move the throttle lever and check the ECO motor for smooth operation. If binding is noted, disconnect the throttle linkage to determine the source of the binding; correct as necessary.

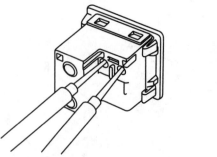
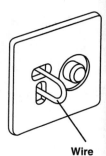

Fig. HA354 – Testing the DC breaker/receptacle for continuity by installing a jumper wire across the outlet terminals. Refer to text for the correct test procedure.

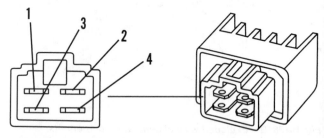

(+)Probe (-)Probe	1	2	3	4
1		∞	∞	∞
2	Continuity		Continuity	Continuity
3	Continuity	∞		∞
4	Continuity	∞	∞	

Fig. HA355 – Terminal test positions for the diode rectifier. Refer to text for the correct test procedure.

Disassembly

Refer to Figs. HA351A and HA351B to view the essential generator components.

To disassemble the generator:

1. Remove the outer sheet metal cover panels.

2. Disconnect the control panel harnesses. Remove the control panel and the inverter unit. Pay close attention to the location of the ground wire locations.

3. Drain the fuel system, then remove the fuel tank, front handle, tank supports, and control panel frame plate.

4. Remove the upper rear shroud frame, then remove the muffler and exhaust system.

5. Remove the flange nuts holding the generator feet to the rubber mounts, then lift the generator from the lower frame, using care when removing the harness wires from the grommeted holes in the lower frame.

6. If necessary, remove the starter assembly at this time. Otherwise, remove the fan cover seal plate and fan cover.

7. Remove the starter clutch, cooling fan and ignition coil, noting the position of the coil mount bolts.

8. The rotor and stator should be removed as an assembly:

NOTE: Striking the rotor will damage it. Handle the rotor carefully.

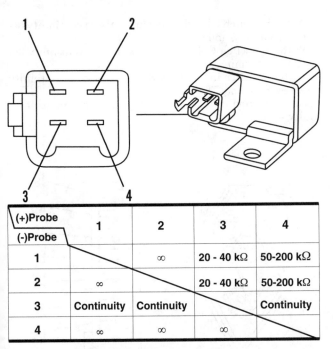

(+)Probe / (-)Probe	1	2	3	4
1		∞	20 - 40 kΩ	50-200 kΩ
2	∞		20 - 40 kΩ	50-200 kΩ
3	Continuity	Continuity		Continuity
4	∞	∞	∞	

Fig. HA356 – Terminal test positions for the regulator/rectifier. Refer to text for the correct test procedure.

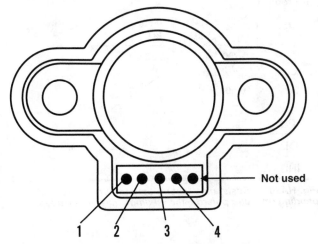

Fig. HA357 – Terminal test positions for the ECO throttle control motor. Refer to text for the correct test procedure.

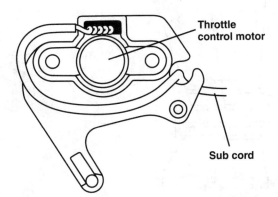

Fig. HA358 – View showing the correct routing of the throttle control motor harness.

a. Using a strap wrench to hold the rotor, remove the 14 mm right hand threaded nut holding the rotor to the engine crankshaft.

b. Rotate the rotor so the stator mount bolts are accessible through the rotor face slots. Remove the bolts.

c. Use a 3-slot puller attached to the threaded holes in the rotor hub face to break the rotor free from the crankshaft. Then remove the rotor/stator assembly. Do not use a jaw style puller to pull on the outer rotor diameter.

Reassembly

To reassemble the generator:

1. Make sure:

a. The matching tapers on the engine crankshaft and inside the rotor shaft are clean and dry.

b. The stator collars and collar mount holes are clean and dry;

c. Be certain there is no metallic foreign material stuck to the rotor magnets; and

d. The crankshaft woodruff key is properly positioned in its slot.

2. To install the rotor and stator as an assembly, fit the rotor over the stator, aligning the rotor face slots with the stator mount bolt holes.

3. Properly position the four stator collars in the rear of the stator:

a. The two collars (4–Fig. HA351B) with shoulders on both sides of the collar go with the holes on either side of the stator harness, positioned toward the engine cylinder head.

b. The two collars (3–Fig. HA351B) with shoulders on only one side of the collar mount in the stator holes away from the harness and head, with the shoulder side toward the engine.

4. Fit the stator/rotor assembly onto the engine, ensuring that the stator mount bolt holes and collars align with the engine posts and the rotor keyway aligns with the crankshaft key.

5. Lightly oil the 14 mm rotor nut threads. Hold the rotor with a strap wrench, and torque the nut 74 N•m (53 ft.-lb.).

6. Install the stator mount bolts. Torque the bolts 11 N•m (95-100 in.-lb.).

7. With the spark plug removed, rotate the rotor to make sure the rotor magnets do not contact the stator. In the event of interference, remove the rotor/stator assembly and install the two double shouldered collars in the two stator locations nearest the point of interference.

8. Install the ignition coil and set the coil gap to 0.3-0.5 mm (.012-.020 in.).

9. Install the cooling fan and starter clutch. For the fan to be properly positioned against the rotor, the small locator fin of the fan must align with the small triangular hole of the rotor.

10. Install the fan cover seal plate and fan cover.

11. If the starter assembly was removed, install it at this time.

12. Carefully set the generator assembly onto the lower frame while feeding the harness wires through their respective grommeted frame holes. Be sure the fan cover seal fits onto the lower cover flange, not against the edge of the flange (Fig. HA359). Torque the generator feet flange nuts 24 N•m (210 in.-lb.).

13. Install the exhaust system and muffler, then install the upper rear shroud frame. Ensure that the fan cover seal fits under the frame, not against the edge of the frame (Fig. HA360).

14. Install the control panel frame plate, fuel tank supports, front handle, and fuel tank.

15. Install the control panel and the inverter unit, then reconnect the control panel harnesses. Insure that the M6x20 flange bolt fastens the single wire ground terminal to the threaded upper boss on the cylinder casting; the M6x28 hex bolt assembly must fasten the double wire ground terminal to the lower frame rail.

16. Install the outer sheet metal cover panels.

ENGINE

To test the ignition control unit
1. Access the control unit and refer to Fig. HA361.
2. Test the resistance of the terminal pairs according to the accompanying chart; note the readings.
3. Depending on meter polarity, the test reading results may be reversed.
4. If resistance readings are beyond the chart specifications, the ignition unit is faulty; replace the unit.

IGNITION SWITCH

To test the ignition switch on electric start units:
1. Disconnect the harness and refer to Fig. HA362.

2. With the switch OFF, there should only be continuity between IG and E, and between G and FS.
3. With the switch ON, there should not be continuity between any terminals.
4. With the switch in Start, there should only be continuity between BAT and ST.
5. Any variation in readings of Steps 2, 3, or 4 indicates a faulty switch.

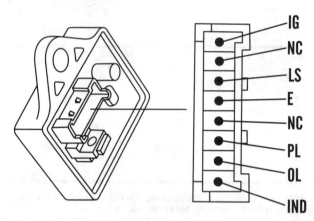

(+)Probe (-)Probe	IG	LS	E	PL	OL	IND
IG		10-50 kΩ	10-50 kΩ	∞	∞	∞
LS	10-200 kΩ		10-200 kΩ	∞	∞	∞
E	10-50 kΩ	10-50 kΩ		∞	∞	∞
PL	∞	∞	∞		∞	∞
OL	∞	∞	∞	∞		∞
IND	∞	∞	∞	∞	∞	

Fig. HA361 – Resistance testing the ignition control unit. Depending on meter polarity, the readings may be reversed.

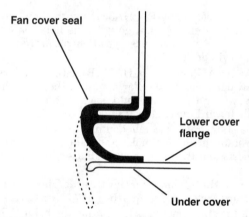

Fig. HA359 – For proper generator cooling, the lip of the fan cover seal must fit flat against the lower frame cover, not against the face edge of the cover.

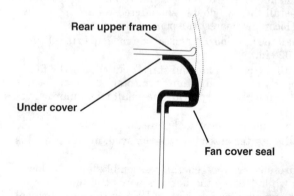

Fig. HA360 – For proper generator cooling, the rear upper frame flange must fit flat over the fan cover seal, not with the flange edge against the back of the seal. against the face edge of the cover.

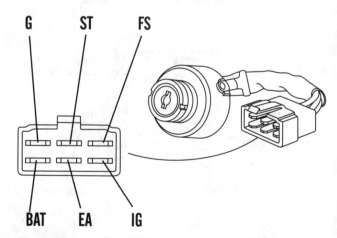

Fig. HA362 – Ignition switch test terminal identification. Refer to text for the correct test procedure.

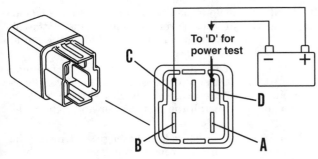

Fig. HA363 – *Starter relay test terminal identification. Refer to text for the correct test procedure.*

STARTER RELAY

To test the starter relay on electric start units:

1. Access the relay, disconnect the harness, and refer to Fig. HA363, but do not connect the battery until Step 4.

2. Test continuity between Terminals A and B; there should be no continuity.

3. Test continuity between Terminals C and D; there should be 89-111 ohms.

4. With a fully charged 12-volt test battery connected to the relay – (+) to C, (-) to D. Retest for continuity between A and B. There should now be continuity.

5. Any test results except those noted in Steps 2, 3, or 4 indicates a faulty relay.

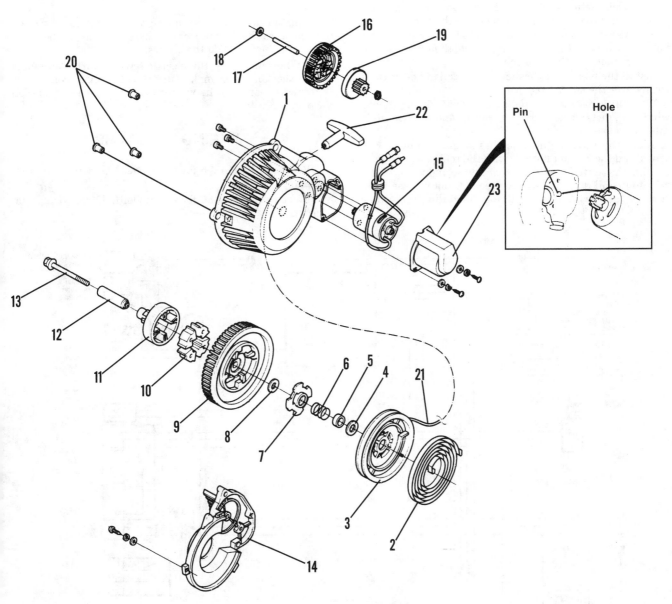

Fig. HA364 – *Exploded view of combination electric and recoil starter assembly.*

1. Starter case	9. Rewind gear	17. Ring gear and bendix shaft
2. Rewind spring	10. Damper (4)	18. Ring gear and bendix shaft washers (2)
3. Reel/pulley	11. Drive cam	19. Bendix gear
4. Reel washer	12. Cam collar	20. Starter mount collars
5. Reel collar	13. Left-handed rewind gear bolt	21. Rope
6. Friction spring	14. Gear cover	22. Grip
7. Ratchet guide	15. 12V starter motor	23. Motor cover
8. Rewind-gear washer	16. Motor to rewind ring gear	

ELECTRIC/REWIND STARTER COMBINATION SERVICE

1. Remove the starter assembly and refer to Fig. HA364. Do not lose the three mount collars.

2. Disassemble and clean as necessary.

NOTE: The center bolt is left hand thread.

3. Inspect all of the gear ratchet components for wear, and the ratchet assemblies for proper operation. Renew as necessary.

4. Inspect the return spring end hooks and the hook tangs on the pulley reel and inside the starter case.

5. With the rope and spring fitted to the reel, lightly coat the rotating friction surfaces of the reel bushing and collar with a light grease, then fit the reel into the case.

6. Set the return spring tension approximately two turns. Feed the rope end through the case hole; then tie a temporary slip knot into the rope end.

7. Install the metric flat washer, metric collar, and friction spring over the starter reel post.

8. With the convex side of the ratchet guide facing the rewind gear, install the metric flat washer and ratchet guide into the gear.

9. Align the ratchet guide notches with the pulley reel tangs, and the ratchet notches with the reel notches; while turning the rewind gear counterclockwise, lightly push the gear to install it onto the pulley reel. Do not release pressure on the rewind gear until after installing the center bolt.

10. Position the four dampers onto the rewind gear. Hand press the drive cam over the dampers; install the collar into the drive cam; then install and tighten the center bolt into the starter case.

11. Locate and identify the alignment hole in the gear end faceplate of the starter motor. Position the hole over the locator pin on the starter case. Apply RTV silicone sealer to the flange. Then install the motor cover and tighten the fasteners.

12. While rotating the gears to align the ratchet, install the bendix gear into the starter motor ring gear.

13. Install the bendix gear shaft into the case, followed by one 6 mm flat washer, the ring and bendix assembly (bendix gear in toward the case), and the second 6 mm washer.

14. Install the gear cover and tighten the six cover screws. Make sure the spring lock washers and flat washers are correctly mounted on the screws.

15. When installing the electric/rewind starter assembly onto the fan cover, make sure that the three collars are properly positioned between the starter case and the fan cover.

WIRING DIAGRAMS

Refer to Fig. HA365 for the model EU2600 wiring diagram.

Refer to Fig. HA366 for the model EU3000 wiring diagram.

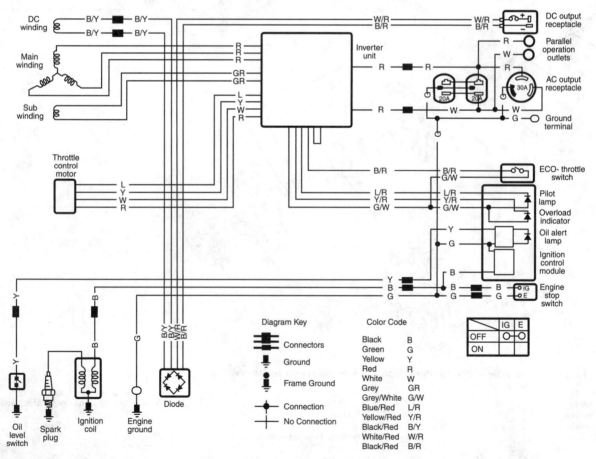

Fig. HA365 – Wiring diagram for model EU2600i, Type A generator.

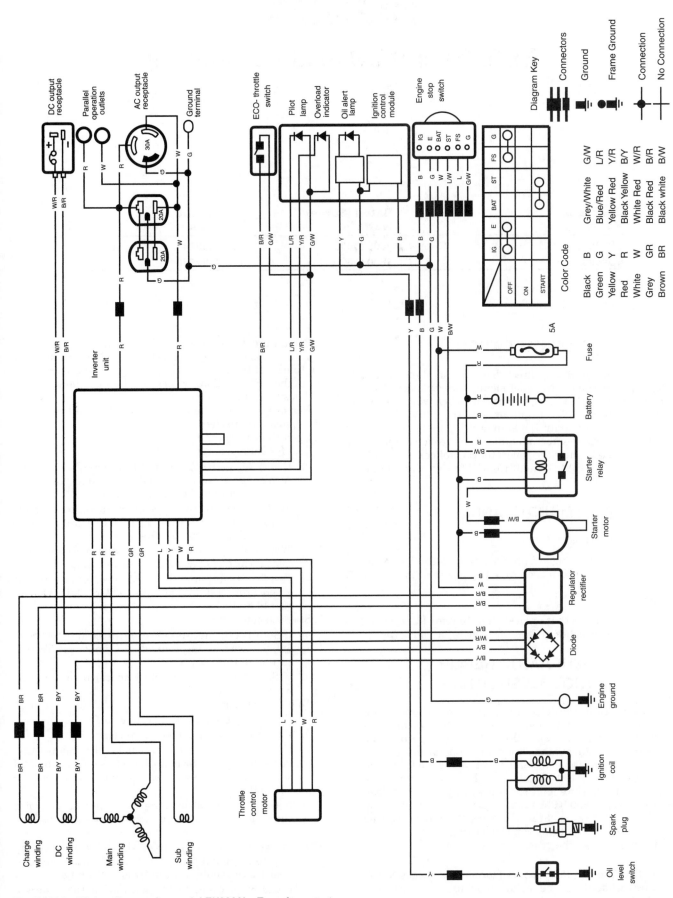

Fig. HA366 – Wiring diagram for model EU3000is, Type A generator.

HONDA

EB3000 and EB4000 Generators

Generator Model	Rated (Surge) Watts	Output Volts
EB3000	2800 (3000)	120/240
EB4000	3800 (4000)	120/240

IDENTIFICATION

These generators are direct drive, brush style, 60 Hz, single phase units with panel mounted receptacles. They are equipped with a voltage selection switch which allows combined 120/240V output or strictly 120V output. auto throttle (idle control) is standard.

The unit serial number is stamped on the side of the engine crankcase, below the generator control box.

Refer to Figs. HA401A and HA401B to view the generator components.

MAINTENANCE

ROTOR BEARING

The endbell bearing is a prelubricated, sealed ball bearing. Inspect for smooth operation. Replace as needed.

BRUSHES

Replace brushes that are less than 13 mm (1/2 in.) long. To access the brushes, remove the generator end cover.

GENERATOR REPAIR
TROUBLESHOOTING

Before testing any electrical components, check the integrity of the terminals and connectors. While testing individual components, also check the wiring integrity. Resistance readings may vary with temperature. The resistance specifications listed in this chapter are for a temperature of 68° F (20° C). All operational testing must be performed with the engine running at its governed no load speed of 3720-3850 RPM, unless specific instructions state otherwise. Always load test the generator after repairing defects discovered during troubleshooting.

To troubleshoot the generator:

NOTE: The positive (+) brush terminal is always the one closest to the rotor bearing, and has a light green with white wire.

1. Circuit breaker must be ON. To test the AC breaker, access the breaker wiring and test the breaker terminals for continuity. If the breaker will not stay set, shows continuity in the OFF position, or does not show continuity in the ON position, the breaker is faulty.

2. Remove the generator end cover, but leave all wires and components connected.

3. Start the generator.

4. Test for voltage output:

   a. Using the 4-wire AC output terminal block, test for AC voltage at the main winding 1 (red to white) connections. Output should be 110-130V.

   b. Again using the 4-wire terminal block, test for AC voltage at the main winding 2 (blue to gray) connections. Output should be 110-130V.

   c. Test for AC voltage at the sub winding connections (black/white to black/white). Output should be 4-6V.

   d. Test for DC output at the brush wire terminals (light green/white [+] to light green/black [−]). Output should be 12-18V.

5. Stop the generator:

   a. If Step 4 tests met specification, proceed to test Step 6-J, followed by the Individual Component Testing section in this chapter.

   b. If Step 4 tests did not meet specification, proceed with Step 6-A.

6. To test the rotor:

   a. Disconnect the AVR 4-pin connector.

   b. Disconnect the brush wire connectors.

   c. Using an ohmmeter, measure field coil winding resistance through the brushes. Resistance should be 71.8 ohms on the EB3000 series, 50.3 ohms on the EB4000 series.

   d. Measure field coil winding resistance at the rotor slip rings. The resistance should be the same as Step c.

   e. Measure field coil winding resistance from each slip ring to a ground on the rotor shaft or laminations. Resistance should be infinity/no continuity.

   f. If test Steps c and d met specification, the rotor and brushes are good.

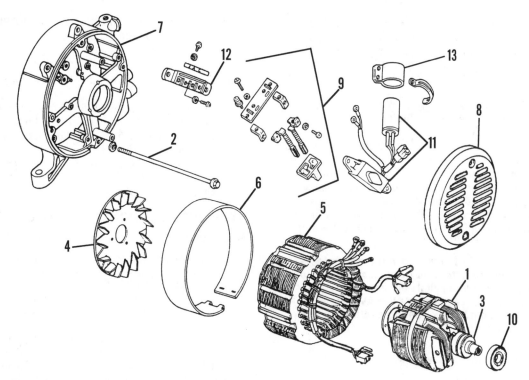

Fig. HA401A – Exploded view of generator components. See Fig. HA401B also.

1. Rotor assembly
2. Rotor through-bolt and washer
3. Slip rings
4. Fan

5. Stator assembly
6. Stator cover
7. Rear housing
8. End cover

9. Brush-holder assembly
10. Rotor bearing
11. AVR assembly with capacitor/condenser
12. AC output terminal strip
13. Capacitor clamp

g. If test Step c failed specification but Step D met specification, the rotor is good. Test and inspect the brushes; replace as necessary.

h. If both test Steps c and d failed specification, the rotor is faulty and must be replaced.

i. If Step e test reads any continuity, the field coil winding is shorted. Replace the rotor.

7. To test the stator:

a. Connect a 12-volt battery to the brush connector terminals.

b. With the terminals still disconnected from the Step 6 test, connect the positive battery post to the light green/white (+) brush holder terminal, then connect the negative battery post to the light green/black (-) brush holder terminal.

NOTE: Improper brush connections will damage the generator.

c. Start the generator, verifying the correct 3720-3850 RPM no load speed.

d. Test the stator output AC voltages using the following test connections. Note the readings.

• Red to white main winding 1 and blue to gray main winding 2 should each read 75-95 volts on the EB3000 series and 85-105 volts on the EB4000 series.

• Black/white to black/white sub winding should read 3-5 volts.

• Light green to white sensor winding should read 10-16 volts.

• Light green/red to green exciter winding should read 35-65 volts.

e. Stop the generator.

f. Using an ohmmeter, measure the resistance of the following test connections. Note the readings.

• The red to white and the blue to gray main winding connections should each read 1.15 ohms on the EB3000 series and 0.55 ohm on the EB4000 series.

• The black/white to black/white sub winding terminals in the 4-pin connector should read 0.23 ohm on the EB3000 series and 0.18 ohm on the EB4000 series.

• The light green/red to green exciter winding terminals in the 4-pin connector should read 2.01 ohms on the EB3000 series and 2.40 ohms on the EB4000 series.

• Connect each red, white, blue, and gray main winding to a good ground on the stator laminations. Resistance should be infinity/no continuity.

• Connect each black/white sub winding terminal to a good ground on the stator laminations. Resistance should be infinity/no continuity.

• Each light green/red and green exciter winding terminal to a good ground on the stator laminations. Resistance should be infinity/no continuity.

g. If all of the tests in Steps d and f met specification, the stator is good, proceed with further component testing. If the component tests meet specification, the automatic voltage regulator (AVR) is faulty and must be replaced. See the NOTE which follows Step k.

h. If any individual test in Steps d or f failed to meet specification, the stator is faulty and must be replaced.

i. Reconnect all previously disconnected wiring and components.

j. Attach DC voltmeter probes to the brush leads

k. Start the generator and immediately test brush voltage; voltage should test 21-29 VDC. If voltage is outside this range, immediately stop the generator. The AVR is faulty. Replace the AVR and repeat this step.

NOTE: If the AVR is faulty and requires replacement, make sure that the voltage selector switch is tested prior to generator start up, as a faulty voltage selector switch can damage the AVR. Refer to the *Individual Component Testing* section for the selector switch test.

8. To test the auto throttle system (Fig. HA402):

NOTE: The auto throttle system will not detect loads below 1.0 amp. If less than a 120-watt load is applied to the generator at 120V (240-watt load at 240V), the auto throttle switch must be OFF so the unit will run at full throttle governed RPM to produce proper voltage.

a. Turn the auto throttle switch OFF.

b. Start the generator; make sure all loads are disconnected.

c. Move the switch to AUTO. After a moment, the engine should idle down to 2100-2300 RPM.

d. Return the switch to OFF. The engine should return to no load full throttle.

e. If the auto throttle system fails Steps c or d, proceed to the Individual Component Testing section.

f. Stop the generator.

INDIVIDUAL COMPONENT TESTING

Receptacles

The generator must be shut down to test the receptacles.

1. Access the receptacle connections inside the control box.

2. Disconnect the RETURN (white) feed wire from the receptacle being tested.

3. Install a jumper wire between the HOT and RETURN outlet terminals (Fig. HA403).

   a. On 120V straight pin and twistlok receptacles, HOT terminals will be either red or blue feed wires. RETURN terminals will be the white feed wire.

   b. On 240V twistlok receptacles, HOT terminals will be the red and blue feed wires. RETURN terminals will be the white feed wire.

4. Test for continuity across the feed terminals. No continuity indicates a faulty receptacle.

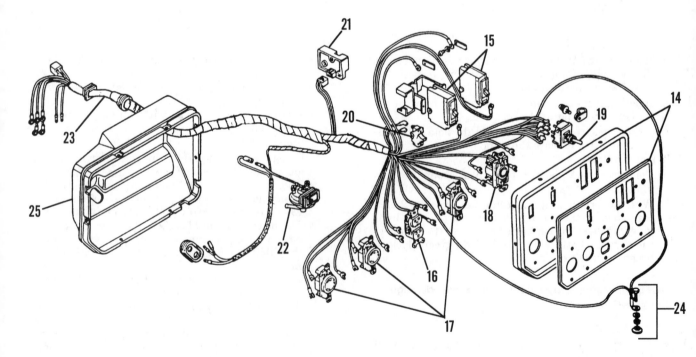

Fig. HA401B– Exploded view of generator components. See Fig. HA401A also.

14. Control panel and faceplate	18. 240V AC Twistlok receptacle	22. Engine switch
15. AC circuit breakers	19. Voltage selector switch	23. Control box harness
16. 120V AC receptacle	20. Auto throttle switch	24. Ground terminal
17. 120V AC Twistlok receptacles	21. Auto throttle control unit	25. Control box rear cover

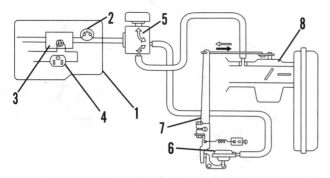

Fig. HA402 – System function view of auto throttle control components.

1. Control box
2. Auto-throttle switch
3. Auto-throttle control unit
4. AC receptacle
5. 3-way solenoid valve
6. Idle-down diaphragm
7. Governor arm
8. Carburetor

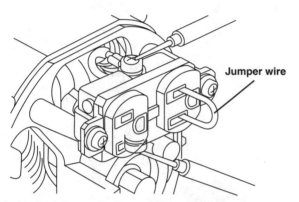

Fig. HA403 – To test the receptacle, disconnect the white feed wire and install a jumper wire across the output terminals. There should be continuity across the feed terminals.

Voltage Selector Switch

Using the switch positions and numbered switch terminals in Fig. HA404, test for continuity between the connected terminals. The failure of any terminal pair to show continuity indicates a faulty switch.

The stator's two main windings, MW1 and MW2 (Figs. HA405 and HA406), are independent of each other but equal in capacity. With the voltage selector switch in the 120/240V position, the output is arranged in series circuits (Fig. HA405). With the switch in 120V only, the output is arranged in a parallel circuit (Fig. HA406). The AVR sensor circuit is also controlled by the switch, so the AVR will accurately control voltage in either switch position. When doing generator output testing, keep in mind that a faulty voltage selector switch could also possibly damage the AVR.

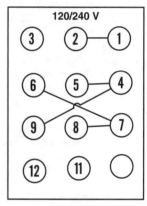
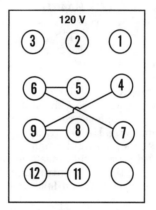

Fig. HA404 – Continuity test positions for the voltage selector switch. There should be continuity between the noted terminals.

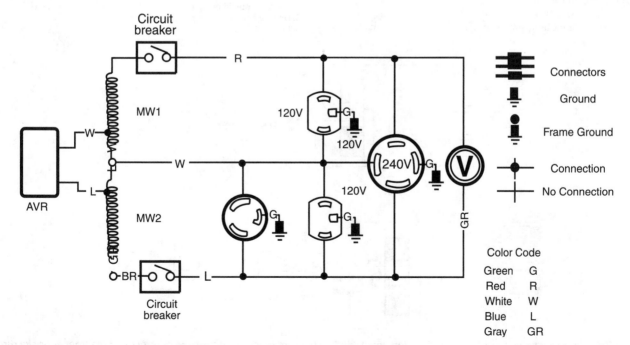

Fig. HA405 – Output wiring with the voltage selector switch in the 120/240V position.

Auto Throttle System

1. To test the auto throttle switch:
   a. Access the switch terminals inside the control box.
   b. With the switch lever UP in the AUTO position, there should be continuity between the two lower terminals on the back of the switch.
   c. With the switch lever DOWN in the OFF position, the two bottom switch terminals should show no continuity.
   d. Any results different from those specified indicate a faulty switch.
2. To test the diaphragm assembly (Fig. HA407):
   a. Apply suction to the vacuum line fitting; if the diaphragm is good, the link will draw down.
   b. Continue to hold suction for approximately one minute to check for pinhole leakdown.
   c. Inspect all vacuum hoses for leaks.
3. To test the three-way solenoid valve (Fig. HA407):
   a. Disconnect the black and yellow harness wires from the solenoid valve.
   b. Test the solenoid wires for continuity. No continuity indicates a faulty solenoid.
4. To test the control unit:
   a. Access the control unit inside the control box.
   b. Verify that the blue and white wires from the stator main windings are passing through the control unit hole from opposite directions (Fig. HA408). If necessary, redirect the wires. Wires passing through the control unit from the same direction will cancel each other's magnetic field, preventing the control unit from sensing voltage differences between load and no load.
   c. If all other auto throttle components test good, and the control unit wires are routed correctly, but the auto throttle still does not operate, the control unit is faulty.

Engine Switch

To test the switch:
1. Access the switch inside the control box.
2. Disconnect the black harness wire from the switch.

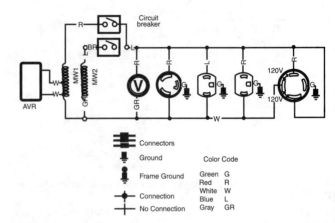

Fig. HA406 – Output wiring with the voltage selector switch in the 120V-only position.

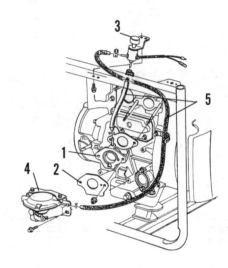

Fig. HA407 – View of the auto throttle control vacuum components.

1. Intake manifold/insulator
2. One-way manifold gasket
3. 3-way solenoid valve
4. Idle-down diaphragm assembly
5. Vacuum lines

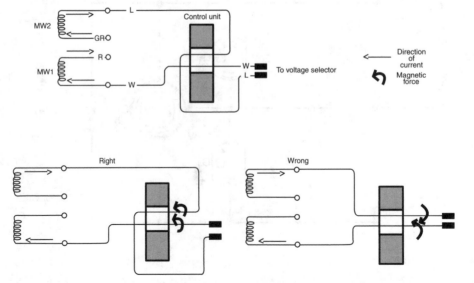

Fig. HA408 – For proper auto throttle control function, main winding wires must pass through the control unit hole from opposite directions.

L. Blue   W. White

3. Test for continuity between the black switch wire and ground. There should be continuity in the OFF position only.

4. Turn the switch ON.

5. Lightly apply air pressure to the switch diaphragm fitting; the diaphragm should trip the switch OFF. (During engine operation, the Oil Alert unit inside the engine crankcase provides crankcase pressure to activate the switch if the oil level becomes low.)

6. Failure of either test Step 3 or 5 indicates a faulty switch.

Condenser/Capacitor

The condenser is a part of the AVR and is not repairable separately.

Automatic Voltage Regulator (AVR)

There is no test procedure available for the AVR. If the rotor, stator, receptacles, breakers, and harnesses all meet specification in the **Troubleshooting** tests, but there is still faulty output, replace the AVR.

DISASSEMBLY

Refer to Figs. HA401A and HA401B for component identification. Always allow the generator to cool down before disassembly. Note that the rotor cannot be removed without first removing the stator. Note the wire routing orientation and the location of any wiring retainers. Exercise caution when handling the stator and rotor so the windings and winding coatings are not damaged.

1. Place a temporary support, such as a 2 × 4, under the engine stator mount housing.

2. Remove the generator end cover.

3. Remove the nuts holding the rear housing feet to the rubber mounts. Note the position of the grounding cable on the foot below the control box.

4. Disconnect the wiring from the rear housing components.

5. Remove the control box from the side of the generator cradle frame.

6. Separate the control box face panel from the rear cover.

7. If replacing the stator, unbend the tabs holding the stator cover to the stator and remove the cover.

8. Remove the brush holder assembly from the rear housing.

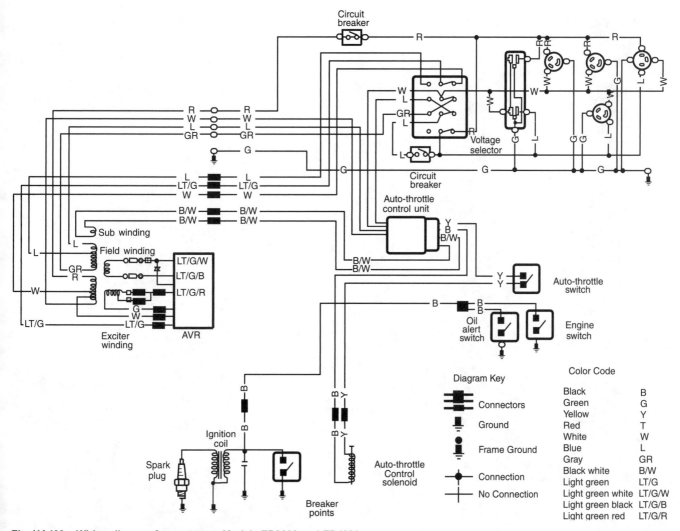

Fig. HA409 – Wiring diagram for generator Models EB3000 and EB4000.

9. Remove the rear housing/stator throughbolts.
10. Remove the rear housing.
11. Remove the stator, noting the wiring orientation.
12. Remove the rotor throughbolt.
13. Using a slide hammer and a threaded adapter, break the rotor free from the taper on the engine crankshaft. Do not allow the rotor to fall once it is loose.
14. Remove the cooling fan from the rotor.

REASSEMBLY

1. Place a temporary support, such as a 2 × 4, under the engine stator mount housing.
2. Service the rotor bearing, if needed.
3. Mount the cooling fan to the rotor.
4. Insure that the matching tapers on the engine crankshaft and inside the rotor shaft are clean and dry. Install the rotor onto the crankshaft; install the rotor throughbolt and washer, then torque the rotor bolt 43-47 N•m (32-34 ft.-lb.).
5. Slide the stator into position over the rotor and against the engine stator mount housing. Ensure that the stator fits snugly against the housing. Ensure the proper wiring orientation.

6. Install the rear housing over the rotor bearing and against the stator.
7. Install and tighten the rear housing/stator throughbolts.
8. Remove the engine spark plug. Using the recoil starter, rotate the rotor to check for interference between the rotor and stator. Correct any misalignment before proceeding. Reinstall the spark plug.
9. Install the brush holder assembly onto the rear housing.
10. If removed during disassembly, install the stator cover over the stator. Bend the tabs to lock the cover into position.
11. Reassemble and install the control box onto the side of the generator cradle frame.
12. Reconnect the wiring to the rear housing components.
13. Reconnect the ground cable to the rear housing foot below the control box. Install and tighten the rubber mount to housing feet nuts.
14. Install the generator end cover.
15. Remove the temporary support.

WIRING DIAGRAMS

Refer to Fig. HA409 for the complete wiring diagram for models EB3000 and EB4000 generators.

HONDA
EB, EG, and EM 3.5 and 5 kW Models

Generator Model	Rated (Surge) Watts	Output Volts
EB3500X, K1	3000 (3500)	120/240
EG3500X, K1	3000 (3500)	120/240
EM3500(S)X, K1	3000 (3500)	120/240 AC/12 DC
EB5000X, K1	4500 (5000)	120/240
EG5000X, K1	4500 (5000)	120/240
EM5000(S)X, K1	4500 (5000)	120/240 AC/12 DC

NOTES: 1. (S) in the model number denotes electric start option.

2. K1 suffix denotes later model generators with frame serial numbers as follows:

EB3500	EA6 3100001 and higher
EG3500	EA6 4100001 and higher
EM3500	EA6 1100001 and higher
EB5000	EA7 3100001 and higher
EG5000	EA7 4100001 and higher
EM5000	EA7 1100001 and higher

Some K1 generator service procedures differ from standard procedures and will be identified as such in this chapter.

IDENTIFICATION

These generators are direct drive, brush style, 60 Hz, single phase units with integral electronic voltage regulation and panel mounted receptacles.

Refer to Figs. HA451-HA455 to view the generator components.

The frame serial numbers on EB and EM series units are located on the engine mount frame crossmember below the carburetor. The frame serial numbers on EG series units are located at the bottom right corner of the back panel of the control box cover.

For the location of engine model and serial numbers, refer to Fig. HA456.

WIRE COLOR CODE

The following color code abbreviations are used to identify the generator wiring in this series:

Bl – Black	Bu – Blue	W – White
R – Red	G – Green	Lb – Light blue
Y – Yellow	Br – Brown	Lg – Light green
Gr – Gray	P – Pink	O – Orange

MAINTENANCE

ROTOR BEARING

The brush end rotor bearing is a pre lubricated, sealed bearing. Inspect and replace as necessary. If a seized bearing is encountered, inspect the rotor journal and the end cover housing bearing cavity for scoring.

BRUSHES

Replace the brushes when the brush length protruding from the brush holder is 5.0 mm (0.20 in.) or less. New brush length is 9.0 mm (0.35 in.).

GENERATOR REPAIR

TROUBLESHOOTING

Before testing any electrical components, check the integrity of the terminals, connectors, and fuse(s). While testing individual components, also check the wiring integrity. Use a digital multimeter for testing. Resistance readings

223

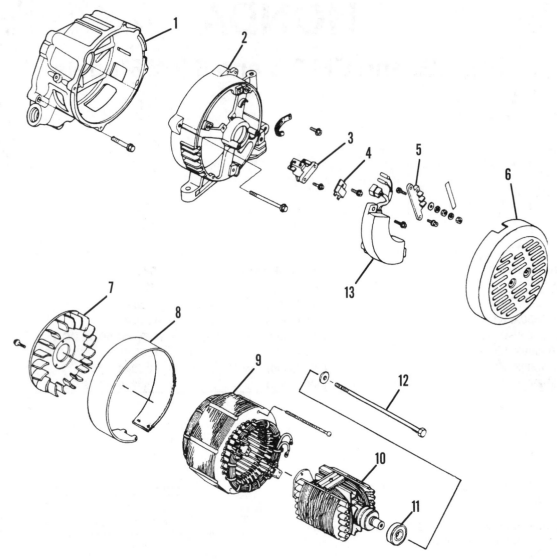

Fig. HA451 – Exploded view of generator head components.

1. Engine PTO housing/adapter
2. Rear housing
3. Brush-holder assembly
4. DC diode (where used)
5. AC terminal strip

6. End cover
7. Cooling fan with three screws
8. Stator cover
9. Stator assembly
10. Rotor assembly

11. Rotor bearing
12. Rotor through-bolt and washer
13. AVR

may vary with temperature and the accuracy of the test equipment.

For access to the components listed in this **Troubleshooting** section, refer to Figs. HA451-HA455 and the later **Disassembly** section. All operational testing must be performed with the engine running at its governed no load speed of 3600-3900 RPM, unless specific instructions state otherwise. Always load test the generator after repairing defects discovered during troubleshooting.

To troubleshoot the generator:

> **NOTE: The positive (+) brush terminal on these generators is always the one closest to the rotor windings. It has a light green with white tracer wire (K1 units use a red wire). The negative (-) brush terminal on these generators is closest to the rotor bearing and has a light green black tracer wire (K1 units use a white wire).**

1. Circuit breakers must be ON. To test either the AC or DC (if equipped) breakers, access the breaker wiring and continuity test the breaker terminals. The breaker is faulty if it will not stay set, shows continuity in the OFF position, or does not show continuity in the ON position.

2. Remove the generator end cover, but leave all wires and components connected.

3. Select the proper wiring test point diagram from the following list and proceed with testing generator output.

EB3500 – Fig. HA457	EB5000 – Fig. HA460
EG3500 – Fig. HA458	EG5000 – Fig. HA458
EM3500 – Fig. HA459	EM5000 – Fig. HA461
EB3500 K1 – Fig. HA462	EB5000 K1 – Fig. HA465
EG3500 K1 – Fig. HA463	EG5000 K1 – Fig. HA466
EM3500 K1 – Fig. HA464	EM5000 K1 – Fig. HA467

4. Start the generator.

5. Test AC output at the following color coded and numbered terminals, then compare to the specifications given:

All EB3500,
  All EG3500,
  All EM3500,
  All EB5000,
  All EG5000, and
  All EM5000   red (2) to white (1),
                     and blue (4)
                     to gray (3)   102-138 volts

6. Test DC voltage at the following color coded and numbered terminals and compare to the specifications given:

EB3500   light green/white (8)
                     to light green/black (7),
EG3500,
  EM3500,
  EG5000, and
  EM5000   light green/white (6)
                     to light green/black (5),
EB5000   light green/white (5)
                     to light green/black (6),
EB3500 K1   red (7) to white (6)
EG3500 K1   red (6) to white (5),
EM3500 K1   red (6) to white (5),
EB5000 K1   red (5) to white (6),
EG5000 K1   red (6) to white (5), and
EM5000 K1   red (6) to white (5)   21-29 volts

7. Stop the generator:

a. If Step 5 and 6 tests met specification, check continuity between generator terminal and receptacles and repair as necessary.

b. If Step 5 and 6 tests did not meet specification, proceed with Step 8.

8. Test the rotor:

a. Unplug the brush wire connections and remove the brush holder assembly.

b. Using an ohmsmeter, test rotor winding resistance at the slip rings. Resistance should be 45-55 ohms on the 3500W generators, 55-65 ohms on the 5000W generators.

c. Measure rotor winding resistance from each slip ring to a ground on the rotor shaft or laminations. Resistance should be infinity/no-continuity.

d. If test Steps b and c met specification, the rotor is good.

e. If any test in Step b or c failed to meet specification, the rotor is faulty. Replace the rotor.

9. Inspect the brush holder and brushes.

a. Check brush length. Exposed brush length should be 5.0 mm (0.20 in.) minimum.

b. Test for continuity between the brush tip and the brush terminal. Continuity should be present.

c. Check for freedom of movement, damage, or unusual wear.

Replace the brush holder assembly if brushes do not meet specification.

10. Unplug and remove the automatic voltage regulator (AVR). Leave all other wires connected.

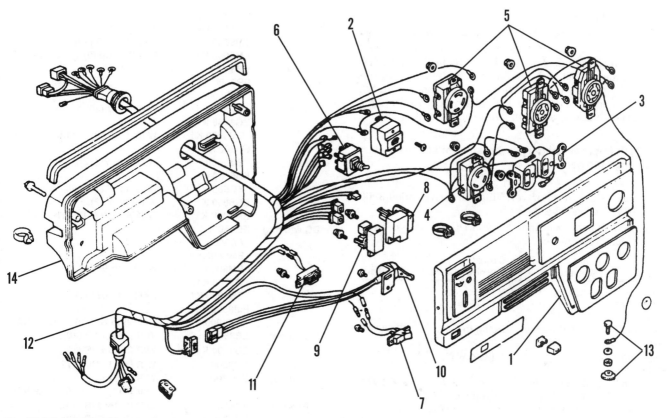

Fig. HA452 – *Exploded view of early style EB series control box assembly.*

1. Control panel
2. AC circuit breaker
3. 120V AC receptacle
4. 120V AC Twistlok receptacle
5. 240V AC Twistlok receptacle
6. Voltage-selector switch
7. Auto-throttle switch
8. Auto-throttle control unit
9. DC diode stack
10. Engine switch
11. Oil alert unit
12. Control box harness
13. Ground terminal assembly
14. Control panel rear cover

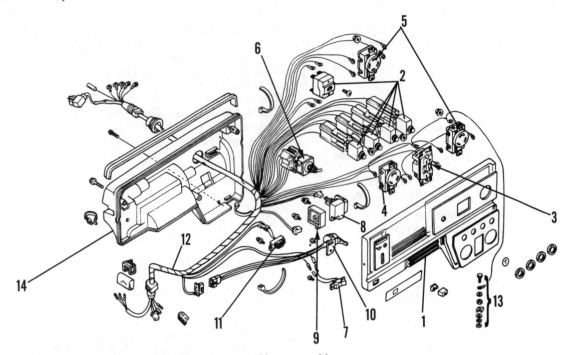

Fig. HA453 – Exploded view of K1-style EB series control box assembly.

1. Control panel
2. AC circuit breakers (5)
3. 120V AC GFCI receptacle
4. 120V AC Twistlok receptacle
5. 240V AC Twistlok receptacle
6. Voltage-selector switch
7. Auto-throttle switch
8. Auto-throttle control unit
9. DC diode stack
10. Engine switch
11. Oil alert unit
12. Control box harness
13. Ground terminal assembly
14. Control panel rear cover

11. To test the stator:

  a. Connect a 12-volt battery to the brush connector terminals.

  b. With the brushes installed and the brush leads disconnected, connect the positive battery post to the light green/white (+) terminal (red on K1 units), then connect the negative battery post to the light green/black (-) terminal (white on K1 units).

NOTE: Improper battery to brush connections will damage the generator.

  c. Start the generator, verifying the correct 3600-3900 RPM no load speed.

  d. Test the stator output AC and DC voltages connections. Note the readings.

Test	Generator	Color/ Number Code	Output
Main winding 1		R (2) to W (1),	
Main winding 2	All 3500 Series	Bu (4) to Gr (3)	86-98 VAC
Main winding 1		R (2) to W (1),	
Main winding 2	All 5000 Series	Bu (4) to Gr (3)	63-73 VAC
Exciter Winding	EB3500	Lg/R (5) to G (6),	
	EB3500 K1	Bu (5) to Bu (5),	
	EG3500, EM3500	Lg/R (7) to G (8),	
	EG3500 K1, EM3500 K1	Bu (7) to Bu (7)	51-61 VAC
	EB5000, EG5000	Lg/R (7) to G (8),	

	EB5000 K1, EG5000 K1	Bu (7) to Bu (7),	
	EM5000	Lg/R (9) to G (10),	
	EM5000 K1	Bu (9) to Bu (9)	37-43 VAC
Sensor winding	EB3500	W (11) to W (12),	
	EB3500 K1, EG3500, EM3500	W (9) to W (10),	
	EG3500 K1, EM3500 K1	W (8) to W (9)	11.6-15.6 VAC
	EB5000, EM5000	W (11) to W (12),	
	EB5000 K1, EG5000	W (9) to W (10),	
	EG5000 K1	W (8) to W (9),	
	EM5000 K1	W (10) to W (11)	9.1-11.1 VAC
DC winding	EB3500, EB5000, EM3500, EM5000	Br (13) to Br (13),	
	EB3500 K1, EB5000 K1	Br (11) to Br (11),	
	EM3500 K1, EM5000 K1	Br (12) to Br (12)	17-23 VAC
DC output	EB3500	W/R (9) to Bl/R (10),	
	EB3500 K1	W/R (8) to Bl/R (13),	
NOTE:	EM3500	W/R (11) to Bl/R (12),	
W/R=(+)	EM3500 K1	W/R (10) to Bl/R (11)	9.9-11.9 VDC
Bl/R=(-)	EB5000	W/R (9) to Bl/R (10),	
	EB5000 K1	W/R (8) to Bl/R (13),	
	EM5000, EM5000 K1	W/R (7) to Bl/R (8)	6.3-8.3 VDC

  e. Stop the generator.

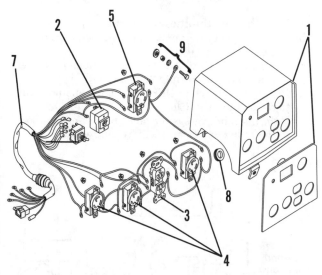

Fig. HA454 – Exploded view of EG series control box assembly.

1. Control box and faceplate
2. AC circuit breaker
3. 120V AC receptacle
4. 120V AC Twistlok receptacle
5. 240V AC Twistlok receptacle
6. Voltage selector switch
7. Control box harness
8. Control box harness grommet

f. Disconnect:
- The 12V test battery.
- The red, white, blue, and gray main winding stator wires from the terminal strip.
- The 4-pin connector with the light green/red and green exciter winding wires.
- On EB and EM models, the stator winding connector with the two brown wires.

g. Using an ohmmeter, measure resistance between the red to white and the blue to gray main winding stator wire terminals. Resistance should be 0.4 ohm on 3500 series units, and 0.2 ohm on 5000 series units.

h Measure resistance between the light green/red and the green exciter winding stator wire terminals; resistance should be 1.2 ohms on all units.

i. Measure resistance between the two brown DC winding stator wire terminals. Resistance should be 0.5 ohms on 3500 series units, and 0.4 ohms on 5000 series units.

j. If all of the tests met specification, the stator is good; proceed with further component testing.

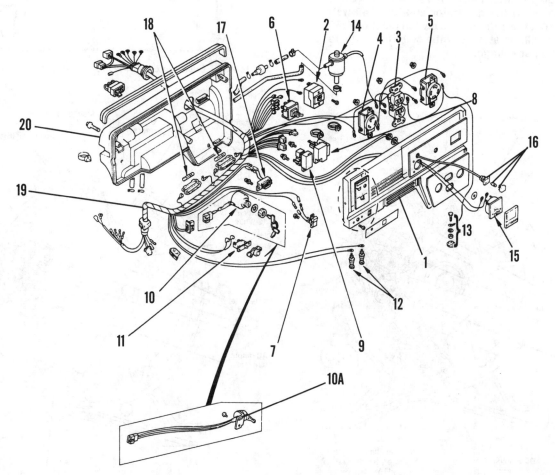

Fig. HA455 – Exploded view of EM series control box assembly.

1. Control panel
2. AC circuit breaker
3. 120V AC receptacle
4. 120V AC Twistlok receptacle
5. 240V AC Twistlok receptacle
6. Voltage-selector switch
7. Auto-throttle switch
8. Auto-throttle control unit
9. DC diode stack
10. Engine switch (EMS)
10A. Engine switch (EM)
11. DC circuit breaker
12. DC terminals
13. Ground terminal
14. Auto-choke solenoid
15. Voltmeter
16. Pilot lamp assembly
17. Oil-alert unit
18. Fuses (EMS)
19. Control-box harness
20. Control-box rear cover

k. If any individual test in Steps D, G, H, or I failed to meet specification, the stator is faulty and must be replaced.

l. Reconnect all previously disconnected wiring and components.

m. Attach DC voltmeter probes to the brush leads.

n. Start the generator and immediately test brush voltage. Voltage should test 21-29 VDC. If voltage is outside this range, immediately stop the generator. The AVR is faulty. Replace the AVR and repeat this Step N.

NOTE: If the AVR is faulty and requires replacement, make sure that the voltage selector switch is tested prior to generator startup, as a faulty voltage selector switch can damage the AVR. Refer to the Individual Component Testing section for the selector switch test.

12. To test the auto throttle system on units so equipped (Fig. HA468):

NOTE: The auto throttle system will not detect loads below 1.0 amp. If less than a 120-watt load is being applied to the generator at 120V (240-watt load at 240V), the auto throttle switch must be OFF so the unit will run at full throttle governed RPM to produce proper voltage.

a. Turn the auto throttle switch OFF.

b. Start the generator; make sure all loads are disconnected.

c. Move the switch to AUTO. After a moment, the engine should idle down to 2000-2400 RPM.

d. Return the switch to OFF; the engine should return to no load full throttle.

e. If the auto throttle system fails Steps c and d, proceed to the Individual Component Testing section.

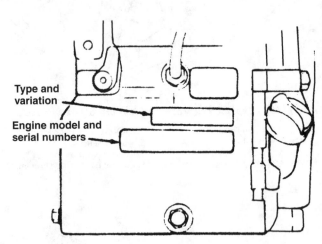

Fig. HA456 – Location of engine model and serial numbers.

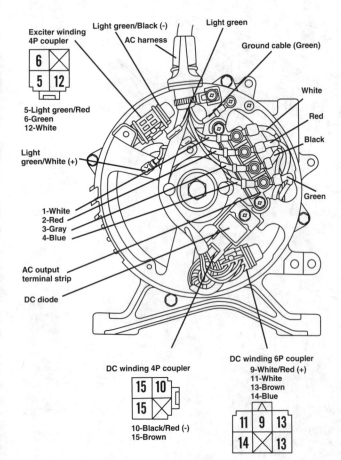

Fig. HA457 - View of end housing wiring connection test points for EB3500 generators.

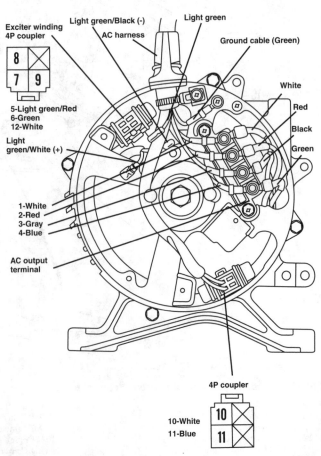

Fig. HA458 - View of end housing wiring connection test points for EG3500 and EG5000 generators.

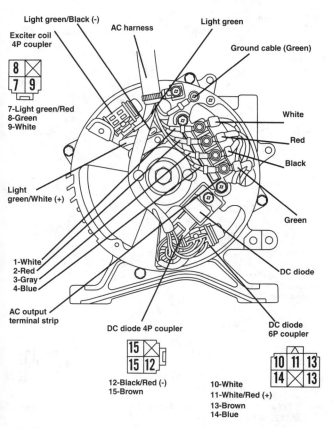

Light green/Black (-)
AC harness
Light green

Exciter coil
4P coupler

8	☒
7	9

7-Light green/Red
8-Green
9-White

Light
green/White (+)

Ground cable (Green)

White

Red

Black

Green

1-White
2-Red
3-Gray
4-Blue

AC output
terminal strip

DC diode

DC diode 4P coupler

15	☒
15	12

12-Black/Red (-)
15-Brown

DC diode
6P coupler

10	11	13
14	☒	13

10-White
11-White/Red (+)
13-Brown
14-Blue

Fig. HA459 – View of end housing wiring connection test points for EM3500 generators.

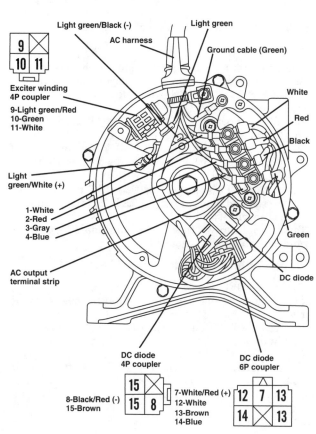

Light green/Black (-)
AC harness
Light green

9	☒
10	11

Exciter winding
4P coupler

9-Light green/Red
10-Green
11-White

Light
green/White (+)

Ground cable (Green)

White

Red

Black

Green

1-White
2-Red
3-Gray
4-Blue

AC output
terminal strip

DC diode

DC diode
4P coupler

15	☒
15	8

8-Black/Red (-)
15-Brown

7-White/Red (+)
12-White
13-Brown
14-Blue

DC diode
6P coupler

12	7	13
14	☒	13

Fig. HA461 – View of end housing wiring connection test points for EM5000 generators.

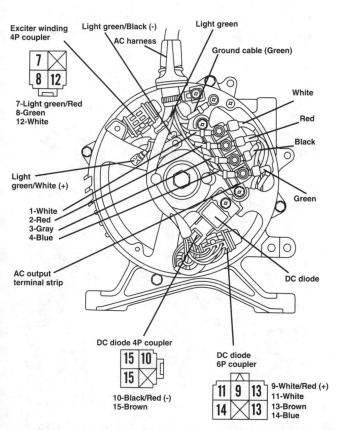

Exciter winding
4P coupler

Light green/Black (-)
AC harness
Light green

7	☒
8	12

7-Light green/Red
8-Green
12-White

Light
green/White (+)

Ground cable (Green)

White

Red

Black

Green

1-White
2-Red
3-Gray
4-Blue

AC output
terminal strip

DC diode

DC diode 4P coupler

15	10
15	☒

10-Black/Red (-)
15-Brown

DC diode
6P coupler

11	9	13
14	☒	13

9-White/Red (+)
11-White
13-Brown
14-Blue

Fig. HA460 – View of end housing wiring connection test points for EB5000 generators.

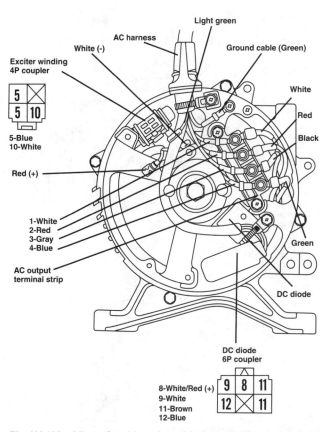

Light green
AC harness
White (-)

Exciter winding
4P coupler

5	☒
5	10

5-Blue
10-White

Red (+)

Ground cable (Green)

White

Red

Black

Green

1-White
2-Red
3-Gray
4-Blue

AC output
terminal strip

DC diode

DC diode
6P coupler

9	8	11
12	☒	11

8-White/Red (+)
9-White
11-Brown
12-Blue

Fig. HA462 – View of end housing wiring connection test points for EB3500 K1 generators.

INDIVIDUAL COMPONENT TESTING

Receptacles

The generator must be shut down to test the receptacles.
1. Access the receptacle connections inside the control box.
2. Disconnect the RETURN (white) feed wire from the receptacle being tested.
3. Install a jumper wire between the HOT and RETURN outlet terminals.

   a. On 120V straight pin and Twistlok receptacles, HOT terminals are either red or blue feed wires. RETURN terminals will be the white wire.

   b. On 240V Twistlok receptacles, HOT terminals are the red and blue feed wires. RETURN terminals will be the white feed wire.

4. Test for continuity across the feed terminals. No continuity indicates a faulty receptacle.

Ground Fault Circuit Interrupter (GFCI)

To test the GFCI:
1. Start the generator and allow a brief no load warm up period.
2. Depress the TEST button; the RESET button should extend. If it does, proceed to Step 3. If the RESET button does not extend, the GFCI is faulty.
3. Press the RESET button. The RESET button should then be flush with the TEST button. If the RESET button does not remain flush with the TEST button, the GFCI is faulty.
4. Stop the generator.

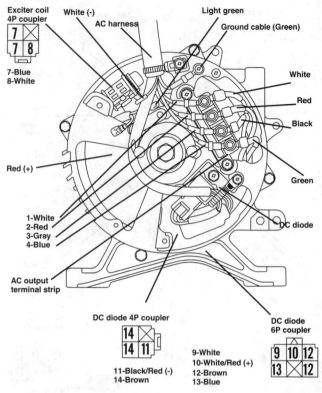

Fig. HA464 – View of end housing wiring connection test points for EM3500 K1 generators.

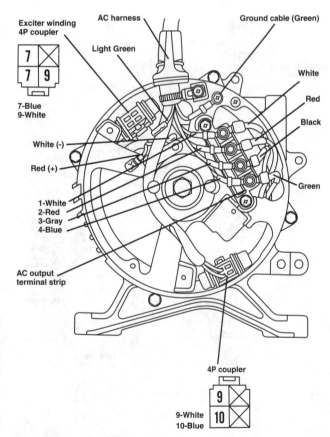

Fig. HA463 – View of end housing wiring connection test points for EG3500 K1 generators.

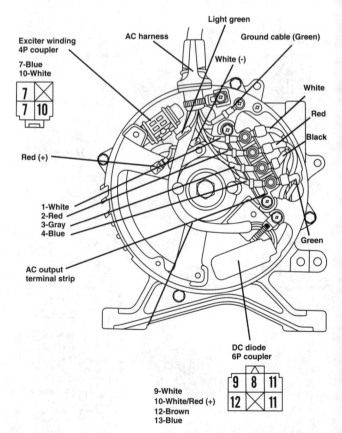

Fig. HA465 – View of end housing wiring connection test points for EB5000 K1 generators.

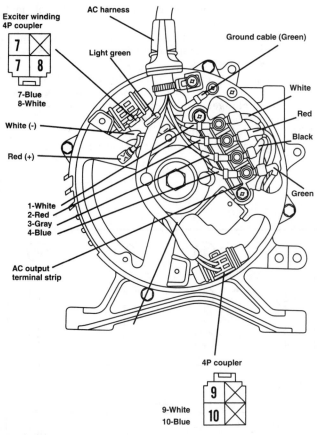

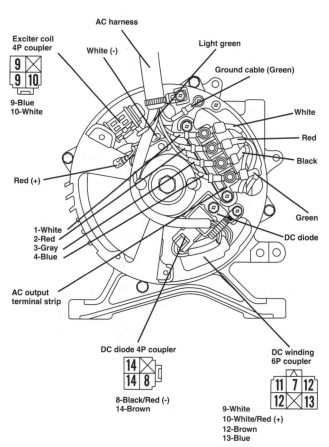

Fig. HA466 – View of end housing wiring connection test points for EG5000 K1 generators.

Fig. HA467 – View of end housing wiring connection test points for EM5000 K1 generators.

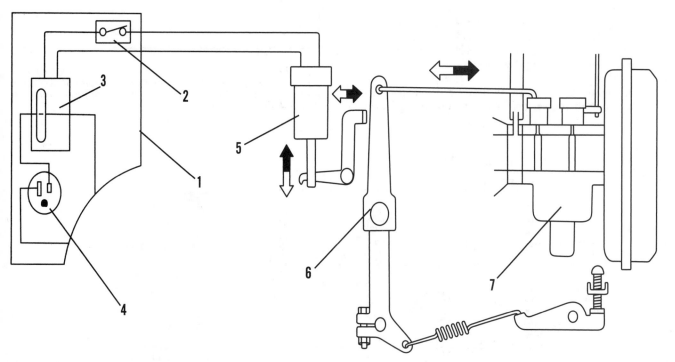

Fig. HA468 – Diagram of auto throttle system components and connections.

1. Control box
2. Auto-throttle switch
3. Auto-throttle control unit

4. AC receptacle
5. Auto-throttle solenoid

6. Governor arm
7. Carburetor

NOTE: If the GFCI seems to trip for no apparent reason, make sure the generator is not vibrating excessively due to faulty rubber mounts or debris between the generator and frame. Excessive vibration can trip the GFCI.

Voltage Selector Switch

Note the orientation of the wiring connector(s) prior to removal.

Using the switch positions and numbered switch terminals in Fig. HA469, test for continuity between the connected terminals. The failure of any terminal pair to show continuity indicates a faulty switch.

On K-1 suffix units, the switch terminals are grouped into two connectors –black and white. When reconnecting the connectors, note that the white connector always fits onto terminals 1-6, while the black connector always fits onto terminals 7-12. Reversing the connectors will cause the switch to operate in reverse.

The stator's two main windings, MW1 and MW2 (Figs. HA470 and HA471), are independent of each other but equal in capacity. With the voltage selector switch in the 120/240V position, the output is arranged in series circuits (Fig. HA470). With the switch in 120V only, the output is arranged in a parallel circuit (Fig. HA471). The AVR sensor circuit is also controlled by the switch, so the AVR will accurately control voltage in either switch position. When doing generator output testing, keep in mind that a faulty voltage selector switch could also possibly damage the AVR.

Auto Throttle System

1. To test the auto throttle switch:
   a. Access the switch terminals inside the control box.
   b. With the switch lever in the AUTO position, there should be continuity between the two lead terminals on the back of the switch.
   c. With the switch lever in the OFF position, the two switch lead terminals should show no continuity.
   d. Any results different from those specified in Steps b and c indicate a faulty switch.
2. To test the solenoid:
   a. Disconnect the two-wire W/R and W/R harness connector from the solenoid valve mounted on the en-

gine governor linkage bracket. Remove the solenoid from the engine.
   b. Test the solenoid wires for continuity. No continuity indicates a faulty solenoid. If continuity is observed, proceed to Step c.
   c. Using a fully charged 12V battery, connect the solenoid terminals to the battery while observing the solenoid plunger. If the plunger activates when battery current is applied, the solenoid is good.
3. To test the control unit:
   a. Access the control unit inside the control box.

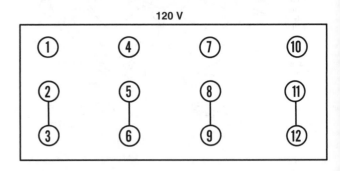

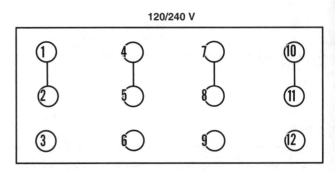

Fig. HA469 - Continuity test positions for the 120-120/240 voltage selector switch. There should be continuity between the noted terminals. On K1 series, refer to text for proper terminal connector installation.

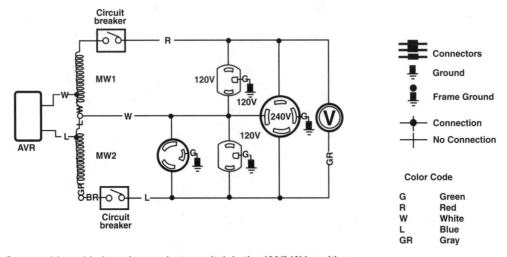

Fig HA470 – Output wiring with the voltage selector switch in the 120/240V position.

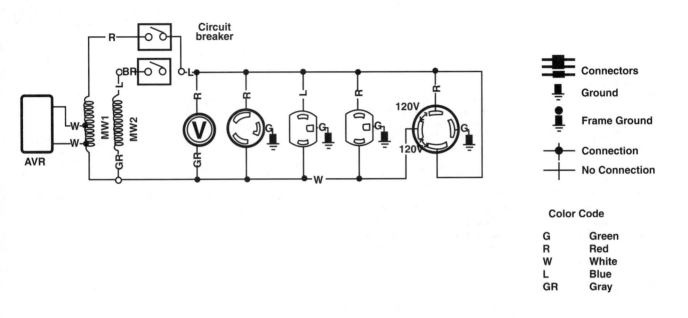

Color Code

G	Green
R	Red
W	White
L	Blue
GR	Gray

Fig. HA471 – Output wiring with the voltage selector switch in the 120V-only position.

Auto-throttle Control Unit
Route the two wires through the control unit opening as shown. Note the direction of the wires.

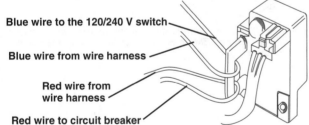

Blue wire to the 120/240 V switch

Blue wire from wire harness

Red wire from wire harness

Red wire to circuit breaker

Fig. HA472 – View of red and blue wires from the Main AC stator winding properly passing through the auto throttle control unit sensor loop from opposite directions. Failure to route the wires as shown will cause the auto throttle system not to function.

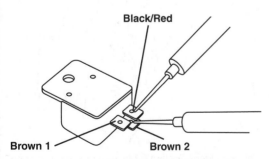

Black/Red

Brown 1

Brown 2

Fig. HA473 – Terminal identification for testing the three-terminal DC diode. Refer to the text for the correct test procedure.

b. Verify that the blue and red wires from the stator main windings are passing through the control unit loop from opposite directions (Fig. HA472). If necessary, redirect the wires. Wires passing through the control unit from the same direction will cancel each other's magnetic field, preventing the control unit from sensing voltage differences between load and no load.

c. If all other auto throttle components test good, and the control unit loop wires are routed correctly, but the auto throttle still does not operate, the control unit is faulty. Rreplace the control unit.

DC Diode

To test the 3-terminal DC diode:

1. Access the diode inside the control box. Unplug the 3-terminal (black/red, brown, and brown) connector from the diode.

2. Refer to Fig. HA473. Using an ohmmeter set on the R × 1 scale, test continuity between black/red to brown 1, black/red to brown 2, and brown 1 to brown 2. Note the readings.

3. Reverse the meter leads and retest the three terminal pairs. Note the readings.

4. There should be continuity in one direction of the black/red to brown 1 and 2 tests but no continuity when the leads are reversed.

5. There should be no continuity in either direction on the brown-to-brown test.

6. Any readings other than those specified in Steps 4 and 5 indicate a faulty DC diode; replace the diode.

Diode Stack

To test the diode stack (Fig. HA474):

1. Access the diode stack inside the control box. Unplug the harness connector(s).

   a. EB Models use one 8-terminal connector.

   b. EM Models use one 8-terminal connector and one 6-terminal connector.

2. Set the ohmmeter on the R × 1 scale.

3. Holding the positive (+) meter probe on the first numbered terminal in each of the following terminal number pairs, continuity test the terminal pairs where applicable (EB models do not have any terminals numbered higher than 9). Note the readings:

1-4 1-7 1-8 3-2 4-8 7-8 9-6 10-11 12-15 14-13

4. Reverse the meter leads and repeat Step 3.

5. One set of Step 3 and 4 readings should have shown continuity and one set of readings should have shown infinity/no continuity. If any terminal pair tested either continuity or infinity in both directions, the diode stack is faulty. Replace the stack.

Pilot Lamp (EM Series)

To test the pilot lamp, access the lamp wiring inside the control box and test the two lamp wires for continuity. A lack of continuity indicates a faulty lamp. Replace the lamp.

Voltmeter

To test the voltmeter:

1. Access the voltmeter wiring inside the control box.

2. Disconnect one voltmeter lead.

3. Use an ohmmeter to test for continuity between the voltmeter posts.

4. A lack of continuity indicates a faulty voltmeter.

Engine Switch
(EB and EM Recoil Start Models)

Test the switch as follows:

1. Access the switch wiring inside the control box.

2. Unplug the switch harness connector.

3. Referring to Fig. HA475, test for continuity between every terminal pair in both the OFF and ON positions; note the readings.

4. Continuity should only be found between IG and E (black and green) and between FS and G (green/white and blue) with the switch OFF.

5. No continuity between the terminals noted in Step 4 or continuity between any other terminal pair in either position indicates a faulty switch.

Engine Switch (EM Electric Start Models)

Test the switch as follows:

1. Access the switch wiring inside the control box.

2. Unplug the switch harness connector.

3. Referring to Fig. HA476, test for continuity between every terminal pair in each of the three positions – OFF, ON and START. Note the readings.

4. Continuity should only be noted between the following terminals in the specified positions:

IG to E (black to green) – OFF
FS to G (green/white to blue) – OFF
BAT to ST (white to black/white) – START

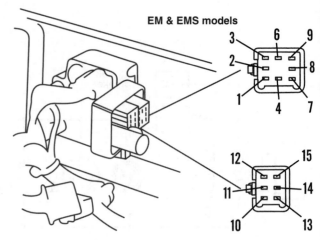

Fig. HA474 – Terminal identification for testing the diode stack. Refer to the text for the correct test procedure.

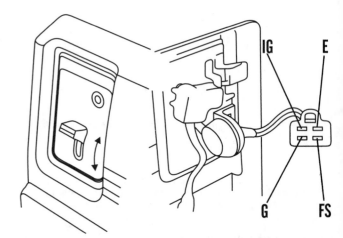

Fig. HA475 – Terminal identification for testing the engine switch on recoil start EB and EM units. Refer to the text for the correct test procedure.

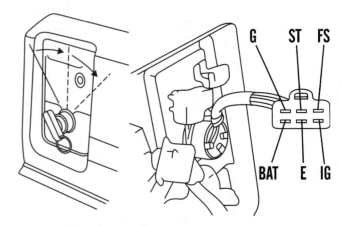

Fig. HA476 – Terminal identification for testing the engine switch on electric start EM units. Refer to the text for the correct test procedure.

5. No continuity between the terminals noted in Step 4 or continuity between any other terminal pair in any switch position indicates a faulty switch.

Engine Switch (EG Models)

To test the switch:

1. Access the switch wiring inside the control box.
2. Disconnect the black switch wire.
3. Test for continuity between the black switch terminal and any switch case ground in both the ON and OFF positions. There should be continuity in the OFF position only. Any other readings indicate a faulty switch.

Oil Alert Switch
(Inside Engine Crankcase)

To test the switch:

1. Access and disconnect the yellow and green two-wire switch harness on the engine crankcase.
2. With the engine oil level full and the generator sitting level, test continuity between the yellow and green leads.
3. If there is continuity, the oil level switch is faulty and must be replaced.
4. Drain engine oil completely, then retest continuity between the yellow and green switch leads.
5. If there is no continuity, the oil level switch is faulty and must be replaced.

Oil Alert Unit Lamp (EG Models)

To test the lamp:

1. Access the lamp wiring inside the control box.
2. Disconnect the black and yellow lamp harness wires.
3. Using a test battery (1.5 volts minimum, 6 volts maximum), connect the battery positive lead to the black pigtail and the battery negative lead to the yellow pigtail. A lamp that does not light is faulty and must be replaced.

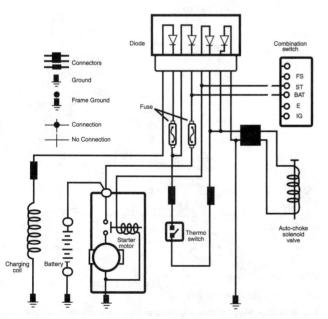

Fig. HA477 – Schematic diagram of the auto choke system components on electric start EM units.

Oil Alert Unit Lamp
(EB and EM Models)

To test the lamp:

1. Access the lamp wiring inside the control box.
2. Disconnect the lamp unit wires.
3. Using a test battery (1.5 volts minimum, 6 volts maximum), connect the battery to the lamp pigtails.
   a. Connect the battery positive terminal to the black lamp pigtail.
   b. On EB and EM3500 models, connect the battery negative terminal to the yellow lamp pigtail.
   c. On EB and EM5000 models, connect the battery negative terminal to the green lamp pigtail.
4. A lamp that does not light is faulty and must be replaced.

Auto Choke Systems

Some electric start EM models are equipped with an electrically operated automatic choke system to aid cold starts (Fig. HA477).

NOTE: Manually operating the choke lever will override the auto choke.

1. When the engine switch is turned to START, battery current energizes the choke solenoid and activates a diaphragm to close the choke shutter.
2. When the engine starts and the switch is moved to RUN, the thermoswitch in the cylinder head allows diode rectified current produced by the charging coil to maintain solenoid control choking.
3. When the head reaches a preset temperature of 25-30° C (77-86° F) on 3500 series units or 30-40° C (86-104° F) on 5000 series units, the thermoswitch opens the circuit and de-energizes the solenoid, allowing the diaphragm to open the choke shutter.
4. To test the EM Auto Choke solenoid:
   a. Access the solenoid inside the control box (Fig. HA478).
   b. Disconnect the two-wire (yellow and green) solenoid harness and the vacuum hoses from Tubes A and B.
   c. Prior to connecting a test battery, manually apply vacuum to Tube A. Air should pass from Tube B out through Tube A.
   d. Connect a 12-volt test battery to the solenoid harness terminals. The solenoid should 'click.'
   e. Again, manually apply vacuum to Tube A. No air should flow through the solenoid.
   f. If the solenoid does not test as specified, the solenoid is faulty.
5. To test the diaphragm:
   a. Manually apply vacuum to the diaphragm fitting.
   b. Observe the linkage. If the linkage does not activate the diaphragm it is faulty.
   c. Make sure that the vacuum hoses are not leaking.
6. To test the charging coil (located on the engine, behind the flywheel):
   a. Disconnect the white charge coil wire terminal from the green harness wire terminal.
   b. Measure resistance between the white wire and ground. Resistance should be 3.0-4.0 ohms. Any reading beyond these values indicates a faulty coil. Replace the coil.
7. To test the thermoswitch, start the engine and allow a brief (3-5 minute) warm-up period with the auto throttle

switch OFF. If the choke has not opened up by the end of the warm up period, the thermoswitch is faulty. Stop the engine, allow it to cool, and replace the thermoswitch.

Some EB series use intake manifold vacuum to operate a diaphragm that closes or opens the choke shutter (Fig. HA479). With the engine stopped, the internal diaphragm spring pushes the choke linkage to close the choke. When the engine starts, intake manifold vacuum draws the choke diaphragm against the spring, opening the choke. As with the EM system, manually operating the choke lever will override the auto choke.

To test the diaphragm:

1. Manually apply vacuum to the diaphragm fitting.
2. Observe the linkage. If the linkage fails to operate, the diaphragm is faulty.
3. Make sure that the vacuum hoses are not leaking.
4. Make sure that the dashpot passes air in both directions – rapidly toward the manifold, slowly toward the diaphragm.
5. Make sure that the carburetor is mounted securely to the engine. Make sure the carburetor mount gaskets are not deteriorated and that the air filter element is clean.

Remote Control Start

Some electric start units use a remote control start option which plugs into the normally blind 8-pin coupler at the back of the generator control box (Fig. HA480). Test the remote control box as follows:

1. Unplug the 6-pin connector from the rear of the box.
2. Test for continuity between the following color coded terminals and box control positions:
   a. Light green to red to white/black – engine switch ON.
   b. Red/white to green – pilot lamp.
   c. Light green to red – starter button IN.
3. No continuity in any of the Step 2 tests indicates a faulty remote control box. Replace the box.
4. If the remote control box meets specification and the remote control harness integrity is secure, but the remote control will not activate the electric start, operate the electric start with the control panel mounted engine switch. If the panel engine switch operates the electric start, the remote control relay box is faulty and must be replaced.

DISASSEMBLY

Refer to Figs. HA451 HA455 for component identification.

1. On EB and EM models:
   a. Remove the upper frame rail assembly.
   b. Drain the fuel and remove the fuel tank.
   c. Remove the muffler and muffler shields.
2. Remove the generator end cover.
3. Noting the wiring orientation and the location of any wiring retainers. Disconnect and unplug all control box harness connectors. Remove the control box.
4. Remove the flange nuts holding the generator rear housing feet to the rubber mounts. Note the location of the ground wire attached to the left foot.
5. Lift the rear housing so the feet clear the rubber mount studs and place a temporary support, such as a 2 × 4, under the engine PTO housing.
6. Disconnect and remove the brush holder assembly.
7. Disconnect any remaining wiring, noting the wiring orientation.
8. Noting the position of the stator cover, straighten the tabs holding the stator cover to the stator. Remove the cover.
9. Remove the four rear housing bolts. Separate the rear housing from the stator. Remove the housing.

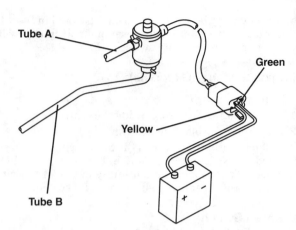

Fig. HA478 – Test diagram for the EM series auto choke solenoid. Refer to the text for the correct test procedure.

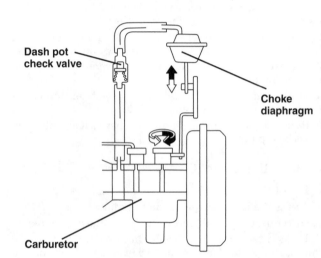

Fig. HA479 – Diagram of vacuum operated EB Auto Choke system.

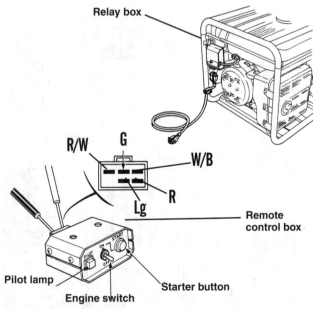

Fig. HA480 – View of remote control box and terminal identification for testing the box. Refer to the text for the correct test procedure.

10. Remove the five stator lamination screws, if necessary. Noting the stator wire orientation, separate the stator from the engine PTO housing. Remove the stator.

11. Remove the rotor throughbolt, then remove the rotor.

   a. The recommended rotor removal method is with a Honda rotor puller (#07HPC-ZB4010A for 3500 series; #07HPC-ZC2010A for 5000 series). Place a support under the rotor. Apply a light amount of penetrating oil through the rotor shaft onto the crankshaft to rotor taper. Insert the appropriate puller. Block the flywheel from turning. Torque the puller 69 N•m (50 ft.-lb.). The rotor should break free from the crankshaft. If the rotor does not loosen, lightly tap the end of the puller with a brass hammer.

   b. An alternate rotor removal method would be to use a slide hammer with a threaded adapter to break the rotor free from the taper on the engine crankshaft. Remove the three screws holding the fan to the rotor.

REASSEMBLY

1. If the engine PTO housing was removed for service:

   a. Make sure that the gasket surfaces and thread holes are clean and dry.

   b. Make sure that the two locator dowels are properly positioned.

   c. Install a new gasket.

   d. Incrementally and alternately torque the seven housing bolts 22-26 N•m (190-225 in.-lb.).

2. Place a temporary support, such as a 2 × 4, under the engine PTO housing so that, when the rear housing is installed onto the stator, the housing feet will clear the rubber mount studs.

3. Service the rotor bearing, if needed.

4. Mount the cooling fan to the rotor.

5. Ensure that the matching tapers on the engine crankshaft and inside the rotor shaft are clean and dry. Install the rotor onto the crankshaft. Install the rotor throughbolt and washer, then torque the rotor bolt 43-47 N•m (375-405 in.-lb./31-34 ft.-lb.).

6. Noting the stator wire orientation, fit the stator over the rotor and against the engine PTO housing. Reinstall the five stator lamination screws, if previously removed.

7. Fit the end housing onto the stator and rotor, being careful not to pinch or kink any wiring. Install four end housing bolts, then incrementally and alternately torque the bolts 8-12 N•m (70-105 in.-lb.).

8. Remove the spark plug and rotate the engine crankshaft to check for interference between the rotor and stator. There should be a consistent gap around the rotor laminations inside the stator. If there is any rubbing or scraping, correct any misalignment prior to proceeding. Reinstall the spark plug.

9. Remove the temporary support from under the end housing and set the housing feet onto the rubber mount studs. Reinstall the ground wire onto the left foot stud. Install the two flange nuts holding the feet to the studs. Torque the nuts 30-40 N•m (260-345 in-lb./22-29 ft.-lb.).

10. Install the stator cover over the stator in its proper position (noted during disassembly). Insert the holding tabs into the slots. Bend the tabs as necessary to snugly hold the cover.

11. Install the brush holder assembly.

12. Install the control box assembly, correctly routing the wiring to prevent kinking or pinching. Replace any wiring retainers removed during disassembly.

13. Refer to Figs. HA457-HA467 as applicable and properly reconnect all wiring.

14. Install the generator end cover.

15. On EB and EM models, reinstall the muffler shields and muffler assembly, fuel tank, and upper frame rail assembly.

FASTENER TORQUE CHART

Fastener torque values are listed in the approximate order of reassembly.

Component Fastener	Torque
Engine PTO housing	22-26 N•m (190-225 in.-lb./16-19 ft-lb)
Rotor throughbolt	43-47 N•m (375-405 in.-lb./31-34 ft-lb)
Generator rear end housing	8-12 N-m (70-105 in.-lb.)
Rear-housing feet mount nuts	30-40 N•m (260-345 in.-lb./22-29 ft-lb)

WIRING DIAGRAMS

To view the necessary model wiring diagram, refer to the appropriate figure as follows:

Model	Figure
EB3500X	HA481
EB5000X	HA488
EB3500X K1	HA482
EB5000X K1	HA489
EG3500X	HA483
EG5000X	HA490
EG3500X K1	HA484
EG5000X K1	HA491
EM3500X	HA485
EM5000X	HA492
EM3500SX	HA486
EM5000SX	HA493
EM3500SX K1	HA487
EM5000SX K1	HA494

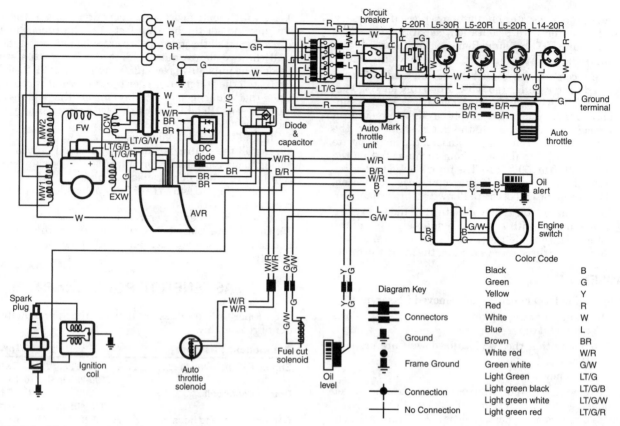

Fig. HA481 – Wiring diagram for the EB3500X generator.

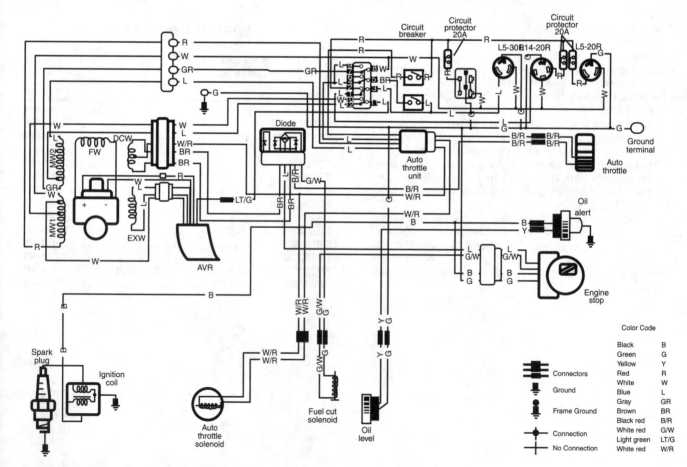

Fig. HA482 – Wiring diagram for the EB3500X K1 generator.

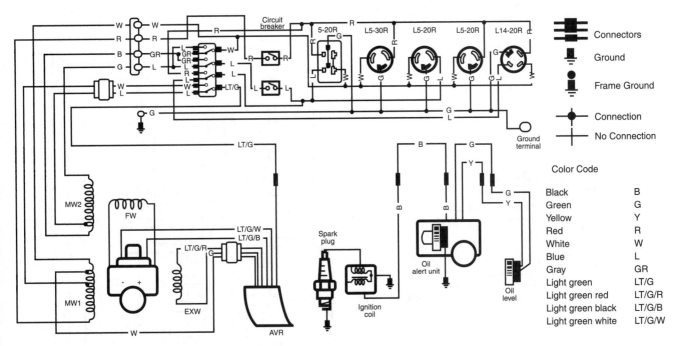

Fig. HA483 – Wiring diagram for the EG3500X generator.

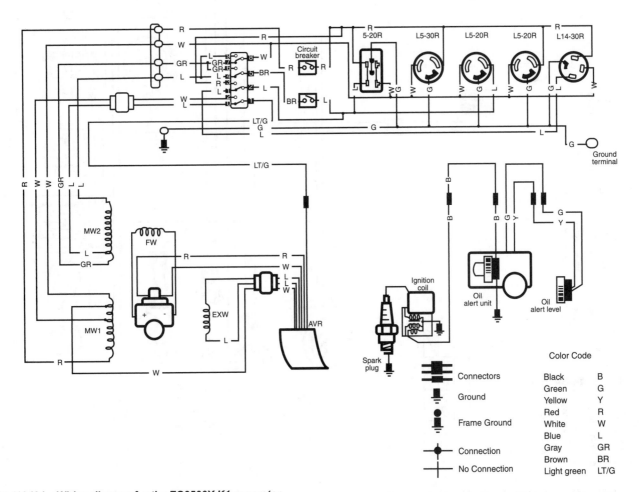

Fig. HA484 – Wiring diagram for the EG3500X K1 generator.

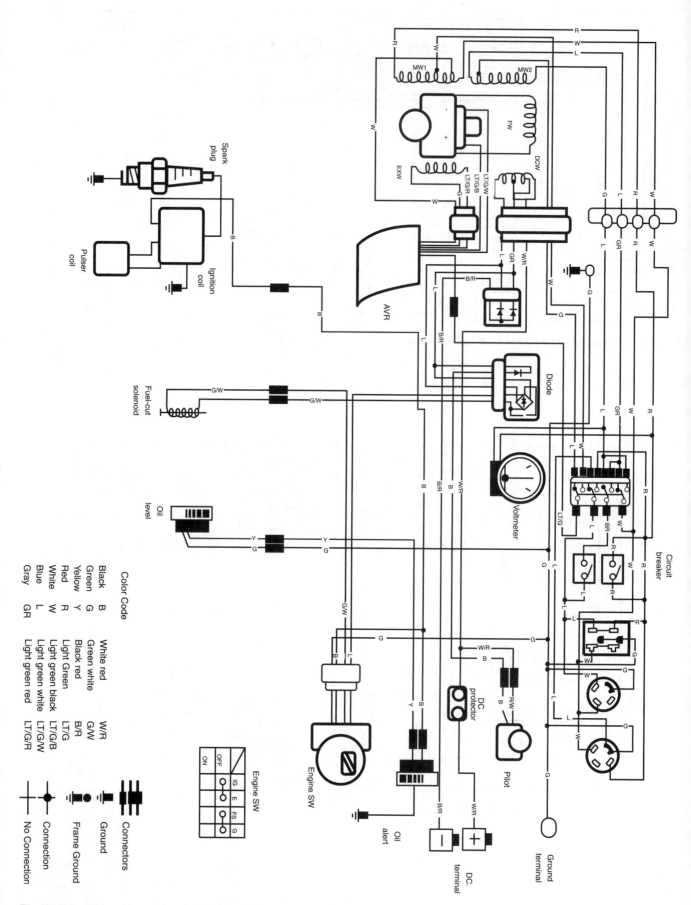

Fig. HA485 – Wiring diagram for the EM3500X generator.

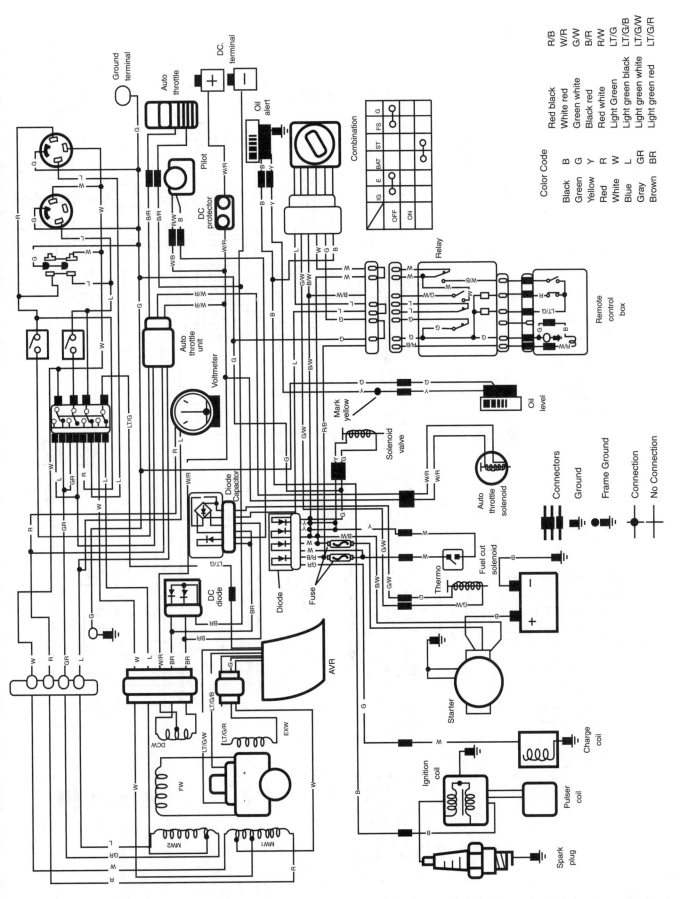

Fig. HA486 – Wiring diagram for the EM3500SX generator.

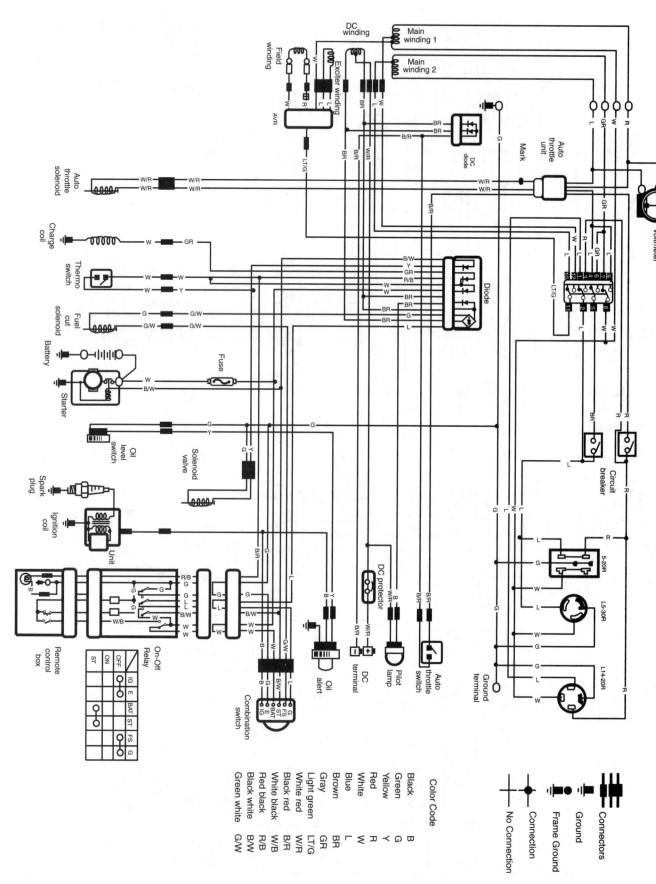

Fig. HA487 – Wiring diagram for the EM3500SX K1 generator.

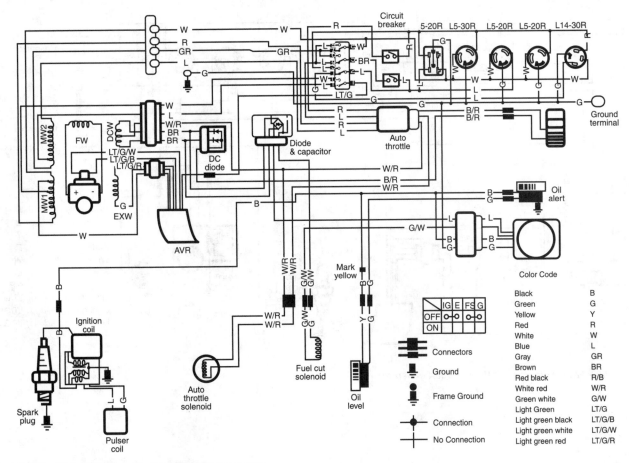

Fig. HA488 – Wiring diagram for the EB5000X generator.

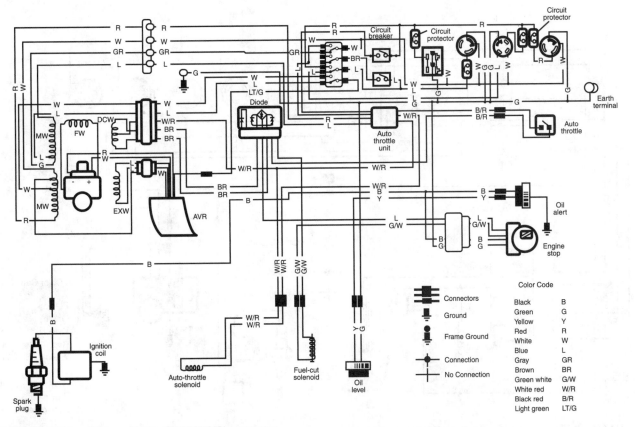

Fig. HA489 – Wiring diagram for the EB5000X K1 generator.

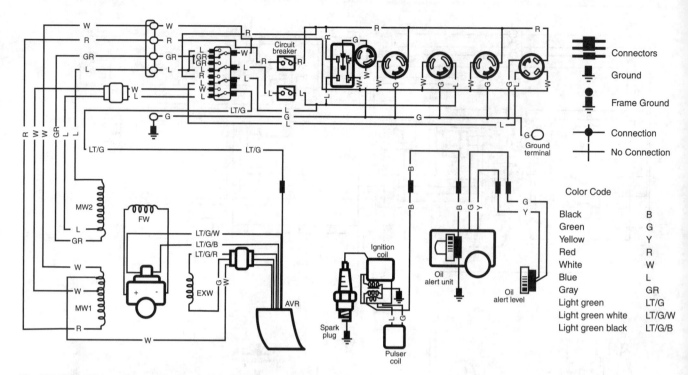

Fig. HA490 – Wiring diagram for the EG5000X generator.

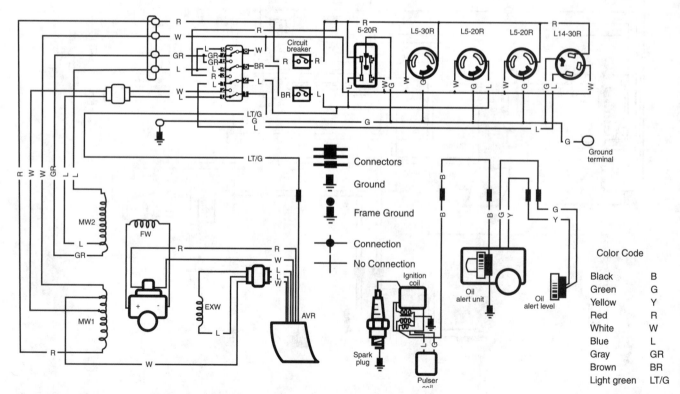

Fig. HA491 – Wiring diagram for the EG5000X K1 generator.

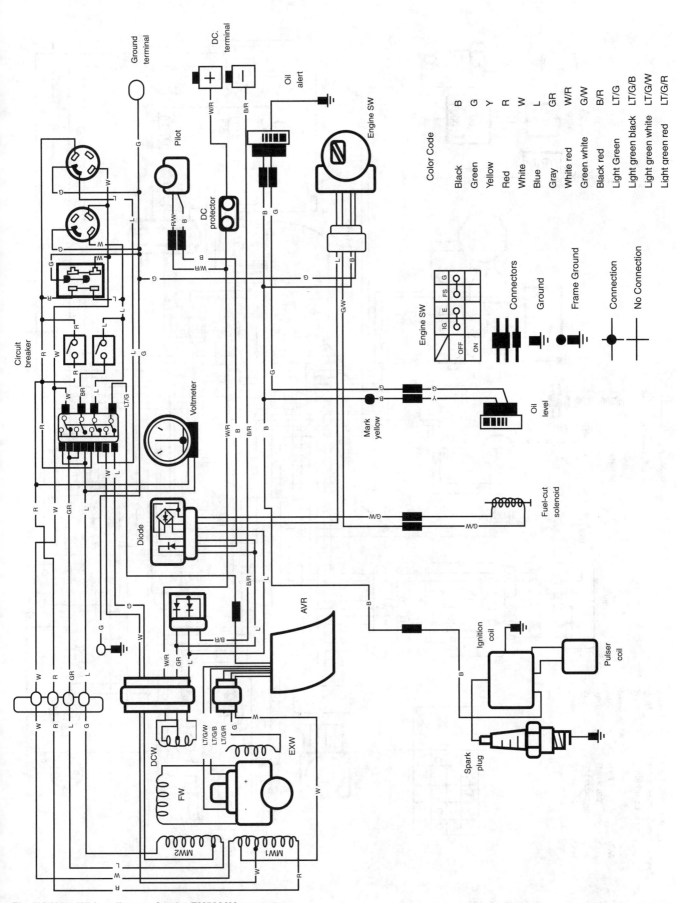

Fig. HA492 – Wiring diagram for the EM5000X generator.

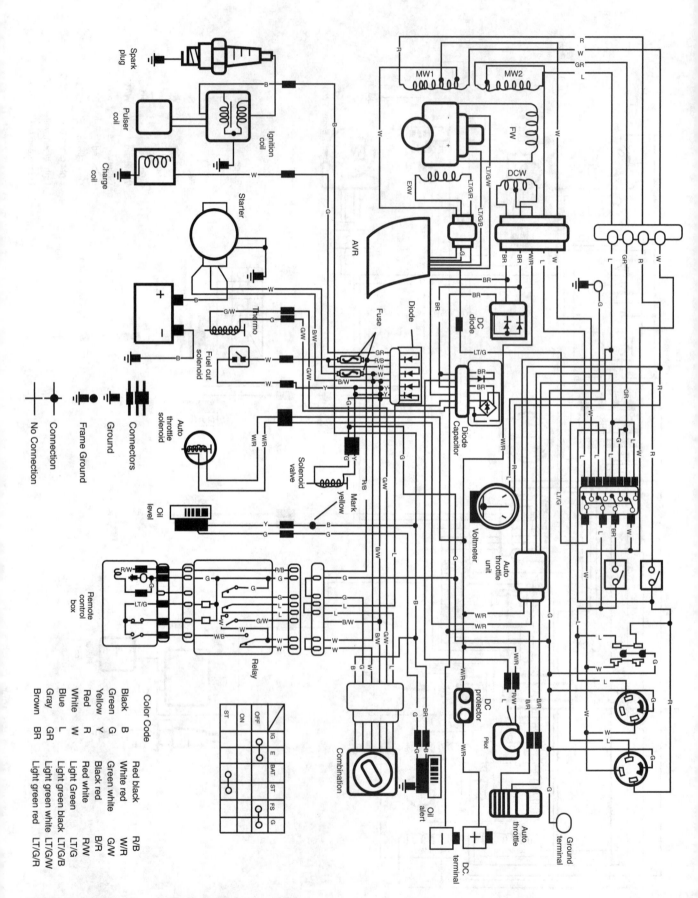

Fig. HA493 – Wiring diagram for the EM5000SX generator.

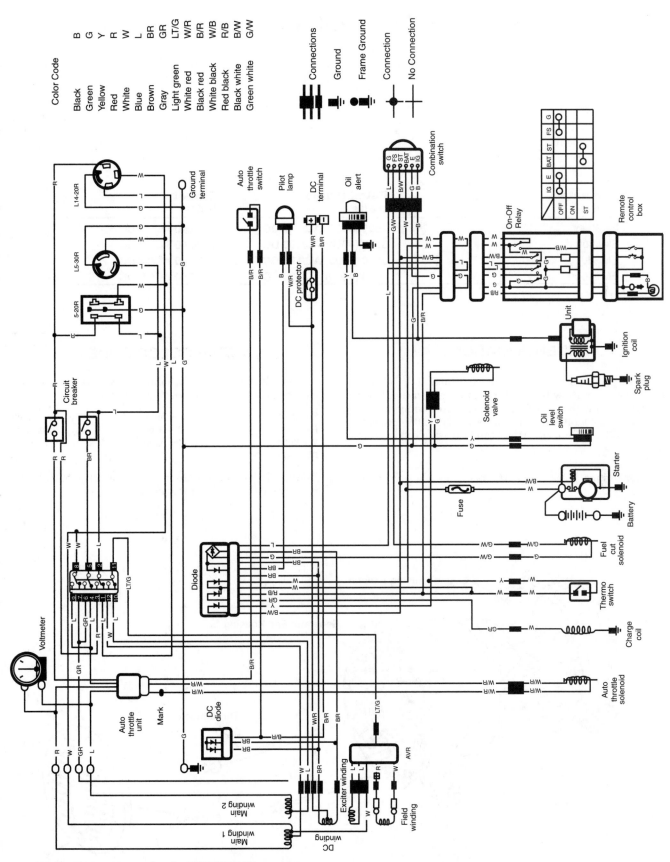

Fig. HA494 – Wiring diagram for the EM5000SX K1 generator.

HONDA
EM3.0, 4.0, EMS4.5 Generators

Generator Model	Rated (Surge) Watts	Output Volts
EM3000	2800 (3000) AC/100 DC	120/240 AC/12 DC
EM4000	3800 (4000) AC/100 DC	120/240 AC/12 DC
EMS4000	3800 (4000) AC/100 DC	120/240 AC/12 DC
EMS4500	4000 (4500) AC/100 DC	120/240 AC/12 DC

IDENTIFICATION

These generators are direct drive, brush style, 60 Hz, single phase units with panel mounted receptacles. They are equipped with a voltage selection switch which allows combined 120/240V output or strictly 120V output. auto throttle (idle control) is standard on electric start units.

Refer to Figs. HA501 and HA502 to view the generator components.

WIRE COLOR CODE

The following color code abbreviations are used to identify the generator wiring in this series:

B – Black	L – Blue	W – White
R – Red	G – Green	Lg – Light green
Y – Yellow	Br – Brown	P – Pink
Gr – Gray		

Multiple wire colors are indicated by a slash. (R/W is a red wire with a white tracer stripe).

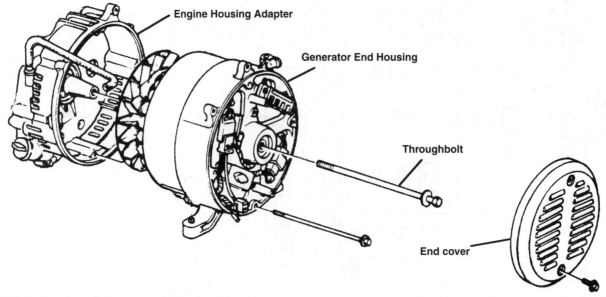

Fig. HA501 – Location of generator model and serial number.

MAINTENANCE

ROTOR BEARING

The brush end rotor bearing is a pre-lubricated, sealed bearing. Inspect and replace as necessary. If a seized bearing is encountered, inspect the rotor journal and the rear housing bearing cavity for scoring.

BRUSHES

The brushes should be replaced when the brush length reaches 13 mm (0.5 in.). The light green/white wire always goes to the positive (+) brush.

GENERATOR REPAIR

TROUBLESHOOTING

Before testing any electrical components, check the integrity of the terminals, connectors, and fuse(s). While testing individual components, also check the wiring integrity. Use a digital multimeter for testing. Resistance readings may vary with temperature and test equipment accuracy. All operational testing must be performed with the engine running 3720 RPM unless specific instructions state other-

wise. Always load test the generator after repairing defects discovered during troubleshooting.

To troubleshoot the generator:

NOTE: The positive (+) brush terminal on these generators is always the one closest to the rotor windings. It has a Lg/W wire. The negative (-) brush terminal on these generators is closest to the rotor bearing and has a Lg/B wire.

1. Circuit breakers must be ON. To test either the AC or DC breakers, access the breaker wiring and test the breaker terminals for continuity. The breaker is faulty if it will not stay set, shows continuity in the OFF position, or does not show continuity in the ON position.

To test the EMS4500 series AC circuit breaker (Fig. HA504):

a. Turn the breaker ON.

b. Test for continuity between the following terminal pairs. red (1) to red (3); red (1) to red (4); and Brown to blue. There should be continuity in all three tests.

c. Turn the breaker OFF and repeat Step b. There should be no continuity in any of the three tests.

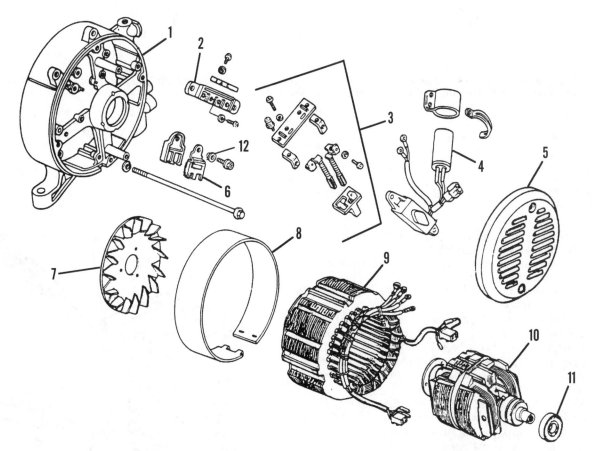

Fig. HA502 – Exploded view of generator head components.

1. Rear housing
2. AC output terminal strip
3. Brush-holder assembly

4. AVR assembly with capacitor/
      condenser
5. End cover
6. DC diodes

7. Cooling fan
8. Stator cover
9. Stator assembly
10. Rotor assembly
11. Rotor bearing
12. Diode insulators

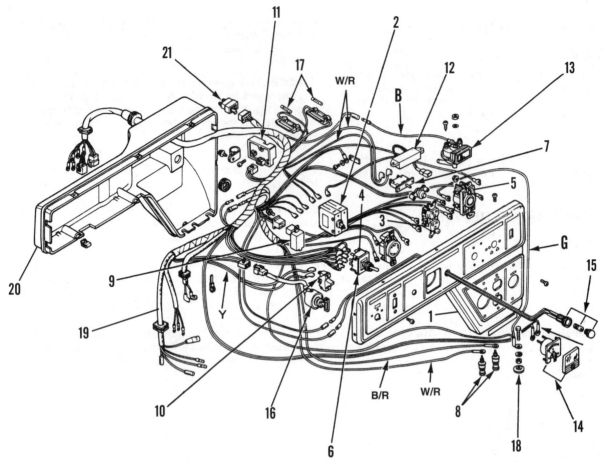

Fig. HA503 – Exploded view of EMS4500 generator control box and components. EM3000 and EM(S) 4000 are similar, using individual AC circuit breakers and a toggle style engine switch for recoil start units.

1. Control panel
2. AC circuit breaker
3. 120V AC receptacle
4. 120V AC Twistlok receptacle
5. 240V AC Twistlok receptacle
6. Voltage-selector switch
7. DC circuit breaker
8. DC output terminals

9. Diode stack
10. Auto-throttle switch
11. Auto-throttle control unit
12. Voltage limiter
13. Oil-alert switch
14. Voltmeter
15. Pilot lamp
16. Engine switch

17. Fuses
18. Ground terminal
19. Control-box harness
20. Control-box cover
21. Remote control connector

d. Any results different from those specified indicate a faulty AC breaker. Replace the breaker.

2. Remove the generator end cover, but leave all wires and components connected.

3. Start the generator.

4. Test voltage output:

   a. Using the 4-wire AC output terminal block, test for AC voltage at the main winding 1 (red to white) connections. Output should be 110-130V.

   b. Again using the 4-wire terminal block, test for AC voltage at the main winding 2 (blue to gray) connections. Output should be 110-130V.

   c. Test for AC voltage at the sub-winding connections (black/white to black/white); output should be 4-6V.

   d. Test for DC output at the DC connections (white/red [+] to black/red [-]); voltage should be 16-22 volts.

   e. Test for DC output at the brush wire terminals (light green/white [+] to light green/black [-]); output should be 12-18V.

5. Stop the generator:

   a. If Step 4 tests met specification, proceed to the Individual Component Testing section in this chapter.

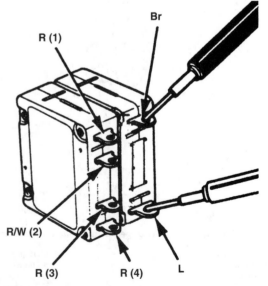

Fig. HA504 – Continuity test points for the AC circuit breaker used on EMS4500 generators.

b. If Step 4 tests did not meet specification, proceed with Step 6.

6. To test the rotor:

   a. Disconnect the AVR 4-pin connector.

   b. Disconnect the brush wire connectors.

   c. Using an ohmmeter, measure field coil winding resistance through the brushes; resistance should be 71.8 ohms on the EM3000 series, 50.3 ohms on the EM and EMS4000 series.

   d. Measure field coil winding resistance at the rotor slip rings. Resistance should be the same as Step c.

   e. Measure field coil winding resistance from each slip ring to a good ground on the rotor shaft or laminations. Resistance should be infinity/no continuity.

   f. If test Steps c and d met specification, the rotor and brushes are good.

   g. If test Step c failed specification but Step d met specification, the rotor is good. Test and inspect the brushes and replace if faulty.

   h If both test Steps C and D failed specification, the rotor is faulty. Replace the rotor.

   i. If either Step E test read any continuity, the field coil winding is shorted. Replace the rotor.

7. To test the stator:

   a. Obtain a fully charged 12-volt battery and a pair of test leads to connect the battery terminals to the brush connector terminals.

   b. With the terminals still disconnected from the Step 6 test, connect the positive battery post to the light green/white (+) brush holder terminal, then connect the negative battery post to the light green/black (-) brush holder terminal.

NOTE: Improper connections will damage the generator.

   c. Start the generator, verifying the correct 3720 RPM no load speed.

   d. Test the stator output voltages using the following color coded test connections; note the readings:

   • Red to white main winding 1 and blue to gray main winding 2 should each read 75-95 VAC on the EM3000 series and 85-105 VAC on the EM and EMS4000 and 4500 series.

   • Black/white to black/white sub-winding should read 3-5 VAC.

   • Light green to white sensor winding should read 10-16 VAC.

   • Light green/red to green exciter winding should read 35-65 VAC.

   • White/red to black/red DC winding output should read 13-17 VDC.

   e. Stop the generator. Disconnect all previously tested connections plus the DC diode connector with the two brown wires.

   f. Using an ohmmeter, measure the resistance of the following color coded test connections. Note the readings.

   • The red to white and the blue to gray main winding connections should each read 1.15 ohms on the EM3000 series and 0.55 ohm on the EM and EMS4000 and 4500 series.

   • The black/white to black/white sub winding terminals in the 4-pin connector should read 0.23 ohm on the EM3000 series and 0.18 ohm on the EM and EMS4000 and 4500 series.

   • The light green/red to green exciter winding terminals in the 4-pin connector should read 2.01 ohms on the EM3000 series and 2.40 ohms on the EM and EMS4000 and 4500series.

   • The brown to brown DC winding terminals in the diode connector should read 1.86 ohms on the EM3000 series and 1.14 ohms on the EM and EMS4000 and 4500 series.

   • Each red, white, blue, and gray main winding connection to a ground on the stator laminations. Resistance should be infinity/no continuity.

   • Each black/white sub winding terminal to a ground on the stator laminations. Resistance should be infinity/no continuity.

   • Each light green/red and green exciter winding terminal to a ground on the stator laminations. Resistance should be infinity/no continuity.

   • Each brown DC winding terminal to a ground on the stator laminations. Resistance should be infinity/no continuity.

   g. If all of the tests in Steps d and f met specification, the stator is good. Proceed with further component testing. If the component tests meet specification, the automatic Voltage Regulator (AVR) is faulty. Replace the AVR.

   h. If any individual test in Steps d or f failed to meet specification, the stator is faulty. Replace the stator.

   i. Reconnect all previously disconnected wiring and components.

   j. Attach DC voltmeter probes to the brush leads.

   k. Start the generator and immediately test brush voltage. Voltage should test 21-29 VDC. If voltage is outside this range, immediately stop the generator. The AVR is faulty and must be replaced.

NOTE: If the AVR is faulty and requires replacement. Be sure the voltage selector switch is tested prior to generator start up, as a faulty voltage selector switch can damage the AVR. Refer to the *Individual Component Testing* section for the selector switch test.

8. To test the auto throttle system (Fig. HA505):

NOTE: The auto throttle system will not detect loads below 1.0 amp. If a load less than 120-watt is being applied to the generator at 120V (240-watt load at 240V), the auto throttle switch must be OFF so the unit will run at full throttle governed RPM to produce proper voltage.

   a. Repair any previously located faults, then properly reconnect all wiring.

   b. Turn the auto throttle switch OFF.

   c. Start the generator; make sure all loads are disconnected.

   d. Move the switch to AUTO. After a moment, the engine should idle down to 2100-2300 RPM.

   e. Return the switch to OFF. The engine should return to no load full throttle.

   f. If the auto throttle system fails Steps d and/or e, proceed to the **Individual Component Testing** section.

   g. Stop the generator.

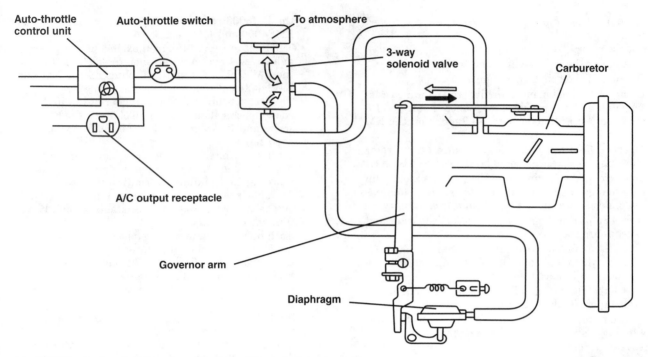

Fig. HA505 – System function view of auto throttle control components.

INDIVIDUAL COMPONENT TESTING

Receptacles

The generator must be shut down to test the receptacles.

1. Access the receptacle connections inside the control box.

2. Disconnect the RETURN (white) feed wire from the receptacle being tested.

3. Install a jumper wire between the HOT and RETURN *outlet* terminals.

   a. On 120V straight pin and twistlok receptacles, HOT terminals will be either red or blue feed wires. RETURN terminals will be the white feed wire.

   b. On 240V twistlok receptacles, HOT terminals will be the red and blue feed wires. RETURN terminals will be the white feed wire.

4. Test for continuity across the *feed* terminals. No continuity indicates a faulty receptacle.

Voltage Selector Switch

Using the switch positions and numbered switch terminals in Fig. HA506, test for continuity between the connected terminals. The failure of any terminal pair to show continuity indicates a faulty switch.

The stator's two main windings, MW1 and MW2 (Figs. HA507 and HA508), are independent of each other but equal in capacity. With the voltage selector switch in the 120/240V position, the output is arranged in series circuits (Fig. HA507); with the switch in 120V only, the output is arranged in a parallel circuit (Fig. HA508). The AVR sensor circuit is also controlled by the switch, so the AVR will accurately control voltage in either switch position. When doing generator output testing, keep in mind that a faulty voltage selector switch could also possibly damage the AVR.

Auto Throttle System

1. To test the auto throttle switch:

   a. Access the switch terminals inside the control box.

   b. With the switch lever UP in the AUTO position, there should be continuity between the two lower terminals on the back of the switch.

   c. With the switch lever DOWN in the OFF position, the two bottom switch terminals should show no continuity.

   d. Any results different from those specified in Steps B and C indicate a faulty switch.

2. To test the diaphragm assembly (Fig. HA505):

   a. Apply suction to the vacuum line fitting. If the diaphragm is good, the link will draw down.

   b. Continue to hold suction for approximately one minute to check for pinhole leakdown.

   c. Inspect all vacuum hoses for leaks.

3. To test the 3-way solenoid valve (Fig. HA505):

   a. Disconnect the black and yellow harness wires from the solenoid valve.

   b. Test the solenoid wires for continuity. No continuity indicates a faulty solenoid.

4. To test the control unit:

   a. Access the control unit inside the control box.

   b. Verify that the blue and white wires from the stator main windings are passing through the control unit hole from opposite directions (Fig. HA509). If necessary, redirect the wires. Wires passing through the control unit from the same direction will cancel each other's magnetic field, preventing the control unit from sensing voltage differences between load and no load.

   c. If all other auto throttle components test good, and the control unit through wires are routed correctly, but the auto throttle still does not operate, the control unit is faulty. Replace the control unit.

Condenser/Capacitor

The condenser is a part of the AVR and cannot be repaired separately.

DC Diode

To test the DC diode on the generator rear housing:
1. Unplug the diode connector.
2. Set the ohmmeter to the R × 1 scale.
3. Touch the positive meter lead to one diode terminal. Touch the negative meter lead to the diode case. Note the reading.
4. Reverse the leads; again note the reading.
5. Repeat Steps 3 and 4 with the other diode terminals.
6. In one direction, resistance should have read approximately 5-30 ohms. In the reverse direction, resistance should have read infinity/no continuity.
7. Any forward reverse test which read ohms both ways or infinity both ways indicates a faulty diode assembly. Replace the diode.

Diode Stack (EMS Series)

To test the diode stack:
1. Unplug the diode stack connector, then refer to Fig. HA510 and the following terminal pair list:

W to Lg	Y to W/B	G to Y/B
P to Y/B	R/B to G	R/B to P

2. With the ohmmeter set to the R × 1 scale, apply the positive (+) meter probe to the W terminal of the first color coded terminal pair, and apply the negative meter probe to the Lg terminal. Note the reading.

3. Reverse the meter leads and again note the reading.

4. Repeat Steps 2 and 3 with the other five color coded terminal pairs, always starting with the positive meter probe on the first terminal of the listed pair.

5. In Steps 2-4, one meter lead test should have shown continuity and the reverse test should have shown no continuity for each of the six color coded terminal pairs, with the first test result of each pair being the same as the first test result of all other pairs.

6. If any terminal pair tested continuity both ways or no continuity both ways, or if any terminal pair tested the reverse of the rest of the pairs, the diode stack is faulty. Replace the stack.

Automatic Voltage Regulator (AVR)

There is no test procedure available for the AVR. If the rotor, stator, receptacles, breakers, and harnesses all meet specification in the **Troubleshooting** tests, but there is still faulty output, replace the AVR.

Pilot Lamp

To test the pilot lamp, disconnect the black/red and white/red lamp pigtails in the control box. There should be continuity. No continuity indicates a faulty lamp. Replace the lamp assembly.

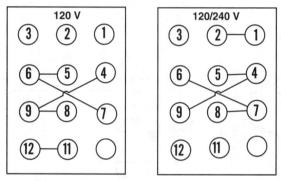

Fig. HA506 – Continuity test positions for the voltage selector switch. Continuity should exist between the noted terminals.

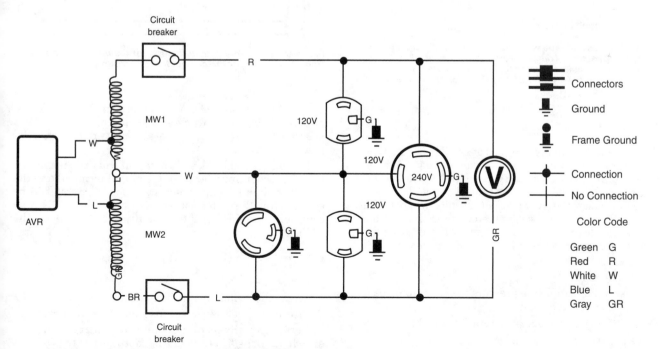

Fig. HA507 – Output wiring with the voltage selector switch in the 120/240V position.

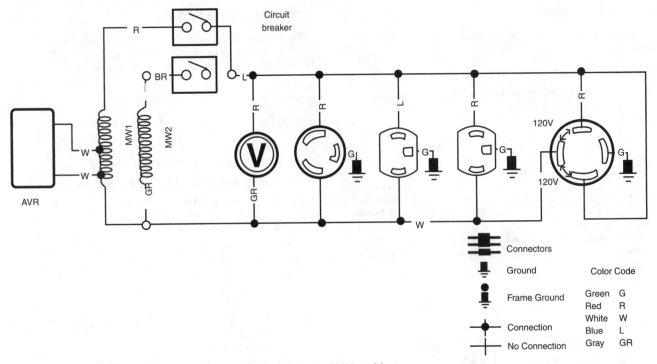

Fig. HA508 – Output wiring with the voltage selector switch in the 120V position.

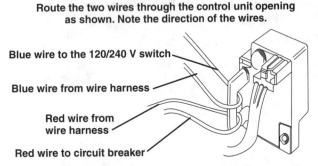

Auto-throttle Control Unit
Route the two wires through the control unit opening
as shown. Note the direction of the wires.

Blue wire to the 120/240 V switch

Blue wire from wire harness

Red wire from wire harness

Red wire to circuit breaker

Fig. HA509 – Red and blue wires from the main AC stator winding properly passing through the auto throttle control unit sensor loop from opposite directions. Failure to route the wires as shown will cause the auto throttle system not to function.

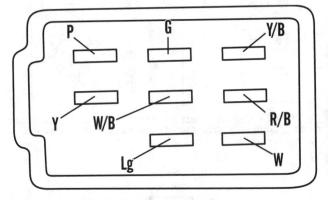

Fig. HA510 – Terminal identification for testing the diode stack. Refer to the text for the correct test procedure.

Oil Alert System

The oil alert system is comprised of an oil cup inside the engine crankcase, a tube connecting the cup to the outside of the crankcase, and a hose connecting the tube to the diaphragm switch on the generator control panel. When engine oil falls below the ADD mark on the dipstick, the tube fitting in the cup is exposed to crankcase pressure. This pressure travels through the tube and hose, activating the switch diaphragm and grounding out the ignition primary circuit.

To test the system:

1. Make sure that crankcase pressure is not escaping past the crankcase tube grommet.

2. Inspect the diaphragm hose for cracks or leaks.

3. Move the switch arm to the ON position.

4. Access the switch connections inside the control box.

5. Manually apply light air pressure to the switch diaphragm fitting. The switch arm should move to the OFF position. Failure to do so indicates a faulty diaphragm. Replace the switch.

6. Disconnect the black switch pigtail from the main harness. Test continuity between the pigtail terminal and the switch ground in both switch positions. There should be continuity in the OFF position only. Continuity in ON or no continuity in OFF indicates a faulty switch. Replace the switch.

Engine Switch (EM Series)

To test the switch:

1. Access the switch inside the control box.

2. Disconnect the switch pigtails from the main harness.

3. Test continuity between the switch pigtails in both switch positions. There should be continuity in the OFF position only.

4. Continuity in the ON position or no continuity in the OFF position indicates a faulty switch. Replace the switch.

Engine Switch (EMS Series)

To test the switch:
1. Access the switch inside the control box.
2. Unplug the 4-wire switch connector from the main harness.
3. Test for continuity:

a. Black to green – switch OFF.
b. White to black/white – switch START.

4. No continuity in Steps 3-A and 3-B indicates a faulty switch; replace the switch.

Voltmeter

To test the voltmeter:
1. Access the voltmeter wiring inside the control box.
2. Disconnect one voltmeter lead.

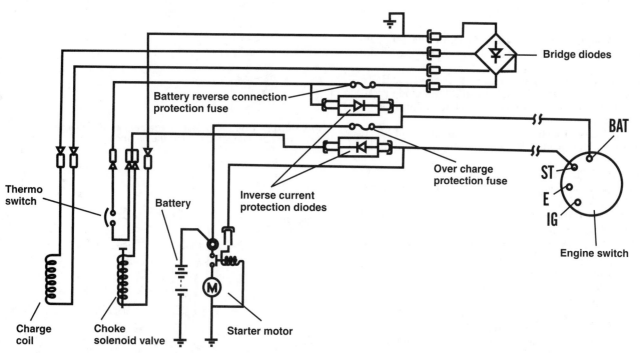

Fig. HA510 – *Schematic diagram of the auto choke system components on EMS series units.*

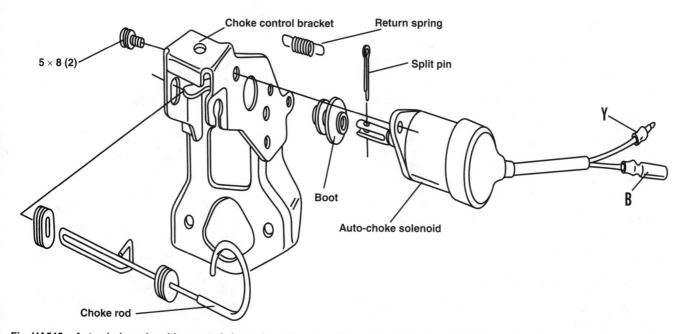

Fig. HA512 – *Auto choke solenoid mounted above the carburetor. Solenoid pigtails should be both continuity tested and power tested: Refer to the text for the correct test procedure.*

3. Using an ohmmeter, test for continuity between the voltmeter posts.
4. A lack of continuity indicates a faulty voltmeter.

Auto Choke Systems

Electric start EMS models are equipped with an electrically operated automatic choke system to aid cold starts (Fig. HA511).

> **NOTE: Manually operating the choke lever will override the auto choke.**

1. When the engine switch is turned to START, battery current energizes the choke solenoid and closes the choke shutter.
2. When the engine starts and the switch is moved to RUN, the thermoswitch in the cylinder head allows diode rectified current produced by the charging coil to maintain solenoid control choking.
3. When the head reaches a preset temperature of 30° C (85-90° F), the thermoswitch opens the circuit, opening the choke shutter.
4. To test the EM auto choke solenoid:
    a. Access the solenoid mounted above the carburetor (Fig. HA512).
    b. Disconnect the yellow and black solenoid pigtail wires from the main harness.
    c. Test the yellow and black solenoid pigtail terminals. Continuity should be present.
    d. Connect a 12-volt test battery to the solenoid pigtail terminals. The solenoid should click, visibly moving the choke linkage.
    e. If the solenoid does not test as specified in Steps C and D, the solenoid is faulty.
5. To test the charge coil (located on the engine, behind the flywheel – Fig. HA513):
    a. Disconnect the pink and green charge coil wire terminals from the main harness.
    b. Test for continuity between the pink and green wire terminals. There should be continuity. A lack of continuity indicates a faulty charge coil. Replace the coil.
6. To test the thermoswitch:
    a. Remove the thermoswitch from the cylinder head.
    b. Suspend the thermoswitch in a pan of water. Do not allow the switch to contact the pan. Connect ohmmeter leads to the thermoswitch wires.
    c. Place a thermometer in the water, then heat the water, observing both the thermometer and the ohmmeter. There should be continuity below the 30° C (85-90° F) cutoff temperature and no continuity above that temperature. The thermoswitch should be replaced if it does not trip within specification.
7. An alternate method would be to start the engine and allow a brief (3-5 minute) warm up period with the auto throttle switch OFF. If the choke has not opened up by the end of the warm up period, the thermoswitch is faulty. Stop the engine, allow it to cool, and replace the thermoswitch.

Remote Control Start

Some electric start units use a remote control start option (Fig. HA514) that plugs into the normally blind 8-pin coupler at the back of the generator control box. To test the remote control box:
1. Unplug the 6-pin connector from the rear of the remote control box.

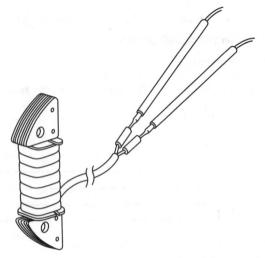

Fig. HA513 – Continuity testing the auto choke charge coil pigtails; the coil is located behind the engine flywheel.

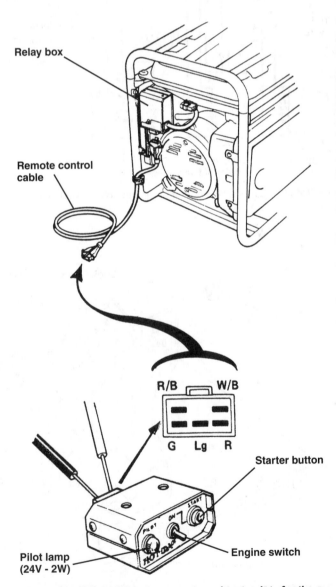

Fig. HA514 - View of the components and test points for the remote control start option.

2. Test for continuity between the following color coded terminals and box control positions:

Light green to red to white/black . . . . engine switch ON
Red/black to green . . . . pilot lamp
Light green to red . . . . start button IN

3. No continuity in any of the Step 2 tests indicates a faulty remote control box. Replace the box.

4. If the remote control box meets specification and the remote control harness integrity is secure, but the remote control will not activate the electric start, operate the electric start with the control panel mounted engine switch. If the panel engine switch operates the electric start, the remote control relay box is faulty. Replace the relay box.

DISASSEMBLY

Refer to Figs. HA501-HA503 for component identification. Always allow the generator to cool down before disassembly. Note that the rotor cannot be removed without first removing the stator. Exercise caution when handling the stator and rotor so the windings and winding coatings are not damaged.

1. If necessary:

a. Remove the upper frame rail assembly.

b. Carefully drain, then remove, the fuel tank.

c. Remove the muffler assembly and muffler shields.

2. Remove the generator end cover.

3. Noting the wiring orientation and the location of any wiring retainers, disconnect and unplug all control box harness connectors; remove the control box.

4. Remove the flange nuts holding the generator rear housing feet to the rubber mounts. Note the location of the ground wire attached to the left foot.

5. Carefully lift the rear housing so the feet clear the rubber mount studs. Place a temporary support, such as a 2 × 4, under the engine PTO housing.

6. Disconnect and remove the brush holder assembly.

7. Disconnect any remaining wiring, carefully noting the wiring orientation.

8. If replacing the stator, note the position of the stator cover wrap, then unbend the two tabs holding the stator cover to the stator. Remove the cover.

9. Remove the four rear housing bolts. Separate the rear housing from the stator.

10. Noting the stator wire orientation, carefully separate the stator from the engine PTO housing.

11. Remove the rotor throughbolt, then remove the rotor.

a. Place a soft support under the rotor to prevent damage upon removal.

b. Squirt a light amount of penetrating oil through the rotor shaft onto the taper.

c. Use a slide hammer with a threaded adapter to break the rotor free from the taper on the engine crankshaft.

12. Remove the screws holding the fan to the rotor.

REASSEMBLY

Refer to Figs. HA501-HA503 for component identification. Note that the rotor must be installed before the stator.

Exercise caution when handling the stator and rotor so the windings and winding coatings are not damaged.

1. If the engine PTO housing was removed for service:

a. Make sure that the gasket surfaces and thread holes are clean and dry.

b. Make sure that the two locator dowels are properly positioned.

c. Install a new gasket.

d. Align the balancer gear as follows:

• Remove the 8 mm blind plug bolt from the outside of the housing.

• Position the piston at TDC.

• Insert a pin through the blind plug hole into the hole in the balance gear.

• Being careful not to damage the crankshaft seal, install the housing onto the engine, meshing the balance gear with the crank gear.

• Incrementally and alternately torque the eight housing bolts 22-26 N•m (190-225 in.-lb.).

• Remove the alignment pin. Install and tighten the blind plug bolt.

2. Place a temporary support, such as a 2 × 4, under the engine PTO housing so that when the rear housing is installed onto the stator, the housing feet will clear the rubber mount studs.

3. Service the rotor bearing, if needed.

4. Mount the cooling fan to the rotor.

5. Make sure that the matching tapers on the engine crankshaft and inside the rotor shaft are clean and dry. Install the rotor onto the crankshaft. Install the rotor throughbolt and washer, then torque the rotor bolt 43-47 N•m (31-34 ft.-lb.).

6. Noting the stator wire orientation, fit the stator over the rotor and against the engine PTO housing.

7. Fit the end housing onto the stator and rotor, being careful not to pinch or kink any wiring. Install the end housing bolts, then incrementally and alternately torque the bolts 8-12 N•m (70-105 in.-lb.).

8. Remove the spark plug and rotate the engine crankshaft to check for interference between the rotor and stator. There should be a consistent gap around the rotor laminations inside the stator. If any rubbing or scraping is evident, correct any misalignment prior to proceeding. Reinstall the spark plug.

9. Remove the temporary support from under the end housing and set the housing feet onto the rubber mount studs. Reinstall the ground wire onto the left foot stud. Install the flange nuts holding the feet to the studs. Torque the nuts 30-40 N•m (260-345 in.-lb./22-29 ft.-lb.).

10. If removed during disassembly, install the stator cover over the stator in its proper position (noted during disassembly). Insert the holding tabs into the slots. Bend the tabs as necessary to snugly hold the cover.

11. Install the brush holder assembly.

12. Install the control box assembly, correctly routing the wiring to prevent kinking or pinching. Replace any wiring retainers removed during disassembly.

13. Refer to Figs. HA515-HA517 as applicable and properly reconnect all wiring.

14. Install the generator end cover.

15. If removed during disassembly, reinstall the muffler shields and muffler assembly, fuel tank, and upper frame rail assembly.

FASTENER TORQUE CHART

Fastener torque values are listed in the approximate order of reassembly.

Component Fastener	Torque
Engine PTO housing	22-26 N•m (190-225 in.-lb./16-19 ft.-lb.)

Rotor throughbolt . . . . 43-47 N•m (375-405 in.-lb./31-34 ft.-lb.)
Generator rear end housing. . . . . 8-12 N.m (70-105 in.-lb.)
Rear housing feet mount nuts . . . . 30-40 N•m (260-345 in.-lb./22-29 ft.-lb.)

WIRING DIAGRAMS

Generator	Diagram
EM3000, EM4000	Fig. HA515
EMS4000	Fig. HA516
EMS4500	Fig. HA517

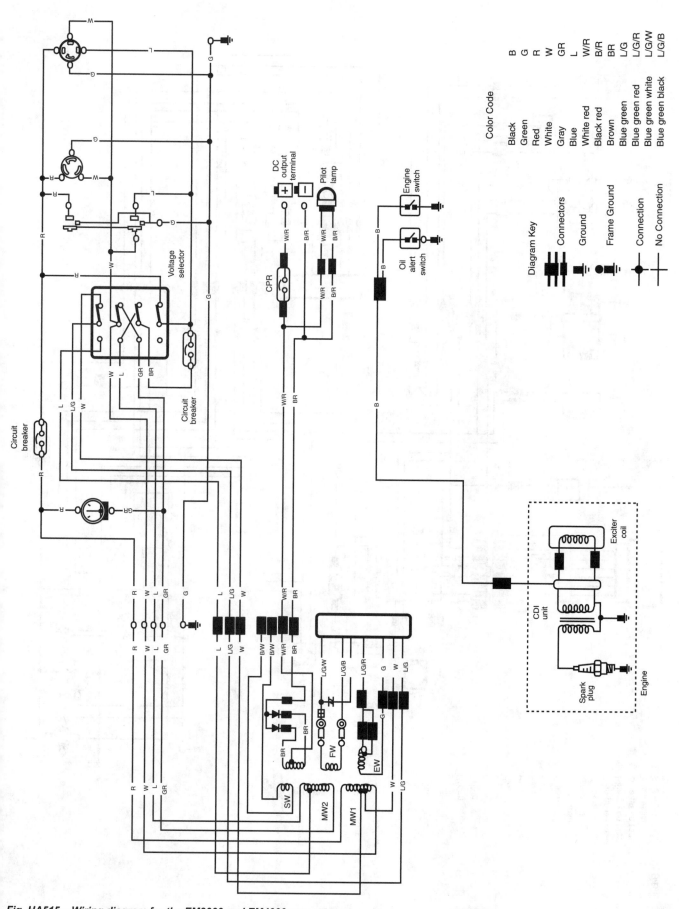

Fig. HA515 – Wiring diagram for the EM3000 and EM4000 generators.

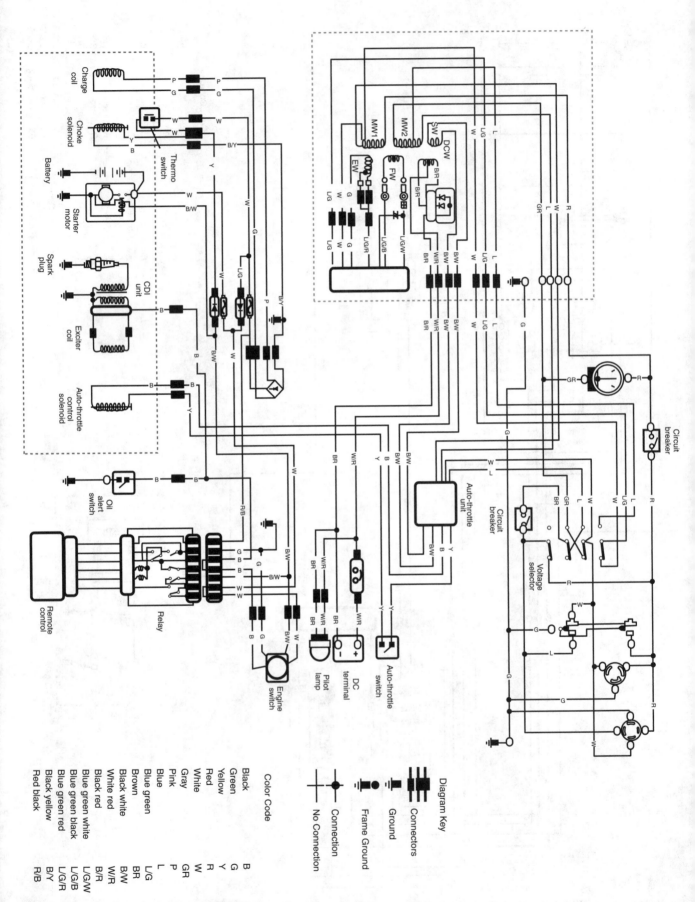

Fig. HA516 – Wiring diagram for the EMS4000 generator.

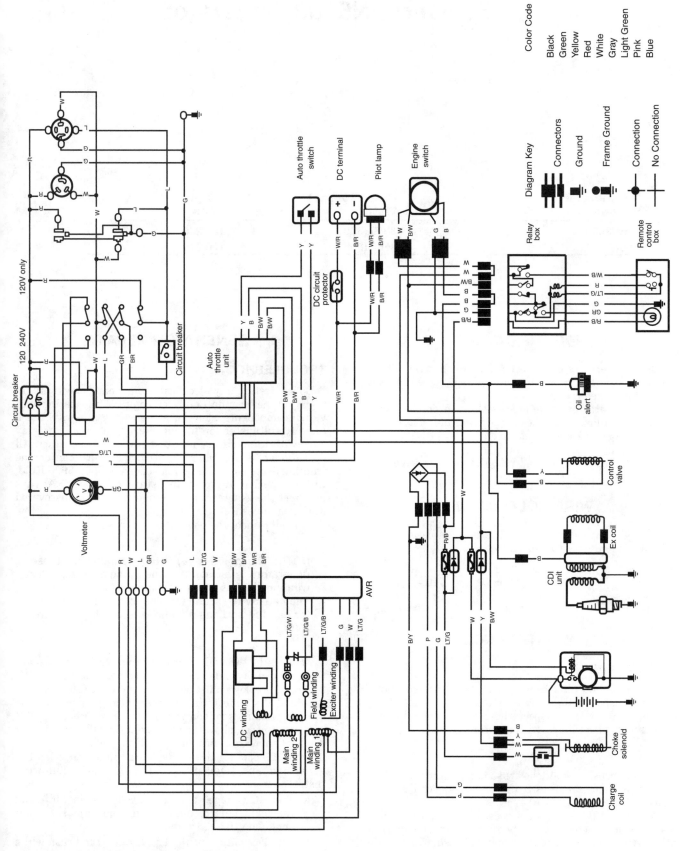

Fig. HA517 – Wiring diagram for the EMS4500 generator.

HONDA
EN3500 and EN5000 Generators

Generator Model	Rated (Surge) Watts	Output Volts
EN3500	3000 (3500)	120/240
EN5000	4500 (5000)	120/240

IDENTIFICATION

These generators are direct drive, brush style, 60 Hz, single phase units with automatic voltage regulation and endbell mounted receptacles.

The generator serial number is located on the frame crossmember just below the engine carburetor. The engine serial number is stamped on the crankcase just above the oil drain plug.

Refer to Figs. HA601A and HA601B to view the generator components.

WIRE COLOR CODE

The following color code abbreviations are used to identify the generator wiring in this series:

Bl – Black	Bu – Blue	W – White
R – Red	G – Green	Lb – Light blue
Y – Yellow	Br – Brown	Lg – Light green
Gr – Gray	P – Pink	O – Orange

MAINTENANCE

ROTOR BEARING

The brush end rotor bearing is a pre-lubricated, sealed bearing. Inspect and replace as necessary. If a seized bearing is encountered, inspect the rotor journal and the rear housing bearing cavity for scoring.

BRUSHES

Replace brushes when the brush length protruding from the brush holder reaches 8.5 mm (0.33 in.) or less. New brush length is 12.0 mm (0.47 in.).

The red AVR wire always goes to the positive brush, which is the brush closest to the rotor windings. The white AVR wire always goes to the negative brush, which is the brush closest to the rotor end bearing.

GENERATOR REPAIR

TROUBLESHOOTING

Before testing any electrical components, check the integrity of the terminals and connectors. While testing individual components, also check the wiring integrity. Use a digital multimeter for testing. Resistance readings may vary with temperature and test equipment accuracy. All operational testing must be performed with the engine running at its governed no load speed of 3600-3900 RPM, unless specific instructions state otherwise. Always load test the generator after repairing defects discovered during troubleshooting.

To troubleshoot the generator:

NOTE: The positive (+) brush terminal on these generators is always the one closest to the rotor windings; it has a red wire coming from the AVR. The negative (-) brush terminal is closest to the rotor bearing and has a white wire coming from the AVR.

1. Remove the two end cover mounting screws and sleeves then remove the end cover. Do not lose the sleeves.
2. Circuit breakers must be ON. To test the AC breaker, access and disconnect the breaker wiring, then test the breaker terminals for continuity. The breaker is faulty if it will not stay set, shows continuity in the OFF position, or does not show continuity in the ON position.
3. Leave all components and wiring connected, start the generator, and refer to Fig. HA602 for the following tests.

   a. Test AC voltage at the main red to main white and the main blue to main white connections. Output should be 102-138 volts.

   b. Test DC voltage at the brush red [+] to brush white [-] connections. Output should be 11-19 volts.

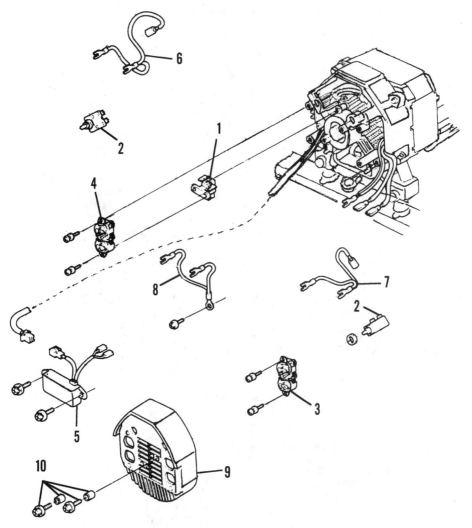

Fig. HA601A – Exploded view of generator components.

1. Brush-holder assembly
2. AC circuit breaker
3. 120V AC receptacle
4. 240V AC receptacle
5. AVR
6. Terminal wire A (Blue)
7. Terminal wire B (Red)
8. Internal ground wire
9. End cover
10. End cover sleeves and screws (2 each)

4. Stop the generator.

   a. If Step 3 tests met specification, proceed to the Individual Component Testing section in this chapter.

   b. If Step 3 tests did not meet specification, proceed with Step 5.

5. To test the rotor:

   a. Disconnect the red and white AVR brush wire connectors.

   b. Using an ohmmeter, measure field coil winding resistance through the brushes. Resistance should be 38-47 ohms on the EN3500 series, 46-57 ohms on the EN5000 series.

   c. Measure field coil winding resistance at the rotor slip rings. Resistance should be the same as Step B.

   d. Measure field coil winding resistance from each slip ring to a ground on the rotor shaft or laminations. Resistance should be infinity/no continuity.

   e. If test Steps b and c met specification, the rotor and brushes are good.

   f. If test Step b failed specification but Step c met specification, the rotor is good. Inspect the brushes and test each brush for continuity through its terminal. If faulty, replace the brush holder assembly.

   g. If both test Steps B and C failed specification, the rotor is faulty. Replace the rotor, then inspect and test the brushes.

   h. If either Step D test reads any continuity, the field coil winding is shorted. Replace the rotor.

6. To test the stator:

   a. Obtain a fully charged 12-volt battery and a pair of test leads to connect the battery terminals to the brush connector terminals.

   b. Disconnect and remove the AVR. Connect the positive battery post to the red (+) brush holder terminal, then connect the negative battery post to the white (-) brush holder terminal.

NOTE: Improper connections will damage the generator.

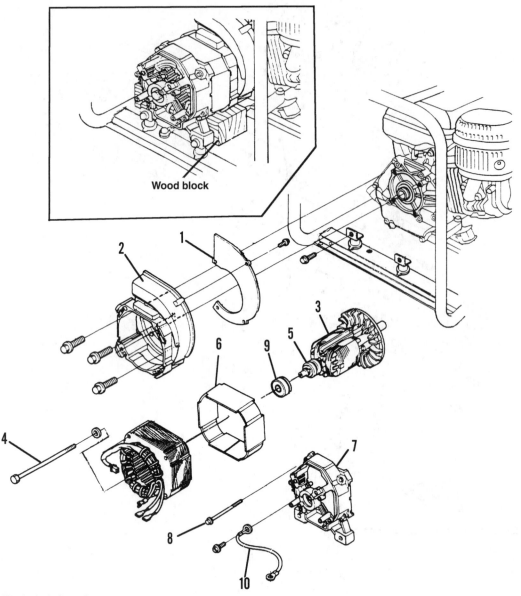

Fig. HA601B – Exploded view of generator components.

1. Front housing duct plate
2. Engine adapter/front housing
3. Rotor assembly
4. Rotor through-bolt and washer

5. Slip rings
6. Stator cover
7. Endbell/rear housing

8. Rear housing/stator through-bolts (4)
9. Rotor bearing
10. External ground wire

c. Start the generator, verifying the correct 3600-3900 RPM no load speed.

d. Test the stator output voltages using the following color coded test connections (Fig. HA602). Note the readings.

   • Red to white main winding 1 and blue to white main winding 2 should each read 93-105 VAC on the EN3500 series and 80-90 VAC on the EN5000 series.

   • Red to white sensor winding should read 93-105 VAC on the EN3500 series and 84-86 VAC on the EN5000 series.

   • Yellow to yellow exciter winding should read 50-60 VAC on the EN3500 series and 41-47 VAC on the EN5000 series.

e. Stop the generator and disconnect the test battery.

f. Using an ohmmeter, measure the resistance of the following test connections. Note the readings.

   • Resistance reading for the red to white and the blue to white main winding connections should be 0.9-1.2 ohms on the EN3500 series and 0.5-0.7 ohm on the EN5000 series.

   • Resistance reading for the yellow to yellow exciter winding terminals in the 4-pin connector should read 1.2-1.4 ohms on the EN3500 series and 0.8-1.0 ohm on the EN5000series.

   • Resistance reading for each red, white, and blue main winding connection to a ground on the stator laminations should be infinity/no continuity.

   • Resistance reading for each yellow exciter winding terminal to a ground on the stator laminations should be infinity/no continuity.

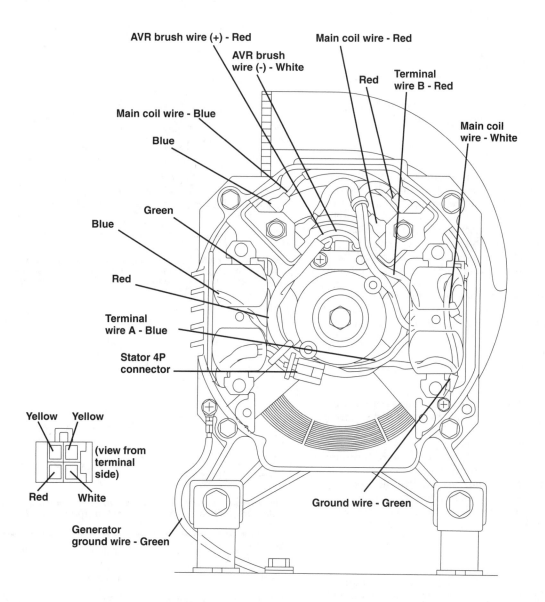

AVR brush wire (+) - Red

AVR brush wire (-) - White

Main coil wire - Red

Red

Terminal wire B - Red

Main coil wire - Blue

Blue

Main coil wire - White

Green

Blue

Red

Terminal wire A - Blue

Stator 4P connector

Yellow   Yellow

(view from terminal side)

Red   White

Ground wire - Green

Generator ground wire - Green

Fig. HA602 - End view of rear housing showing the generator wiring and test points. The AVR has been removed in this view. It mounts at the bottom of the housing and connects to the 4-terminal connector and the brush assembly.

g. If all of the tests in Steps d and f met specification, the stator is good. Proceed with further component testing. If the component tests meet specification, the automatic voltage regulator (AVR) is faulty and must be replaced.

h. If any individual test in Steps D or F failed to meet specification, the stator is faulty and must be replaced.

INDIVIDUAL COMPONENT TESTING

Receptacles

The generator must be shut down to test the receptacles.

1. Remove the generator end cover to access the receptacle(s).

2. Disconnect one feed wire from the receptacle being tested.

3. Install a jumper wire between the HOT and RETURN *outlet* terminals (Fig. HA603).

4. Test for continuity across the *feed* terminals. No continuity indicates a faulty receptacle.

5. If replacing a receptacle, do not tighten the receptacle mounting screws until after the end cover is installed, so the receptacle can be properly aligned with the end cover opening.

Automatic Voltage Regulator (AVR)

There is no test procedure available for the AVR. If the rotor, stator, receptacles, breakers, and harnesses all meet specification in the **Troubleshooting** tests, but there is still faulty output, replace the AVR.

Engine Switch

The engine switch is mounted on the side of the engine, below the gas tank. To test the switch:
1. Disconnect the 2-lead switch wire from both engine wiring terminals.
2. Test continuity between the switch pigtail and any good ground on the engine blower housing. There should be continuity in the OFF position only.

Oil Alert System

The oil alert system works in conjunction with the engine switch, and is comprised of two components—the oil level switch, mounted inside the engine crankcase, and the stop switch module, mounted next to the engine switch.

To test the oil alert stop switch module (Fig. HA604).
1. Disconnect the yellow and black module wires.
2. Using a 100 kΩ/VDC impedance tester, alternately test between the yellow wire terminal, black wire terminal, and module case.
3. Compare the Step 2 test readings to the value given in Table HA605.

Any readings other than those specified indicate a faulty module.

To test the oil level switch:
1. Verify that the engine oil level is full and the generator is sitting level.
2. Disconnect the yellow switch wire.
3. Test for continuity between the yellow wire terminal and any good engine crankcase ground. There should be infinity/no continuity.
4. Drain the engine oil completely.
5. Retest for continuity between the yellow wire terminal and ground. There should be continuity.
6. Any test results other than those specified in Steps 3-5 indicate a faulty oil level switch. Replace the switch.

Anytime the oil alert wiring is disconnected, always refer to the color coded wiring diagram to insure proper reconnection.

DISASSEMBLY

Refer to Figs. HA601A and HA601B for component identification. Note that the rotor cannot be removed without first removing the stator. Note the wire routing orientation and the location of any wiring retainers. Exercise caution when handling the stator and rotor so the windings and winding coatings are not damaged.
1. Remove the end cover mounting screws and sleeves and remove the end cover. Do not lose the sleeves.
2. Disconnect and remove the AVR.
3. Remove the brush holder assembly.
4. Disconnect the red and blue stator wires from the circuit breakers. Disconnect the white stator wire from the 120V receptacle.
5. Disconnect the ground wire.
6. Remove the stator throughbolt and washer.
7. Unbolt the rear housing feet from the rubber mount brackets.
8. Carefully lift the rear housing so the feet clear the brackets. Place a temporary support, such as a 2 × 4, under the front housing.
9. Remove the four rear housing/stator throughbolts.
10. Carefully remove the rear housing assembly.
11. Carefully remove the stator, noting the wiring orientation. If the stator is being replaced, remove the stator cover for reuse.
12. Using a slide hammer with the proper adapter threaded into the rotor shaft, carefully remove the rotor.

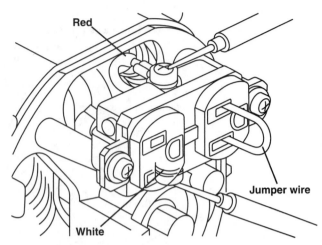

Fig. HA603 - To test the receptacle, install a jumper wire across one pair of output terminals, and disconnect at least one feed wire. There should be continuity across the feed terminals.

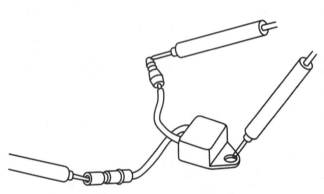

Fig. HA604 - View of the oil alert module showing the probe points for testing. Refer to the text for the correct test procedure.

(+)Probe ⟍ (-)Probe	Black	Yellow	Body
Black		500-10000Ω	∞
Yellow	500-10000Ω		∞
Body	∞	∞	

Table HA605

REASSEMBLY

Refer to Figs. HA601A and HA601B for component identification. Note that the rotor must be installed before installing the stator. Exercise caution when handling the stator and rotor so the windings and winding coatings are not damaged.

1. If the generator front housing was removed for service, install it with the air duct at the top. Torque the four bolts incrementally and alternately to 26 N•m (240 in.-lb.)
2. Place a temporary support, under the front housing so that when the rear housing is installed onto the stator, the housing feet will clear the rubber mount brackets.
3. Service the rotor bearing, if needed.
4. Insure that the mating surfaces of the engine crankshaft and rotor shaft are clean and dry.
5. Install the rotor. Install and finger tighten the rotor throughbolt and washer.

6. If the stator cover was removed, install it onto the stator at this time. Fit the stator over the rotor, noting the stator wiring orientation.

7. Install the rear housing over the rotor bearing and against the stator. Install and finger tighten the housing bolts.

8. Remove the temporary front housing support, then lower the rear housing feet onto the rubber mounts and bolt the mount brackets to the feet.

9. Holding the engine flywheel with a strap wrench, torque the rotor throughbolt 44 N•m (33 ft.-lb.).

10. Torque the four rear housing/stator through bolts incrementally and alternately 10 N•m (85-90 in.-lb.).

11. Remove the spark plug and rotate the engine crankshaft to check for interference between the rotor and stator. There should be a consistent gap around the rotor laminations inside the stator. If any rubbing or scraping is evident, correct any misalignment prior to proceeding. Reinstall the spark plug.

12. Reconnect the rear housing to frame ground wire.

13. Connect the white stator wire to the 120V receptacle. Connect the red and blue stator wires to their respective circuit breakers.

14. Install the brush holder assembly.

15. Install and connect the AVR. Ensure that the AVR harness is UP and that the AVR mounting tabs are aligned inside the rear housing casting tabs.

16. Install the end cover, cover mounting sleeves, and cover screws. If the receptacles were removed from the rear housing, do not tighten the receptacle mounting screws until the end cover is installed, so proper receptacle/end cover alignment can be achieved.

WIRING DIAGRAM

Refer to Fig. HA606 for the wiring diagram for these generators.

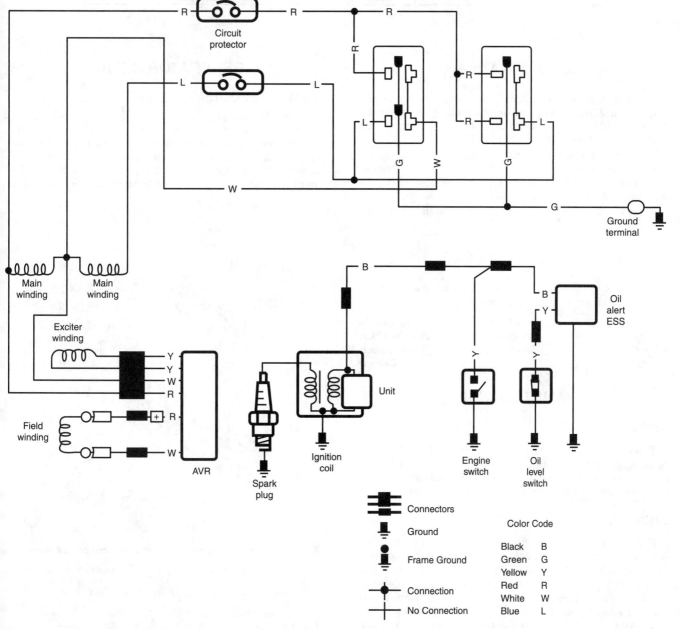

Fig. HA606 – Wiring diagram for the EN3500 and EN5000 generators.

HONDA
EB6500 Series

Generator Model	Rated (Surge) Watts	Output Vollts
EB6500	5500 (6500)	120/240

IDENTIFICATION

These generators are direct drive, brush style, 60 Hz, single phase units with automatic voltage Regulation, auto throttle idle control, voltage selector, and panel mounted receptacles.

The generator serial number is located on the frame crossmember just below the engine carburetor. The engine identification numbers are stamped on the crankcase just below the generator control panel.

Refer to Figs. HA651 and HA652 to view the generator components.

WIRE COLOR CODE

The following color code abbreviations are used to identify the generator wiring in this series:

Bl – Black	Bu – Blue	W – White
R – Red	G – Green	Lb – Light blue
Y – Yellow	Br – Brown	Lg – Light green
Gr – Gray	P – Pink	O – Orange

MAINTENANCE

ROTOR BEARING

The brush end rotor bearing is a pre-lubricated, sealed bearing. Inspect and replace as necessary. If a seized bearing is encountered, inspect the rotor journal and the rear housing bearing cavity for scoring.

BRUSHES

Replace the brushes when the brush length protruding from the brush holder reaches 8.5 mm (0.33 in.) or less. New brush length is 12.0 mm (0.47 in.).

The light green/white AVR wire always goes to the positive (+) brush, which is the brush closest to the rotor windings. The light green/Black AVR wire always goes to the negative (-) brush, which is the brush closest to the rotor end bearing.

GENERATOR REPAIR

TROUBLESHOOTING

Before testing any electrical components, check the integrity of the terminals and connectors. While testing individual components, also check the wiring integrity. The use of a digital multimeter is recommended for testing. Resistance readings may vary with temperature and test equipment accuracy. All operational testing must be performed

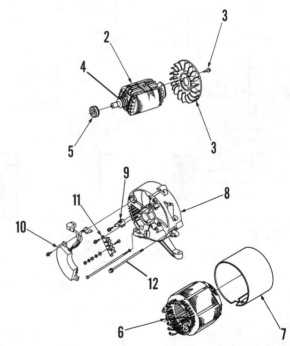

Fig. HA651 – Exploded view of generator head components.

1. Crankcase cover/front adapter
2. Rotor assembly
3. Rotor fan with three screws
4. Slip rings
5. Rotor bearing
6. Stator assembly
7. Stator cover
8. Endbell/rear housing
9. Brush-holder assembly
10. AVR assembly
11. AC output terminal strip
12. Rear housing/stator through-bolts

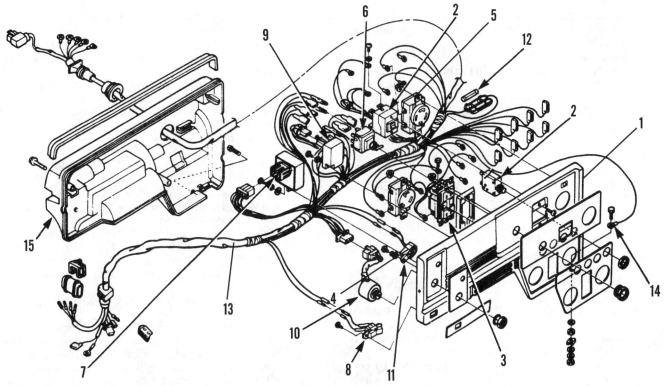

Fig. HA652 – Exploded view of generator control box and components.

1. Control panel
2. AC circuit breakers
3. 120V AC GFCI receptacle
4. 120V AC Twistlok receptacle
5. 240V AC Twistlok receptacle

6. Voltage-selector switch
7. Diode stack
8. Auto-throttle switch
9. Auto-throttle control unit
10. Engine ignition switch

11. Oil-alert unit
12. Fuse
13. Control-box harness
14. Ground terminal
15. Control-box cover

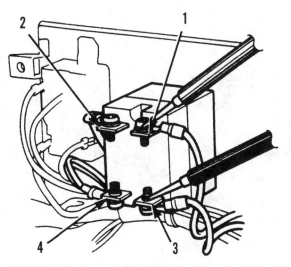

Fig. HA653 – Test connections for the 4-terminal AC circuit breaker. Refer to the text for the proper test procedure.

with the engine running at its governed no load speed of 3600-3900 RPM, unless specific instructions state otherwise. Always load test the generator after repairing defects discovered during troubleshooting.

To troubleshoot the generator:

NOTE: The positive (+) brush terminal on these generators is always the one closest to the rotor

windings. It has a light green/white wire coming from the AVR. The negative (-) brush terminal is closest to the rotor bearing and has a light green/Black wire coming from the AVR.

1. Remove the end cover mounting screws then remove the end cover.

2. Circuit breakers must be ON. To test the AC breakers, disconnect the breaker wiring, then continuity test the breaker terminals. Refer to Fig. HA653 for the 4-terminal breaker. The breaker is faulty if it will not stay set, shows continuity in the OFF position, or does not show continuity in the ON position.

3. Leave all components and wiring connected, start the generator, and refer to Fig. HA654 for the following color coded and numbered tests.

   a. Test AC voltage at the main red to main white and the main blue to main gray connections. Output should be 102-138 volts.

   b. Test DC voltage at the brush light green/white to brush light green/black connections. Output should be 21-29 volts.

4. Stop the generator:

   a. If Step 3 tests met specification, proceed to test Step 5-I, followed by the Individual Component Testing section in this Chapter.

   b. If Step 3 tests did not meet specification, proceed with Step 5-A.

5. To test the rotor:

   a. Disconnect the light green/white and light green/black AVR brush wire connectors.

b. Using an ohmmeter, measure field coil winding resistance through the brushes; resistance should be 57-67 ohms.

c. Measure field coil winding resistance at the rotor slip rings. Resistance should be the same as Step B.

d. Measure field coil winding resistance from each slip ring to a ground on the rotor shaft or laminations. Resistance should be infinity/no continuity.

e. If test Steps B and C met specification, the rotor and brushes are OK.

f. If test Step B failed specification but Step C met specification, the rotor is good. Inspect the brushes and test each brush for continuity through its terminal. If faulty, replace the brush holder assembly.

g. If both test Steps B and C failed specification, the rotor is faulty. Replace the rotor, then inspect and test the brushes.

h. If either Step D test read any continuity, the field coil winding is shorted; replace the rotor.

i. The rotor can also be dynamically tested by performing a current draw test.
- With the VOM on the DCV scale, note the battery voltage.
- With the VOM on the R × 1 scale, note the rotor winding resistance across the slip rings.
- Using ohm's Law applicable to this test (I=V÷R), calculate the current flow (I) by dividing the rotor winding resistance (R) into the battery voltage (V). Note the calculated amperage.
- Disconnect the brush feed wires and the 4-terminal AVR connector.
- Connect the positive terminal of the test battery to the positive side of the VOM/ammeter (Fig. HA655).
- Connect the ammeter negative lead to one terminal of the 3-amp breaker.
- Connect a test wire from the positive brush terminal to the other terminal of the 3-amp breaker.
- Connect a test wire from the negative brush terminal to the negative battery terminal.
- Note the field winding current draw on the ammeter.
- Start the engine. With the engine running at rated no load RPM, note the amperage draw on the ammeter.
- Compare the engine running ammeter reading to both the engine stopped ammeter reading and the calculated amperage.

j. If the running amperage is significantly higher than the static or calculated amperage, or if the circuit breaker trips while running, the rotor is shorted; replace the rotor. If there is NO amperage draw during the running test, the rotor has an open winding and must be replaced. If the running amperage is approximately the same as the static and the calculated amperage, and if the circuit breaker does not trip while running, the rotor is OK.

6. To test the stator:

a. Connect a 12-volt battery to the brush connector terminals.

b. Disconnect and remove the AVR. Connect the positive battery post to the light green/white (+) brush holder terminal, then connect the negative battery post to the light green/black (-) brush holder terminal.

NOTE: Improper battery connections will damage the generator.

c. Start the generator, verifying the correct 3600-3900 RPM no load speed.

d. Test the stator output voltages using the following test connections (Fig. HA654); note the readings:
- Red to white main winding 1 and blue to gray main winding 2 should each read 61-69 VAC.
- Light green (10)-to white (9) sensor winding should read 8-12 VAC.
- Light green/red (7) to green (8) exciter winding should read 39-51 VAC.
- Brown (11) to brown (12) DC sub winding should read 12-18 VAC.

e. Stop the generator and disconnect the test battery.

f. Using an ohmmeter, measure the resistance of the following test connections:
- The red to white and the blue to gray main winding connections should each read 0.1-0.3 ohm.
- The light green/red (7) to green (8) exciter winding terminals in the 4-terminal connector should read 0.7-1.1 ohms.
- The brown (11) to brown (12) DC sub winding terminals should read 0.4 ohm.
- The reading for each red, white, gray, and blue main winding connection to a ground on the stator laminations should be infinity/no continuity.
- The reading for each light green/red and green exciter winding terminal to a ground on the stator laminations should be infinity/no continuity.
- The reading for each brown DC sub winding terminal to a ground on the stator laminations should be infinity/no continuity.

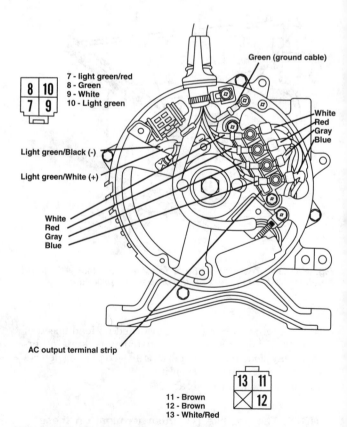

Fig. HA654 – View of end housing wiring connection test points.

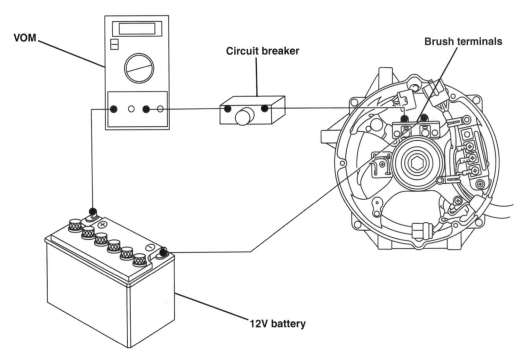

Fig. HA655 – Component connections for performing the dynamic rotor test. Refer to the text for the proper test procedure.

g. If all of the tests in Steps D and F met specification, the stator is good; proceed with further component testing. If the component tests meet specification, the automatic voltage Regulator (AVR) is faulty and must be replaced.

h. If any individual test in Steps D or F failed to meet specification, the stator is faulty and must be replaced.

i. Reconnect all previously disconnected wiring and components.

j. Attach DC voltmeter probes to the brush leads.

k. Start the generator and immediately test brush voltage; voltage should test 21-29 VDC. If voltage is outside this range, immediately stop the generator – the AVR is faulty and must be replaced.

NOTE: If the AVR is faulty and requires replacement, make sure that the voltage selector switch is tested prior to generator start up, as a faulty voltage selector switch can damage the AVR. Refer to the *Individual Component Testing* section for the selector switch test.

7. To test the auto throttle system (Fig. HA656):

NOTE: The auto throttle system will not detect loads below 1.0 Amp. If less than a 120-watt load is being applied to the generator at 120V (240-watt load at 240V), the auto throttle switch must be OFF so the unit will run at full throttle governed RPM to produce proper voltage.

a. Turn the auto throttle switch OFF.

b. Start the generator, making sure all loads are disconnected.

c. Move the switch to AUTO. After a moment, the engine should idle down to 2000-2400 RPM.

d. Return the switch to OFF. The engine should return to no load full throttle.

e. If the auto throttle system fails Steps C and D, proceed to the **Individual Component Testing** section.

INDIVIDUAL COMPONENT TESTING

GFCI Receptacle

To test the GFCI:

1. Start the generator and allow a brief no load warm up period.

2. Press the TEST button. The RESET button should extend. If it does, proceed to Step 3. If the RESET button does not extend, the GFCI is faulty.

3. Press the RESET button. The RESET button should then be flush with the TEST button. If the RESET button does not remain flush with the TEST button, the GFCI is faulty.

4. Stop the generator.

NOTE: If the GFCI seems to trip for no apparent reason, check to make sure the generator is not vibrating excessively due to faulty rubber mounts or debris between the generator and frame. Excessive vibration can trip the GFCI.

Twistlok Receptacles

The generator must be shut down to test the receptacles.

1. Access the receptacle connections inside the control box.

2. Disconnect the RETURN (white) feed wire from the receptacle being tested.

3. Install a jumper wire between the HOT and RETURN *outlet* terminals.

a. On 120V twistlok receptacles, HOT terminals will be either red or blue feed wires. RETURN terminals will be the white feed wire.

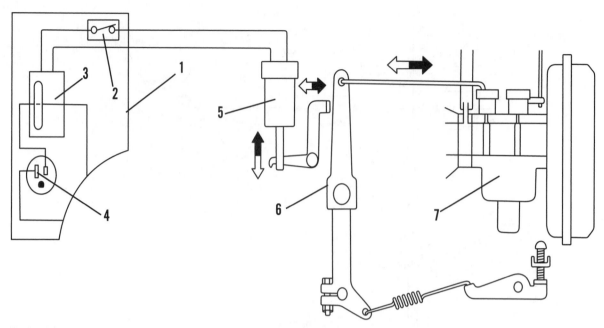

Fig. HA656 – Diagram of auto throttle system components and connections.

1. Control box
2. Auto throttle switch
3. Auto throttle control unit

4. AC output receptacle
5. Solenoid

6. Governor arm
7. Carburetor

b. On 240V twistlok receptacles, HOT terminals will be the red and blue feed wires. RETURN terminals will be the white feed wire.

4. Test for continuity across the feed terminals; no continuity indicates a faulty receptacle.

Voltage Selector Switch

Using the switch positions and numbered switch terminals in Fig. HA657, test for continuity between the following numbered terminals in the positions noted. The failure of any terminal pair to show continuity indicates a faulty switch.

Position	Terminal Pair Continuity
120V only	2-3, 5-6, 11-12
120/240V	1-2, 4-5, 10-11

The stator's two main windings, are independent of each other but equal in capacity. With the voltage selector switch in the 120/240V position, the output is arranged in series circuits. with the switch in 120V only, the output is arranged in a parallel circuit. The AVR sensor circuit is also controlled by the switch, so the AVR will accurately control voltage in either switch position. When doing generator output testing, keep in mind that a faulty voltage selector switch could also possibly damage the AVR.

Diode Stack

To test the diode stack:

1. Disconnect the diode stack connector (Fig. HA658) inside the control box.

2. With the VOM set to the R × 1 scale:

a. Place the positive (+) meter probe on the first terminal pair position indicated in the following list. Place the negative (-) meter probe on the second terminal pair position indicated in the following list.

1-2   3-4   5-4   6-4   7-4

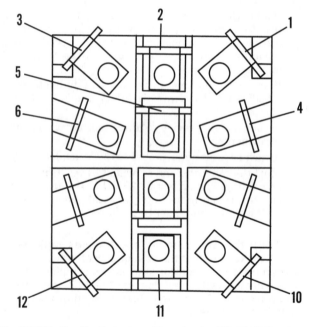

Fig. HA657 - Continuity test positions for the 120-120/240 voltage selector switch. Refer to the text for the proper test procedure.

b. Note the meter reading.

c. Reverse the meter probes and again note the reading.

d. One test should have shown continuity, and one test should have shown infinity/no continuity.

e. Repeat the A-B/C-D test with the other terminal pairs. All tests should show continuity/infinity in the same order.

f. Continuity test all the other terminal pairs (1-3, 1-4, 1-5, 1-6, 1-7, 2-1, 2-3, etc.) in both directions. There should be no continuity between any other terminal pairs.

3. Any test results other than those indicated in Steps A-F indicate a faulty diode stack which must be replaced.

Auto Throttle System

1. To test the auto throttle switch (Fig. HA656):
    a. Access and disconnect the switch terminals inside the control box.
    b. With the switch lever in the ON position, there should be continuity between the two pigtails.
    c. With the switch lever in the OFF position, the two pigtails should show no continuity.
    d. Failure of the switch to test as specified indicates a faulty switch. Replace the switch.
2. To test the solenoid:
    a. Disconnect the solenoid harness connector from the solenoid assembly on top of the engine.
    b. Using a fully charged 12V battery, connect the battery terminals to the solenoid terminals. The solenoid plunger should pull the throttle lever to idle position.

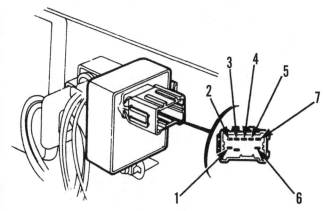

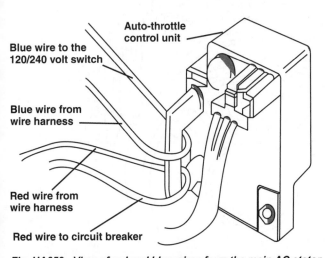

Fig. HA658 - Terminal identification for testing the diode stack. Refer to the text for the correct test procedure.

Fig. HA659 - View of red and blue wires from the main AC stator winding properly passing through the auto throttle control unit sensor loop from opposite directions. Failure to route the wires as shown will cause the auto throttle system not to function.

Blue wire to the 120/240 volt switch

Auto-throttle control unit

Blue wire from wire harness

Red wire from wire harness

Red wire to circuit breaker

c. Disconnect the test battery; return spring action should bring the throttle lever back to non idle position.
    d. Failure of the solenoid to activate as described in Steps 2 and 3 indicate a faulty solenoid. Replace the solenoid assembly.
3. To test the control unit:
    a. There is no specific test for the auto throttle control unit. If the blue and red generator main winding leads are correctly routed through the sensor loop on the back of the control unit (Fig. HA659), and if the generator DC sub winding, diode stack, and auto throttle solenoid and switch all test to specification, but the auto throttle still does not function, the control unit is faulty. Replace the control unit.
    b. If replacing the control unit, make sure that the blue and red main winding leads pass through the sensor loop from opposite directions as shown in Fig. HA659. If the wires pass through the loop from the same direction, they will cancel each other's magnetic field, preventing the new control unit from sensing voltage differences, thereby causing the new unit to malfunction.

Oil Alert

To test the oil alert indicator lamp:

1. Access the lamp unit inside the generator control box.
2. Disconnect the black and yellow harness connectors.
3. Using a test battery (.5 volts minimum; 6 volts maximum):
    a. Connect the positive battery post to the black lamp unit pigtail.
    b. Connect the negative battery post to the yellow pigtail.
4. Failure of the lamp to light when connected to the test battery indicates a faulty oil alert unit; replace the unit.

To test the engine mounted oil alert switch:

1. Verify that the engine oil level is full and the generator is sitting level.
2. Disconnect the yellow switch wire.
3. Start the engine.
4. Ground the yellow wire terminal against the engine.
    a. The oil alert lamp should blink, and
    b. The engine should quit.
5. Also disconnect the green switch wire.
6. Test for continuity between the yellow and green wire terminals. There should be none.
7. Any test results other than those specified indicate a faulty oil alert switch. Replace the switch.

Engine Ignition Switch

To test the switch:

1. Access and unplug the switch connector inside the generator control box (Fig. HA660).
2. Test for continuity between the named and color coded terminals in the following list with the switch in the position noted:

OFF position . . . . IG (black) to E (green), and FS (green/white) to G (blue)
ON position . . . . No continuity between any terminals
'Start' position . . . . BAT (white) to ST (Black/white)

3. Any readings other than those specified indicate a faulty switch; replace the switch.

Auto Choke System

To test the auto choke system (Fig. HA661):

1. Disconnect the vacuum hose from the carburetor mounted diaphragm assembly.

2. Manually apply vacuum to the diaphragm fitting and observe the choke rod. The rod should pull toward the diaphragm housing.

3. Hold manual vacuum for approximately one minute to test for leakdown.

4. Release manual vacuum and observe the choke rod for smooth operation.

5. Inspect the vacuum hoses for cracks or leaks.

6. Make sure that the in line dashpot check valve is mounted with the black end toward the diaphragm.

7. Make sure that the carburetor to block insulator and insulator gaskets are in good condition, preventing any air leaks.

DISASSEMBLY

Refer to Figs. HA651 and HA652 for component identification. Note that the rotor cannot be removed without first removing the stator. Exercise caution when handling the stator and rotor so the windings and winding coatings are not damaged.

1. If necessary:

   a. Remove the upper frame rail assembly.

   b. Carefully drain, then remove, the fuel tank.

   c. Remove the muffler assembly and muffler shields.

2. Remove the generator end cover.

3. Noting the wiring orientation and the location of any wiring retainers, disconnect and unplug all control box harness connectors; remove the control box.

4. Remove the flange nuts holding the generator rear housing feet to the rubber mounts. Note the location of the ground wire attached to the left foot.

5. Lift the rear housing so the feet clear the rubber mount studs and place a temporary support, such as a 2 ×4, under the engine PTO housing.

6. Disconnect and remove the brush holder assembly.

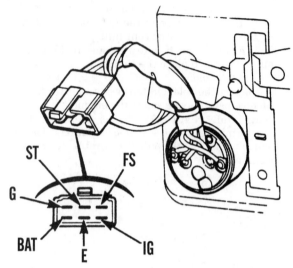

Fig. HA660 – View of terminal test points on the engine ignition switch. Refer to the text for the proper test procedure.

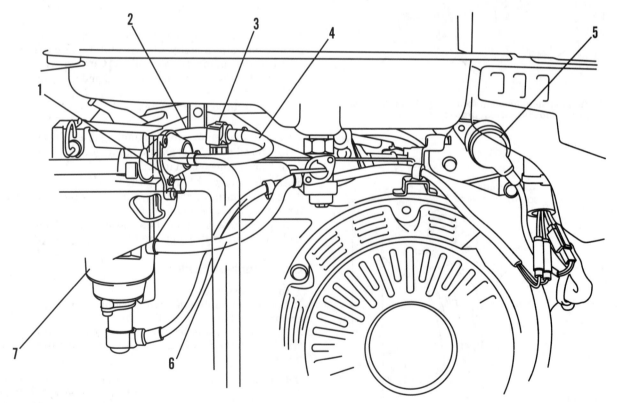

Fig. HA661 – View of the auto choke system components.

1. Choke diaphragm	4. Vacuum hose	6. Fuel hose
2. Vacuum hoses	5. Auto-throttle solenoid	7. Carburetor
3. Dashpot check valve		

7. Disconnect any remaining wiring, carefully noting the wiring orientation.

8. If replacing the stator, note the position of the stator cover wrap. Straighten the tabs holding the stator cover to the stator; remove the cover.

9. Remove the rear housing bolts. Separate the rear housing from the stator.

10. Noting the stator wire orientation, separate the stator from the engine PTO housing.

11. Remove the rotor throughbolt, then remove the rotor.

   a. Place a soft support under the rotor to prevent damage upon removal.

   b. Apply a light amount of penetrating oil through the rotor shaft onto the crankshaft to rotor taper.

   c. Use a slide hammer with a threaded adapter to break the rotor free from the taper on the engine crankshaft.

12. Remove the screws holding the fan to the rotor.

REASSEMBLY

Refer to Figs. HA651 and HA652 for component identification. Note that the rotor must be installed before the stator. Exercise caution when handling the stator and rotor so the windings and winding coatings are not damaged.

1. If the engine PTO housing was removed for service:

   a. Insure that the gasket surfaces and thread holes are clean and dry.

   b. Make sure that the two locator dowels are properly positioned.

   c. Install a new gasket.

   d. Carefully align the governor components.

   e. Incrementally and alternately torque the seven housing bolts to 24 N•m (205-210 in.-lb.).

2. Place a temporary support, such as a 2 × 4, under the engine PTO housing so that, when the rear housing is installed onto the stator, the housing feet will clear the rubber mount studs.

3. Service the rotor bearing, if needed.

4. Mount the cooling fan to the rotor.

5. Insure that the matching tapers on the engine crankshaft and inside the rotor shaft are clean and dry. Install the rotor onto the crankshaft. Install the rotor throughbolt and washer, then torque the rotor bolt to 45 N•m (32. ft.-lb.).

6. Noting the stator wire orientation, fit the stator over the rotor and against the engine PTO housing. Make sure that the grommet has been correctly installed in the top of the housing prior to tightening the stator against the housing.

7. Fit the end housing onto the stator and rotor, being careful not to pinch or kink any wiring. Install the four end housing bolts, then incrementally and alternately torque the bolts 10 N•m (85 in-lb).

8. Remove the spark plug and rotate the engine crankshaft to check for interference between the rotor and stator. There should be a consistent gap around the rotor laminations inside the stator. If any rubbing or scraping is evident, correct any misalignment prior to proceeding. Reinstall the spark plug.

9. Remove the temporary support from under the end housing and set the housing feet onto the rubber mount studs. Install the ground wire onto the left foot stud. Install the flange nuts holding the feet to the studs and torque the nuts 35 N•m (25 ft.-lb.).

10. If removed during disassembly, install the stator cover over the stator in its proper position (noted during disassembly). Insert the holding tabs into the slots. Bend the tabs as necessary to snugly hold the cover.

11. Install the brush holder assembly.

12. Install the control box assembly, correctly routing the wiring to prevent kinking or pinching, and replacing any wiring retainers removed during disassembly.

13. Refer to Figs. HA662-HA664 as applicable and properly reconnect all wiring.

14. Install the generator end cover.

15. If removed during disassembly, reinstall the muffler shields and muffler assembly, fuel tank, and upper frame rail assembly. When reinstalling the fuel tank, insure that the tank cushion grommets are installed in the tank sides with the thick side of the grommet DOWN to support the weight of the tank.

FASTENER TORQUE CHART

Fastener torque values are listed in the approximate order of reassembly.

Component Fastener	Torque
Engine PTO Housing	24 N•m
	(205-210 In-lb/17.4 ft.-lb)
Rotor Throughbolt	45 N•m
	(390 In-lb/32 ft.-lb)
Generator Rear End Housing	10 N•m (85 in.-lb.)
Rear-housing Feet Mount Nuts	35 N•m
	(300-305 in-lb/25 ft.-lb.)

WIRING DIAGRAMS

Refer to Fig. HA662 to view the output wiring with the voltage=selector switch in the 120/240V position.

Refer to Fig. HA663 to view the output wiring with the voltage-selector switch in the 120V-only position.

Refer to Fig. HA664 to view the complete wiring diagram for this generator series.

120/240V Position

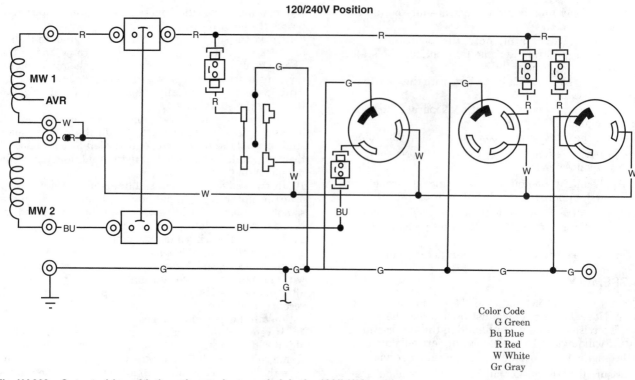

Fig. HA662 – Output wiring with the voltage selector switch in the 120/240V position.

Color Code
G Green
Bu Blue
R Red
W White
Gr Gray

120V Position

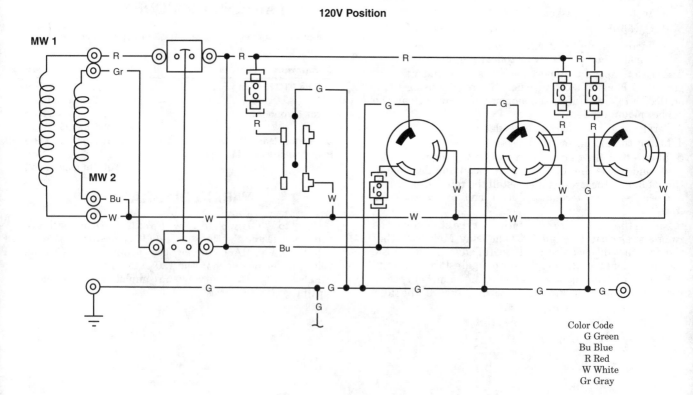

Fig. HA663 –Output wiring with the voltage selector switch in the 120V only position.

Color Code
G Green
Bu Blue
R Red
W White
Gr Gray

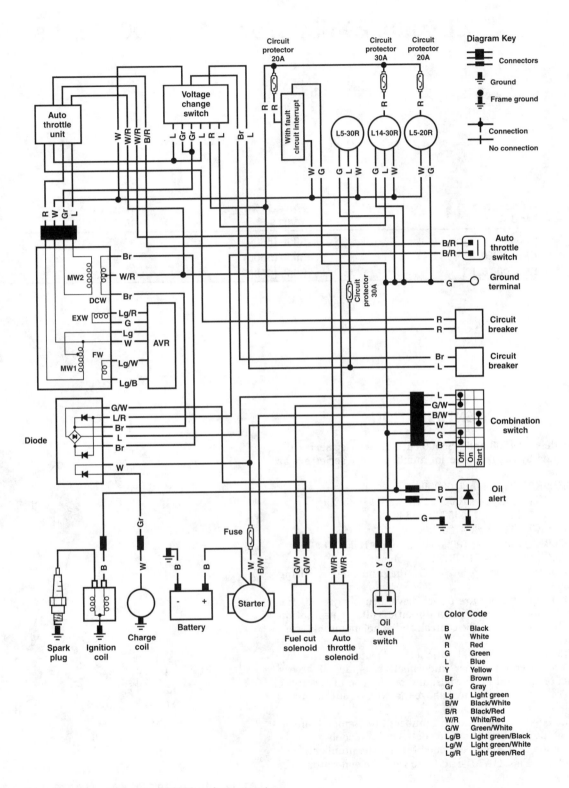

Fig. HA664 – Complete wiring diagram for the EB6500 series generator.

HONDA
EL5000, EX5500, and ES6500 Series

Generator Model	Rated (Surge) Watts	Output Volts
EL5000	5000 (5000)	120/240
EX5500K0, K1	5000 (5500)	120/240
ES6500K0, K1	6000 (6500)	120/240

IDENTIFICATION

Models EX5500 and ES6500 each have three series, determined by the frame serial number. These series are as follows:

EX5500K0 – Through #1000000
EX5500K1 (Early) – #1000001 through #1035729
EX5500K1 (Late) – #1035730 and higher
ES6500K0 – Through #1000000
ES6500K1 (Early) – #1000001 through #1035730
ES6500K1 (Late) – #1035731 and higher

Refer to the appropriate wiring diagram when servicing these units.

These generators are direct drive, brush style, 60 Hz, single phase units with automatic voltage regulation (AVR), auto throttle idle control, and panel mounted receptacles.

The generator serial number is located on the upper frame crossmember underneath the hinged access panel on EL and ES models, and on the end of the outer generator cover panel, just above the exhaust outlet, on EX models.

The engine identification numbers are stamped on the side of the crankcase on all models. On EX models, the side inspection door must be opened to view the numbers.

Refer to Figs. HA701-HA704 to view the generator components.

MAINTENANCE

ROTOR BEARING

The brush end rotor bearing is a pre-lubricated, sealed bearing. Inspect and replace as necessary. If a seized bear-

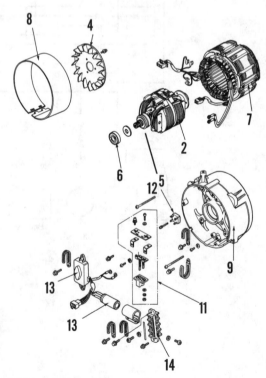

Fig. HA701 – Exploded view of generator head components.

1. Engine adapter/front housing
2. Rotor assembly
3. Rotor through-bolt and washer
4. Cooling fan and three mount screws
5. Sliprings
6. Rotor bearing
7. Stator assembly
8. Stator cover
9. Rear housing
10. Rear housing/stator through-bolts (4)
11. Brush-holder assembly
12. Diode stack
13. AVR assembly with capacitor/condenser
14. AC output terminal strip

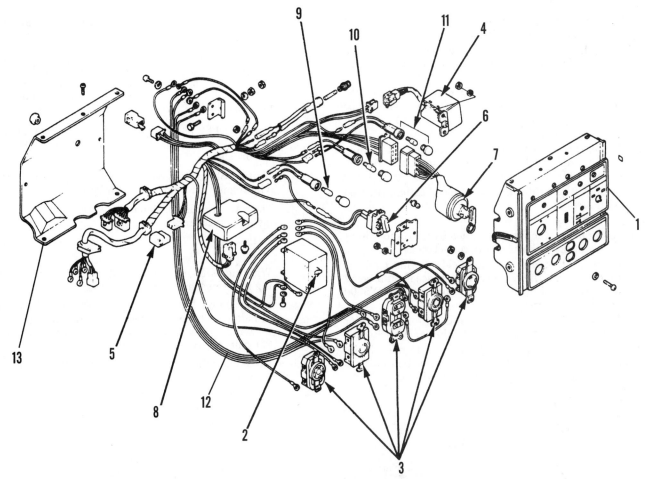

Fig. HA702 – Typical exploded view of series EL5000 and ES6500 control panel. ES6500K1 view shown; non K1 panel uses dual individual circuit breakers. EL5000 panel is similar.

1. Control panel
2. AC circuit breaker
3. 120V and 240V AC receptacles
4. Reset switch
5. Diode stack

6. Auto-throttle switch
7. Engine switch
8. Idle-control unit
9. Pilot lamp

10. Low-oil lamp
11. Coolant lamp
12. Control-panel harness
13. Panel rear cover

ing is encountered, inspect the rotor journal and the rear housing bearing cavity for scoring.

BRUSHES

The brushes should be replaced when the brush length reaches 13 mm (0.5 in.) with the brushes removed from the brush holder

The light green/white AVR wire always goes to the positive (+) brush. The light green/black AVR wire always goes to the negative (-) brush.

GENERATOR REPAIR

TROUBLESHOOTING

Before testing any electrical components, check the integrity of the terminals and connectors. While testing individual components, also check the wiring integrity. Use a digital multimeter for testing: Resistance readings may vary with temperature and test equipment accuracy. All operational testing must be performed with the engine running at its governed no load speed of 3750 RPM, unless

specific instructions state otherwise. Always load test the generator after repairing defects discovered during troubleshooting. For access to the components listed in this Troubleshooting section, refer to Figs. HA701-HA704 and the later Disassembly section.

To troubleshoot the generator:

> **NOTE: The positive (+) brush terminal on these generators is always the one with the light green/white wire coming from the AVR. The negative (-) brush terminal has a light green/black wire coming from the AVR.**

1. Remove the generator rear housing end cover. On EL5000 models, remove the recoil starter assembly.
2. Leave all components and wiring connected, start the generator, and refer to Fig. HA705 for the following color coded tests:

   a. Test AC voltage at the white to brown (red or black depending on model) main winding 1 connections, and at the white to blue main winding 2 connections. Output should be 114-137 volts on each winding.

b. Test AC voltage at the gray to gray sub-winding 1 terminals. Output should be 14-18 volts.

c. Test DC voltage at the blue/red to black/red sub winding 2 terminals. Output should be 10-14 volts.

d. Test DC voltage at the light green/white to light green/black brush connections. Output should be 22-28 volts.

3. Stop the generator:

a. If Step 2 tests met specification, proceed to test Step 4I, followed by the Individual Component Testing section in this chapter.

b. If Step 2 tests did not meet specification, proceed with Step 4A.

4. To test the rotor:

a. Disconnect the light green/white and light green/black AVR brush wire connectors.

b. Using an ohmmeter, measure field coil winding resistance through the brushes. Resistance should be 52 ohms on the EL5000 series, 57 ohms on the EX5500 and ES6500 series.

c. Measure field coil winding resistance at the rotor slip rings. Resistance should be the same as Step B.

d. Measure field coil winding resistance from each slip ring to a ground on the rotor shaft or laminations. Resistance should be infinity/no continuity.

e. If test Steps B and C met specification, the rotor and brushes are good.

f. If test Step B failed specification but Step C met specification, the rotor is good. Inspect the brushes and test each brush for continuity through its terminal. If faulty, replace the brush holder assembly or brushes as necessary.

g. If both test Steps B and C failed specification, the rotor is faulty. Replace the rotor, then inspect and test the brushes.

h. If either Step D test read any continuity, the field coil winding is shorted. Replace the rotor.

i. The rotor can also be dynamically tested by performing a current draw test.

• With the VOM on the DCV scale, note the battery voltage.

• With the VOM on the R × 1 scale, note the rotor winding resistance across the slip rings.

• Using Ohm's Law applicable to this test ($I=V÷R$), calculate the current flow (I) by dividing the rotor winding resistance (R) into the battery voltage (V). Note the calculated amperage.

• Disconnect the brush feed wires and the 4-terminal AVR connector.

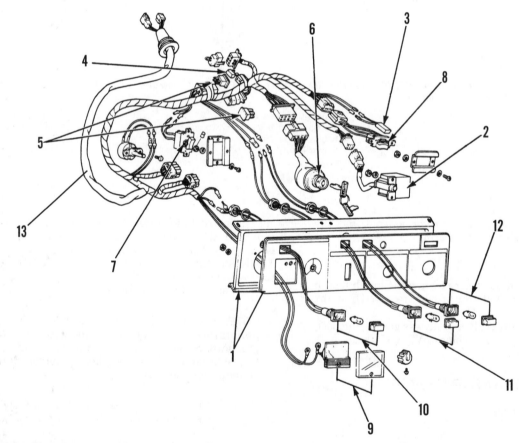

Fig. HA703 – Exploded view of EX5500 control panel.

1. Control panel and faceplate
2. Reset switch
3. Diode
4. Silicon rectifier
5. Diode stacks (2)

6. Engine switch
7. Auto-throttle switch
8. Remote-control switch
9. Voltmeter

10. Pilot lamp
11. Low-oil lamp
12. Coolant lamp
13. Control-panel harness

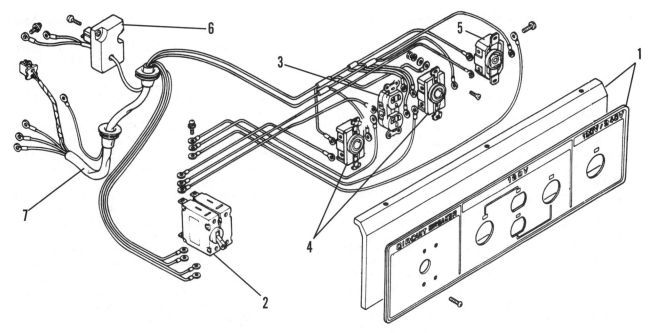

Fig. HA704 – Exploded view of EX5500 receptacle panel.

1. Receptacle panel and faceplate
2. AC circuit breaker assembly
3. 120V AC receptacle

4. 120V AC Twistlok receptacle
5. 240V AC Twistlok receptacle

6. Idle-control unit
7. Panel harness

6-P connector: EX5500
4-P connector: EL5000, ES6500

AC output terminal block

Red, brown, or black

White

Blue

Grommet

Wire clamps

Green

Fig. HA705 – View of rear housing wiring connections showing test points and proper wire routing.

- Connect the positive (+) terminal of the test battery to the positive side of the VOM/ammeter (Fig. HA706).
- Connect the ammeter negative lead to one terminal of a 3-amp circuit breaker.
- Connect a test wire from the positive brush terminal to the other terminal of the 3-amp breaker.
- Connect a test wire from the negative brush terminal to the negative battery terminal.
- Note the field winding current draw on the ammeter.
- Start the engine; with the engine running at rated no load RPM, note the amperage draw on the ammeter.

j. Compare the ammeter reading with the engine running to both the engine stopped ammeter reading and the calculated amperage.

- If the running amperage is significantly higher than the static or calculated amperage, or if the circuit breaker trips while running, the rotor is shorted; replace the rotor.
- If there is NO amperage draw during the running test, the rotor has an open winding and must be replaced.
- If the running amperage is approximately the same as the static and the calculated amperage, and if the circuit breaker does not trip while running, the rotor is good.

5. To test the stator:

a. Obtain a fully charged 12-volt battery and a pair of test leads to connect the battery terminals to the brush connector terminals.

b. Disconnect and remove the AVR. Connect the positive battery post to the light green/white positive brush holder terminal, then connect the negative battery post to the light green/black negative brush holder terminal.

NOTE: Improper connections will damage the generator.

c. Start the generator, verifying the correct 3750 RPM no load speed.
d. Test the stator output voltages using Fig. HA705 and the following color coded test connections.
- White to brown (or red, or black, depending on model) main winding 1 and white to blue main winding 2 should each read 60-70 VAC.
- Light green to white sensor winding should read 8-10 VAC.
- Light green/red to green exciter winding should read 30-36 VAC.
- Gray to gray sub winding 1 should read 8.5-10.5 VAC.
- Blue/red (+) to black/red (-) sub winding 2 should read 6.5-8.5 VDC.
e. Stop the generator and disconnect the test battery.
f. Using an ohmmeter, measure the resistance of the following test connections.
- Resistance for the blue to white and the white to brown (depending on model) main winding connections should each be 0.32 ohm on the EL5000 series or 0.24 ohm on the EX5500 and ES6500 series.
- Resistance for the light green/red to green exciter winding terminals in the 4-terminal AVR connector should be 1.92 ohms on the EL5000 series, 1.45 ohms on the EX5500 series, or 1.65 ohms on the ES6500 series.
- Resistance for the gray to gray DC sub winding terminals in the 4-terminal or 6-terminal stator harness connector should be 0.27 ohm on the EL5000 series or 0.22 ohm on the EX5500 and ES6500 series.
- Resistance for each white, blue, and brown, red or black (depending on model) main winding connec-

tion to a ground on the stator laminations should be infinity/no continuity.
- Resistance for each light green/red and green exciter winding terminal to a ground on the stator laminations should be infinity/no continuity.
- Resistance for each gray DC sub winding terminal to a ground on the stator laminations should be infinity/no continuity.
g. If all of the tests in Steps D and F met specification, the stator is good; proceed with further component testing. If the component tests meet specification, the automatic voltage regulator (AVR) is faulty and must be replced.
h. If any individual test in Steps D or F failed to meet specification, the stator is faulty. Replace the stator.
i. Reconnect all previously disconnected wiring and components.
j. Attach DC voltmeter probes to the brush leads.
k. Start the generator and immediately test brush voltage. Voltage should test 22-28 VDC. If voltage is outside this range, immediately stop the generator. The AVR is faulty and must be replaced.
6. To test the auto throttle system:

NOTE: The auto throttle system will not detect loads below 1.0 amp. If load less than 120 watt is being applied to the generator at 120V (240-watt load at 240V), the auto throttle switch must be OFF so the unit will run at full throttle governed RPM to produce proper voltage.

a. Turn the auto throttle switch OFF.
b. Start the generator; make sure all loads are disconnected.
c. Move the switch to AUTO. After a moment, the engine should idle down to 2100-2300 RPM.

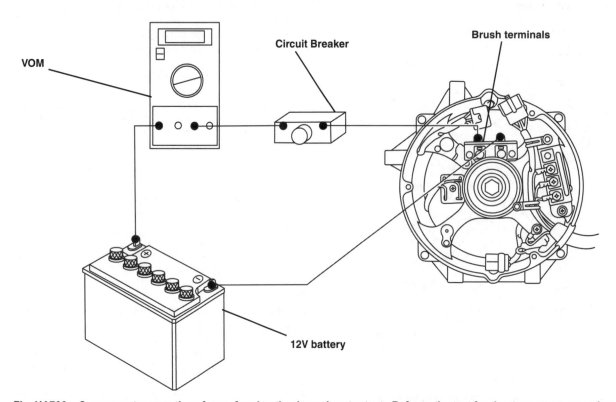

Fig. HA706 – Component connections for performing the dynamic rotor test. Refer to the text for the proper test procedure.

d. Return the switch to OFF. The engine should return to no load full throttle.

e. If the auto throttle system fails Steps C and D, proceed to the **Individual Component Testing** section.

INDIVIDUAL COMPONENT TESTING

Circuit Breakers

To test the breaker:
1. Access the breaker wiring inside the control box.
2. Disconnect the wiring from one breaker terminal.
3. Continuity test the breaker terminal pairs. If the breaker will not stay set, shows continuity in the OFF position, or does not show continuity in the ON position the breaker is faulty.

Receptacles

The generator must be shut down to test the receptacles.
1. Access the receptacle connections inside the control box.
2. Disconnect the RETURN (white) feed wire from the receptacle being tested.
3. Install a jumper wire between the HOT and RETURN *outlet end* terminals.

   a. On 120V straight pin and twistlok receptacles, HOT terminals will be either brown, red, black *or* blue feed wires. RETURN terminals will be the white feed wire.

   b. On 240V twistlok receptacles, HOT terminals will be the brown, red, black *and* blue feed wires; RETURN terminals will be the white feed wire.

4. Test for continuity across the *feed terminals*. No continuity indicates a faulty receptacle.

Auto Throttle System

1. To test the auto throttle switch:

   a. Access the switch terminals inside the control box.

   b. With the switch rocker UP in the AUTO position, there should be continuity between the two pigtails on the back of the switch.

   c. With the switch rocker DOWN in the OFF position, the two switch pigtails should show no continuity.

   d. Any results different from those specified in Steps B and C indicate a faulty switch.

2. To test the diaphragm assembly (Fig. HA707):

   a. Apply suction to the vacuum line fitting. If the diaphragm is good, the link will draw down.

   b. Continue to hold suction for approximately one minute to check for pinhole leakdown.

   c. Release suction and verify that the return spring pushes the link back out.

   d. Inspect all vacuum hoses for leaks.

3. To test the control unit:

   a. Access the control unit inside the control box.

   b. Verify that the blue and brown, red or black (depending on model) wires from the stator main windings are passing through the control unit hole from opposite directions (Fig. HA708) on their way to the circuit breakers. If necessary, redirect the wires. Wires passing through the control unit from the same direction will cancel each other's magnetic field, preventing the control unit from sensing voltage differences between load and no load.

   c. If all other auto throttle components test good, and the control unit wires are routed correctly, but the

auto throttle still does not operate, the control unit is faulty and must be replaced.

Condenser/Capacitor

The condenser is a part of the AVR and cannot be repaired separately.

Diode Stack

To test the diode stack:
1. Access and unplug the diode stack connector located inside the control box.
2. Set the VOM to the R × 1 scale, then refer to Fig. HA709.
3. Placing the positive (+) meter probe on the first numbered terminal in each of the following terminal pairs, continuity test between the terminals. Note the readings:

1-2 4-1 3-2 4-3

4. Reverse the leads and repeat the Step 3 test.
5. In one direction, the meter should have shown continuity, with no continuity shown in the reverse test. If both tests of one terminal pair showed either continuity or no continuity, the diode stack is faulty. Replace the diode stack.

Voltmeter (Applicable Units Only)

To test the voltmeter:
1. Access the meter terminals inside the control box.
2. Disconnect one meter lead.
3. Test for continuity between the meter terminals. A lack of continuity indicates a faulty meter.

Variable Resistor (Applicable Units Only)

To test the resistor:
1. Access and disconnect the yellow resistor wires inside the control box.
2. Turn the resistor knob to its maximum clockwise position; do not force.
3. Continuity test between the yellow pigtails. Resistance should be 400 ohms. Any other unreasonable value indicates a faulty resistor.

Engine Switch – EL5000 Series

To test the switch:
1. Access and disconnect the switch wires inside the control box.
2. With the switch ON, there should not be continuity between any of the switch wire terminals.
3. With the switch OFF, there should be continuity between the green/white to red/white terminals and between the black to black/yellow terminals.
4. Any other test results indicate a faulty switch.

Engine Switch – EX5500 and ES6500 Series

To test the switch:
1. Access and disconnect the 8 wire switch harness connector inside the generator control box.
2. Refer to Fig. HA710; test continuity between the terminals in each switch position noted:

Position	Terminal Continuity
OFF	Red to green/white, and black to black/yellow
ON	White to light green to black/white
START	White to light green to yellow to black/white

3. If any switch positions fail to show continuity as specified, the switch is faulty and must be replaced.

Indicator Lamps

The indicator lamps are OK if continuity exists between the lamp pigtail terminals.

Auto Choke (Applicable Units Only)

To test the auto choke solenoid:

1. Access and disconnect the yellow and black solenoid wires.

2. Using a known good, fully charged 12V test battery, connect the battery terminals to the solenoid terminals. The solenoid plunger should extend.

3. Disconnect the test battery; the plunger should retract.

4. If the solenoid does not test to specification in Steps 2 and 3, the solenoid is faulty.

Remote Control Start

Some electric start 5500 series units use a remote control start option (Fig. HA711) which plugs into a 6-terminal coupler underneath the generator control panel.

To test the panel mounted remote control switch:

1. Access and disconnect the switch harness connector behind the control panel.

2. Continuity test the following color coded terminal pairs in the switch position noted:

OFF – Green to green/white, and white to white/green
ON – Black to green, and white to white/black

3. Failure of any terminal pair to continuity test as specified indicates a faulty switch.

To test the remote control box:

1. Unplug the 6-pin connector from the rear of the remote control box.

2. Test for continuity between the following color coded terminals and box control positions:

Light green to red to white/black – Engine switch ON
Red/black to green – Pilot lamp
Light green to red – Start button IN

3. No continuity in any of the Step 2 tests indicates a faulty remote control box; replace the box.

4. If the remote control switch and box meet specification and the remote control harness integrity is secure, but the remote control will not activate the electric start, operate the electric start with the control panel mounted engine switch. If the panel engine switch operates the electric start, the remote control relay box is faulty and must be replaced.

DISASSEMBLY

Refer to Figs. HA701-HA704 for generator component identification. Note that the rotor cannot be removed without first removing the stator. Exercise caution when handling the stator and rotor so the windings and winding coatings are not damaged.

To disassemble the generator:

1. On the EL5000 and ES6500 series:

 a. Disconnect and remove the battery, if applicable.

 b. Remove the top cover, upper frame side covers, and upper frame assembly.

 c. Carefully drain and then remove the fuel tank and heat shield.

 d. If necessary, drain the cooling system, then remove the radiator debris shield, radiator, and radiator upper mount.

 e. Remove the outer muffler heat screen, muffler assembly, and inner muffler heat shield.

 f. Remove the generator rear cover (recoil starter assembly on EL5500).

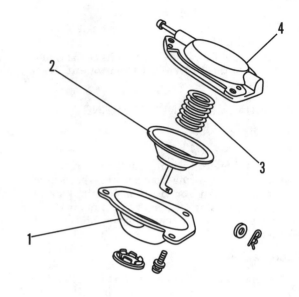

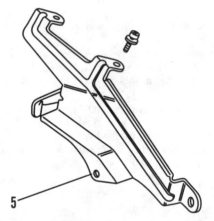

Fig. HA707 – Exploded view of auto throttle diaphragm. Refer to the text for the correct test procedure.

1. Diaphragm base
2. Diaphragm
3. Diaphragm spring
4. Diaphragm cover
5. Diaphragm mount bracket

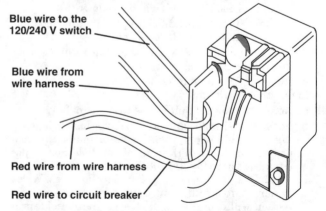

Auto-throttle Control Unit
Route the two wires through the control unit opening as shown. Note the direction of the wires.

Blue wire to the 120/240 V switch

Blue wire from wire harness

Red wire from wire harness

Red wire to circuit breaker

Fig. HA708 – View of red (or brown, or black, depending on model) and blue wires from the main AC stator winding properly passing through the auto throttle control unit sensor loop from opposite directions. Failure to route the wires as shown will cause the auto throttle system not to function.

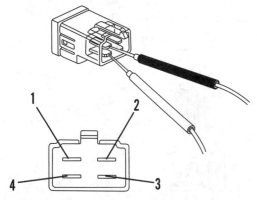

Fig. HA709 – View of diode stack terminals numbered for testing. Refer to the text for the proper test procedure.

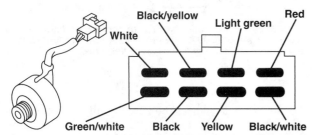

Fig. HA710 – View of color coded ignition switch connector terminals for testing the switch. Refer to the text for the correct test procedure.

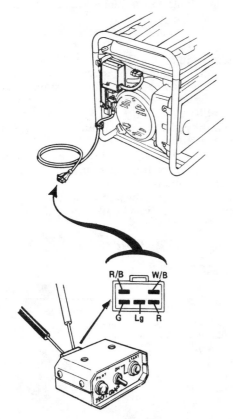

Fig. HA711 – View of the components and switch test points for the EX5500 Remote control start option. The relay unit on this series is mounted behind the control panel.

g. Disconnect the control box harness terminals from the rear of the generator.

h. Disconnect and remove the control box assembly. Note the wiring routing and orientation.

2. On the EX5500 series:

a. Disconnect and remove the battery.

b. Remove the generator hood, fuel tank hood, and upper crossmember beam.

c. If necessary, drain, then remove, the fuel tank and heat shield.

d. Remove the air filter assembly.

e. Remove and disconnect the control panel assembly. Note the wiring routing and orientation.

f. Remove and disconnect the receptacle panel assembly. Note the wiring routing and orientation.

g. Drain the cooling system. Unplug the thermoswitch and fan motor connections, then unbolt the radiator mounts from the side panel.

h. Remove the rear generator ventilation shield and the upper rear handle crossmember and hooks.

i. Undo the AC terminal block from inside the side panel. Remove the radiator side outer panel, then remove the radiator.

j. Remove the muffler side outer panel.

k. Remove the outer muffler heat shield, muffler assembly, and inner muffler heat shield.

l. Remove the rear generator frame shield, then remove the top generator frame shield.

m. Remove the generator end cover.

3. Disconnect and remove the AVR assembly from the generator rear housing.

4. Disconnect the control box harness. Note the location and orientation of the wire clamps and ties. Slide the harness grommet from the housing slot and set aside the harness.

5. Remove the brush holder assembly.

6. Unscrew the AC output terminal block, leaving the stator wires connected to the block, unless the stator is being replaced.

7. Disconnect any remaining housing component wiring.

8. Remove the two mount bolts.

9. Place a temporary support, such as a 2 × 4, under the engine stator housing.

10. Remove the four rear housing mount bolts. Separate the rear housing from the stator while sliding the stator wires through the housing openings.

11. Separate the stator from the engine housing. If replacing the stator, note the position of the stator cover wrap, then unbend the two tabs and remove the cover.

12. Place a padded board under the rotor to prevent damage in case it is dropped during removal.

13. Remove the rotor throughbolt and washer.

14. Using a slide hammer with the proper threaded adapter (recommended adapter is OTC-33-12 × 1.25), break the rotor free from the engine crankshaft taper. Remove the rotor.

15. Remove the rotor fan.

REASSEMBLY

Refer to Figs. HA701-HA704 for generator component identification. The rotor must be installed before the stator. Exercise caution when handling the stator and rotor so the windings and winding coatings are not damaged.

1. Place a temporary support, such as a 2 × 4, under the engine stator housing.

2. Service the rotor bearing, if needed.

3. Mount the cooling fan to the rotor.

4. Make sure that the matching tapers on the engine crankshaft and inside the rotor shaft are clean and dry. Install the rotor onto the crankshaft. Install the rotor throughbolt and washer, then torque the rotor bolt 60-70 N•m (43-50 ft.-lb.).

5. If the stator cover wrap was removed, install the cover. Bend the two tabs to hold the cover in position.

6. Noting the stator wire orientation, place the stator over the rotor, positioning the stator tightly against the engine housing.

7. Slide the stator wires through the proper rear housing openings, then fit the housing tightly against the stator.

8. Install the housing mount bolts. Incrementally and alternately torque the bolts to 20-28 N•m (180-240 in.-lb.).

9. Rotate the engine crankshaft to check for interference between the rotor and stator. There should be a consistent gap around the rotor laminations inside the stator. If any rubbing or scraping is evident, correct any misalignment prior to proceeding.

10. Make sure that the stator cover wrap is snug against the engine housing–there should be no clearance or gap.

11. Remove the temporary engine housing support; install and tighten the two rubber mount bolts.

12. Fasten the AC output terminal block to the rear housing.

13. Install the brush holder assembly.

14. Properly insert the control box harness grommet into the rear housing slot, then install the AVR assembly into the housing.

15. Refer to Fig. HA705. Connect all rear housing wiring, making sure that the wiring is routed properly and the wire clamps are in their proper positions.

16. On the EX5500 series:

   a. Install the generator end cover.

   b. Install the top generator frame shield, followed by the rear generator frame shield.

   c. Install the inner muffler heat shield, muffler assembly, and outer muffler heat shield. A new gasket is recommended when installing the muffler, and the rounded edge of the gasket outer diameter must face the muffler.

   d. Install the muffler side outer panel.

   e. Position the radiator assembly in the generator compartment with the hoses loosely connected to the hose necks. Install the radiator side outer panel, making sure that the panel properly aligns with the frame locator tabs. Fasten the radiator to the side panel, then tighten the hoses and plug in the thermoswitch and fan motor connectors.

   f. Mount the AC terminal block inside the side panel (engine end – Fig. HA712) and connect the harness wires.

   g. Install the upper rear handle crossmember and hooks, and the rear generator ventilation shield.

   h. Noting the wiring routing and orientation from the disassembly procedure, reconnect and install the receptacle panel assembly, then reconnect and install the control panel assembly.

   i. Install the air filter assembly.

   j. Install the fuel tank heat shield and fuel tank.

   k. Install the upper crossmember beam, fuel tank hood, and generator hood.

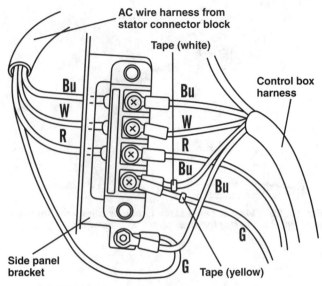

Fig. HA712 – *View of stator harness to control box harness connections on the EX5500 AC terminal block which mounts next to the engine, inside the radiator side cover panel.*

   l. Install and connect the battery. Fill the cooling system and fuel tank.

17. On the EL5000 and ES6500 series:

   a. Noting the wiring routing and orientation from the disassembly procedure, reconnect and install the control box assembly.

   b. If not already done, reconnect the control box harness terminals to the rear housing terminal block (Fig. HA705).

   c. Install the generator rear cover (recoil starter assembly on EL5000).

   d. Install the inner muffler heat shield, muffler assembly, and outer muffler heat screen. A new gasket is recommended when installing the muffler, and the rounded edge of the gasket outer diameter must face the muffler.

   e. Install the radiator upper mount, radiator, and radiator debris shield.

   f. Install the fuel tank heat shield and fuel tank.

   g. Install the upper frame assembly, two upper frame side covers, and the top cover.

   h. Install and connect the battery, if applicable. Fill the cooling system and fuel tank.

ENGINE

WIRING DIAGRAMS

Generator	Diagram
EL5000	HA713
EX5500K0	HA714
EX5500K1 (Early)	HA715
EX5500K1 (Late)	HA716
ES6500K0	HA717
ES6500K1 (Early)	HA718
ES6500K1 (Late)	HA719

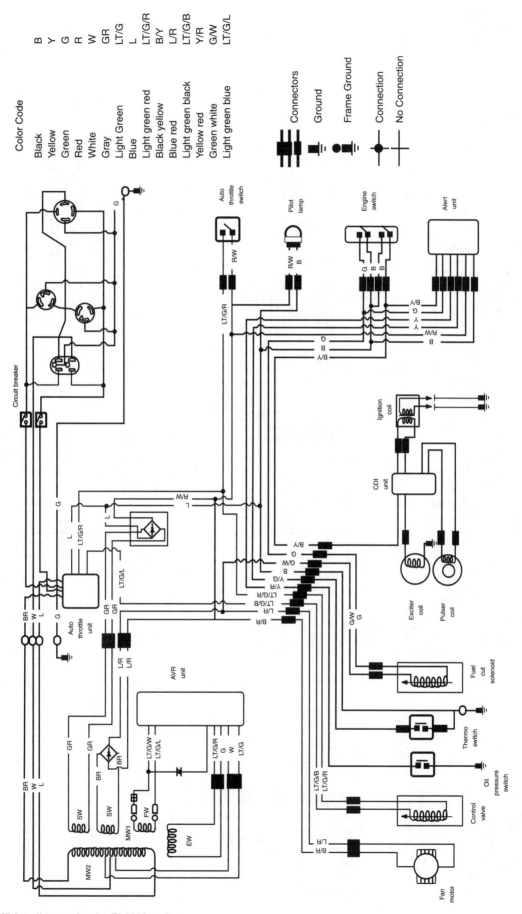

Fig. HA713 – Wiring diagram for the EL5000 series.

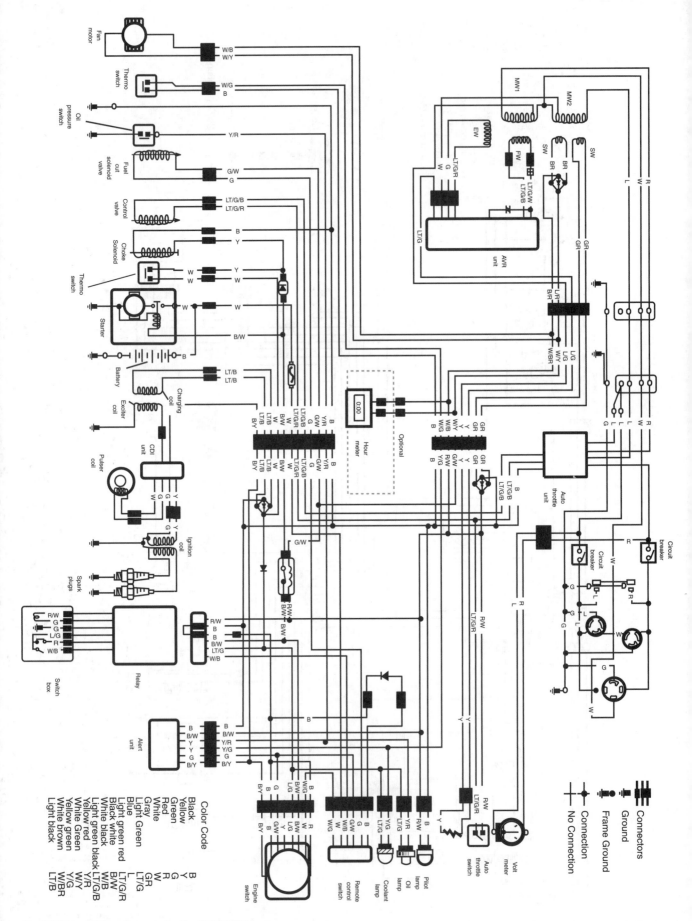

Fig. HA714 – Wiring diagram for the EX5500K0 series.

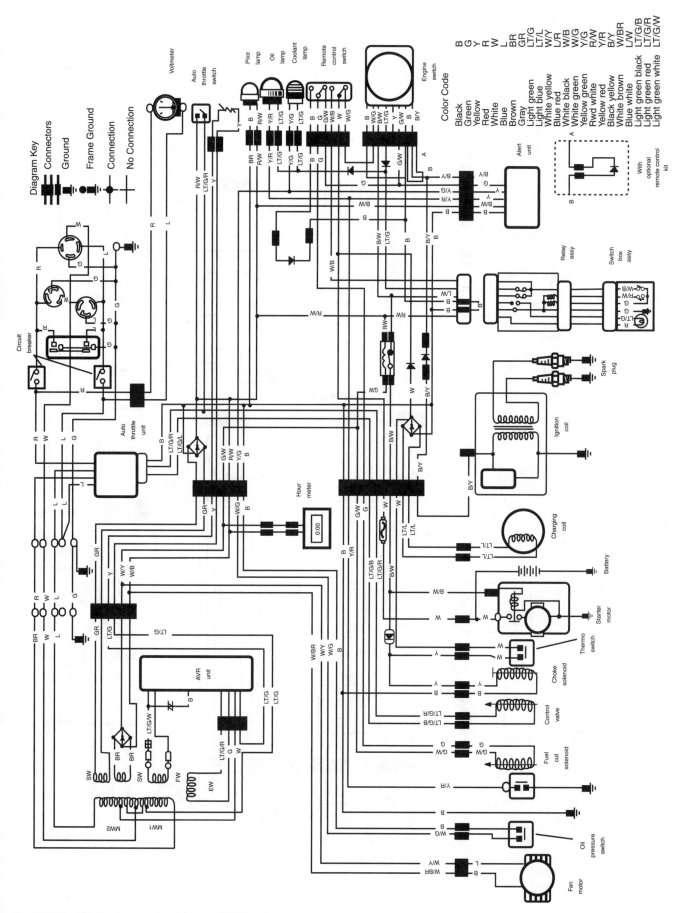

Fig. HA715 – Wiring diagram for the early EX5500K1 series.

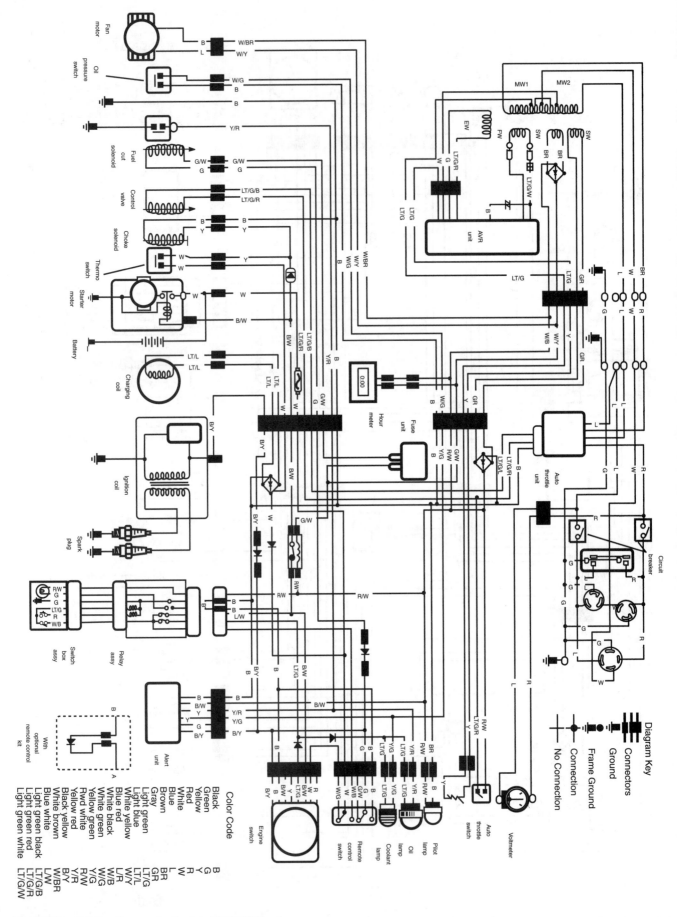

Fig. HA716 – Wiring diagram for the late EX5500K1 series.

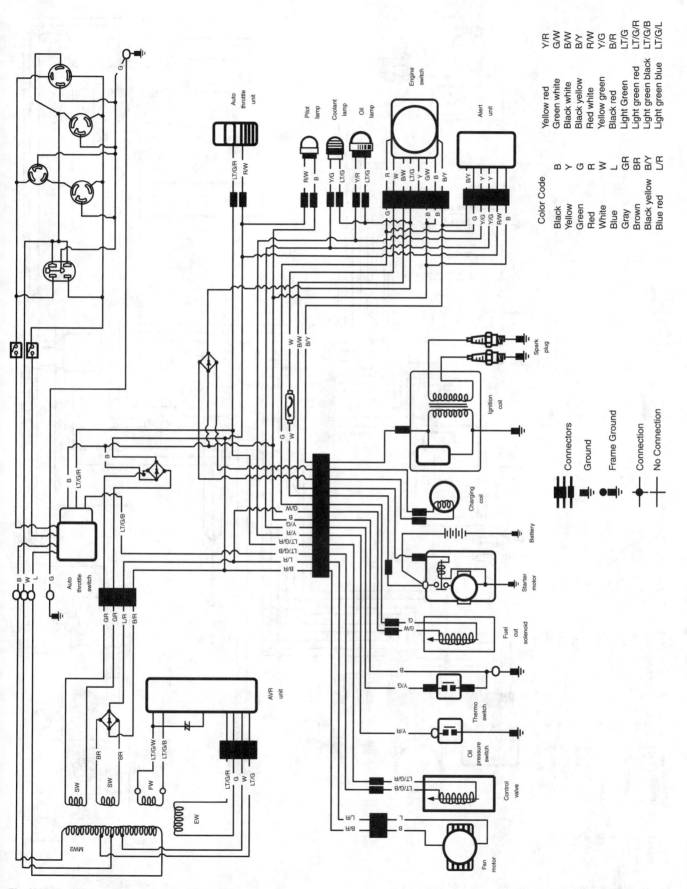

Fig. HA717 – Wiring diagram for the ES6500K0 series.

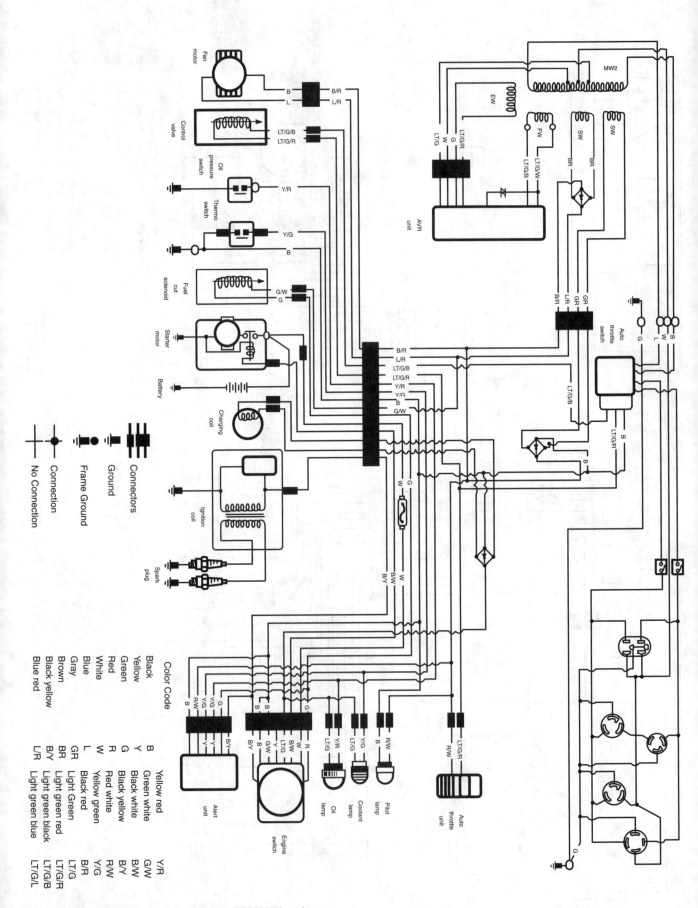

Fig. HA718 – Wiring diagram for the early ES6500K1 series.

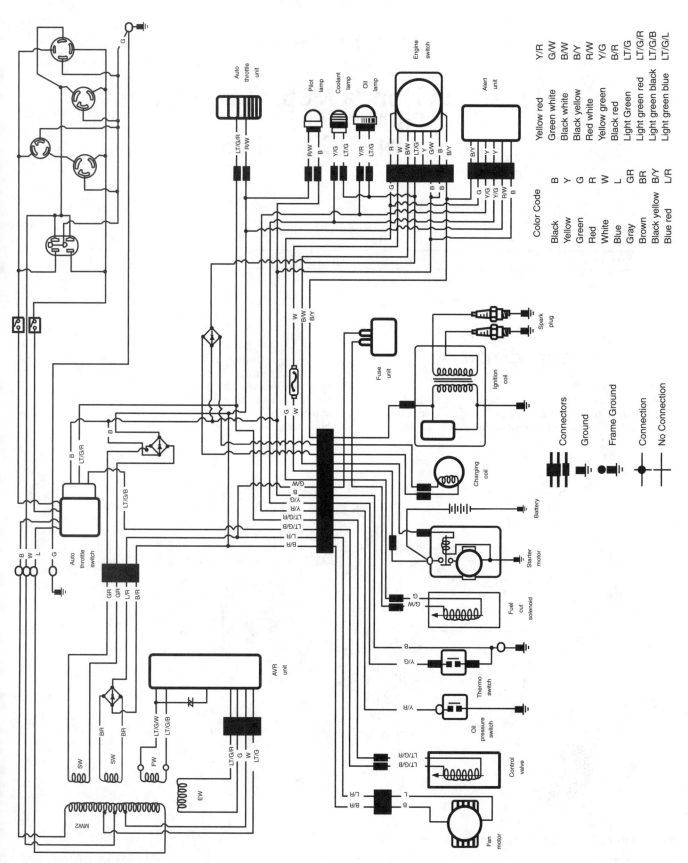

Color Code

Yellow red	Y/R	Black	B
Green white	G/W	Yellow	Y
Black white	B/W	Green	G
Black yellow	B/Y	Red	R
Red white	R/W	White	W
Yellow green	Y/G	Blue	L
Black red	B/R	Gray	GR
Light Green	LT/G	Brown	BR
Light green red	LT/G/R	Black yellow	B/Y
Light green black	LT/G/B	Blue red	L/R
Light green blue	LT/G/L		

Connectors   Ground   Frame Ground

Connection   No Connection

Fig. HA719 – Wiring diagram for the late ES6500K1 series.

HONDA

EB11000 Series

American Honda Motor Co., Inc.
4900 Marconi Dr.
Alpharetta, GA 30005

Generator Model	Rated (Surge) Watts	Output Volts
EB11000	9500 (11000)	120/240

IDENTIFICATION

These generators are direct drive, brush style, 60 Hz, single phase units with automatic voltage regulation, auto throttle idle control, and panel mounted receptacles.

The generator serial number is located on the engine mount frame crossmember just to the right of the battery. The engine identification numbers are stamped on the crankcase just below the electric starter motor.

Refer to Figs. HA1001A, HA1001B, HA1001C and HA1002 to view the generator components.

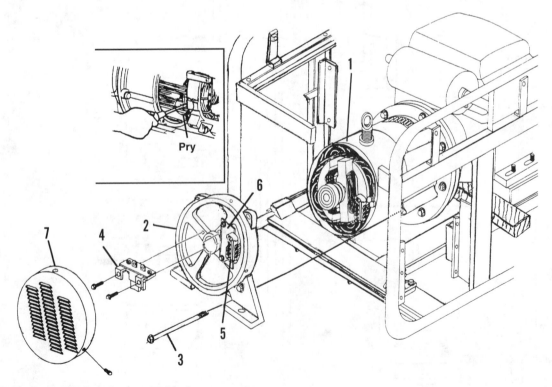

Fig. HA1001A – Exploded view of generator head components.

1. Stator assembly
2. Rear housing
3. Rear housing/stator through-bolt (4)

4. Brush assembly
5. Harness terminal block

6. Terminal-block mount plate
7. End cover

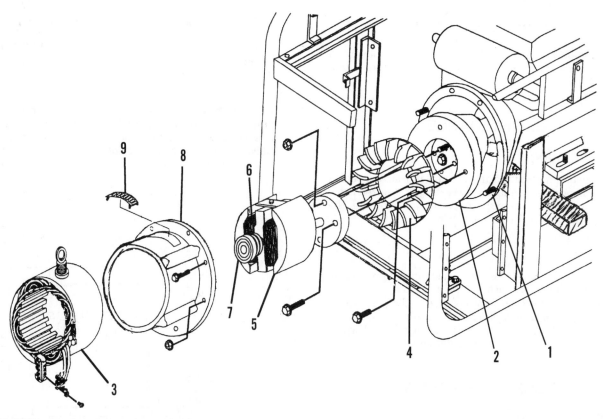

Fig. HA1001B – Exploded view of generator head components.

1. Engine adapter/front housing
2. Coupling
3. Stator assembly
4. Cooling fan

5. Rotor assembly
6. Slip rings
7. Rotor bearing

8. Stator-mount housing
9. Cover plate

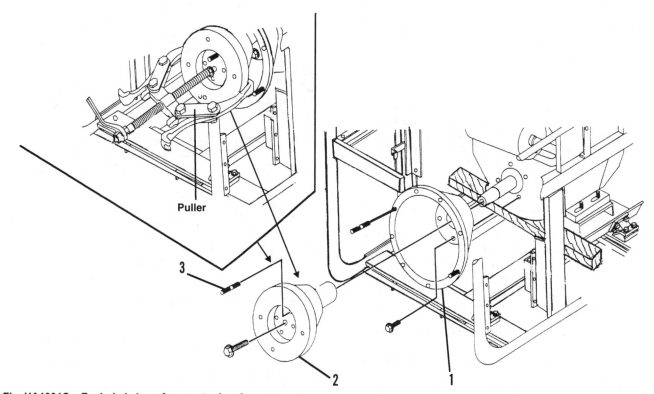

Fig. HA1001C – Exploded view of generator head components.

1. Engine adapter/front housing
2. Rotor coupling

3. Coupling locator stud

4. Puller

WIRE COLOR CODE

A wire identified by two colors separated by a slash is a two-color wire. For example, red/white is a red wire with a white tracer stripe.

MAINTENANCE

ROTOR BEARING

The brush end rotor bearing is a pre lubricated, sealed bearing. Inspect and replace as necessary. If a seized bearing is encountered, inspect the rotor journal and the rear housing bearing cavity for scoring.

BRUSHES

Replace the brushes when the brush length protruding from the brush holder reaches 7.0 mm (0.28 in.).

The red/white AVR wire always goes to the positive (+) brush, which is the brush closest to the rotor windings. The black/white AVR wire always goes to the negative (-) brush, which is the brush closest to the rotor end bearing.

GENERATOR REPAIR

TROUBLESHOOTING

Before testing any electrical components, check the integrity of the terminals and connectors. While testing individual components, also check the wiring integrity. Use a digital multimeter to test. Resistance readings may vary with temperature and test equipment accuracy. All operational testing must be performed with the engine running at 3750 RPM, unless specific instructions state otherwise. Always load test the generator after repairing defects discovered during troubleshooting.

To troubleshoot the generator:

NOTE: The positive (+) brush terminal on these generators is always the one closest to the rotor windings. It has a red/white wire coming from the AVR. The negative (-) brush terminal is closest to the rotor bearing and has a black/white wire coming from the AVR.

1. Remove the three end cover mounting screws then remove the end cover.
2. Circuit breakers must be ON. To test the AC breakers, disconnect the breaker wiring and continuity test between the top and bottom terminals of each breaker or breaker section (the toggle type breaker is a two-section breaker). The breaker is faulty if it will not stay set, shows continuity in the OFF position, or does not show continuity in the ON position.
3. Ground fault circuit interrupters (GFCIs) must be reset. To test the GFCI:

    a. Start the generator and allow a brief no load warm up period.

    b. Press the TEST button. The RESET button should extend. If it does, proceed to Step C. If the RESET button does not extend, the GFCI is faulty or it is not receiving input from the generator.

    c. Press the RESET button. The RESET button should then be flush with the TEST button. If the RESET button does not remain flush with the TEST button, the GFCI is faulty.

    d. Stop the generator.

NOTE: If the GFCI seems to trip for no apparent reason, make sure the generator is not vibrating excessively due to faulty rubber mounts or debris between the generator and frame. Excessive vibration can trip the GFCI.

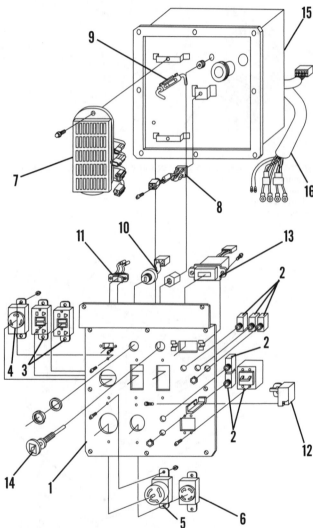

Fig. HA1002 – Exploded view of generator control box and components.

1. Control panel
2. AC circuit breakers
3. 20A 120V AC GFCI receptacles
4. 30A 120V AC Twistlok receptacle
5. 50A 120/250V AC Twistlok receptacle
6. 30A 125/250V AC Twistlok receptacle
7. AVR
8. Full-wave rectifier
9. Fuse
10. Engine switch
11. Auto-throttle switch
12. Auto-throttle control unit
13. Hourmeter
14. Engine choke cable
15. Control box
16. Control-box wire harness

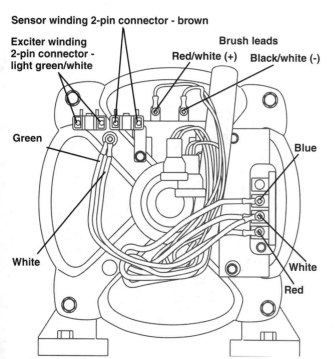

Fig. HA1003 – View of rear housing wiring connections and test points.

4. Leave all components and wiring connected, start the generator, and refer to Fig. HA1003 for the following tests.

a. Test AC voltage at the main red to main white and the main blue to main white connections. Output should be 105-135 volts.

b. Test DC voltage at the brush red/white [+] to brush black/white [-] connections. Output should be 34-42 volts.

5. Stop the generator.

a. If Step 4 tests met specification, proceed to test Step 6-I, followed by the Individual Component Testing section in this chapter.

b. If the tests in Step 4 did not meet specification, proceed with Step 6-A.

6. To test the rotor:

a. Disconnect the red/white and black/white AVR brush wire connectors.

b. Using an ohmmeter, measure field coil winding resistance through the brushes. Resistance should be 22-35 ohms.

c. Measure field coil winding resistance at the rotor slip rings. Resistance should be the same as Step B.

d. Measure field coil winding resistance from each slip ring to a ground on the rotor shaft or laminations. Resistance should be infinity/no continuity.

e. If test Steps B and C met specification, the rotor and brushes are good.

f. If test Step B failed specification but Step C met specification, the rotor is good. Inspect the brushes and test each brush for continuity through its terminal. If faulty, replace the brush holder assembly.

g. If both test Steps B and C failed specification, the rotor is faulty. Replace the rotor, then inspect and test the brushes.

h. If either Step D test read any continuity, the field coil winding is shorted. Replace the rotor.

i. The rotor can also be dynamically tested by performing a current draw test. The equipment required for this test is a fully charged, known good condition 12V battery; a digital VOM capable of reading at least 3 amps and a 3-amp circuit breaker (Honda P/N 2652931).

j. With the VOM on the DCV scale, note the battery voltage. With the VOM on the R × 1 scale, note the rotor winding resistance across the slip rings. Using ohm's law applicable to this test (I=V÷R), calculate the current flow (I) by dividing the rotor winding resistance (R) into the battery voltage (V). Note the calculated amperage.

k. Disconnect the brush feed wires and the two two terminal AVR connectors brown and brown and light green/white and light green/white. Connect the positive (+) terminal of the test battery to the positive (+) side of the VOM/ammeter (Fig. HA1004).

l. Connect the ammeter negative (-) lead to one terminal of the 3-amp circut breaker. Connect a test wire from the positive (+) brush terminal to the other terminal of the 3-amp breaker.

m. Connect a test wire from the negative brush terminal to the negative battery terminal. Note the field winding current draw on the ammeter.

n. Start the engine. With the engine running at rated no load RPM, note the amperage draw on the ammeter.

o. Compare the ammeter reading with the engine running to both the reading with the engine stopped and the calculated amperage:

• If the running amperage is significantly higher than the static or calculated amperage, or if the circuit breaker trips while running, the rotor is shorted; replace the rotor.

• If there is NO amperage draw during the running test, the rotor has an open winding. Replace the rotor.

• If the running amperage is approximately the same as the static and the calculated amperage, and if the circuit breaker does not trip while running, the rotor is good.

7. To test the stator:

a. Obtain a fully charged 12-volt battery and a pair of test leads to connect the battery terminals to the brush connector terminals.

b. Disconnect and remove the AVR. Connect the positive battery post to the red/white (+) brush holder terminal, then connect the negative battery post to the black/white (-) brush holder terminal.

NOTE: Improper battery connections will damage the generator.

c. Start the generator, verifying the correct 3750 RPM no load speed.

d. Test the stator output voltages using the following test connections:

• Blue to white main winding 1 and red to white main winding 2 should each read 51-69 VAC.

• Brown to brown sensor winding should read 8-12 VAC.

• Light green/white to light green/white exciter winding should read 45-55 VAC.

e. Stop the generator and disconnect the test battery.

f. Using an ohmmeter, measure the resistance of the following test connections:

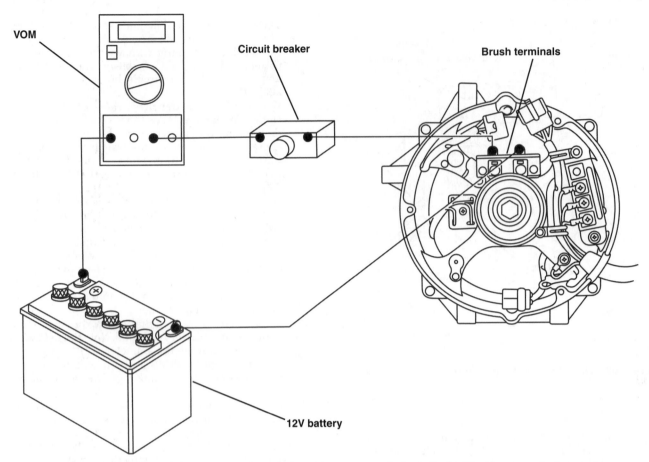

Fig. HA1004 – Component connections for performing the dynamic rotor test. Refer to the text for the proper test procedure.

- The blue to white, and the red to white main winding connections should each read 0.1-0.3 ohms. The blue to red main winding connection should read 0.2-0.6 ohms.
- The light green/white to light green/white exciter winding terminals in the 2-pin connector should read 0.7 1.1 ohms.
- The brown to brown sensor winding terminals in the 2-pin connector should read 0.03 ohm maximum.
- Resistance for each red, white, and blue main winding connection to a ground on the stator laminations should be infinity/no continuity.
- Resistance for each light green/white exciter winding terminal to a ground on the stator laminations should be infinity/no continuity.
- Resistance for each brown sensor winding terminal to a ground on the stator laminations should be infinity/no continuity.

g. If all of the tests in Steps D and F met specification, the stator is good. Proceed with further component testing. If the component tests meet specification, the automatic voltage regulator (AVR) is faulty and must be replaced.

h. If any individual test in Steps D or F failed to meet specification, the stator is faulty. Replace the stator.

i. Reconnect all previously disconnected wiring and components.

j. Attach DC voltmeter probes to the brush leads.

k. Start the generator and immediately test brush voltage. Voltage should test 34-42 VDC. If voltage is out-side this range, immediately stop the generator. The AVR is faulty and must be replaced.

12. To test the auto throttle system:

NOTE: The auto throttle system will not detect loads below 1.0 Amp. If less than a 120-watt load is being applied to the generator at 120V (240-watt load at 240V), the auto throttle switch must be OFF so the unit will run at full throttle governed RPM to produce proper voltage.

a. Turn the auto throttle switch OFF.

b. Start the generator; make sure all loads are disconnected.

c. Move the switch to AUTO. After a moment, the engine should idle down to 2000-2400 RPM.

d. Return the switch to OFF. The engine should return to no load full throttle.

e. If the auto throttle system fails Steps C and D, proceed to the **Individual Component Testing** section.

INDIVIDUAL COMPONENT TESTING

Automatic Voltage Regulator (AVR)

There is no test procedure available for the AVR. If the rotor, stator, receptacles, breakers, and harnesses all meet specification in the **Troubleshooting** tests, but there is still faulty output, replace the AVR.

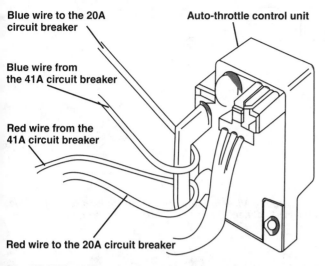

Blue wire to the 20A circuit breaker

Auto-throttle control unit

Blue wire from the 41A circuit breaker

Red wire from the 41A circuit breaker

Red wire to the 20A circuit breaker

Fig. HA1005 – View of red and blue wires from the main AC stator winding properly passing through the auto throttle control unit sensor loop from opposite directions. Failure to route the wires as shown will cause the auto throttle system not to

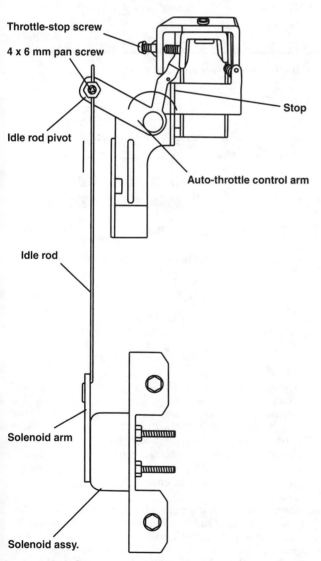

Throttle-stop screw

4 x 6 mm pan screw

Stop

Idle rod pivot

Auto-throttle control arm

Idle rod

Solenoid arm

Solenoid assy.

Fig. HA1006 – View of auto throttle linkage. Refer to the text for the correct adjustment procedure.

Auto Throttle System

1. To test the auto throttle switch:
   a. Access and disconnect the two black/red to green switch harness terminals inside the control box.
   b. With the switch lever in the ON position, there should be continuity between the two pigtails.
   c. With the switch lever in the OFF position, the two pigtails should show no continuity.
   d. Failure of the switch to test as specified indicates a faulty switch; replace the switch.
2. To test the solenoid:
   a. Disconnect the solenoid harness connector from the solenoid assembly on the back of the engine, next to the oil filter. The harness wires are white/blue and the solenoid leads will be white.
   b. Using a fully charged 12V test battery, connect the battery terminals to the solenoid terminals. The solenoid arm should pull the throttle linkage to idle position.
   c. Disconnect the test battery. The throttle linkage should return back to non idle position.
   d. Failure of the solenoid to activate as described in Steps B and C indicate a faulty solenoid. Replace the solenoid assembly.
3. To test the control unit:
   a. There is no specific test for the auto throttle control unit. If the blue and red generator main winding leads are correctly routed through the sensor loop on the back of the control unit (Fig. HA1005), and if the generator DC sub winding, diode stack, and auto throttle solenoid and switch all test to specification, but the auto throttle still does not function, the control unit is faulty and must be replaced.
   b. If replacing the control unit, make sure that the blue and red main winding leads pass through the sensor loop from opposite directions as shown in Fig. HA1005. If the wires pass through the loop from the same direction, they will cancel each other's magnetic field, preventing the new control unit from sensing voltage differences, thereby causing the new unit to malfunction.
4. To adjust the idle speed to 2000-2400 RPM (2200 RPM recommended):
   a. With the engine and the auto throttle switch both OFF, verify that the auto throttle control arm contacts the arm stop (Fig. HA1006).
   b. Loosen the 4 mm screw in the idle rod pivot.
   c. Hold the control arm against the stop.
   d. Move the rod toward the control arm/pivot to its maximum position. Do not force.
   e. Remove the 6 mm shoulder screws and the fan cover hood from over the air filter tube.
   g. Start the engine. Allow it to warm up to normal operating temperature.
   g. Turn the auto throttle switch to ON/AUTO.
   h. Note the engine RPM. If necessary, adjust with the throttle stop screw.

Full Wave Rectifier

To test the rectifier:
1. Disconnect the 4-wire (gray, gray, green, and white) rectifier harness connector inside the generator control box.
2. Using a VOM set to the R × 1 scale and Fig. HA1007 indicated in Fig. HA1007A, test for continuity between the color coded terminals noted:

3. A reverse polarity meter may show readings opposite those indicated. If so, reverse the probes.

4. Reverse the Step 2-3 probes and retest the terminal pairs that tested *Continuity*. They should now test *Infinity*.

5. Any test results other than those specified in Steps 2 and 4 indicate a faulty rectifier. Replace the rectifier.

Engine Switch

To test the switch:

1. Access and disconnect the 8-wire switch harness connector inside the generator control box.

2. Refer to Fig. HA1008. Test continuity between terminals in each switch position as follows:

Position	Terminal Continuity
OFF	G (red) to 'FS' (green/white), and E (black) to IG (black/yellow)
ON	B (white) to AS (light green)
START	B (white) to AS (light green) to S (yellow) to 'ST' (black/white)

3. If any switch positions fail to show continuity as specified, the switch is faulty and must be replaced.

Hourmeter

To test the hourmeter:

1. Obtain a fully charged, 12V test battery.

2. Disconnect the two-wire meter connector inside the generator control box.

3. Connect the battery positive to the white/yellow meter lead.

4. Connect the battery negative to the yellow/green meter lead.

5. Replace the meter if it fails to operate.

DISASSEMBLY

Refer to Figs. HA1001 and HA1002 for component identification. The rotor cannot be removed without first removing the stator and front housing. Exercise caution when handling the stator and rotor so the windings and winding coatings are not damaged.

To access the generator components:

1. Remove the louvered cover, upper side plates, and the hanger beam cross member from the top of the cradle frame.

2. Drain as necessary, then remove the fuel tank, hoses, and filter/shutoff valve.

3. Disconnect and remove the battery.

4. Remove the engine heat shield.

5. Remove the generator end cover.

6. Remove the engine fan cover hood over the air filter tube. Disconnect the choke cable.

7. Disconnect the wiring and remove the control box harness from the generator rear housing, then disconnect the 8-terminal engine harness.

8. Remove the fasteners holding the generator control box and control panel faceplate to the cradle frame, then remove the control box assembly.

9. If control box component access is necessary, remove the remaining upper and lower center faceplate screws; separate the faceplate from the control box.

10. Remove the brush holder assembly from the rear housing.

11. Unbolt the terminal block assembly from the rear housing mount plate, then remove the mount plate.

12. Remove the rear housing mount bolts and the engine mount nuts and washers.

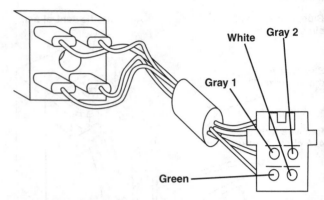

Fig. HA1007 – View of the terminal connections for testing the full wave rectifier. Refer to the text for the correct test procedure.

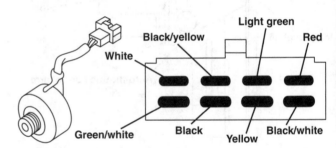

	Tester (+)			
Tester (+)		GRAY 1	GRAY 2	WHITE	GREEN
GRAY 1		∞	Continuity	∞
GRAY 2	∞		Continuity	∞
WHITE	∞	∞		∞
GREEN	Continuity	Continuity	Continuity	

Fig. HA1007A–Rectifier continuity chart.

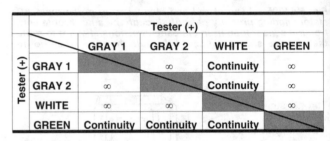

Fig. HA1008 – View of engine switch connector terminals. Refer to the text for the proper test procedure.

13. Lift the rear housing slightly in order to place a temporary support, such as a 2×4, under the engine side cover so that the rear housing feet are not contacting the generator lower cross member. The stator top has provision for a lift ring, if desired.

14. Remove the four bolts that mount the rear and front housings.

15. Prying gently in the slots provided, remove the rear housing from the stator, allowing the terminal block to slide through the housing hole and remain with the stator.

16. Noting the wiring orientation, remove the stator.

17. Remove the six bolts and two stud nuts holding the generator front housing to the engine front housing; remove the generator housing.

18. Using care not to break any fan blade fins, remove the three bolts and stud nut holding the rotor flange to the engine coupler and remove the rotor. It may be necessary to apply a light amount of penetrating oil to the flange to coupler area. Gently rock the rotor, to free it from the coupler.

19. Remove the cooling fan from the coupler.

20. Loosen the coupler center bolt, backing it out from the crankshaft 3 or 4 threads. Using a two-prong puller, break the coupler free from the engine crankshaft. Remove the center bolt and coupler.

21. If necessary, remove the engine front housing, noting UP housing orientation.

REASSEMBLY

Refer to Figs. HA1001 and HA1002 for component identification, and Figs. HA1003 and HA1009 for proper wiring connections. The rotor must be installed before the front housing and stator. Exercise caution when handling the stator and rotor so the windings and winding coatings are not damaged.

To reassemble the generator components:

1. If removed during disassembly, install the engine front housing, noting UP housing orientation. Incrementally and alternately torque the housing bolts 24 N•m (205-210 in.-lb.).

2. Place a temporary support, such as a 2 × 4, under the engine side cover so the generator rear housing will clear the lower crossmember during installation.

3. Make sure that the matching tapers on the engine crankshaft and inside the rotor coupler are clean and dry. Install the coupler onto the crankshaft. Install the coupler flange bolt, then torque the bolt 40 N•m (350-360 In-lb).

4. Install the cooling fan onto the coupler.

5. Make sure that the rotor to coupler mount faces are clean and dry. Fit the rotor onto the coupler using the coupler stud for alignment. Using care not to break any fan blade fins and install the bolts and stud nut. Then incrementally and alternately torque the fasteners to 40 N•m (350-360 in-lb.).

6. Using the two engine front housing studs for alignment, mount the generator front housing to the engine housing. incrementally and alternately torque the fasteners 24 N•m (205-210 in.-lb./17 ft.-lb.).

7. Noting the wiring orientation, fit the stator over the rotor and against the front housing.

8. Carefully feed the stator terminal block through the proper housing hole, and install the rear housing onto the rotor bearing and against the stator.

9. Install the housing mount bolts. Incrementally and alternately torque the bolts 40 N•m (350-360 in-lb.).

10. Remove the spark plugs and rotate the engine crankshaft to check for interference between the rotor and stator. There should be a consistent gap around the rotor laminations inside the stator. If any rubbing or scraping is

evident, correct any misalignment prior to proceeding. Reinstall the spark plugs and torque to 18 N•m (155 in.-lb.).

11. Lift the rear housing slightly and remove the temporary engine support.

12. Install the generator rear housing mount bolts and the engine mount washers and flange nuts. Torque the engine nuts 40 N•m (350-360 in-lb.), and torque the rear housing bolts to 60 N•m (43 ft.-lb.).

13. Install the terminal block mount plate onto the rear housing, then bolt the terminal block onto the plate.

14. Install the brush holder assembly.

15. Assemble the control box faceplate to the box, being careful not to pinch or kink any component wiring. Install the control box assembly onto the cradle frame.

16. Being careful not to kink or pinch any wires, reconnect the control box harness wiring to the proper rear housing points, then reconnect the 8-terminal engine harness.

17. Reconnect the engine choke cable. Install the engine fan cover hood over the air filter tube.

18. Install the generator end cover.

19. Install the engine heat shield.

20. Install and reconnect the battery.

21. Install and reconnect the fuel tank, hoses, and filter/shutoff valve.

22. Install the hanger beam crossmember, upper side plates, and louvered cover onto the top of the cradle frame. Torque the crossmember bolts 60 N•m (43 ft.-lb.).

FASTENER TORQUE CHART

Fastener torque values are listed in the approximate order of reassembly.

Component Fastener	Torque
Engine front housing	24 N•m (205-210 in.-lb./17 ft.-lb.)
Crankshaft coupler;	
Rotor to coupler	40 N•m (350-360 in-lb./29 ft.-lb.)
Generator front housing	24 N•m (205-210 in.-lb./17 ft.-lb.)
Rear housing to front housing;	
engine mount flange nut	40 N•m (350-360 in.-lb./29 ft./lb.)
Spark plug	18 N•m (155 in.-lb.)
Rear housing foot mount bolt;	
Hanger beam crossmember	60 N•m (515 in.-lb./43 ft.-lb.)

WIRING DIAGRAM

Refer to Fig. HA1010 for the EB11000 series wiring diagram.

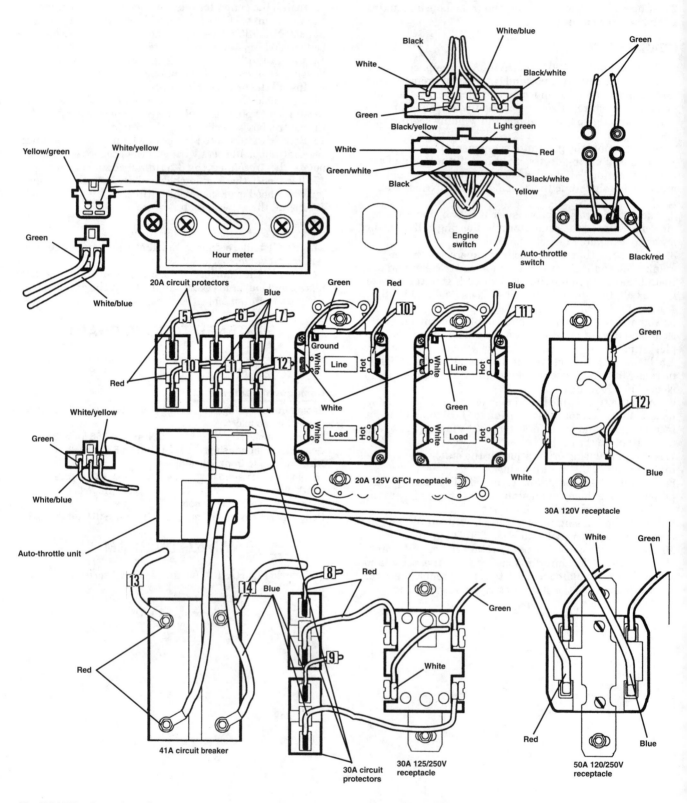

Fig. HA1009 – Rear view of generator control panel showing components installed and wired.

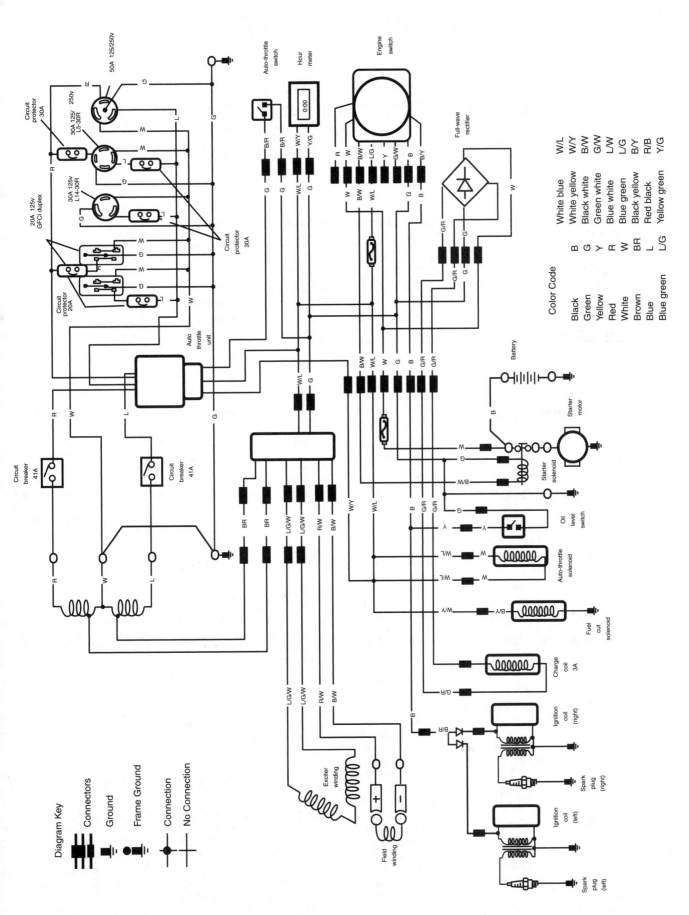

Fig. HA1010 – Wiring diagram for the EB11000 series generator.

JOHN DEERE

Worldwide Commercial and Consumer Equipment Division
14401 Carowinds Blvd.
Charlotte, NC 28241-7047

Generator Model	Rated (Surge) Watts	Output Volts	Engine Make	HP
250G	2300 (2500)	120	BandS	5
440G	4000 (4400)	120/240	BandS	8
550GE	5000 (5500)	120/240	B&S	11
G2500K	2300 (2500)	120	Kawasaki	5.3
G4400K	4000 (4400)	120/240	Kawasaki	7.9
G5500K/KE	5000 (5500)	120/240	Kawasaki	11.1

IDENTIFICATION

John Deere generators are brush style, single phase, 60 Hz units with integral electronic voltage regulation.

WIRING COLOR CODES

John Deere wiring schematics use the following color code abbreviations to identify the generator wiring:

Blk	Black	Blk/Wht	Black w/White tracer stripe
Blu	Blue	Blu/Wht	Blue w/White tracer stripe
Brn	Brown	Brn/Wht	Brown w/White tracer stripe
Grn	Green	Brn/Yel	Brown w/Yellow tracer stripe
Gry	Gray	Dk Brn/Lt Grn	Dark Brown w/Light Green tracer stripe
Org	Orange	Dk Brn/Red	Dark Brown w/Red tracer stripe
Pnk	Pink	Dk Brn/Yel	Dark Brown w/Yellow tracer stripe
Pur	Purple	Org/Wht	Orange w/White tracer stripe
Red	Red	Pnk/Blk	Pink w/Black tracer stripe
Tan	Tan	Pur/Wht	Purple w/White tracer stripe
Wht	White	Red/Blk	Red w/Black tracer stripe
Yel	Yellow	Red/Wht	Red w/White tracer stripe
Dk Blu	Dark Blue	Wht/Blk	White w/Black tracer stripe
Lt Blu	Light Blue	Wht/Red	White w/Red tracer stripe
Dk Grn	Dark Green	Yel/Blk	Yellow w/Black tracer stripe
Lt Grn	Light Green	Yel/Red	Yellow w/Red tracer stripe
		Yel/Wht	Yellow w/White tracer stripe

MAINTENANCE

ROTOR BEARING

The endbell bearing and race should be inspected, cleaned, and relubricated any time the endbell is removed. If the bearing runs hot to the touch, replacement may be necessary.

BRUSHES

Brushes must be at least 3/8 in. long. Shorter brushes require replacement. Brushes can only be accessed by removing the endbell. Refer to Disassembly section in this chapter.

TROUBLESHOOTING

A generator analyzer tool #08371 and a field flasher tool #UP00457 are available from John Deere to troubleshoot generator malfunctions. Prior to any disassembly, initial testing should be performed with the two tools.

Output voltage should be 113-127V on the 120-volt receptacles and 226-254V on the 240-volt receptacles. No load voltage with the idle control activated should be 80-95V. If little or no output is generated, the following items could be potential causes:

1. Engine RPM–The engine must maintain a minimum 3550 RPM under load. Normal no load speed is 3750-3800 RPM. No load speeds with idle control activated should be 2600-2800 RPM on models 250G, 440G, G2500K, and G4400K, and 2400-2600 RPM on models 550GE and G5500K/KE. Refer to the appropriate engine section for engine service. Refer to the **idle control** section in this chapter for idle control service.

2. Loose component mounting fasteners (low or no voltage).

3. Circuit breakers or GFCI switch breaker (low or no voltage).

4. Insecure wiring connections (low or no voltage).

5. Faulty receptacle (low or no voltage).

6. Capacitor (low or no voltage).

7. Brushes (no voltage).

8. Rotor:

   a. Loss of residual magnetism (no voltage).

NOTE: Residual magnetism can be restored without generator disassembly, and is the recommended first step if the generator has no output. Refer to Field Flashing section.

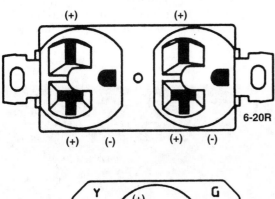

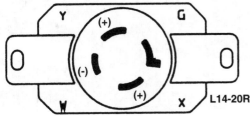

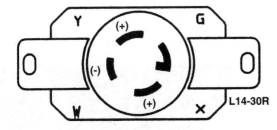

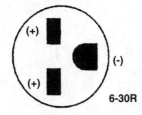

Fig. JD100 – Test connections for the 240V receptacles. Refer to text for correct test procedure.

b. Open or shorted windings (no or low voltage, respectively).

c. Dirty, broken, or disconnected slip rings (no voltage).

9. Stator windings (low voltage – shorted; no voltage – open).

10. Idle control malfunction (erratic or no idle).

11. Insufficient cooling system ventilation (overheating).

TESTING AND TROUBLESHOOTING

RECEPTACLES

NOTE: The generator should be properly grounded to earth prior to testing the receptacle output.

To test the receptacles:

1. Start the generator, verifying proper no load speed of 3750-3800 RPM.

2. Make sure that the circuit breakers are set. If the circuit breakers fail to set, refer to the Circuit Breaker section in this chapter.

3. Set the VOM scale to AC volts.

4. For the 120V outlets, place the black VOM to neutral/ground and the red lead to the small flat leg of each duplex outlet. The reading should be 117-130 volts.

5. For the 240V receptacles, refer to the polarity markings of Fig. JD100.

   a. Positive (+) slots are *hot*.

   b. Negative (-) slots are *neutral* or *return*.

   c. Lock slots, where used, are *frame ground*.

6. Test between the positive (+) slots and both the negative (-) slots and the ground (G) slots. the reading should be 234-260 volts.

   a. If only one receptacle shows no voltage, the receptacle may be faulty or may not be receiving output. Inspect the receptacle and wiring.

   b. A no voltage reading on all receptacles usually indicates a loss of residual magnetism: Refer to the following **Field Flashing** section in this chapter.

   c. A reading in the 1-3 volt range on all receptacles is a measurement of residual magnetism only. It indicates a fault in either the brushes, rotor, voltage regulator, bridge rectifier, or quad winding. Proceed with further testing.

   d. A half voltage reading (60V at the 120V receptacles; 120V at the 240V receptacles) usually indicates a faulty bridge rectifier or rectifier terminal connections. Proceed to **Bridge Rectifier** section in this chapter for further testing.

   e. A less than half voltage reading (30-50V) usually indicates a blown capacitor (capacitor bottom is bulged). a 3/4-voltage reading (90V of 120V; 180V of 240V) usually indicates a faulty capacitor. Proceed with further testing.

NOTE: If the 120V receptacle is being replaced on 120/240V units, the connector tab on the 'Hot' side must be broken off: Failure to remove the tab will give an indication of overload upon generator start up. Torque terminal screws to 14 in.-lb. On duplexes, torque ground screws to 10 in.-lb.

NOTE: When replacing receptacles, always make sure that the circuit breaker capacity matches the receptacle. If upgrading from an original 15-amp receptacle or GFCI to a 20-amp receptacle or GFCI, replace the breaker with a matching unit.

Field Flashing

If there is no output at any receptacle, stop the generator and proceed as follows:

Using the #08371 Analyzer and the #UP00457 Field Flasher

1. Disconnect the large main harness connector and the small excitation harness connector from the back of the control panel (Fig. JD101).
2. Connect the main and excitation connectors to analyzer #08371.
3. Start the generator and observe the analyzer.

NOTE: Do not unplug the analyzer while the generator is running.

  a. If both neon lights are OFF:
- Turn the field flasher (#UP00457) switch ON.
- Plug the flasher into one of the analyzer receptacles.
- When the flasher switch trips, the residual magnetism should be restored.
- If both neon lights remain OFF, there is usually a problem in the excitation circuit.
- If both lights light after flashing the fields, stop the generator, then restart it. If both lights are OFF again, the rotor will not hold residual magnetism. Replace the rotor.

  b. If one neon light is ON (models 250G and G2500K) or if both neon lights are ON (models 440G, 550GE, G4400K, and G5500K/KE), the brushes, slip rings, rotor, and stator are OK, but the voltage regulator board or the control box wiring is faulty.

  c. If one green light is out, one output leg is open (4000 and 5000W units).

  d. If one green light is dim, one output leg is partially shorted.

4. Continue to run the generator for three minutes. If there is a layer short, the windings should begin to smoke. Stop the generator. The stator is faulty and must be replaced.

Using Only the #UP00457 Field Flasher

1. Start the generator and allow a brief warm up period.
2. Turn the field flasher switch ON.
3. Plug the flasher into one of the generator's 120V outlets.
4. When the flasher switch trips, residual magnetism should be restored.
5. Stop the generator, remove the flasher, then restart the generator.
6. Test the output at the 120V outlet. If there is no output, the rotor will not hold residual magnetism. Replace the rotor.

IDLE CONTROL

NOTE: On Model 440G, if the engine surges or hunts with a load of between 500 and 1000 watts, and the problem is not caused by the carburetor or governor, replace the paddle and electromagnet.

Switch Test

To test the idle control switch:
1. Disconnect the spark plug wire.
2. Remove the Torx head screws from the control panel faceplate, then separate the faceplate from the rear cover.
3. Disconnect the wiring terminals from the back of the idle start switch.
4. Using a VOM set to the R × 1 scale, test for continuity between the switch terminals. Switch should have continuity in the IDLE position only. Any other test results indicate a faulty switch.

Paddle Setting

Refer to Fig. JD102 for model 250G, Fig. JD103 for models 440G and 550GE, and Fig. JD104 for models G4400 and G5500.

For proper paddle function:
1. Manually move the paddle against the electromagnet contact. Make sure that the flat face of the paddle (A) is parallel with the face of the electromagnet. Bend the paddle as necessary.
2. Connect a tachometer to the engine. Start the generator and allow a brief warm up period.
3. Manually hold the paddle toward the electromagnet until the idle control linkage stop prevents further move-

Fig. JD101 – Typical rear view of the control panel showing the wiring connectors.

Fig. JD102 – View of idle control electromagnet and paddle on model 250G generator.

Fig. JD103 – View of idle control electromagnet and paddle on generator models 440G and 550GE.

Fig. JD104 – View of idle control electromagnet and paddle on generator models G4400 and G5500.

Fig. JD105 – With the idle control electromagnet energized, the carburetor throttle arm should not contact the slow idle speed adjusting screw. Refer to text.

ment. Do not force. Observe the engine RPM. Idle control speed should be 2600-2800 RPM on models 250G, 440G, G2500K, and G4400K, and 2400-2600 RPM on models 550GE and G5500K/KE. To adjust idle control speed to this range, refer to the electromagnet test and adjustment.

4. Release the paddle.

5. Move the idle control switch to the IDLE position. The paddle should pull to the electromagnet. If the paddle does not pull to the electromagnet, proceed with the Electromagnet Test and Adjustment.

6. Turn the idle control switch OFF and stop the generator.

Electromagnet Test and Adjustment

1. To test the electromagnet:
   a. With the generator stopped and the idle control switch OFF, disconnect the two black wires from the idle control connector terminals at the rear of the control panel (Fig. JD101).
   b. Using the VOM, test the resistance between the two black wires. Resistance should test 250 ohms on 2300W units and 285 ohms on 4000 and 5000W units.
   c. Test for continuity between each black wire and the electromagnet body. There should be no continuity.
   d. Failure of the continuity tests indicates a faulty electromagnet. Replace the electromagnet.
   e. If the electromagnet tests good, and the idle control switch passes the switch test, but the electromagnet still does not energize when the switch is moved to the IDLE position, check the continuity of the wiring between the idle control switch and the circuit board. Refer to **Disassembly, Control Panel**.
   f. If the wiring continuity tests OK, reassemble, then test the voltage to the electromagnet.
   g. Start the generator and, with the VOM set to read DC voltage, connect the VOM leads to the idle control connector terminals exposed in Step 1. VOM should read 35-45 VDC. Insufficient DC voltage indicates a faulty circuit board. Replace the board.

2. To adjust the electromagnet:
   a. Connect a tachometer to the engine, then start the generator and allow a brief warm up period.
   b. Move the idle control switch to IDLE.
   c. Observe the tachometer for the proper idle control speed: 2600-2800 RPM on models 250G, 440G, G2500K, and G4400K, and 2400-2600 RPM on models 550GE and G5500K/KE.
   d. Make sure that the electromagnet is positioned close enough to the paddle to exert a sufficient pull so that the paddle holds the idle control RPM. If necessary, loosen the electromagnet lock nuts and move the electromagnet for a stronger pull.
   e. On units with Briggs and Stratton engines, the paddle will need to contact the electromagnet so that the carburetor idle speed adjusting screw (C – Fig. JD105) does not contact the throttle arm.
   f. On units with Kawasaki engines, the paddle does not necessarily have to contact the electromagnet, as the idle control RPM is adjusted with the adjusting screw (C – Fig. JD106, or Fig. JD107, depending on engine). The throttle arm should not be touching the carburetor idle speed adjusting screw.

GROUND FAULT CIRCUIT INTERRUPTER (GFCI)

To test the GFCI:

1. Start the generator and allow a brief warm up period.

2. Depress the TEST button. The RESET button should extend. If the RESET button does not extend, the GFCI is faulty.

3. With the RESET button extended, plug a test light into the GFCI receptacle. If the light illuminates with the RESET button extended, the GFCI line and load terminals

have been wired backwards. To correct, stop the generator, then:

   a. Refer to Fig. JD108.

   b. Access the GFCI connections (refer to Disassembly – Control panel section in this chapter).

   c. Determine whether the GFCI receptacle is a Bryant or an Arrow Hart.

   d. Make sure that the input power wires connect to the proper *line* screws, and, when applicable, any protected receptacles downstream of the GFCI connect to the proper *load* screws. When replacing a GFCI receptacle, always note the *line* and *load* connection orientation.

   e. Make sure that the White (neutral) wires connect to the correct silver colored screws, and the black or red (hot) wires connect to the correct brass screws. The Green (ground) wire always connects to the green screw.

4. To restore power, restart the generator and depress the RESET button firmly into the GFCI until a definite click is heard. If properly reset, the RESET button is flush with the surface of the TEST button. When the RESET button stays in, the GFCI is on.

5. Stop the generator.

> NOTE: If the GFCI seems to trip for no apparent reason, check to make sure the shipping block (packing material) was removed from under the engine when the generator was initially prepared for service. If the block was not removed, the vibration isolation mounts will not properly prevent engine vibration from reaching the control panel, and the vibration will trip the GFCI.

CIRCUIT BREAKERS

To test each circuit breaker:

1. Make sure the breaker is set. If tripped, reset.

2. Disassemble as necessary to access the breaker, (refer to Disassembly, Control Panel then remove the wires from the rear of the breaker.

3. Using the R × 1 scale on the VOM, test for continuity across the breaker terminals.

4. If no continuity reading is noted, or if the breaker fails to reset, the breaker is faulty. Replace the breaker.

DISASSEMBLY

> NOTE: Before beginning any disassembly, allow the generator to cool.

Refer to Figs. JD109 and JD110 for component identification.

Generator Head

1. On electric start models, disconnect the battery cables and remove the battery.

2. Shut off the fuel valve, disconnect the fuel hose, and remove the fuel tank.

3. Disconnect the wiring connectors from the rear of the control panel (Fig. JD101).

4. Remove the hardware which fastens the generator support bracket to the mounts on the generator cradle frame.

5. Place a temporary support block under the engine adapter, if necessary.

6. Remove the four stator throughbolts from the endbell flanges.

Fig. JD106 – View of the fast idle adjustment screw on 2300W units with Kawasaki engines. Refer to text for adjustment procedure.

Fig. JD107 – View of the fast idle adjustment screw on 4000 and 5000W units with Kawasaki engines. Refer to text for adjustment procedure.

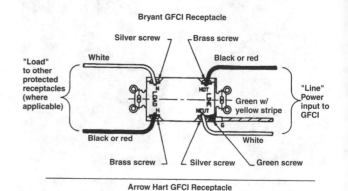

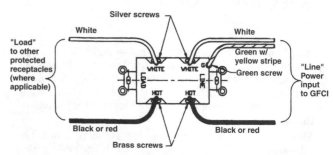

Fig. JD108 – Wire side views of Bryant and Arrow Hart GFCI receptacles showing wiring connection differences.

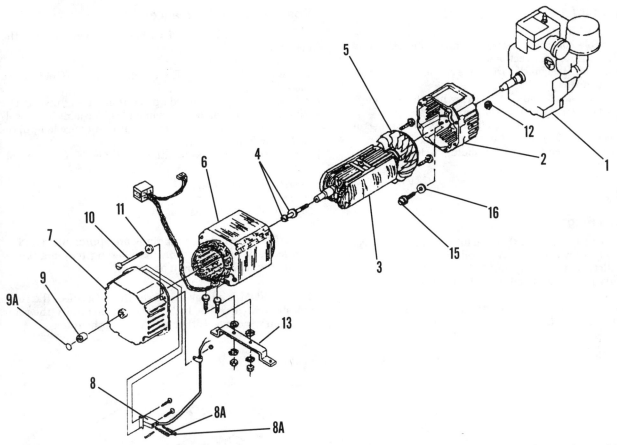

Fig. JD109 – Typical view of John Deere generator head components. The support bracket will vary depending on the model.

1. Engine
2. Engine adapter/front housing
3. Rotor assembly
4. Rotor through bolt and washer
5. Fan with four screws

6. Stator assembly
7. Endbell/rear housing
8. Brush holder
8A. Brush (2)
9. Rotor bearing

9A. Plug
10. Endbell/stator through-bolt (4)
11. Flat washer (4)
12. Hex nut (4)
13. Stator mount

7. Gently pull the stator/endbell assembly straight out from the generator.

NOTE: After the endbell is partially pulled out, the brushes will spring out of their holder.

8. Remove the endbell from the stator by sliding the stator harness and its protective rubber grommet from the lower slot in the endbell.

9. To inspect and service the endbell bearing, remove the bearing cap plug from the endbell.

10. Remove the rotor throughbolt and lock washer.

11. Remove the rotor using one of the following two methods:

NOTE: To prevent rotor damage, do not let the rotor fall from the engine crankshaft upon removal.

a. The recommended method is to use a slide hammer with a 3/8-16 end threaded into the rotor shaft. Screw the slide hammer into the end of the rotor shaft approximately 1/2 in. Tap the slide hammer until the rotor is loose from the engine crankshaft.

b. An alternate method is to make a push tool to fit inside the rotor shaft (Fig. JD111):

• Obtain a length of 0.25 in. all thread rod.

• Thread the rod through the rotor shaft and into the engine crankshaft until it bottoms (do not force), then unscrew it one turn.

• Mark the rod flush with the end of the rotor shaft.

• Remove the rod and cut the in rotor length 0.5 in. (13 mm) shorter than the mark.

• Cut a screwdriver slot in the outer end of the rod.

• Lightly screw the rod back through the rotor shaft and into the engine crankshaft, then back it out one turn, as before.

• Using a 3/8-16 × 1 in. bolt as a jack screw, tighten the bolt into the end of the rotor shaft and against the rod. Tighten the screw and gently tap with a brass hammer until the rotor breaks loose.

Control Panel

1. If not previously done, disconnect the wiring connectors from the rear of the control panel (Fig. JD101).

2. Remove the four Torx head screws that hold the rear cover of the control panel to the generator cradle frame (two on each side).

3. Carefully remove the control panel assembly.

4. Unscrew the Torx screws holding the control panel faceplate to the rear cover, then carefully separate the faceplate assembly from the rear cover.

Control Board

Refer to Fig. JD112.

1. Disassemble control panel as per preceding **Control Panel** disassembly instructions.

2. Disconnect the two black wire spade connectors (A) from the circuit board.

3. Disconnect the red and black wires which feed through the circuit board coil (B) from the circuit breakers.

4. Disconnect the red white and black wire 4-pin connector (C) from the side of the circuit board.

5. Remove the circuit board by sliding it from its control panel slots.

ROTOR

Rotor Slip Rings

The slip rings (Fig. JD113) are an integral part of the rotor assembly; they cannot be replaced.

Clean dirty or rough sliprings with a suitable solvent, then lightly sand with ScotchBrite. Do not use emery cloth or crocus cloth.

Cracked, chipped, broken, or grooved slip rings are not repairable. In these cases, the rotor must be replaced.

Rotor Winding Resistance

The rotor can be tested without removing it from the generator.

To test resistance:

1. Use a VOM set on the R × 1 scale. hold the VOM leads to the rotor slip rings.

2. If no reading is observed upon initial testing, inspect the winding to slip ring connection and solder joint. If the solder joint is faulty, and if there is sufficient wire length, re-solder the joint and retest.

3. Rotor resistance specifications at 77° F (25° C) are as follows:

2300W units – 46.7 ohms
4000W units – 67.0 ohms
5000W units – 76.0 ohms

NOTE: All resistance readings are approximate and may vary depending on the winding temperature and the test equipment accuracy.

4. If the VOM readings are lower than specified, the rotor has shorted windings. If the VOM reads infinity, the rotor has open windings. Either way, replace the rotor.

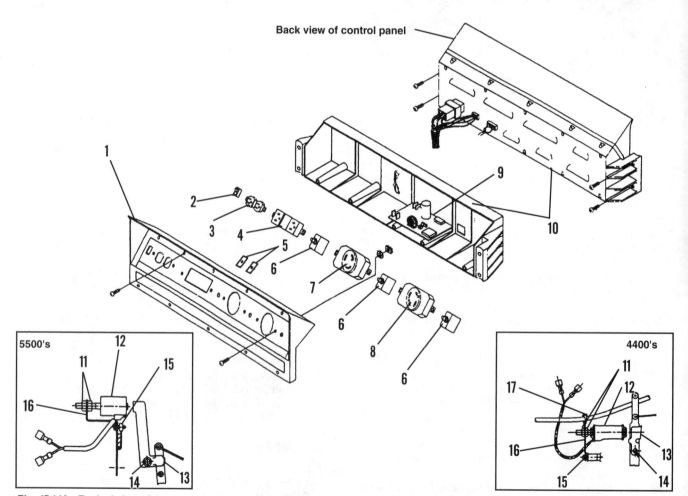

Back view of control panel

5500's

4400's

Fig. JD110 – Typical view of the control panel used on John Deere generators. Pictured is the panel used on the 4000 and 5000W models. the panel on the 2300W models has fewer receptacles and breakers.

1. Face plate
2. Rocker switch
3. 120V AC receptacle
4. 120V AC GFCI receptacle
5. Rubber grommets
6. Circuit breakers
7. 120V AC Twistlok receptacle
8. 240V AC Twistlok receptacle
9. Control board
10. Control panel
11. Nut
12. Electromagnet
13. Paddle
14. Screw
15. Nut
16. Bracket
17. Clamp

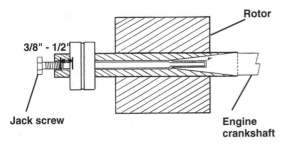

Fig. JD111 – Alternate push rod tool method of rotor removal. Tool can be constructed in the shop.

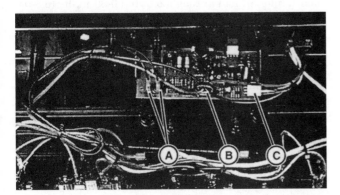

Fig. JD112 – View showing the control panel faceplate separated from the back cover, as well as most of the components mounted inside the control panel assembly.

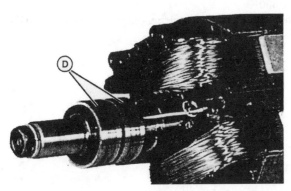

Fig. JD113 – View showing the brush slip rings (D) on the rotor.

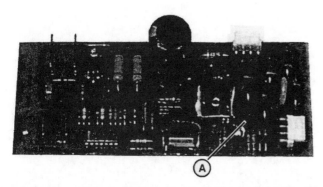

Fig. JD114 – View of a replacement control board showing the 120/240V jumper connector (A). Refer to text for jumper application.

5. Test for continuity between each slip ring and the rotor shaft. If either test shows continuity, the rotor windings are shorted to the rotor shaft. Replace the rotor.

STATOR

Winding Resistance

To test the stator windings:

1. If not previously done, disconnect the wiring connectors from the rear of the control panel (Fig. JD101).

2. With the VOM set to the R × 1 scale, and using the unplugged connector blocks, measure winding resistances and compare to the following specifications:

   a. On 2300W units:

White to black (main coil) . . . . 0.450-0.506 ohm
White to yellow (1) (quad coil) . . . . 1.568-1.768 ohms
White to yellow (2) (quad coil) . . . . 1.568-1.768 ohms
Yellow to yellow (quad coil) . . . . 3.136-3.536 ohms
Green (ground) to stator body. . . . . < 0.5 ohm
White to stator body,
  Black to stator body, and
  Yellow to stator body . . . . Infinite resistance

   b. On 4000W units:

Black to white (main coil) . . . . 0.350-0.394 ohm
Red to white (main coil) . . . . 0.350-0.394 ohm
White to yellow (1) (quad coil) . . . . 1.343-1.513 ohms
White to yellow (2) (quad coil) . . . . 1.343-1.513 ohms
Yellow to yellow (quad coil) . . . . 2.686-3.026 ohms
Green (ground) to stator body. . . . . < 0.5 ohm
Black to stator body,
  White to stator body,
  Red to stator body, and
  Yellow to stator body. . . . . Infinite Resistance

   c. On 5000W units:

Black to white (main coil) . . . . 0.280-0.316 ohm
Red to white (main coil) . . . . 0.280-0.316 ohm
White to yellow (1) (quad coil) . . . . 1.306-1.473 ohms
White to yellow (2) (quad coil) . . . . 1.306-1.473 ohms
Yellow to yellow (quad coil) . . . . 2.612-2.946 ohms
Green (ground) to stator body . . . . < 0.5 ohm
Black to stator body,
  White to stator body,
  Red to stator body, and
  Yellow to stator body. . . . . Infinite Resistance

3. If test results do not meet specifications, replace the stator assembly.

REASSEMBLY

Control Board

Refer to Fig. JD112.

NOTE: Any plastic cable ties that were cut and removed during disassembly must be replaced during reassembly.

1. Clean all contacts on the board prior to installation.

2. When installing a new control board into generator models 250G and G2500K, remove the two jumpers (A) from the board (Fig. JD114). Jumpers must be left in position on the board for all other installations, or the generator will have no 240V output.

3. Assemble the circuit board to the back cover of the control panel by sliding it into its slots.

4. Connect the R/W/B wire 4-pin connector (C) to the side of the circuit board.

5. Reconnect the red and black wires which feed through the circuit board coil (B) to the circuit breakers.

6. Connect the two black wire spade connectors (A) to their proper board terminals.

Control Panel

1. Carefully mate the rear cover to the control panel faceplate, making sure that no wires are pinched. insert and tighten, but **do not force**, the Torx head screws which hold the faceplate to the rear cover.

2. Fit the control panel to the generator cradle frame, then fasten the panel with the four Torx head screws at the back of the rear cover (two on each side).

3. Reconnect the wiring connectors from the generator head to the rear of the control panel (Fig. JD101).

Generator Head

Refer to Figs. JD109 and JD110 for component identification.

1. If the engine adapter was removed during disassembly, reinstall at this time. The unslotted side is 'Up'. Torque the adapter bolts as follows:

2300W units – 120-150 in.-lb. (13.6-16.9 N•m)

4000 and 5000W units – 240-250 in.-lb. (27.1-28.2 N•m)

2. Place a temporary 2 × 4 support block under the engine adapter, if necessary.

3. Make sure that the matching tapers on the engine crankshaft and inside the rotor shaft are clean and dry. Install the rotor onto the crankshaft. Install the rotor bolt and lockwasher. Tighten the rotor bolt finger tight at this time.

4. Service the rotor bearing in the endbell, if needed.

5. Install the brushes in the brush holder. Hold the brushes in place against their springs by inserting a straightened paper clip or similar tool through the endbell hole next to the bearing. Insert the tool from the outside in, making certain that it passes through the brush holder and over both brushes.

6. Assemble the endbell to the stator by feeding the stator harness through the lower slot in the endbell, making sure that the protective grommet is properly installed in the slot.

7. Carefully install the stator/endbell assembly over the rotor, correctly aligning the rotor endbell bearing and the stator to engine adapter locator pins.

8. Slide the two lower stator throughbolts into position, tightening them finger tight.

9. Slide the two upper stator throughbolts into position, then torque all four throughbolts to 60-80 in.-lb. (6.8-9.0 N•m).

10. Make sure that the rotor turns freely inside of the stator, then torque the rotor bolt to 100-140 in.-lb. (11.3-15.8 N•m).

11. Remove the brush holder tool, and install the rotor bearing plug cap.

12. Install the generator support bracket bolts. Torque the bolts to 145-155 in.-lb. (164.-17.5 N•m).

13. Reconnect the wiring connectors to the rear of the control panel (Fig. JD101).

14. Install the fuel tank and reconnect the fuel hose.

15. On electric start models, install the battery and reconnect the battery cables.

WIRING DIAGRAMS

Refer to Fig. JD115 for wiring diagram for models 250G and G2500K.

Refer to Fig. JD116 for wiring diagram for models 440G, 550G, G4400K, and G5500K/KE.

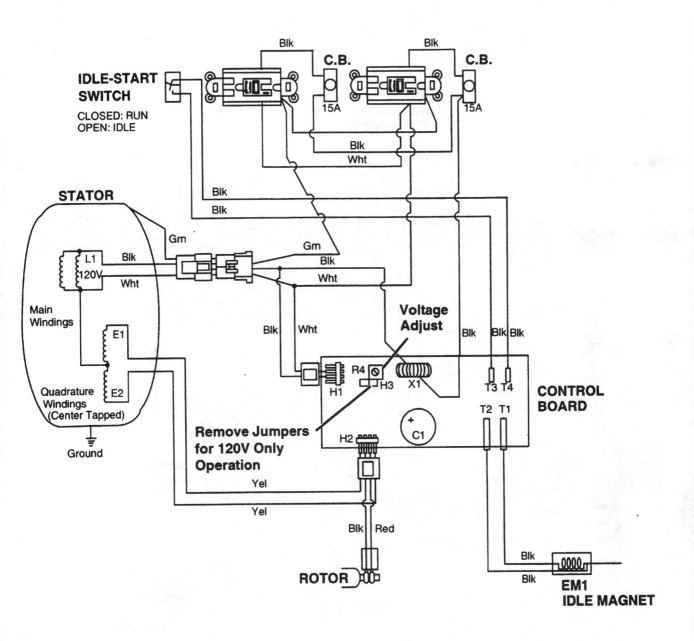

IDLE-START SWITCH

CLOSED: RUN
OPEN: IDLE

C.B. 15A

C.B. 15A

Blk

Blk

Blk
Wht

Blk
Blk

STATOR

Grn

Grn

Blk

L1
120V

Blk

Wht

Wht

Main Windings

Voltage Adjust

Blk

Blk Blk

E1

E2

Blk

Wht

H1

R4
H3

X1

T3 T4

CONTROL BOARD

Quadrature Windings (Center Tapped)

Ground

Remove Jumpers for 120V Only Operation

H2

+ C1

T2 T1

Yel

Yel

Blk Red

ROTOR

Blk

Blk

EM1 IDLE MAGNET

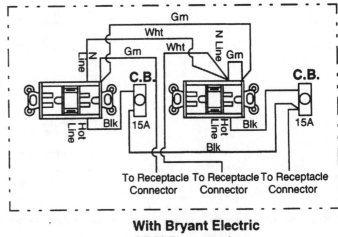

Gm
Wht
N Gm
N Line
Wht
Gm

N Line

C.B. 15A

Hot Line

Blk

C.B. 15A

Hot Line

Blk

Blk

To Receptacle Connector

To Receptacle Connector

To Receptacle Connector

With Bryant Electric GFCI Receptacle

Fig. JD115 – Wiring diagram for generator models 250G and G2500K.

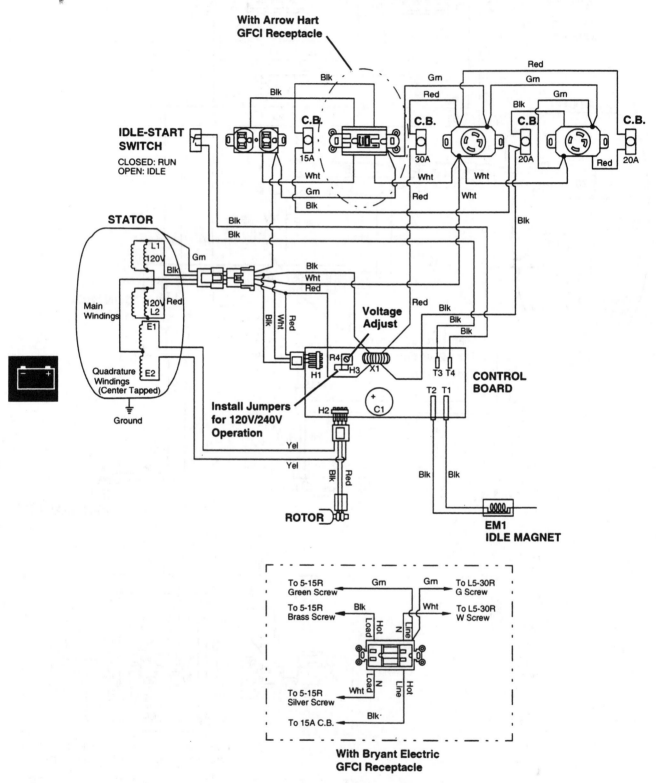

Fig. JD116 – Wiring diagram for generator models 440G, 550GE, G4400K, and G5500K/KE.

MITSUBISHI

MGA, MGE, and MGW Series

Mitsubishi Generator Service and Sales
222 N. LaSalle St., Suite 999
Chicago, IL 60601

Generator Model	Rated (Surge) Watts	Output Volts
MGA1300	1100 (1300)	120
MGA1800	1500 (1800)	120
MGA2900	2500 (2900)	120
MGE1800	1500 (1800)	120
MGE2900	2500 (2900)	120
MGE4000	3300 (4000) AC/100 DC	120/240 AC/12 DC
MGE4800	4100 (4800) AC/100 DC	120/240 AC/12 DC
MGE5800	5000 (5800) AC/100 DC	120/240 AC/12 DC
MGE6700	5800 (6700) AC/100 DC	120/240 AC/12 DC
MGW150A	3600 DC/2400 AC	26 DC/120 AC
MGW200A	5040 DC/3600 AC	28 DC/120 AC

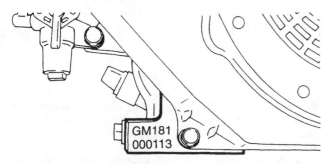

Fig. MB100 – Engine model and serial numbers are located on the side of the base mounting boss.

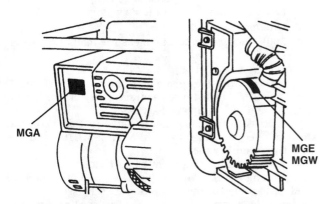

Fig. MB101 – Generator ideification is located on the side of the control box assembly.l

NOTE: MGA series units are brushless generators. MGE series units are brush style generators. MGW series units are brush style DC welders with AC capability.

Some units may be equipped with 12V electric start.

IDENTIFICATION

These generators are direct drive, 60 Hz, single phase AC output units with panel mounted receptacles.

The engine ID numbers are located on the side of the base mounting boss, to the left of the bottom blower housing bolt (Fig. MB100).

MGA series generator identification is located on the side of the control box assembly, next to the exhaust outlet (Fig. MB101).

MGE and MGW series generator model identification is located on a plate on the end of the fuel tank, above the generator end cover. The generator serial number is located on a plate on the top surface of the generator end cover (Fig. MB101).

Refer to Figs. MB102-MB106 to view the generator components.

MAINTENANCE

ROTOR BEARING

The brush end rotor bearing is a pre-lubricated, sealed bearing. Inspect and replace as necessary. If a seized bear-

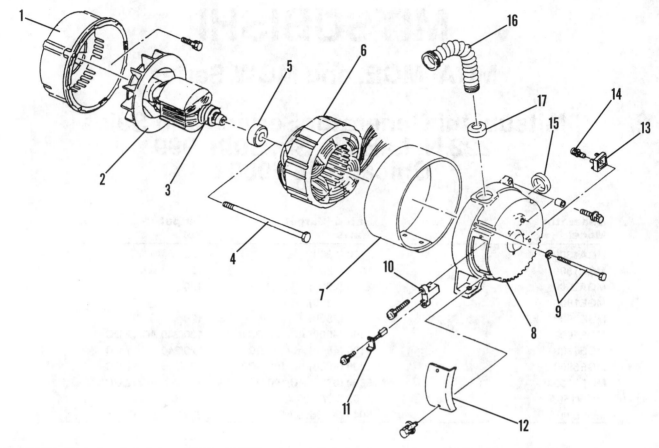

Fig. MB102 – Typical exploded view of MGA and MGE series generator head components. MGE is pictured. MGA is similar, minus the brushes and the diode stack.

1. Engine adapter/front housing
2. Rotor assembly with cooling fan
3. Slip rings
4. Rotor through-bolt
5. Rotor bearing
6. Stator assembly

7. Stator cover
8. Endbell/rear housing
9. Rear housing/stator through-bolts
10. Brush holder
11. Brush (2)
12. Brush cover

13. Diode stack
14. Diode spacer
15. Cap
16. Wire harness protector tube
17. Tube flange seal

ing is encountered, inspect the rotor journal and the rear housing bearing cavity for scoring.

BRUSHES

Replace the brushes when the brush length protruding from the brush holder reaches 5.0 mm (0.20 in.) or less. New brush length is 14 mm (0.55 in.) on MGE1800 and MGE2900 units, 15 mm (0.59 in.) on MGE4000-6700 units. Brushes should move freely within the holder.

When servicing brushes or performing brush tests, always observe proper brush polarity: The positive (+) brush is fed by the light green wire and is the brush closest to the rotor windings (inside brush). The negative (-) brush is fed by the brown wire and is the brush closest to the rotor bearing (outside brush).

GENERATOR REPAIR

TROUBLESHOOTING

Before testing any electrical components, check the integrity of the terminals and connectors.

1. While testing individual components, also check the wiring integrity.

2. The use of a fully charged, 12V test battery and the Mitsubishi tester #KJ10003AA is recommended for testing. If the Mitsubishi tester is not available, use the following test equipment.

    a. AC voltmeter – 0-150V for single voltage generators. 0-300V for dual voltage units.

    b. AC ammeter capable of handling the unit being tested.

    c. Frequency meter capable of reading 45-65 Hz, and able to handle the generator's voltage.

    d. Ohmmeter.

    e. 500V megger tester to measure insulation resistance.

    f. Engine tachometer.

4. Test specifications are given for an ambient temperature of 20° C (68° F). Resistance readings may vary with temperature and test equipment accuracy.

5. For access to the components listed in this **Troubleshooting** section, refer to Figs. MB102-MB106 and the later **Disassembly** section.

6. All operational testing must be performed with the engine running at its governed no-load speed of 3700-3900 RPM, unless specific instructions state otherwise.

7. Always load test the generator after repairing defects discovered during troubleshooting.

Megger Test the Complete Generator System Insulation

1. Make sure that the circuit breakers are *on* and functioning.
2. Connect one megger test probe to any hot terminal of a receptacle.
3. Connect the other megger test probe to any ground terminal.
4. The megger should indicate a resistance of 1 MΩ or more. If less than 1 MΩ, megger test the rotor, stator, and control panel resistance individually. Refer to the subsequent Trouble shooting sections.

Troubleshoot the MGA Series Brushless Generator

1. Circuit breakers must be *on*. To test either the AC or DC (if equipped) breakers, access the breaker wiring and continuity test the breaker terminals with the generator shut down. The breaker is faulty if it will not stay set, shows continuity in the *off* position or does not show continuity in the *on* position,
2. Receptacles must be functioning. To test the receptacle(s):

  a. Access the receptacle connections.
  b. Disconnect the *return* feed wire from the receptacle being tested.
  c. Install a jumper wire between the *hot* and *return* outlet terminals.

  d. Test for continuity across the feed terminals. No continuity indicates a faulty receptacle.
3. Start the generator, allowing it to reach operating temperature.
4. Test the AC voltage at one of the 120V receptacles. Output should be 120-134 volts. AC voltage at the 240V receptacles should test 240-269 volts. DC voltage, if equipped, should test 12-20 volts.
5. Stop the generator.
6. Access and disconnect the stator winding wires. With the meter set to the R × 1 scale, apply the meter leads to the color coded wire terminals noted and test the stator winding resistance using the following specifications:

Model	Winding	Ohms
MGA1300	Red to white AC	0.79-0.97
	Yellow to	
	yellow capacitor	3.91-4.77
MGA1800	Red to white AC	0.69-0.85
	Blue to black AC	0.69-0.85
	Yellow to	
	yellow capacitor	1.90-2.32
	Brown to brown DC	0.20-0.24
MGA2900	Red to white AC	0.50-0.62
	Blue to black AC	0.50-0.62
	Yellow to	
	yellow capacitor	1.50-1.84
	Brown to brown DC	0.20-0.24
ALL	Each wire to ground	Infinity

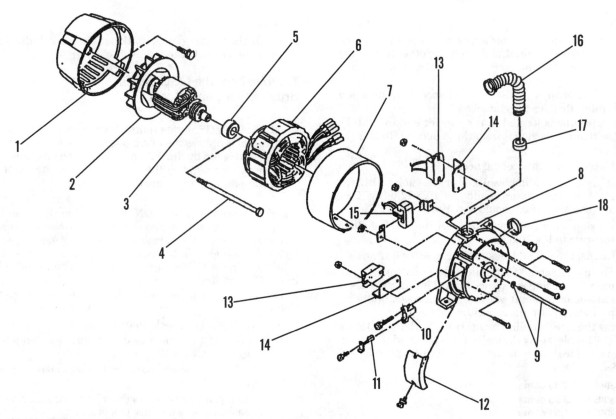

Fig. MB103 - Exploded view of MGW series generator head components.

1. Engine adapter/front housing
2. Rotor assembly with cooling fan
3. Sliprings
4. Rotor through-bolt
5. Rotor bearing
6. Stator assembly
7. Stator cover
8. Endbell/rear housing
9. Rear housing/stator through-bolt (4)
10. Brush holder
11. Brush (2)
12. Brush cover
13. Diode (2; one each polarity)
14. Diode heat sink
15. Current-transfer unit/transformer
16. Wire harness protector tube
17. Tube flange seal
18. Cap

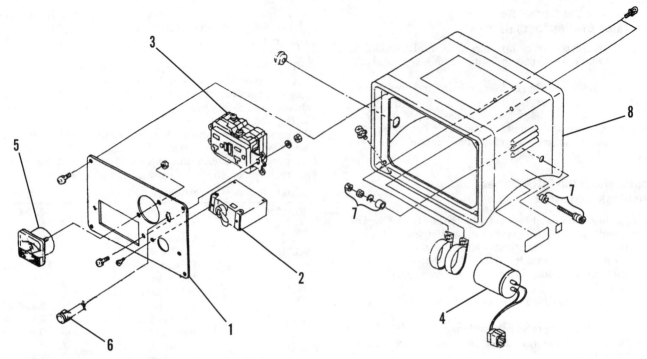

Fig. MB104 – Exploded view of MGA series control box components.

1. Control-panel faceplate
2. AC circuit breaker
3. 120V AC GFCI receptacle
4. Capacitor/condenser
5. Voltmeter
6. Pilot light
7. Ground terminal
8. Control box

If megger testing the stator, apply one megger lead to the red winding wire terminal, and apply the other lead to the lamination core. The megger should read a minimum of 1 MΩ.

The failure of any winding to test as specified indicates a faulty stator. Replace the stator.

7. The capacitor is located in the control box and can be tested without removing it from the generator. To test the capacitor:

    a. Unplug the capacitor harness.

    b. If using the Mitsubishi tester or equivalent, the MGA1300 generator should test 10.5-12.0 (C × ΩF), and the MGA1800 should test 19-22 (C × ΩF).

    c. If using other test equipment, set the VOM ohmmeter scale to R × 10,000.

    d. Connect the VOM leads to the capacitor terminals.

    e. The meter should first indicate low resistance, gradually increasing resistance toward infinity. No meter reading or constant continuity indicates a faulty capacitor. Replace the capacitor.

8. Access the rotor winding connections. With the VOM set to the R × 1 scale, apply the meter leads to the winding connections and test the winding resistance using the following specifications:

MGA1300: 2.31-2.83 ohms
MGA1800: 1.93-2.35 ohms
MGA2900: 1.55-1.91 ohms

If megger testing the rotor, apply one megger lead to one of the rotor's soldered winding terminals, and apply the other lead to any ground on the rotor shaft or laminations. The megger should read a minimum of 1 MΩ.

Failure of the rotor to test as specified indicates a faulty rotor. Replace the rotor.

9. For further testing, refer to the subsequent Individual Component Testing section in this chapter.

Troubleshoot the MGE Series Brush Style Generator

NOTE: On MGE series units, the positive (+) brush terminal is always the one closest to the rotor windings and is fed by the light green wire. The negative (-) brush terminal is the one closest to the rotor bearing and is fed by the brown wire.

1. Circuit breakers must be ON. To test either the AC or DC (if equipped) breakers, access the breaker wiring and continuity test the breaker terminals with the generator shut down. The breaker is faulty if it will not stay set, shows continuity in the OFF position or does not show continuity in the ON position.

2. Receptacles must be functioning. To test the receptacle(s):

    a. Access the receptacle connections.

    b. Disconnect the *return* feed wire from the receptacle being tested.

    c. Install a jumper wire between the *hot* and *return* outlet terminals.

    d. Test for continuity across the feed terminals. No continuity indicates a faulty receptacle.

3. Make sure that the automatic idle down switch is OFF.

4. Remove the brush cover. Start the generator, allowing it to reach operating temperature.

5. Test the generator output:

    a. AC voltage at the 120V receptacles should be 118-130 volts.

b. AC voltage at the 240V receptacles should test 236-260 volts.

c. DC voltage at the red (+) and black (-) control panel terminals should test 12-17 volts. and

d. DC voltage at the light green (+) and brown (-) brush wire terminals should be 12-18 volts.

6. Stop the generator.

7. Test the rotor.

The slip rings are an integral part of the rotor assembly and cannot be replaced.

Clean dirty slip rings with a suitable solvent, then lightly sand with ScotchBrite. Do not use steel wool, emery cloth or crocus cloth.

Cracked, chipped, broken, or grooved slip rings are not repairable. In these cases, the rotor will need replacement.

a. Disconnect the light green and the brown brush wire connectors.

b. With the ohmmeter set to the R × 1 scale, measure rotor winding resistance through the brushes. Resistance specifications are as follows:

MGE1800: 8.3-9.2 ohms
MGE2900: 6.8-7.5 ohms
MGE4000: 6.8-7.8 ohms
MGE4800: 5.7-6.5 ohms
MGE5800: 6.0-6.9 ohms
MGE6700: 6.3-7.3 ohms

c. Remove the brush assembly and measure resistance at the rotor slip rings using the specifications listed in Step B.

d. Measure resistance from each slip ring to a ground on the rotor shaft or laminations. Resistance should be infinity/no continuity. Any continuity reading from this test indicates a shorted rotor winding.

e. If using a megger, measure the insulation resistance between each slip ring and a ground on the rotor shaft or laminations. Resistance should be more than 3 MΩ.

f. If test Steps B, C, D, and E meet specification, the rotor and brushes are good.

g. If test Step B failed specification but Steps C, D, and E meet specification, the rotor is OK. Inspect the

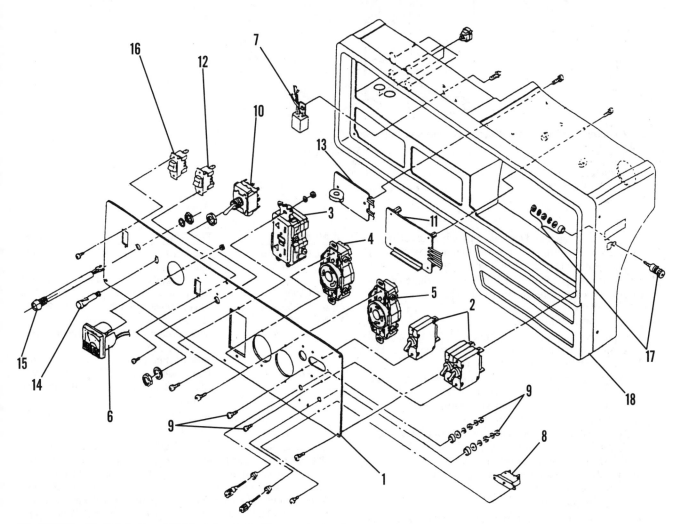

Fig. MB105 – Exploded view of MGE series control box components.

1. Control-panel faceplate
2. AC circuit breakers
3. 120V AC GFCI receptacle
4. 120V AC Twistlok receptacle
5. 240V AC Twistlok receptacle
6. AC voltmeter
7. Rectifier
8. DC circuit breaker
9. DC terminals (2)
10. Full-power switch
11. AVR unit
12. Idle-down switch
13. Idle-control unit
14. Pilot lamp
15. Low-oil lamp
16. Engine switch
17. Ground terminal
18. Control box

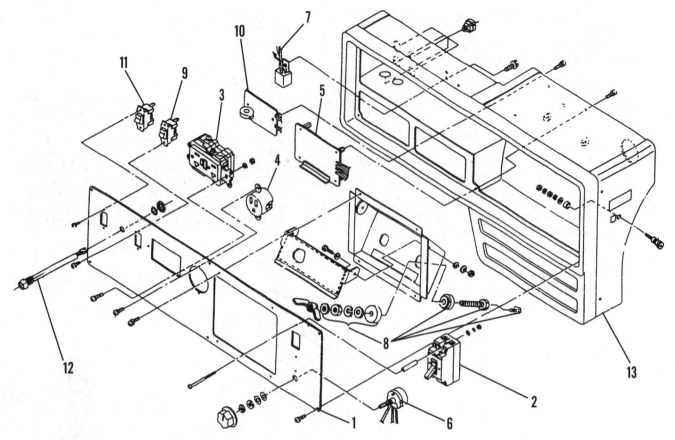

Fig. MB106 – Exploded view of MGW series control box components.

1. Control panel faceplate
2. AC circuit breaker
3. 120V AC GFCI receptacle
4. 120V AC individual receptacle
5. AVR unit

6. Voltage-adjust rheostat
7. Rectifier
8. DC welding terminal (2)
9. Idle-down switch

10. Idle control unit
11. Engine switch
12. Low-oil light
13. Control box

brushes for length and test each brush for continuity through its terminal and for freedom of movement. If faulty, replace the brush holder assembly.

  h. If test Steps B, C, D, and/or E failed specification, the rotor is faulty. Replace the rotor, then inspect and test the brushes.

8. Test the stator dynamically:

  a. Remove the control panel assembly from the control box. Position it so it will not cause any short circuits during the following operational tests.

  b. Remove the brush cover. Disconnect the light green and the brown brush wire connectors, then disconnect the AVR coupler from the stator wiring harness.

  c. Connect the positive test battery terminal to the positive brush terminal, the terminal closest to the rotor windings. Connect the negative battery terminal to the negative brush terminal, the terminal nearest the rotor bearing.

  d. Start the generator.

  e. Measure the output voltages at the color coded circuits noted in the following specification list. Note the readings.

WARNING: To prevent shock or electrocution while testing, only touch the insulated parts of the tester probes. Insulated gloves are recommended.

Model	Winding	Voltage
MGE1800 and	Red to white main	110-126 VAC
MGE2900	Light green to	
	white exciter	120-140 VAC
MGE4000,	Red to white main 1	110-126 VAC
MGE4800, and	Blue to black main 2	110-126 VAC
MGE5800	Light green to	
	white exciter	120-140 VAC
	Orange (+) to	
	Brown/white (-) DC*	14-20 VDC
MGE6700	Red to white main 1	102-118 VAC
	Blue to black main 2	102-118 VAC
	Light green to	
	white exciter	120-140 VAC
	Orange (+) to	
	brown/white (-) DC*	14-20 VDC

*DC output should be measured at the outside control panel posts first. If output is low there, then measure at the inside color coded connections to isolate the panel harness. If output is low inside, stop the generator, then test the DC winding diode. If faulty, replace. If the diode tests OK, the stator DC winding is faulty. Replace the stator.

  f. Stop the generator and disconnect the 12V test battery.

  g. If any test results were marginally questionable, proceed to Step 9 and test the stator winding resistances. If all test results were within specification, proceed to Step H.

h. Reconnect all previously disconnected wires and components, except leave the brush cover off and the control panel open.

i. Start the generator.

j. Observing proper brush polarity (light green is positive, brown is negative), test DC voltage at the brushes. Output should be 12-20 volts.

k. If the rotor is good and stator output meets Step E specification, but brush output does not meet Step J specification, the AVR is faulty. Stop the generator and replace the AVR.

l. Stop the generator.

9. To static test the stator windings:

a. Access and disconnect the stator winding wires.

b. With the VOM set to the R × 1 scale, apply the meter leads to the color coded winding terminals noted and test the stator winding resistance using the following specifications:

Model	Winding	Ohms
MGE1800	Red to white AC	0.3-0.5
	Light green to white exciter	2.4-3.0
MGE2900	Red to white AC	0.2-0.3
	Light green to white exciter	2.1-2.5
MGE4000	Red to white AC	0.3-0.5
	Blue to black AC	0.3-0.5
	Light green to white exciter	1.7-2.1
MGE4800	Red to white AC	0.2-0.3
	Blue to black AC	0.2-0.3
	Light green to white exciter	1.4-1.7
MGE5800	Red to white AC	0.2-0.3
	Blue to black AC	0.2-0.3
	Light green to white exciter	1.3-1.5
MGE6700	Red to white AC	0.1-0.2
	Blue to black AC	0.1-0.2
	Light green to white exciter	1.0-1.2
ALL	Each wire to ground	Infinity

c. If megger testing the stator, apply one megger lead to the red winding wire terminal, and apply the other lead to the lamination core. The megger should read a minimum of 3 MΩ. When testing, do not allow any other winding wire terminals to contact any part of the stator.

d. The failure of any winding to test as specified indicates a faulty stator. Replace the stator.

10. To test the automatic idle down system:

a. To test the switch.

• Disconnect the switch wires behind the control panel.

• Test the switch terminals for continuity – ON should show continuity. OFF should show no continuity. Any other results indicate a faulty switch.

b. To test the solenoid.

• Disconnect the two black idle down solenoid wire terminals.

• Test resistance across the black to black terminals. Resistance should be 15-19 ohms.

• A solenoid which does not meet the test specification is faulty.

c. To test the solenoid feed voltage.

• Access the idle controller inside the control box.

• Disconnect the two light blue controller wires. Use caution to prevent the loose wires from shorting out against any control box components.

• Start the engine, verifying proper 3700-3900 rpm no load speed.

• Test controller output DC voltage between each light blue wire and the green controller wire terminal. Voltage should be 30-36 volts on MGE4000 and 4800 units, or 21-25 volts on MGE5800 and 6700. If the light blue to green voltage is not in this range, proceed to Step D.

d. To test the regulator feed voltage.

• Make sure that the idle down switch is off.

• With the engine still at no load RPM, test DC voltage between the red rectifier output wire and the green ground wire terminal. If the rectifier resembles #1, Fig. MB107, voltage should be 14-16 VDC. If the rectifier resembles #2, Fig. MB107, voltage should be 12-15 VDC.

e. Stop the generator. If Step 2 test voltage does not meet specification, refer to the Diode Stack test and check the diodes. If the diode stack tests good, the charging coil behind the engine flywheel is faulty. Replace the charge coil.

f. If regulator feed voltage meets specification but preceding solenoid feed voltage does not meet specification, the idle controller is faulty. Replace the controller.

Troubleshoot the MGW Series Welder Generator

1. Make sure that the automatic idle down switch is off.

2. Circuit breakers must be on. To test either the AC or DC (if equipped) breakers, access the breaker wiring and continuity test the breaker terminals with the generator shut down. The breaker is faulty if it will not stay set, shows continuity in the *off* position or does not show continuity in the *on* position, the breaker is faulty.

3. Receptacles must be functioning. To test the receptacle(s) (except the GFCI; refer to Step 4):

a. Access the receptacle connections.

b. Disconnect the return feed wire from the receptacle being tested.

c. Install a jumper wire between the hot and return outlet terminals.

d. Test for continuity across the feed terminals. No continuity indicates a faulty receptacle.

4. Test the GFCI under power. If the generator has no output, disconnect the GFCI generator feed wires and carefully run a feed jumper from a wall outlet to test the GFCI:

a. Start the generator and allow a brief no load warm up period.

b. Depress the TEST button. the RESET button should extend. If it does, proceed to Step C. If the RESET button does not extend, the GFCI is faulty.

c. Depress the RESET button. The RESET button should then be flush with the TEST button. If the RESET button does not remain flush with the TEST button, the GFCI is faulty.

NOTE: If the GFCI seems to trip for no apparent reason, check to make sure the generator is not vibrating excessively due to faulty rubber mounts or debris between the generator and frame. Excessive vibration can trip the GFCI.

5. Test the DC voltage at the welding terminal. Output should be 50-57 volts.

6. Test the AC voltage at the receptacles. Output should be 118-124 volts at the 120V receptacles.

7. If Steps 5 and 6 do not test to specification, remove the brush cover and measure the DC voltage at the red (+) and brown (-) brush terminals. Output should be 16-23 volts.

8. Stop the generator.

9. Test the rotor.

The slip rings are an integral part of the rotor assembly. they are not replaceable.

Clean dirty or rough slip rings with a suitable solvent, then lightly sand with ScotchBrite. Do not use steel wool emery cloth or crocus cloth.

Cracked, chipped, broken, or grooved slip rings are not repairable. In these cases, the rotor will need replacement.

a. Disconnect the red and brown brush wire connectors.

b. If megger testing the rotor, measure the insulation resistance between each slip ring and a ground on the rotor shaft or laminations. Resistance should be more than 3 MΩ.

c. With the ohmmeter set to the R × 1 scale, measure rotor winding resistance through the brushes. Resistance should be 5.5-6.7 ohms on the MGW150 or 6.2-7.6 ohms on the MGW200.

d. Remove the brush assembly and measure resistance at the rotor slip rings using the specifications listed in Step C.

e. Measure resistance from each slip ring to a ground on the rotor shaft or laminations. Resistance should be infinity/no continuity. Any continuity reading from this test indicates a shorted rotor winding.

f. If test Steps B, C, D, and E met specification, the rotor and brushes are good.

g. If test Step C failed specification but Steps B, D, and E met specification, the rotor is good. Inspect the brushes for length and test each brush for continuity through its terminal and for freedom of movement. If faulty, replace the brush holder assembly.

h. If test Steps B, C, D, and/or E failed specification, the rotor is faulty. Replace the rotor, then inspect and test the brushes.

10. To test the stator dynamically:

a. With the brush wires disconnected, connect the positive test battery terminal to the positive brush terminal, the one closest to the rotor windings. Connect the negative battery terminal to the negative brush terminal, the one nearest the rotor bearing.

b. Start the generator.

c. Measure the DC output voltage at the welding terminals. Output should be 40-47 volts.

d. Measure the AC output voltage at the receptacles. Output should be 98-107 volts at the 120V receptacles.

e. Stop the generator.

f. Disconnect the test battery.

g. If test Steps C and D met specification, reconnect the brush wire terminals. The generator is good. If either test Step C or D failed specification, proceed to the next test.

11. To static test the stator windings:

a. Disconnect the stator winding wires. On the AC winding, disconnect both white wires on units with two white wires, even though it will only be necessary to test between one white wire and the red wire.

b. With the VOM set to the R × 1 scale, apply the meter leads to the color coded winding terminals noted and test the stator winding resistance using the following specifications. (Refer to Fig. MB108 for the DC winding arrangement)

Model	Winding	Ohms
MGW150	Red to white AC	0.31-0.39
	White 1 to white 2 DC,	
	White 2 to white 3 DC,	
	White 3 to white 1 DC,	
	White 4 to white 5 DC,	
	White 5 to white 6 DC, and	
	White 6 to white 4 DC	0.061-0.075
MGW200	Red to white AC	0.24-0.30
	White 1 to white 2 DC,	
	White 2 to white 3 DC,	
	White 3 to white 1 DC,	
	White 4 to white 5 DC,	
	White 5 to white 6 DC, and	
	White 6 to white 4 DC	0.041-0.051
ALL	Each wire to ground	Infinity

c. If megger testing the stator, apply one megger lead to the red winding wire terminal, and apply the other lead to the lamination core. The megger should read a minimum of 3 MΩ.

d. If any winding test in Steps B or C failed specification, the stator is faulty. If the windings met Steps B and C specification, the stator is OK, but the AVR unit is faulty. Replace the AVR unit.

12. Test the diode stack (Fig. MB109):

a. Access and unplug the diode stack harness connection, but leave the stack mounted in the rear housing.

NOTE: The following continuity test specifications are for the Mitsubishi tester or for most analog circuit testers. If the polarity of the meter being used is opposite that of the Mitsubishi tester, the test results will be reversed.

b. Set the ohmmeter to the R × 1 scale and test continuity between terminals shown in Fig. MB109, carefully noting the terminal identification.

c. Place the red (+) tester probe on the positive (+) diode terminal, then place the black (-) tester probe on each (~) diode terminal. Continuity should be noted. Place the red (+) tester probe on the negative (-) diode terminal, then place the black (-) tester probe on each (~) diode terminal. No continuity/infinity should be noted.

d. Place the black (-) tester probe on the negative (-) diode terminal, then place the red (+) tester probe on each (~) diode terminal. Continuity should be noted.

e. Place the black (-) tester probe on the positive (+) diode terminal, then place the red tester probe on each (~) diode terminal. No-continuity/infinity should be noted.

f. Any test results other than those specified indicate a faulty diode stack. Replace the diode stack.

13. To test the automatic idle down system on units so equipped:

a. To test the switch:

• Disconnect the switch wires behind the control panel.

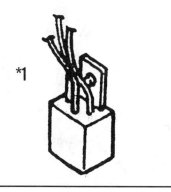

*1

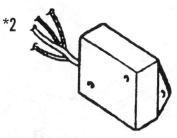

*2

Fig. MB107 – View showing the differences in appearance between the rectifiers used to supply feed voltage to the idle down regulator. Refer to the text for the correct test procedure.

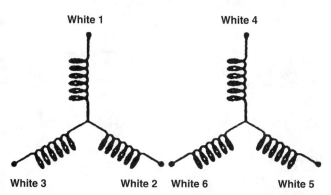

White 1   White 4

White 3   White 2   White 6   White 5

Fig. MB108 – View showing the test points for the star connected DC windings on the MGW series stator.

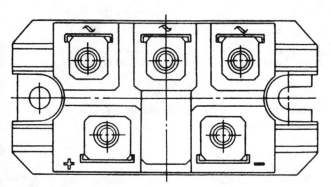

Fig. MB109 – View of the DC welding output diode stack test connections used on the MGW series. Refer to the text for the correct test procedure.

- Test the switch terminals for continuity–ON should show continuity. OFF should show no continuity. Any other results indicate a faulty switch.
  b. To test the solenoid:
- Access and disconnect the two black idle down solenoid wire terminals.
- Test resistance across the black to black terminals. Resistance should be 15-19 ohms.
- A solenoid that does not meet the specification is faulty.
  c. To test the solenoid feed voltage:
- Access the idle controller inside the control box.
- Disconnect the two light blue controller wires. Use caution to prevent the loose wires from shorting out against any control box components.
- Start the engine, verifying proper 3700-3900 rpm.
- Test controller output DC voltage between each light blue wire and the green controller wire terminal. Voltage should be 30-36 volts on MGW150 units, 21-25 volts on MGW200. If the light blue to green voltage is not in this range, proceed to Step D.
  d. To test the regulator feed voltage:
- Make sure that the idle down switch is off.
- With the engine still at no load RPM, test DC voltage between the red rectifier output wire and the green ground wire terminal. If the rectifier resembles #1, Fig. MB107, feed voltage should be 14-16 VDC. If the rectifier resembles #2, voltage should be 12-15 VDC.
  e. If Step D voltage does not meet specification, refer to the **Diode Stack** test and check the diodes. If the diode stack tests OK, the charging coil behind the engine flywheel is faulty. Replace the charge coil.
  f. If regulator feed voltage meets specification but preceding solenoid feed voltage does not meet specification, the idle controller is faulty. Replace the controller.
  g. Adjust the idle down rpm as follows:
- Turn the idle control switch off.
- Start the generator, verifying the correct no load speed of 3700-3900 rpm.
- Turn the idle control switch on. engine speed should drop to 1900-2100 rpm.
  h. If engine speed does not maintain a consistent idle:
- Manually move governor arm to the idle position (Fig. MB111).
- Measure the amount of plunger extended from the solenoid body. The dimension should be approximately 5 mm (0.20 in.).
- If dimension is not to specification, loosen the solenoid mount bolts and slide the solenoid as necessary. Retighten the bolts when the proper dimension is obtained.
- Release the governor arm. Recheck idle rpm. Readjust, if necessary.

INDIVIDUAL COMPONENT TESTING

Full Power Switch
(MGE4000-6700 and MGW Units)

Refer to the respective wiring diagram for the numbered switch points for this test.

To test the switch:

1. Unplug the switch connector behind the control panel, noting connector and wiring orientation.

2. In the 120V position, there should be continuity between the 2-3, 5-6, and 8-9 terminals.

3. In the 120/240V position, there should be continuity between the 2-1, 5-4, and 8-7 terminals.

4. Failure to test continuity between the terminals noted in Steps 2 and 3 denotes a faulty switch.

Diode Bridge For the 12VDC Output Circuit

NOTE: The following continuity test specifications are for the Mitsubishi tester or for most analog circuit testers. If the polarity of the meter being used is opposite that of the recommended tester, the test results will be reversed.

Refer to Fig. MB110 to view the diode bridge for this test.
1. Disconnect the diode bridge connector inside the generator control box.
2. Using the following chart and specifications, test for continuity between the color coded terminals noted in Fig. MB110.

Red (+) Meter Probe	Black (-) Meter Probe	Reading
Brown 1	Brown 2	No continuity
Brown 1	Orange	No continuity
Brown 1	Brown/White	Continuity
Brown 2	Brown 1	No continuity
Brown 2	Orange	No continuity
Brown 2	Brown/White	Continuity
Orange	Brown 1	Continuity
Orange	Brown 2	Continuity
Orange	Brown/White	Continuity
Brown/white	Brown 1	No continuity
Brown/white	Brown 2	No continuity
Brown/white	Orange	No continuity

2. Any test results other then those specified indicate a faulty diode bridge. Replace the bridge.

Pilot Lamp

If the lamp illuminates when rated voltage is applied to the receptacle with the circuit breaker *on*, the lamp is good. There are no other recommended tests for the pilot lamp.

Oil Warning Light

1. Disconnect the light wire terminals inside the control box.
2. Being careful to observe proper polarity, apply 3-5 VDC to the lamp terminals.
3. If the lamp fails to illuminate after Step 2, the lamp is faulty.

Voltmeter

If the voltmeter indicates rated AC voltage when it is applied to the meter, the voltmeter is OK. There are no other recommended tests for the voltmeter.

Control Panel

If the complete generator system megger test at the beginning of the **Troubleshooting** section failed to meet specification, and subsequent megger testing of the rotor and stator met specification, the control panel assembly should be megger tested. To test the panel assembly:
1. Access the interior of the panel.
2. Apply one megger lead to panel ground.
3. Apply the other lead to all current carrying sections of the panel components, one component at a time.

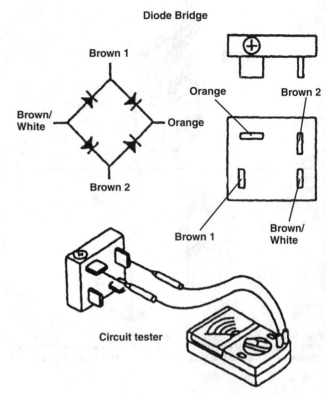

Fig. MB110 – *View of the diode bridge, including test points, for the 12 VDC output circuit on units so equipped.*

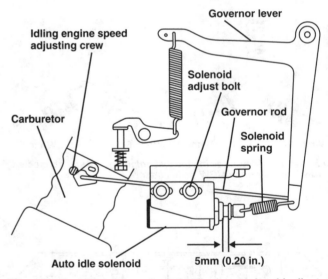

Fig. MB111 – *View of the automatic idle down solenoid adjustment. Refer to the text for the correct test procedure.*

4. Any component reading less than 1 MΩ is faulty. replace the component.

Auto Choke System

Some electric start models are equipped with an automatic choke system to aid cold starts (Fig. MB112). If the choke does not appear to be opening properly after the engine starts, inspect the system components.

1. Check the feed voltage to the bimetal heater. input should be 14 VDC.
2. Disconnect the bimetal heater wires and test the heater element resistance. Resistance should be 26 ohms.
3. Check the diaphragm vacuum tube for cracks and leaks. Manually apply vacuum to the diaphragm to verify linkage movement.
4. Inspect the carburetor vacuum tube fitting for obstructions.
5. Inspect the carburetor choke shaft and shutter for freedom of movement.

DISASSEMBLY

Refer to Figs. MB102-MB106 for component identification. Always allow the generator to cool down before disassembly. The rotor cannot be removed without first removing the stator and rotor so the windings and winding coatings are not damaged.
1. On MGE and MGW series, drain and remove the fuel tank.
2. Loosen the control panel fasteners. access and disconnect the control panel harness wires. Remove the control panel. On electric start models, disconnect and remove the battery.

3. Remove the flexible harness cover tube from the rear control box hole. Remove the control box.
4. Remove the muffler assembly.
5. On brush style units, remove the brush cover, then disconnect the brush wires and remove the brush holder assembly. Note the brush wire polarity.
6. Remove the nuts holding the rear housing feet to the rubber mounts. Note the location of the ground wire.
7. Carefully lift the rear housing slightly in order to place a temporary support, such as a wooden block, under the generator front housing so that the rear housing feet clear the rubber mount studs.
8. Straighten the tabs holding the stator cover to the stator. Remove the cover.
9. Remove the rear housing to front housing bolts and the ground wire. Using a light plastic or rubber mallet, evenly tap and remove the housing so as not to damage the stator harness.
10. Noting the stator orientation, remove the stator. Set the stator and rear housing aside.
11. Remove the rotor throughbolt, then remove the rotor.
   a. Place a soft support under the rotor to prevent damage if the rotor falls.
   b. Apply a light amount of penetrating oil through the rotor shaft onto the crankshaft to rotor mating area.
   c. Use a slide hammer with a threaded adapter to break the rotor free from the engine crankshaft.

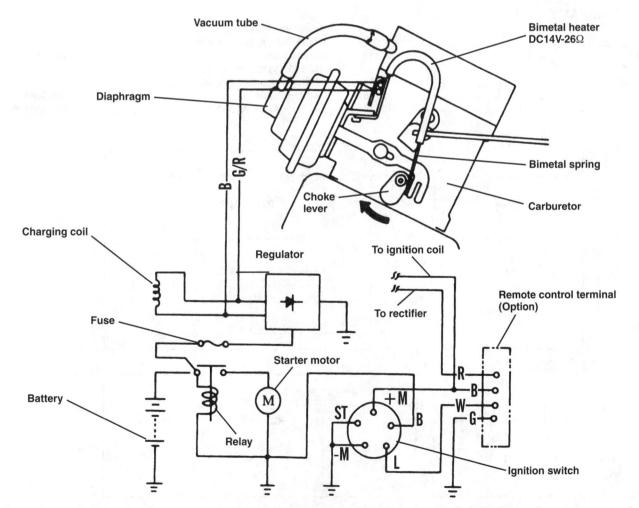

Fig. MB112 – Schematic view of the auto choke system used on some electric start units. Refer to the text for component testing.

REASSEMBLY

Refer to Figs. MB102-MB106 for component identification. The rotor must be installed before the stator. Exercise caution when handling the stator and rotor so the windings and winding coatings are not damaged.

1. If the generator front housing was removed for service, install it at this time. torque the bolts to 13 N•m (105-110 in.-lb.).

2. Service the rotor bearing, if needed.

3. Place a temporary support, such as a wooden block, under the generator front housing so that the rear housing feet will clear the rubber mount studs.

4. Make sure that the matching tapers on the engine crankshaft and inside the rotor shaft are clean and dry. Install the rotor onto the crankshaft. install the rotor throughbolt and washer, then torque the rotor bolt. Torque M8 bolts to 14 N•m (120 in.-lb.), and torque M10 bolts to 30 N•m (265 in.-lb., or 22 ft.-lb.)

5. Align and install the stator, followed by the rear housing. Tap around the rear housing perimeter to seat the rear housing against the stator and the stator against the front housing. Make sure that the rear housing ground wire is correctly reconnected.

6. Install the rear housing to front housing bolts. Torque M8 bolts to 14 N•m (120 in-lb), and torque M6 bolts to 5 N•m (40-45 in-lb).

7. Remove the engine spark plug and rotate the crankshaft to check for interference between the rotor and stator. There should be a consistent gap around the rotor laminations inside the stator. If any rubbing or scraping is evident, correct any misalignment prior to proceeding. Reinstall the spark plug.

8. Install the stator cover. Bend the cover tabs to snugly hold the cover against the stator. If the tabs are broken off, a pair of large plumber's hose clamps will suffice.

9. Lift the rear housing to remove the temporary support from under the front housing. Lower the rear housing feet onto the rubber mount studs. reinstall the ground wire, then tighten the stud nuts.

10. On brush style units, install the brush holder assembly into the rear housing. Observing proper polarity, reconnect the brush wires–light green to positive inside brush, and brown to negative outside brush.

11. Install the muffler assembly.

12. Insert the stator harness and the flexible harness cover tube into the rear control box hole. install the control box.

13. Reconnect the control panel harness wires. Install the panel.

14. Reinstall and reconnect the fuel tank.

15. On electric start models, reinstall and reconnect the battery.

WIRING DIAGRAMS

Model	Diagram
MGA1300	Fig. MB113
MGA1800, MGA2900	Fig. MB114
MGE1800	Fig. MB115
MGE2900	Fig. MB116
MGE4000, MGE4800, MGE5800, MGE6700, with recoil start	Fig. MB117
MGE4000, MGE4800, MGE5800, MGE6700, with electric start	Fig. MB118
MGW150A, MGW200A	Fig. MB119

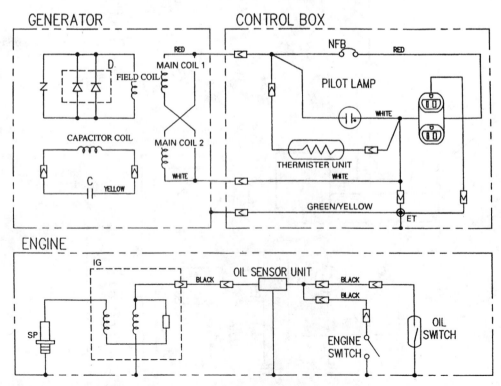

Fig. MB113 – Wiring diagram for the MGA1300 series.

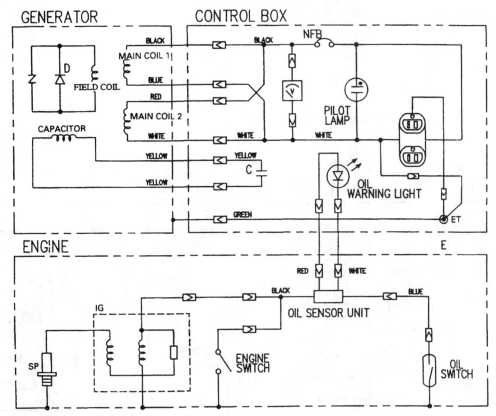

Fig. MB114 – Wiring diagram for the MGA1800 and MGA2900 series.

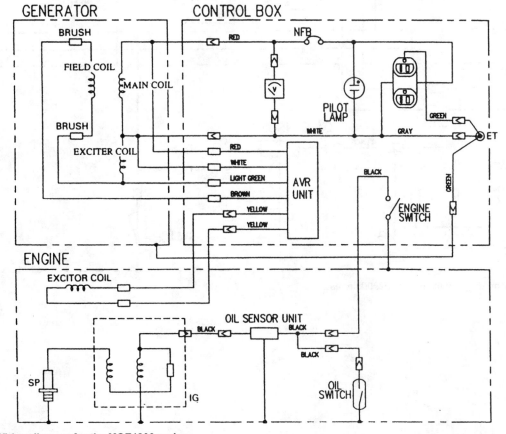

Fig. MB115 – Wiring diagram for the MGE1800 series.

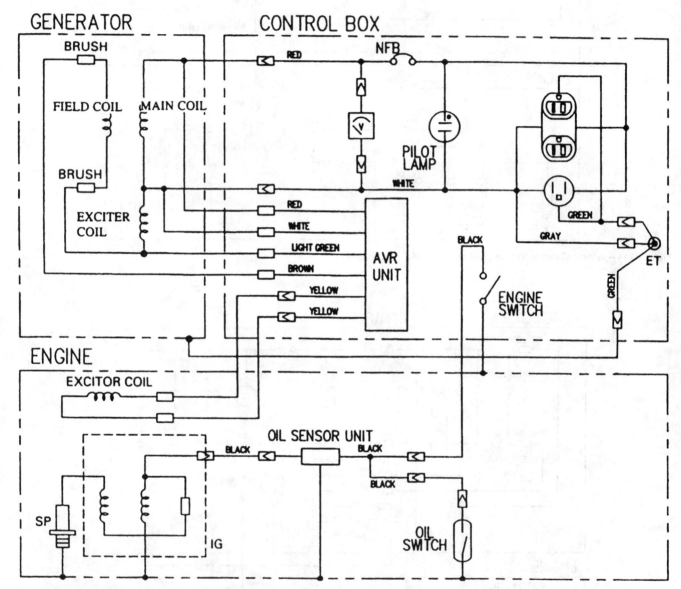

GENERATOR   CONTROL BOX

BRUSH

FIELD COIL   MAIN COIL

BRUSH

EXCITER
COIL

RED   NFB

PILOT
LAMP

WHITE

RED
WHITE
LIGHT GREEN
BROWN

AVR
UNIT

YELLOW
YELLOW

GREEN
GRAY   ET

BLACK

ENGINE
SWITCH

GREEN

ENGINE

EXCITOR COIL

OIL SENSOR UNIT

BLACK   BLACK

BLACK

SP   IG

OIL
SWITCH

Fig. MB116 – Wiring diagram for the MGE2900 series.

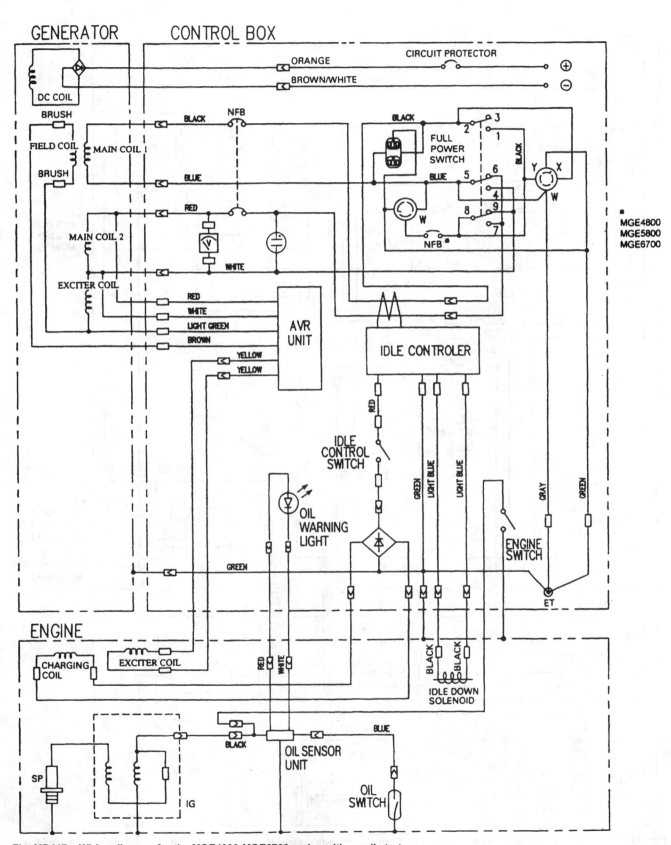

Fig. MB117 – Wiring diagram for the MGE4000-MGE6700 series with recoil start.

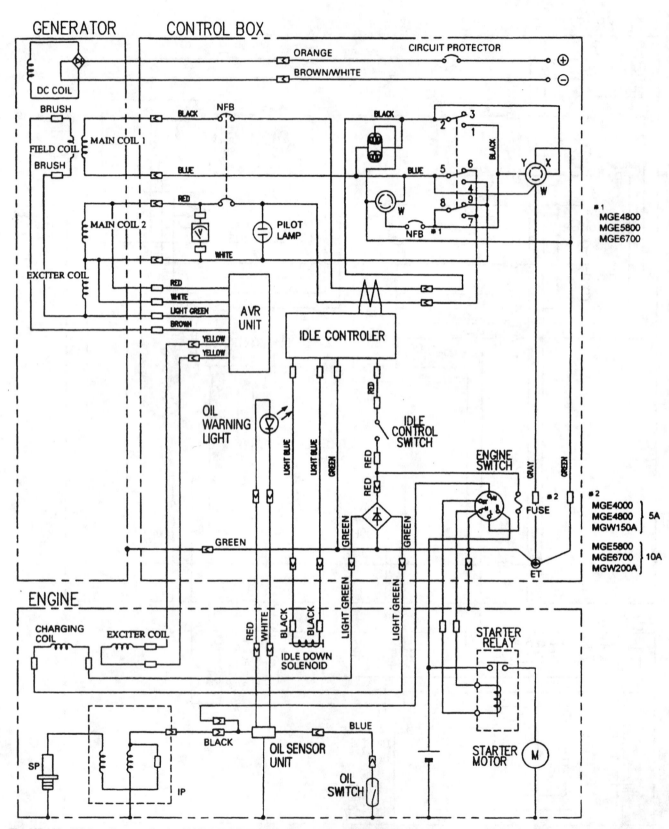

Fig. MB118 – Wiring diagram for the MGE4000-MGE6700 series with electric start. Refer to Fig. MB112 for the auto choke option wiring.

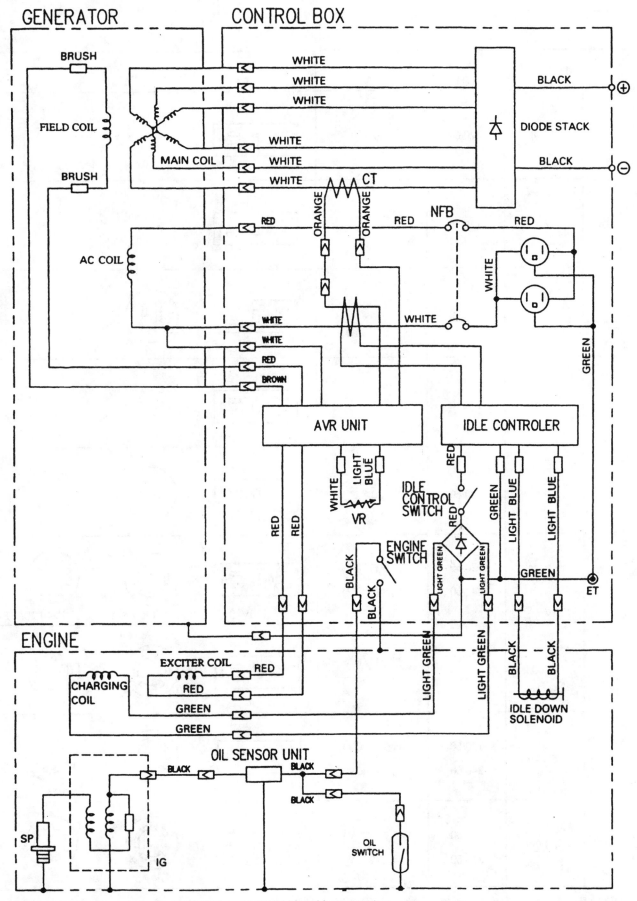

Fig. MB119 – Wiring diagram for the MGW150A and MGW200A welder generators.

Model		Rated V/Hz	Rated A	Receptacle
MGA	1300	120/60	9.2	
	1800	120/60	12.5	
	2900	120/60	20.8	
Also MGW	150A 200A			
MGE	1800	120/60	12.5	
	2900	120/60	20.8	
	4000	120/60 240/60	27.5 (120V) 13.8 (240V)	120V20A 120V30A 120/240 20A4P
	4800	120/60 240/60	34.2 (120V) 17.1 (240V)	
	5800	120/60 240/60	41.7 (120V) 20.8 (240V)	120V30A 120/240 30A4P
	6700	120/60 240/60	48.3 (120V) 24.2 (240V)	

Fig. MB120 – Rated specifications for North American generators.

MITSUBISHI
MGC Series

Generator Model	Rated (surge) Watts	Output Volts
MGC1101-F01	950 (1050) AC/100 DC*	120 AC/12 DC*

*DC output on Canadian units is 106 watts @ 12.78 volts

IDENTIFICATION

These generators are direct drive, 60 Hz, multi-pole, single-phase output, self exciting, permanent magnet, inverter control units.

The generator model and serial numbers are located on a plate at the lower edge of the generator side cover, on the side opposite the receptacle.

Refer to Fig. MB201 to view the generator components.

GENERATOR REPAIR

TROUBLESHOOTING

Before testing any electrical components, check the integrity of the terminals and connectors. While testing individual components, also check the wiring integrity. The use of the Mitsubishi tester #KJ10003AA is recommended for testing. The Mitsubishi tester has the following capabilities:

a. AC voltmeter – 0-500 volts.
b. Frequency – 25-70 Hz.
c. Condenser capacity – 10-100 μF.
d. Analog Ohmmeter – 0.1-1,999 ohms.
e. Insulation resistance – 3 MΩ

Test specifications are given for an ambient temperature of 20° C (68°F). Resistance readings may vary with temperature and test equipment accuracy. All operational testing must be performed with the engine running at its governed no load speed of 4000 RPM, unless specific instructions state otherwise. Always load test the generator after repairing defects discovered during troubleshooting. For access to the components listed in this **Troubleshooting** section, refer to Fig. MB201 and the later **Disassembly** section.

To troubleshoot the generator:

1. Circuit breakers must be *on*. To test either the AC or DC breakers, access the breaker wiring and continuity test the breaker terminals with the generator shut down. The breaker is faulty it will not stay set, shows continuity in the *off* position or does not show continuity in the *on* position.

2. Receptacles must be functioning. To test the receptacle(s):

   a. Access the receptacle connections.
   b. Disconnect the *return* feed wire from the receptacle being tested.
   c. Install a jumper wire between the *hot* and *return* outlet terminals.
   d. Test for continuity across the feed terminals. No continuity indicates a faulty receptacle.

3. Start the generator, allowing it to reach operating temperature.

4. Test the AC voltage at one of the 120V receptacles. Output should be 114-126 volts. DC voltage should test approximately 19 volts on U.S. units, 21 volts on CSA (Canadian) units.

5. Stop the generator.

6. If neither AC nor DC voltage met specification in Step 4, disconnect the 8-pin stator connector from the inverter. With the meter set to the R × 1 scale, apply the meter leads to the three pair of main winding yellow to yellow terminals (Fig. MB202: A-B, B-C, and C-A). Resistance should read 1.35-1.65 ohms.

7. If AC voltage did not meet specification:

   a. Reconnect the 8-pin stator to inverter connector.
   b. Unplug the 4-pin inverter output connector.
   c. Start the generator, insuring that all loads are disconnected.
   d. Measure voltage at the white to brown terminals. Output should be 114-126 volts.
   e. Unplug the 8-pin stator to inverter connector.
   f. Measure voltage at the same three pair of main winding yellow to yellow terminals (A-B, B-C, and C-A) used for the resistance test in preceding Step 6. output should be 185-195 volts.
   g. Measure voltage at the exciter winding red/white to red/white terminals. Output should be 23-27 volts.
   h. Stop the generator.
   i. If test Step 7-D did not meet specification but Steps 7-F and 7-G both met specification, the inverter module is faulty. Replace the module.

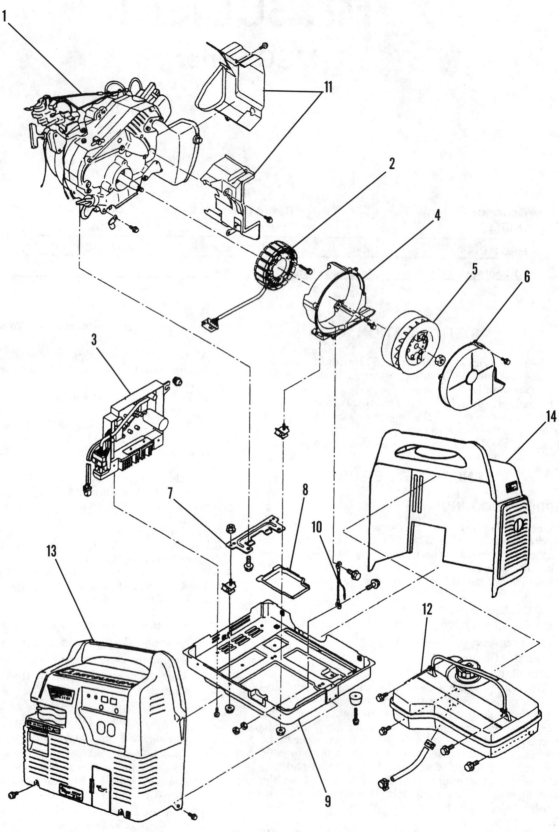

Fig. MB201 – Exploded view of the generator components.

1. Engine assembly
2. Stator assembly
3. Inverter unit
4. Rotor case
5. Rotor assembly

6. Rotor cover
7. Engine-mount bracket
8. Oil pan cover
9. Bottom cover
10. Ground wire assembly

11. Muffler cover
12. Fuel tank
13. Case, control-panel side
14. Case, rear side

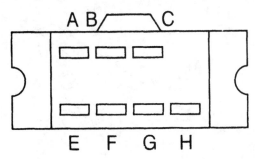

Fig. MB202 – View of the 8-pin stator to inverter connector used for stator testing. Refer to the text for the proper test procedures.

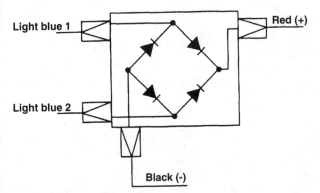

Fig. MB203 – View of the DC-circuit rectifier showing the terminal placement and a superimposed view of the internal circuitry. Refer to the text for the correct test procedure.

j. If test Steps 7-D and either/or both Steps 7-F and 7-G failed to meet specification, the stator is faulty. Replace the stator.

8. If DC voltage did not meet specification:

a. Unplug the 4-pin inverter output connector.

b. Start the generator, making sure that all loads are disconnected.

c. Measure DC voltage at the red to black terminals. Output should be 19 volts on U.S. units, 21 volts on CSA units.

d. Stop the generator, unplug the 8-pin stator to inverter connector, then restart the generator.

e. Measure AC voltage at the light blue to light blue terminals. Output should be 19 volts on U.S. units, 21 volts on CSA units.

f. Stop the generator.

g. Measure resistance between the two light blue wire terminals coming from the stator. Resistance should be 0.12-0.14 ohms

h. If test Step C failed, but Steps E and G met specification, the rectifier is probably faulty. Proceed to Step J.

i. If test Steps C, E, and G failed, the stator DC winding is faulty. Replace the stator.

j. To test the rectifier, remove all four rectifier connector terminals, then continuity test the color coded terminal positions noted in Fig. MB203. With the VOM set to 'Ohms', continuity should test as follows:*

Red (+) Probe	Black (-) Probe	Result
Light blue 1	Light blue 2	No continuity
Light blue 1	Red	No continuity
Light blue 1	Black	Continuity
Light blue 2	Light blue 1	No continuity
Light blue 2	Red	No continuity
Light blue 2	Black	Continuity
Red	Light blue 1	Continuity
Red	Light blue 2	Continuity
Red	Black	Continuity
Black	Light blue 1	No continuity
Black	Light blue 2	No continuity
Black	Red	No continuity

*The above listed specifications are for the Mitsubishi tester or for most analog circuit testers. If the polarity of the meter being used is opposite that of the Mitsubishi tester, the test results will be reversed.

A rectifier not meeting the above specifications is faulty.

DISASSEMBLY

Refer to Fig. MB201 for component identification. Allow the generator to cool down prior to disassembly, but drain the engine oil while the engine is hot. Exercise caution when handling the stator so the windings and winding coatings are not damaged. Exercise caution when handling the rotor–do not drop, heat, or allow debris to accumulate on the magnets.

1. Remove the oil fill cap, tilt the generator and drain the engine oil.

2. Place a drain pan under the hose below the air filter. Loosen the carburetor bowl drain screw and drain the fuel tank.

3. Unscrew and remove the engine switch cap on the control panel.

4. Remove the eight screws securing the generator case halves. Lay the control panel side case down. Tie a slip knot in the starter rope at the starter housing, then remove the starter rope grip.

5. Unplug the control panel to inverter connector. Remove the control panel side case.

6. Disconnect the fuel hose at the carburetor.

7. Remove the four fuel tank bolts. Simultaneously remove the fuel tank and air filter side case half.

8. Disconnect the bottom cover ground wire, and unplug the stator to inverter harness connector.

9. Using a protective pad, place the generator on its side, remove the bottom cover, then place the generator upright again.

10. Remove the rotor cover.

11. Remove the two muffler cover shields.

12. Place alignment marks on the rotor nut and engine crankshaft for later reassembly, then remove the rotor nut.

13. Use a two bolt puller to remove the rotor (Fig. MB204).

14. Again placing the generator on its side, remove the rotor case.

15. Remove the stator, noting the correct wiring orientation.

REASSEMBLY

Refer to Fig. MB201 for component identification.

1. Noting the correct wiring orientation, install the stator to the engine.

2. Using a protective pad, place the generator on its side and install the rotor case.

3. Be sure the rotor magnets are clean. Make sure that the matching tapers on the engine crankshaft and inside the rotor are clean and dry. Install the rotor onto the crankshaft. Install the rotor nut and torque the nut to 45 N•m (400 in.-lb., or 33 ft.-lb.). With the rotor nut properly torqued, the disassembly alignment marks should closely match.

4. Install the two muffler cover shields.

5. Install the rotor cover.

6. Install the bottom cover, then set the generator upright.

7. Reconnect the stator to inverter harness connector and reconnect the bottom cover ground wire, taking care not to pinch or kink the wires.

8. Simultaneously install the air filter side case half and the fuel tank. Install and tighten the four fuel tank bolts.

9. Reconnect the fuel hose to the carburetor.

10. Placing the control panel side case half next to the generator, reconnect the control panel to inverter connector.

11. Feed the starter rope through the case grommet. Install the rope grip onto the rope, then undo the inner rope slip knot.

12. Position the engine switch into the case. screw the switch cap onto the switch.

13. Using the eight case screws, secure the case halves.

WIRING DIAGRAM

Refer to Fig. MB205 to view the generator wiring diagram.

ENGINE

Model	GM82P
No. Cyls	1
Bore	52.0 mm (2.05 in.)
Stroke	38.0 mm (1.49 in.)
Displacement	80 cc (4.88 ci)
Power Rating	1.8 kW (2.44 hp)
Governed RPM	4000

The engine is a 4-stroke design with pushrod operated overhead valves.

Refer to Fig. MB206 to view the engine components.

To access the engine, refer to the **GENERATOR REPAIR, Disassembly** section of this chapter.

MAINTENANCE

Spark Plug

Recommended spark plug is NGK BPR6HS or equivalent. Recommended gap is 0.7-0.8 mm (.028-.032 in.).

CAUTION: Do not use abrasive blasting to clean spark plugs, as this may introduce abrasive material into the engine which could cause extensive damage.

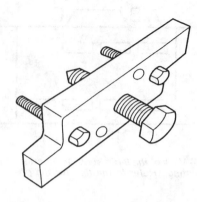

Fig. MB204 – The recommended rotor and flywheel puller for the MGC series generator.

Lubrication

The engine is splash lubricated by a dipper on the connecting rod cap.

Oil capacity is 0.40 litre (13 fl. oz.). The crankcase is full when no more oil can be added to the oil fill hole with the generator sitting level.

Use oil with an API service classification of SH, SJ, or better. Use SAE 30 oil if temperatures are consistently above 10E C (50E F). SAE 40 if temperatures are consistently above 25E C (77EF). SAE 10W30 in any temperature. or SAE 5W30 if temperatures are consistently below freezing. DO NOT USE 10W40 oil.

Drain the old oil when the oil is hot to insure the removal of all impurities. To drain the oil, unscrew the oil fill cap, then tilt the generator.

Whenever the crankcase oil has been drained, dry test the operation of the low oil shutdown switch:
1. Make sure that the generator is sitting level.
2. Remove the spark plug.
3. Connect a spark tester to the plug lead and cylinder block.
4. Pull the starter rope and check for spark. There should be none.
5. If the tester shows spark with the crankcase empty of oil, the shutdown switch is faulty. Refer to the subsequent Ignition System section for testing and replacement.

Air Filter

1. To access the filter, unscrew the filter cover.
2. Gently clean the sponge filter element in a warm water and detergent/water solution. Rinse thoroughly, then allow to dry completely.
3. Pour clean engine oil onto the element, squeezing out any excess oil.
4. Reinstall the element.
5. Reinstall the filter cover, insuring a proper, tight fit.

Valve Clearance

Check and adjust valve clearance with the engine cold.
1. Remove the valve cover.
2. Place the piston at TDC compression stroke by removing the spark plug, rotating the flywheel, and observing the piston head through the spark plug hole.
3. Using a feeler gauge, check valve clearance. Intake and exhaust clearance should each be 0.10-0.12 mm (.004-.005 in.).

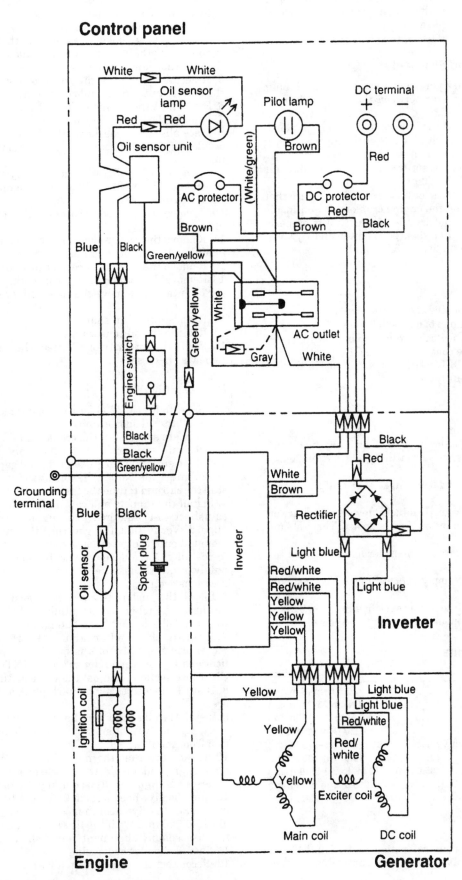

Fig. MB205 – Wiring diagram for the MGC series generator.

4. To adjust valve clearance, loosen the rocker arm adjusting screw locknut, then turn the adjusting screw for more or less clearance. Tighten the locknut, then recheck clearance and readjust, if necessary.

5. Using a new gasket, reinstall the valve cover.

Ignition System and Flywheel

1. Disconnect the spark plug lead, remove the spark plug, and connect a spark tester between the plug lead and a good engine ground.

2. Access and disconnect the ignition module ground lead.

3. Rapidly spin the flywheel while watching the spark tester. No spark indicates a faulty ignition module.

4. If replacing the module, set the flywheel-to-module air gap at 0.3-0.5 mm (.012-0.20 in.).

5. If removing the flywheel for any reason, make sure that the matching tapers on the engine crankshaft and inside the flywheel are clean and dry, the flywheel key is properly aligned and not scored or sheared. Tighten the flywheel nut to 35.0 N•m (310 in.-lb., or 25.9 ft.-lb.).

Engine Switch

To test the switch:

1. Access the switch behind the control panel.

2. Disconnect at least one black switch wire.

3. Continuity test the switch terminals in both switch positions. There should be continuity in the OFF position only. No continuity in the OFF position or continuity in the ON position indicates a faulty switch.

Oil Sensor

To test the oil sensor:

1. With the engine shut down, prior to draining the oil, unplug the blue harness wire from the crankcase oil sensor unit.

2. With the engine sitting level, make sure that the crankcase oil level is *full*.

3. Test continuity between the blue oil level switch wire on the engine and any ground on the crankcase. There should be no continuity.

4. Remove the oil fill cap, tilt the generator, and drain the oil completely.

5. With the crankcase empty, repeat Step 3. There should now be continuity.

6. Failure of the switch to test as specified in Steps 3 and 5 indicate a faulty oil sensor switch. Replace the switch.

Torque Specifications

Fastener	Torque
Standard M3 × 0.5 bolt	0.5 N•m (5 in.-lb.)
Standard M4 × 0.7 bolt	3.0 N•m (25-27 in.-lb.)
Standard M5 × 0.8 bolt	4.0 N•m (35-37 in.-lb.)
Standard M6 × 1.0 bolt	5.0 N•m (45 in.-lb.)
Self-tapping M6 × 1.0 bolt	12.0 N•m (105-110 in.-lb.)
Standard M8 × 1.25 bolt	10.0 N•m (90 in.-lb.)
Flywheel nut M12 × 1.75	35.0 N•m (310 in.-lb., or 25.9 ft.-lb.)
Connecting rod	10.0 N•m (90 in.-lb.)
Cylinder head	25.0 N•m (220 in.-lb.)

Disassembly

1. To access the engine, refer to the **GENERATOR REPAIR, Disassembly** section and disassemble the generator components. Refer to Fig. MB206 to view the engine components.

2. Remove the engine mount bracket.

3. Remove the muffler assembly.

4. Disconnect the choke wires. Remove the air cleaner nuts. While sliding the carburetor off the studs, carefully remove the governor link and spring.

5. If necessary, remove the governor lever. While holding the bottom nut, loosen and remove the top nut.

6. Disconnect the engine switch wires.

7. Remove the blower housing.

8. Place alignment marks on the flywheel nut and engine crankshaft for later reassembly, then remove the flywheel nut.

9. Use a two bolt puller to remove the flywheel and the flywheel key (Fig. MB204). Remove the ignition module and the oil sensor wires.

10. Remove the cylinder covers, valve cover, and spark plug.

11. Remove the rocker arm assembly, then remove the cylinder head assembly and push rods.

12. Remove the crankcase cover, camshaft, and tappets.

13. Remove the connecting rod cap, noting the cap and dipper orientation. Remove the connecting rod/piston assembly.

14. Remove the crankshaft.

15. Remove the oil sensor unit.

16. Verify the integrity of all internal engine components. Renew as necessary.

Reassembly

Refer to Fig. MB206 for component identification.

1. Install the oil sensor unit.

2. Lubricate and install the crankshaft.

3. Lubricate and install the piston/connecting rod assembly. Make sure that the piston ring end gaps are staggered at 120° intervals. Noting the correct rod cap orientation, install the cap and torque the fasteners 10.0 N•m (90 in.-lb.).

4. Install the tappets and camshaft, making sure that the camshaft-to-crank gear timing marks are properly aligned. Verify that the governor thrust cap is properly mounted onto the governor assembly.

5. Install the crankcase cover. Incrementally and alternately torque the fasteners as specified in the Torque Specifications section.

6. Install the push rods, a new head gasket, and the cylinder head assembly. Incrementally and alternately torque the head bolts to 25.0 N•m (220 in.-lb.).

7. Install the rocker arm assembly. Torque the rocker arm assembly bolts as specified in the Torque Specifications section, then refer to the MAINTENANCE, Valve Clearance section to adjust the valve settings.

8. Install the valve cover, spark plug, and cylinder covers.

9. Install the oil sensor wire harness and the ignition module, temporarily tightening the module in the farthest position from the flywheel.

10. Make sure that the matching tapers on the engine crankshaft and inside the flywheel are clean and dry. Insert and align the flywheel key. Install the flywheel. Torque the flywheel nut 35.0 N•m (310 in.-lb., or 25.9 ft.-lb.). Set the flywheel to module air gap at 0.3-0.5 mm (0.012-0.020 in.). With the flywheel nut properly torqued, the disassembly alignment marks should closely match.

11. Install the blower housing.

12. Reconnect the engine switch wires.

13. Install the carburetor, governor linkage and air filter assembly. Reconnect the choke wires.

14. If the governor lever was removed, reinstall and adjust as follows:

    a. Set the governor lever onto the governor shaft post.

    b. Holding the carburetor throttle in the wide open position, rotate the governor shaft to its maximum clockwise position. Do not force.

    c. While holding the throttle and shaft, tighten the top governor shaft nut.

    d. Release the linkage.

15. Install the muffler assembly.

16. Install the engine mount bracket.

17. Refer to the **GENERATOR REPAIR, Reassembly** section to finish the generator reassembly.

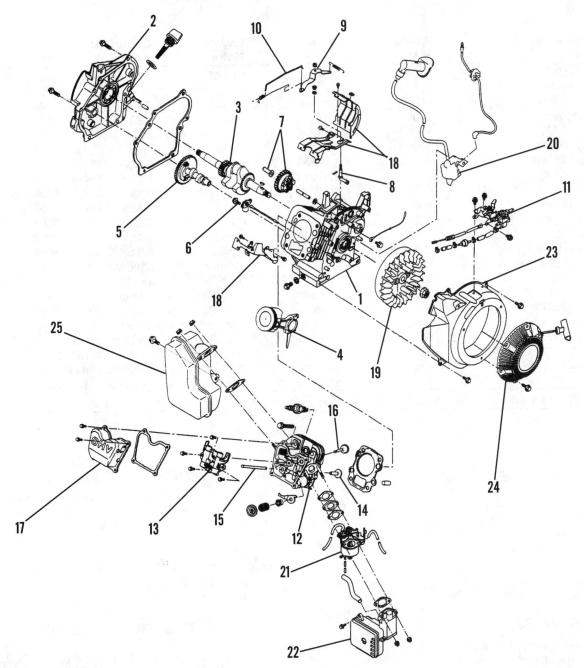

Fig. MB206 – Exploded view of the engine components.

1. Cylinder block
2. Crankcase cover
3. Crankshaft
4. Connecting rod/piston assembly
5. Camshaft
6. Oil-sensor unit
7. Governor assembly
8. Governor shaft
9. Governor arm
10. Governor link
11. Governor control assembly
12. Cylinder head assembly
13. Rocker-arm assembly
14. Valve lifter (2)
15. Pushrod (2)
16. Valve (1 intake, 1 exhaust)
17. Valve cover
18. Cylinder covers (3)
19. Flywheel
20. Ignition unit
21. Carburetor
22. Air filter assembly
23. Fan cover/blower housing
24. Recoil starter assembly
25. Muffler assembly

NorthStar

K-BAR Industries, Inc.
2050 Airtech Rd.
Faribault, MN 55021

Generator Model	Rated Watts	Output Volts	Engine Make	Governed Model	RPM
PG 165912	2300	120	Honda	GC160VXA	3600
PG 165960	4000	120/240	Tecumseh	HM80	3600
IPG 165926	4500	120/240	Honda	GX270VDE2	3600
IPG 165920	6600	120/240	Honda	GX390VXE2	3600
IPG 165922	8500	120/240	B&S Vanguard	303447-1042	3600
PPG 165911	4500	120/240	Honda	GX270 VTN2	3600
PPG 165947	4500	120/240	B&S Vanguard	185432-0606	3600
PPG 165945	4800	120/240	Kohler	CS8.5-9215410	3600
PPG 165914	6600	120/240	Honda	GX390 VAG2	3600
PPG 165957	6600	120/240	Honda	GX390 VXE2	3600
PPG 165917	8500	120/240	B&S Vanguard	303447-1042	3600
PPG 165923	0,500	120/240	Honda	VXA1	3600
PPG 165925	3,500	120/240	Kohler	CH25GS-68617	3600
DPG 165930	6120	120/240	Yanmar	L100ACE-DEG	3600
DPG 165931	9600	120/240	Hatz	1D81Z	3600
BDG 165915	2600	120/240	(N/A: Requires minimum 5HP)		3600
BDG 165913	5000	120/240	(N/A: Requires minimum 10HP)		3600
BDG 165918	7200	120/240	(N/A: Requires minimum 14HP)		3600
BDG 165928	9600	120/240	(N/A: Requires minimum 18HP)		3600
PTOG 165929	2,000	120/240	(PTO drive: Requires min. 24HP)		540
PTOG 165937	4,000	120/240	(PTO drive: Requires min. 48HP)		540
PTOG 165949	4,000	120/240	(PTO drive: Requires min. 108HP)		540

NOTE: Rated output wattage is continuous run. Surge wattage is slightly higher.

IDENTIFICATION

NorthStar generators are single phase, 60Hz, brushless design generators with a 1.0 power factor. No load frequency at 3770-3800 RPM should be 62.8-63.3 Hz. Minimum frequency at 100% load is 56.5 Hz at 3390 RPM.

MAINTENANCE

ON DPG, IPG, PG, & PPG MODELS

On models with a serviceable bearing, the endbell bearing and race should be inspected, cleaned, and relubricated any time the endbell is removed. Some models have a pre-lubricated sealed bearing. If the bearing runs hot to the touch, replacement may be necessary.

ON BDG MODELS

The rotor shaft has sealed roller bearings at both ends. Inspect the bearings whenever servicing the generator. Replace as necessary.

ON PTOG MODELS

The rotor shaft has pre-lubricated sealed ball bearings at both ends. Inspect the bearings whenever servicing the generator. Replace as necessary.

The gearbox requires SAE 90 synthetic gear oil in the following capacities.

165929– 0.86 qt. (0.82 litre)
165937– 0.53 qt. (0.5 litre)
165949 – 2.0 qt. (1.9 litre)

TROUBLESHOOTING

No load voltages on 60Hz units should be 125-127 volts from the 120V receptacles and 250-254 volts from the 240V receptacles at 3770-3800 rpm. If little or no output is generated, the following items could be potential causes.

1. Engine rpm–Engine must maintain 3600 rpm under load, or 3770-3800 rpm no load. Engine rpm could also be set too high, causing overvoltage.
2. Loose component mounting fasteners (low or no voltage).
3. Circuit breakers (low or no voltage).
4. Insecure wiring connections (low or no voltage).

5. Faulty receptacle (low or no voltage).
6. Capacitor (low or no voltage).
7. Diodes (low or no voltage).
8. Stator windings (low voltage – shorted; no voltage – open).
9. Rotor
   a. Open or shorted windings (no or low voltage, respectively).
   b. Loss of residual magnetism (no voltage).

NOTE: Residual magnetism can be restored without generator disassembly, and is the recommended first step whenever generator has no output: Refer to RESIDUAL MAGNETISM Section.

10. Voltage regulator on PPG 165925 out of adjustment (low or high voltage).
11. Idle Control not working (erratic or no idle).
12. Belt or coupler slippage on BDG models (low voltage)
13. Insufficient cooling system ventilation (overheating)

GENERATOR REPAIR

RESIDUAL MAGNETISM

If low residual magnetism is suspected as the cause of no output, the recommended method of remagnetizing the rotor (commonly referred to as "flashing the fields") is by using a commercially available field flasher, such as a

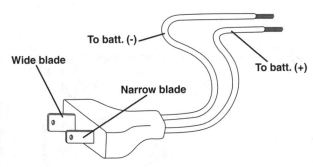

Fig. NS01 – Homemade jumper tool for remagnetizing rotors.

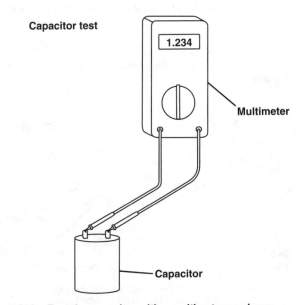

Fig. NS02 – Test the capacitor with a multimeter as shown.

Homelite Tool #UP00457. If such a tool is not available, *carefully* proceed as follows:

1. Construct a jumper tool (Fig. NS01) using a male two-prong 110-volt household plug and two 16-gauge minimum wire leads of sufficient length to reach from the generator's 110V plug in to a nearby fully charged DC power source such as a 12V automotive battery. Alligator clips on the jumper wire ends would be helpful.

NOTE: Mark the jumper wires so that the wire coming from the narrow spade on the 110V plug will connect to the positive (+) battery terminal. The wire from the wide spade must connect to the negative (-) battery terminal.

2. Put on safety glasses.
3. Start the generator.

WARNING: Double check to make sure there is no output at the 240V receptacles to prevent flashback or explosion.

4. With the jumper tool connected to the battery, plug the jumper tool into one 120V receptacle on the generator for 1-2 seconds, then immediately withdraw it.

WARNING: Leaving the jumper tool plugged into the generator too long could cause the battery/power supply to explode.

CAUTION: Do not allow the jumper tool spade prongs to contact any conductive material while the tool is connected to the power supply.

5. Check for voltage at the 240V receptacle (use the 120V receptacle on the PG 165912).
6. If no output voltage is present, recheck the power supply for full voltage, then repeat the field flash for twice as long (2-4 seconds), promptly pulling the jumper plug back out.
7. Recheck for output voltage. If none is present, the rotor is faulty. Replace the rotor.

CAPACITOR TESTING

CAUTION: Even with the generator stopped, the capacitor should have a retained charge if it is working properly. Do not touch the capacitor terminals as electric shock may result. Always discharge the capacitor by shorting across the terminals with a screwdriver or similar tool with an insulated handle.

To test the capacitor:
1. Check the outside of the capacitor for the rating identification.
2. Using a multimeter as shown in Fig. NS02, check the capacitor against its rating.

CIRCUIT BREAKER

To test the circuit breaker on units so equipped:
1. Make sure the breaker is set. If tripped, reset.
2. Remove the wires from the rear of the breaker.
3. Using the R × 1 scale on the VOM, test for continuity across the breaker terminals.

4. If no continuity reading is noted, or if the breaker fails to reset, the breaker is faulty. Replace the breaker.

ROTOR

Diode Test

1. Using caution not to break the diode leads, disconnect at least one end of the diode from the rotor. Fig. NS03 shows both diode leads disconnected.
2. Using the R × 10K scale on the VOM, test the diode in one direction, then reverse the leads and test the diode in the reverse direction. A good diode should show a very high resistance with the leads one direction, and a very low resistance when the leads are reversed.
3. If the diode is faulty, replace the diode, always observing proper polarity.

An alternate method can be used to test the diodes which does not require the removal of the diode from the rotor (Fig. NS04). To use this method:
1. Connect a 45-watt automotive headlight type bulb in series with a 12V battery.
2. Connect the battery and bulb to the diode leads, first one way, then with the leads reversed. The bulb should light in only one direction.

Rotor Winding Resistance

The rotor can be tested without removing it from the generator.

> **NOTE: On diode equipped units, all resistance tests must be performed with the diodes disconnected from the rotor windings.**

To test resistance, use a VOM set on the R × 1 scale, and test the windings according to the specifications in the generator's owner's manual, using Fig. NS03 for reference. Testing with the diodes in place as shown is acceptable.

> **NOTE: All resistance readings are approximate and may vary depending on the winding temperature and the test equipment accuracy.**

STATOR WINDING RESISTANCE

To test the stator windings, set the VOM to the R × 1 scale, then test each winding according to the specifications in the generator's owner's manual, using Fig. NS05 for reference.

> **NOTE: All resistance readings are approximate and may vary depending on the winding temperature and the test equipment accuracy.**

VOLTAGE REGULATOR (PPG 165925)

Only two of the four adjusting screws on the voltage regulator (Fig. NS06) are field adjustable without special test equipment.
1. VOLT–Adjusts voltage. Rotating the *volt* trim screw clockwise increases output voltage, counterclockwise decreases voltage.
2. STAB–Adjusts output voltage stability. Increased voltage stability, or consistency, results in slightly longer voltage response times. Quicker voltage response time comes with larger voltage fluctuations (decreased stability). Rotating the STAB trim screw clockwise decreases voltage

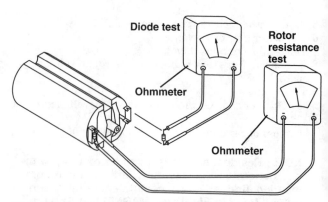

Fig. NS03 – Test the diodes and rotor winding resistances with a VOM as shown.

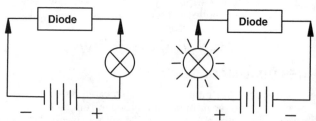

Fig. NS04 – Bulb testing the rotor diodes without removing them from the rotor. Refer to text.

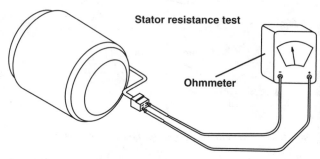

Fig. NS05 – Test the stator windings with a VOM as shown.

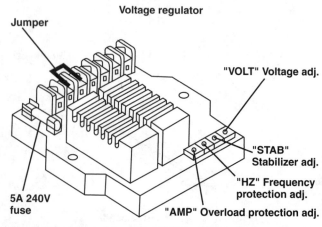

Fig. NS06 – View of the PPG 165925 voltage regulator. HZ and AMP screws are not field adjustable: Refer to text for VOLT and STAB adjustment procedures.

stability. Counterclockwise rotation increases voltage stability. To correctly adjust this setting:

a. Start with the generator *off*.

b. Turn the adjustment screw to the full counterclockwise position.

c. Start the generator and operate it in the normal no load position, except for one 1-filament light bulb.

d. Slowly turn the adjustment screw clockwise until the light starts to oscillate.

e. Slowly turn the adjustment screw counterclockwise until the light becomes stable.

3. HZ – Frequency protection. This trim screw is factory sealed.

4. AMP – Overload protection. This trim screw is factory sealed.

VOLTAGE ADJUST (PTOG 165937)

No Load Voltage Adjust

NOTE: Adjustments must be made with the generator off. Increasing the air gap raises the no load voltage. decreasing the air gap lowers the no load voltage.

To raise the no load voltage:

1. Loosen the three fasteners holding the "I" portion of the transformer in place.

2. Insert a 0.004 in. feeler gauge into the air gap.

3. Press the "I" portion against the feeler gauge and transformer.

4. Lightly snug the fasteners.

5. Remove the feeler gauge.

6. Tighten the fasteners.

7. Start the generator and check the output voltage, then stop the generator again.

8. If the voltage is still too low, repeat the process using a 0.006 in. feeler gauge.

9. Continue adding 0.002 in. clearance at a time until the voltage is acceptable.

10. Apply a 30% rated load for 5 minutes, then recheck no load voltage.

11. Repeat as necessary until no load voltage remains consistent.

12. To lower the no load voltage, measure the air gap, then use the same procedure as used to raise the voltage, except begin with a feeler gauge 0.002 in. less than what was measured.

Full load Voltage Adjust

NOTE: Adjustment must be made with the generator off.

To change full load voltage, the individual brass links on the transformer control board must be aligned according to the load voltage compensation settings. The transformer control board is located next to the transformer, inside the control box.

After compensation adjustments have been made, start the generator and check the output. If the compensation adjustments do not bring the output into specifications, the control board may be faulty.

IDLE CONTROL (DPG 165930)

Adjusting Idle Speed

Check the idle speed against the specification listed in the owner's manual. On these units, use a Vibra-tach to check rpm.

To decrease idle speed:

1. Locate the white UHMW contact block on the idle solenoid shaft.

2. Loosen the adjusting nut to the left (solenoid side) of the block one turn at a time.

3. Tighten the adjusting nut to the right (throttle linkage side) of the block.

4. Check idle speed.

To increase idle speed:

1. Locate the white UHMW contact block on the idle solenoid shaft.

2. Loosen the adjusting nut to the right (throttle linkage side) of the block one turn at a time.

3. Tighten the adjusting nut to the left (solenoid side) of the block.

4. Check idle speed.

Troubleshooting

If idle control difficulties arise, check the in line starter solenoid fuse and the fuse on the time delay relay prior to testing the system components. Replace if faulty. Also check to sure that the 12V battery is fully charged.

If the solenoid does not extend when the key is turned:

1. Remove the black and white wires from the *extend* solenoid coil.

2. Test the coil: Coil resistance should test approximately 0.4 ohm.

3. If coil resistance is acceptable, the black circuit board sending module is faulty. Replace the module.

If the solenoid does not hold after the key returns from the start to the run position:

1. Adjust the engine speed as shown in the preceding adjusting idle speed section. For the solenoid to hold, it must be at the very end of its stroke.

2. Remove the black and red wires from the *hold* solenoid coil.

3. Test the coil. Coil resistance should test approximately 12.0 ohms.

4. Test the red murphy time delay relay in the control box:

   a. Supply 12VDC power to the relay – positive (+) to 'B' and negative (-) to 'G'.

   b. Ground the 'S' terminal.

   c. After approximately 15 seconds, the contacts should transfer position.

   d. Disconnect the 'S' terminal ground. The contacts should return to their original position.

   e. Failure of the relay to perform as noted indicates a faulty relay.

5. If these tests are successful, the black circuit board sending module is faulty. Replace the module.

WIRING DIAGRAMS

Refer to the owner's manual of the unit being serviced for the correct wiring schematic.

PORTER-CABLE

DeVILBISS AIR POWER COMPANY
213 Industrial Drive
Jackson, Tennessee 38301-9615

Generator Model	Rated (Surge) Watts	Output Volts	Engine HP	Wiring Diagram Figure Location
BS500	5000 (6250)	120/240	10	PC121
BS600	6000 (7250)	120/240	11	PC119
BSV800	8000 (9400)	120/240	14	PC114
CDGT3010, CDGT3010-1, CDGT3010-2	3000 (3750)	120	6	PC116
CDGTP3010,				PC115
CDGTP3010-1,				PC116
CDGTP3010-2	3000 (3750)	120	6	PC116
CGBV4000, CGBV4000-1	4000 (5000)	120/240	7.5	PC117
CGTP3000,				PC115
CGTP3000-1,				PC116
CGTP3000-2	3000 (3750)	120	6	PC116
CH250	2500 (3125)	120	5	PC115
CH350CS	3500 (4375)	120/240	6.5	PC124
CTE300	3000 (3750)	120	6	PC116
DGHC6510	6500 (8000)	120/240VAC; 12VDC	13	PC110
DGT5010, DGT5010-1	5250 (6500)	120/240	10	PC121
DTE325	3250 (4000)	120	6	PC116
GB4000, GB4000-1, GB4000-2, GB4000-3 GB4010, GB4010-1, GB4010-2	4000 (5000)	120/240	8	PC112
GB5000, GB5000-1, GB5000-2, GB5000-3, GB5000-4 GB5010, GB5010-1, GB5010-2	5000 (6250)	120/240	10	PC112
GBE4010	4000 (5000)	120/240	8	PC113
GBFE6010,				PC118
GBFE6010-1	6000 (7500)	120/240	11	PC119
GBP4000	4000 (5000)	120/240	8	PC112
GBV4600	4600 (5750)	120/240	9	PC113
GBV5000, GBV5000-1	5000 (6250)	120/240	9	PC113
GBV7000,				PC113
GBV7000-1,				PC113
GBV7000-2,				PC113
GBV7000-3,				PC113
GBV7000-4				PC114
GBV7010,				PC113
GBV7010-1,				PC113
GBV7010-3				PC114
GBVE7010-3	7000 (8400)	120/240	14	PC114
GBVE8000	8000 (9400)	120/240	14	PC114

GBVF5000	5000 (6250)	120/240	9	PC123
GHC6510	6500 (8000)	120/240VAC; 12VDC	13	PC110
GHV4250C	4250 (5250)	120/240	9	PC109
GHV4500, GHV4500-1, GHV4500-2, GHV4500-3, GHV4500-4, GHV4510, GHV4510-1, GHV4510-2, GHV4520	4500 (5500)	120/240	9	PC111
GT5000,				PC120
GT5000C,				PC122
GT5000C-1,				PC122
GT5010	5000 (6250)	120/240	10	PC120
GT5250, GT5250-1, GT5250-2, GT5250-WK, GT5250-WK-1	5250 (6500)	120/240	10	PC121
H450CS	4500 (5500)	120/240	9	PC125
H650CS	6500 (8000)	120/240	13	PC110
T525	5250 (6500)	120/240	10	PC121
T550	5500 (6875)	120/240	10	PC121

NOTE: Wiring diagrams are found at the end of the Porter-Cable section.

IDENTIFICATION

Porter-Cable portable generators are single phase, 60Hz, brushless design generators set to operate at a governed full load speed of 3600 rpm.

WIRING COLOR CODES

All color coded wiring, including gauge size, is identified in each respective wiring diagram. The WIRING DIAGRAMS grouping is at the end of this Porter-Cable section.

MAINTENANCE

The rotor end bearing is a prelubricated, sealed ball bearing. Inspect and replace as necessary.

TROUBLESHOOTING

These generators should produce 60-62.5Hz at a no load speed of 3750 rpm. If little or no output is generated, the following items could be potential causes:
1. Engine rpm–Engine must maintain 3600 rpm under load. Engine rpm could also be set too high, causing overvoltage.
2. Loose component mounting fasteners (low or no voltage).
3. Fuses or circuit breakers/GFCI switch breaker (low or no voltage).
4. Insecure wiring connections (low or no voltage).
5. Faulty receptacle (low or no voltage).
6. Capacitor (low or no voltage).
7. Diodes (low or no voltage).
8. Stator windings (low voltage–shorted; no voltage–open).
9. Rotor
   a. Open or shorted windings (no or low voltage, respectively).
   b. Loss of residual magnetism (no voltage).

NOTE: Residual magnetism can be restored without generator disassembly, and is the recommended first step whenever generator has no output: Refer to RESIDUAL MAGNETISM Section.

10. Idle control not working (erratic or no idle).
11. Bridge rectifier (no DC output).
12. Insufficient cooling system ventilation (overheating).

GENERATOR REPAIR

RESIDUAL MAGNETISM

If low residual magnetism is suspected as the cause of no output, re excite as follows:
1. Start the engine and allow a brief warm up period.
2. Manually apply full throttle to the engine for a maximum of 3 seconds, then release the throttle and allow the engine to resume normal governed full throttle rpm.
3. Check the generator output. If there is still no output, stop the engine.
4. Remove the stator end cover (10-Fig. PC 101 or PC 102).
5. On all models except the GHC6510, disconnect the capacitor leads from the capacitor (7). On the GHC6510, remove the two pink capacitor leads from the wire harness, not the capacitor.
6. Obtain a fully charged 6-volt lantern battery including wire leads, and connect one battery terminal to one capacitor terminal. Momentarily touch the second battery lead to the other capacitor terminal to cause an arc. Disconnect the battery.
7. Reattach the capacitor leads and reinstall the end cover.
8. Restart the engine and check the generator output.
9. If there is still no output, shut the engine off and proceed with further testing and repair.

RECEPTACLES

To test the receptacles:
1. Start the generator.

2. Set the VOM scale to 250VDC, and refer to Fig. PC103.

3. For the 120V outlets, place the black VOM to ground and the red lead to the small flat leg of each duplex outlet. The reading should be 108-132 volts

4. For the 240V receptacle, connect the leads to the outlet as shown.

5. If no reading is obtained at the receptacle, test for voltage at the wire harness supplying the receptacle.

6. If no reading is obtained at the harness end, check the harness wires for continuity between the panel receptacle ends and the stator ends.

7. If continuity tests OK, proceed with further generator component testing.

NOTE: Torque the terminal screws to 14 in.-lb. On duplexes, torque ground screws to 10 in.-lb.

Disassembly

Generator Head Assembly

To disassemble:

1. Place a 2 × 4 under the engine adapter for support.

2. Remove the stator end cover and disconnect the receptacle wires (Fig. PC101 or PC102).

3. Remove the rubber isolators from the end bearing housing feet.

4. Remove the four 7/16 in. nuts from the stator throughbolts.

5. Disconnect the capacitor leads (the capacitor harness on model GHC6510).

6. Remove the stator throughbolt nuts (6-Fig. PC101 or PC102).

7. Remove the end bearing housing (5) from the stator (16). It may be necessary to tap the inner edge of the housing with a mallet to free it from the stator.

8. Noting the stator wire orientation, remove the stator by pulling it off of the stator throughbolts, then remove the throughbolts.

NOTE: The rotor cannot be removed without first removing the stator.

9. Remove the 1/2 in. nut holding the rotor shaft on, then remove the rotor (17-Fig. PC101 or PC102) by pulling it off the engine crankshaft and stud. To break the rotor loose, it may be necessary to use a mallet and tap the rotor on the laminations.

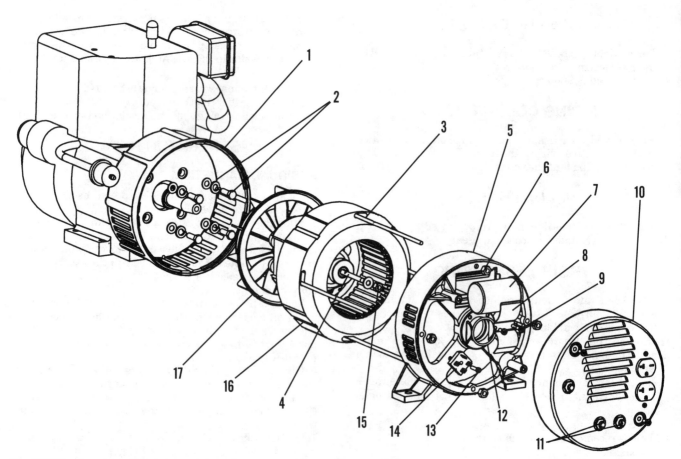

Fig. PC101 – Removing the stator end cover.

1. Adapter	7. Capacitor	13. 30-40 in.-lb. (3.4-4.5 N.m)
2. 204-264 in.-lb. (23.1-29.8 N.m)	8. Capacitor bracket	14. Diode rectifier
3. Stator through-bolt	9. 30-40 in.-lb. (3.4-4.5 N.m)	15. 120-144 in.-lb. (13.6-16.2 N.m)
4. Rotor through-bolt	10. End cover	16. Stator assy.
5. Bearing support	11. Circuit breakers	17. Rotor & fan assy.
6. 60-70 in.lb. (6.8-7.9 N.m)	12. "O" ring	

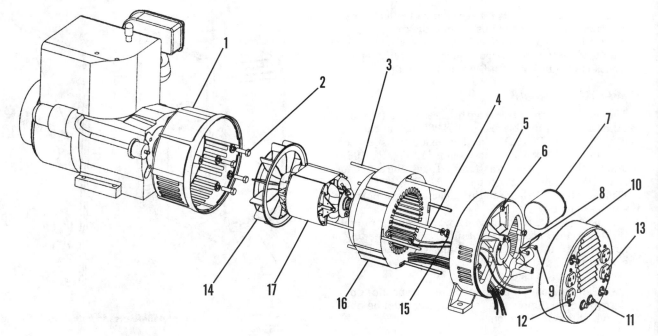

Fig. PC102 – Exploded view of GB4000 series generator.

1. Adapter
2. 204-264 in.-lb. (23.1-29.8 N.m)
3. Stator through-bolt
4. Rotor through-bolt
5. Bearing support
6. 60-70 in.lb. (6.8-7.9 N.m)

7. Capacitor
8. Capacitor bracket
9. 30-40 in.-lb. (3.4-4.5 N.m)
10. End cover
11. Circuit breakers
12. 120V receptacle

13. 240V receptacle
14. Fan
15. 120-144 in.-lb. (13.6-16.2 N.m)
16. Stator assy.
17. Rotor assy.

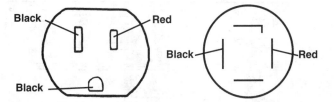

Fig. PC103 – VOM connections for testing output at the 120V and 240V receptacles respectively.

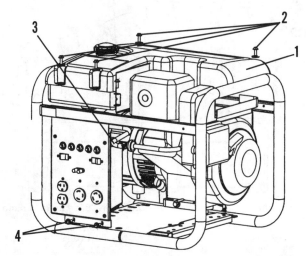

Fig. PC104 – Control-panel removal for the Model CGBV4000.

1. Gas tank
2. Screws

3. Fuel line
4. Screws

CAUTION: The plastic rotor fan is part of the rotor assembly. Handle the rotor carefully to prevent fan damage.

10. If needed, the rotor shaft stud in the engine crankshaft can be removed at this time.

Control Panel Assembly on Model CGBV4000

To disassemble:

1. Shut off the fuel tank valve. Drain and remove the fuel tank, including the fuel line.
2. Disconnect the panel assembly wire harness.
3. Remove the two bottom face panel screws (Fig. PC104) and **carefully** remove the panel assembly.
4. To disassemble the panel assembly:
   a. Remove the four face panel screws holding the back cover (Fig. PC105).
   b. Separate the back cover from the front panel.
   c. Disconnect the harness connector. If needed, the circuit board can be removed from the back cover.

Control Panel Assembly on Model GHC6510

To disassemble:

1. Disconnect the wire harness connector plug from the back of the control panel, then use a 3/8 in. socket to remove the four screws holding the back of the face panel to the generator cradle frame. Carefully remove the panel assembly.
2. Use a T-15 Torx wrench to remove the two screws holding the capacitor bracket to the inside of the panel cover. These are on the left side of the panel back cover.
3. Remove the circuit board from the inside of the panel cover using needle nose pliers. Squeeze the four push posts

on the right side of the cover, then depress the lock tabs on the harness connector and push it into the cover.

4. Use a T-20 Torx wrench to remove the six face-panel screws holding the back cover in place. Carefully separate the back cover (15-Fig. PC106) from the front panel (1).

CAPACITOR TESTING

The capacitor can be tested without removing it from the generator. On all models except GHC6510, the capacitor (7-Fig. PC101 or PC102) is located inside the stator end cover. On the GHC6510, the capacitor (10-Fig. PC 106) is located in the control panel assembly. To test the capacitor:

1. Set VOM ohmmeter scale to R × 10,000.
2. Connect the VOM leads to the capacitor terminals, noting proper polarity.
3. The meter should first indicate low resistance, gradually increasing resistance toward infinity. No meter reading or constant continuity indicates a faulty capacitor. Replace the capacitor.

> **NOTE: If replacing the capacitor, note capacitor polarity prior to removal. Correct polarity must be observed upon installation.**

DIODES

To test the diode(s):
1. Remove the diode from its bracket.
2. With a VOM set on the 5VDC scale, connect the VOM leads to the diode terminals.
3. Attach the positive terminal of a new AA battery to the diode in the direction indicated by the arrow on the diode (Fig. PC107 or PC108). VOM should read 1.5.
4. Reverse the battery: VOM should now read 0.7-0.9.
5. Readings other than those specified indicate a faulty diode.

If, instead of a VOM, a Fluke 77 or 87 meter is used, set it to the diode symbol and listen for a beep from the meter when performing Step 3. No beep indicates a faulty diode.

CIRCUIT BREAKERS

To test the circuit breaker on units so equipped:
1. Make sure the breaker is set. If it is tripped, reset.
2. Remove the wires from the rear of the breaker.
3. Using the R × 1 scale on the VOM, test for continuity across the breaker terminals.
4. If no continuity reading is noted, or if the breaker fails to reset, the breaker is faulty. Replace the breaker.

GROUND FAULT CIRCUIT INTERRUPTER (GFCI)

To test the GFCI:
1. Start the generator and allow a brief warmup period.
2. Depress the TEST button: The RESET button should extend. If the RESET button does not extend, the GFCI is faulty.
3. To restore power, depress the RESET button firmly into the GFCI until a definite click is heard. If properly reset, the RESET button is flush with the surface of the TEST button. When the RESET button stays in, the GFCI is *on*.
4. Stop the generator.

STATOR WINDING RESISTANCE

1. Set the VOM to the R × 1 scale.
2. Test for continuity between the following stator leads:
   a. Black-Orange
   b. Red-Green

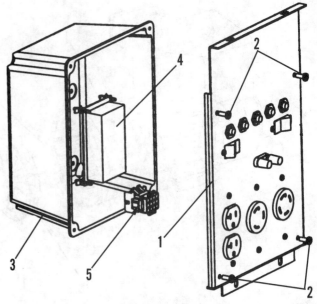

Fig. PC105 – Control-panel disassembly for the Model CGBV4000.

1. Control panel	4. Circuit board
2. Screws	5. Connector
3. Back cover	

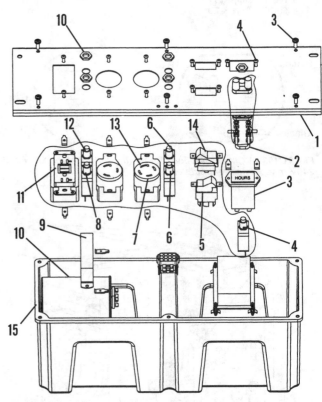

Fig. PC106 – Exploded view of Model DGHC6510 control panel.

1. Front panel	9. Clamp
2. Resistor	10. Capacitor
3. Hour meter	11. 12V GFCI receptacle
4. Circuit breaker 10A	12. Circuit breaker 20A
5. Full power switch	13. Receptacle 4-prong
6. Circuit breaker 25A	14. Idle control switch
7. 125V receptacle	15. Rear cover
8. Circuit breaker 25A	

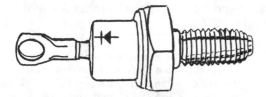

Fig. PC107 – View of diode used on single-diode rotors. Also used on dual-diode rotors in conjunction with diode in Fig. PC108. Direction of arrow marked on side of diode must be observed when testing the diode. Refer to text.

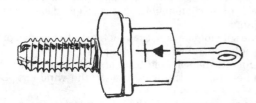

Fig. PC108 – View of diode used on dual-diode rotors in conjunction with diode in Fig. PC107. Direction of arrow marked on side of diode must be observed when testing the diode. Refer to text.

   c. Pink-Pink

3. No continuity/open circuit indicates a faulty stator.
4. Scrape a small amount of varnish from the laminations to make a good test ground.
5. Set the VOM to the R × 1 scale.
6. Test for continuity between the following points:

   a. Black stator lead-stator laminations.
   b. Red stator lead-stator laminations.
   c. Each Pink capacitor lead-stator lamination.
7. No continuity/open circuit indicates a good stator.

To test the capacitor winding: With the VOM still set on the R × 1 scale, connect the VOM leads between the black stator lead and a pink capacitor lead. Then connect the VOM leads between the red stator lead and a pink capacitor lead. The VOM should read infinity. Any continuity reading indicates a faulty stator.

NOTE: The white and yellow windings are for the idle control and DC circuits respectively. Their test procedures will be covered in subsequent sections.

Winding resistance test values for generators requiring such values are listed in the following chart:

ROTOR WINDING RESISTANCE

One Diode Rotors

1. Remove the 7/16 in. nut, and remove the 1/4 in. female connector from the diode bracket.
2. Set the VOM to the R × 1 scale.
3. Attach one VOM lead to the female connector. Attach the other VOM lead to the solder junction on the diode.
4. No reading/infinity indicates an open winding and a faulty rotor.

Resistance Values in Ohms@70° F

Model No.		Main Winding (Red to Green) (Black to Orange)	Aux. Winding (Pink Capacitor Leads)	Rotor	I/C	D/C
GPB4000 BG4000 GB500-1 GB5010 GBV5000 GBE4010	GB5000 BF5000-2 GBV4600 GBV5000-1 GHV4500	.3-.6	1.2-1.6	.3-.8	N/A	N/A
GB400-1 GB4010	GB400-2	.3-.6	1.3-1.8	.6-1.2	N/A	N/A
GB400-3 GB5000-3 GT5000	GBE4010-1 GB5000-4 GBP4000-1	.3-.6	2.2-2.6	2.0-2.3	N/A	N/A
GT5250	GHV4500-3	.3-.7	1.7-2.1	.6-1.6	N/A	N/A
GBFE6010		.4-.7	1.7-2.6	1.7-2.0	N/A	N/A
GBV7000 GBV7000-2GBV7010 GBV7000-3	GBV7000-1	0-.3	.6-1.1	.5-.9	N/A	N/A
GBV7000-4	GBVE8000	.2-.6	.6-1.3	.9-1.7	N/A	N/A
CGPT3000		.4-.7	2.3-2.7	1.9-2.3	N/A	N/A
CGBV4000		.3-.7	2.0-2.3	2.0-2.3	.3-.6	.3-.6

5. Scrape a small amount of varnish from the surface of the laminations to make a good ground for the next test.

6. Attach one VOM lead to the female connector. Attach the other VOM lead to the lamination ground spot.

7. Any continuity reading obtained indicates a shorted winding and a faulty rotor.

Two-Diode Rotors

1. Remove one diode.

2. Set the VOM to the R × 1 scale.

3. Attach one VOM lead to the solder junction of the removed diode. attach the other VOM lead to the solder junction of the remaining diode.

4. No reading/infinity indicates an open winding and a faulty rotor.

5. Scrape a small amount of varnish from the surface of the laminations to make a good ground for the next test.

6. Attach one VOM lead to the solder junction of the removed diode. Attach the other VOM lead to the lamination ground spot.

7. Any continuity reading obtained indicates a shorted winding and a faulty rotor.

Both One- and Two-Diode Rotors

Some generator models require specific rotor winding resistance test values. These are listed in the chart which follows the preceding *stator* section. Any rotor tested which reads beyond the specified range is faulty.

12-VOLT DC/BATTERY CHARGER (MODELS CGBV4000 AND GHC6510)

To test the DC circuit:

1. Remove the stator end cover.

2. Set the VOM on the R × 1 scale.

3. Connect the VOM leads to the two yellow stator leads. VOM should read 0.07 ohms.

4. Scrape a small amount of varnish from the surface of the stator laminations to make a good ground for the next test.

5. Connect one VOM lead to one yellow stator lead. Connect the other VOM lead to stator lamination ground. VOM should read infinity.

6. Readings other than those specified indicates a faulty stator. If the test readings are as specified, the bridge rectifier is faulty (14-Fig. PC101).

IDLE CONTROL

Testing the Switch

> NOTE: Always turn the idle control switch *off* when removing or installing it, or whenever the generator is being test run, unless the test requires the switch *on*.

1. Remove the two screws holding the switch to the front of the control panel.

2. Disconnect the wire harness from the rear of the switch.

3. With the VOM set to the R × 1 scale, attach the VOM leads to the switch terminals.

4. Turn the switch *on*. A low resistance reading indicates a good switch. No reading/infinity indicates a faulty switch.

Testing and Replacing the Throttle Solenoid (Model CGBV4000)

1. Disconnect the solenoid wire harness.

2. Loosen but **do not remove** the solenoid bracket lock nut, then slide the solenoid from the slotted bracket.

3. With the VOM set on ohms, connect the VOM leads to the solenoid leads. The VOM should read 140-160 ohms. A reading beyond this range indicates a faulty solenoid.

4. Reconnect the solenoid to its wire harness, but do not mount it into its bracket slot at this time.

5. Start the generator and turn the idle control switch *on*, then place the solenoid end near the generator cradle frame. If properly magnetized, the solenoid should stick to the frame.

6. If the solenoid does not magnetize to the frame, proceed to the idle control stator winding test.

7. To install the solenoid back onto the generator, thread the inner nut all the way back to the windings: Do not force.

8. Slide the solenoid into the bracket slot.

9. Thread the outer solenoid nut down only until the nut face is flush with the end of the solenoid shaft.

10. Using a VOM able to register Hz, connect the VOM to one 120V outlet.

11. Start the generator and turn the idle control switch *on*. The magnetism of the solenoid will pull it into place. The VOM should read 60 Hz.

12. Thread the outer nut down until the VOM reads 45 Hz.

13. Thread the inner nut out against the bracket, then wrench tighten the nuts, **making sure** the VOM continues to read 45 Hz until the nuts are tight.

14. Turn the idle control switch *off*. The VOM will return to 60 Hz.

Testing and Replacing the Throttle Solenoid (Model GHC6510)

> NOTE: This solenoid test (Steps 1 and 2) also applies to Model BS600.

1. Disconnect the solenoid wire harness.

2. With the VOM set to the R × 1 scale, connect the VOM leads to the brass terminals inside the solenoid harness connector: The VOM should read 25-27 ohms. A reading beyond this range indicates a faulty solenoid. If the solenoid tests OK, proceed to the idle control stator winding test.

3. To replace the solenoid, use a 10 mm socket to remove the three engine heat shield screws. Set the shield aside.

4. Use a Phillips screwdriver to remove the two solenoid screws.

5. Slide the crank arm from the solenoid shaft and remove the solenoid.

6. Install the new solenoid by sliding the shaft over the crank arm, installing the two Phillips screws, then remounting the heat shield.

7. Reconnect the solenoid wire harness.

Testing the Idle Control Stator Winding

If the solenoid ohm tests OK but does not magnetize when the idle control switch is *on*, test the stator winding as follows:

1. Make sure the generator switch is *off* and the generator is cool.

2. Remove the stator housing end cover.

3. With the VOM set to the R × 1 scale, connect the VOM leads to the two white stator wires. The resistance reading

should be 0.05 ohm. Higher resistance or no reading indicates a faulty stator.

4. Scrape a small amount of varnish from the stator laminations to make a good ground for the next test.

5. With the VOM still set to R × 1, connect one VOM lead to one white stator wire. Connect the other VOM lead to the stator lamination ground. Any continuity reading indicates a shorted winding and a faulty stator.

6. If the stator is faulty, replace the stator, then retest the solenoid magnetism. When the stator and solenoid are good, but the solenoid will not magnetize, the circuit board is faulty. Replace the circuit board.

7. To access the circuit board, refer to **Disassembly, Control Panel** in this chapter's **GENERATOR REPAIR** section.

> NOTE: Carefully note the circuit board wiring orientation when replacing the circuit board on Model GHC6510.

Setting the Engine Idle Speed (Model GHC6510)

1. Set the VOM to the Hz scale, and connect the VOM leads to the 120V GFCI outlet.

2. Start the generator, allow a brief warm up period, then turn the idle control switch *on*.

3. Using the Phillips screwdriver, turn the carburetor adjusting screw left or right as necessary until the VOM reads 45 Hz.

4. Turn the idle control switch *off*. the VOM will return to 60 Hz.

5. Turn the idle control switch *on* and *off* a couple of times, allowing the engine to stabilize each time, making certain that 45 Hz is maintained when the switch is *on*. Readjust as necessary.

Reassembly

Control Panel Assembly on Model CGBV4000

To assemble:

1. Make certain that the circuit board and the wiring harness connector are properly fastened into the back cover.

2. Fasten the front panel to the back cover with the four face panel screws.

3. Carefully place the control panel assembly into position in the generator cradle frame, then fasten the face panel to the frame with the two bottom screws.

4. Reconnect the panel assembly wire harness at the rear of the panel.

5. Reinstall and reconnect the fuel tank.

Control Panel Assembly on Model GHC6510

To assemble:

1. Attach the capacitor bracket, harness connector plug, and circuit board to the inside of the back cover.

2. Carefully install the back cover onto the front panel, making sure there are no pinched or kinked wires. Fasten the front panel to the back cover with the six T-20 Torx head screws.

3. Carefully position the panel assembly onto the generator cradle frame, then install the four 3/8 in. head screws to hold the panel in place.

4. Reattach the wire harness connector.

Generator Head Assembly

1. If the rotor shaft stud was removed from the engine crankshaft during disassembly, reinstall it now.

2. Place the 2 × 4 under the engine adapter for support.

3. Make sure that the matching tapers on the engine crankshaft and inside the rotor shaft are clean and dry. fit the rotor onto the crankshaft. Install and tighten the ½ in. rotor stud nut.

> CAUTION: The plastic rotor fan is part of the rotor assembly. Handle the rotor carefully to prevent fan damage.

4. Install the stator throughbolts, then, noting the proper stator wire orientation, fit the stator over the bolts and the rotor. Install it all the way up to the engine adapter.

5. Fit the end bearing housing onto the rotor and stator. It may be necessary to lightly tap the housing with the mallet for the proper snug fit.

> CAUTION: Make sure that the stator wires are not pinched and are properly routed through the housing openings.

6. Reconnect the capacitor leads or the capacitor harness on Model GHC6510.

7. Install and tighten the four 7/16 in. nuts on the stator throughbolts.

8. Install the rubber isolators onto the end bearing housing feet.

9. Reconnect the stator end cover receptacle wires and install the cover.

10. Remove the 2 × 4 support from under the engine adapter.

WIRING DIAGRAMS

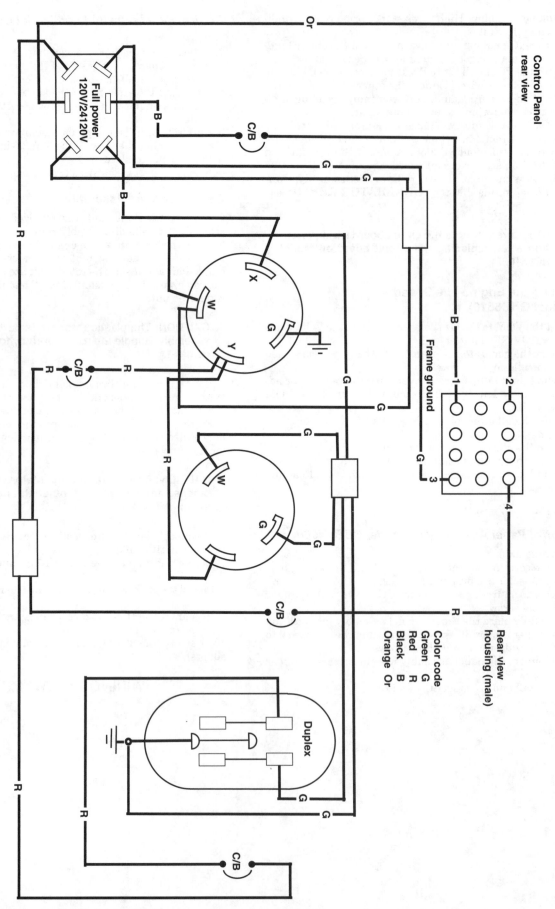

Fig. PC109 – Wiring diagram for Model GHV4250C.

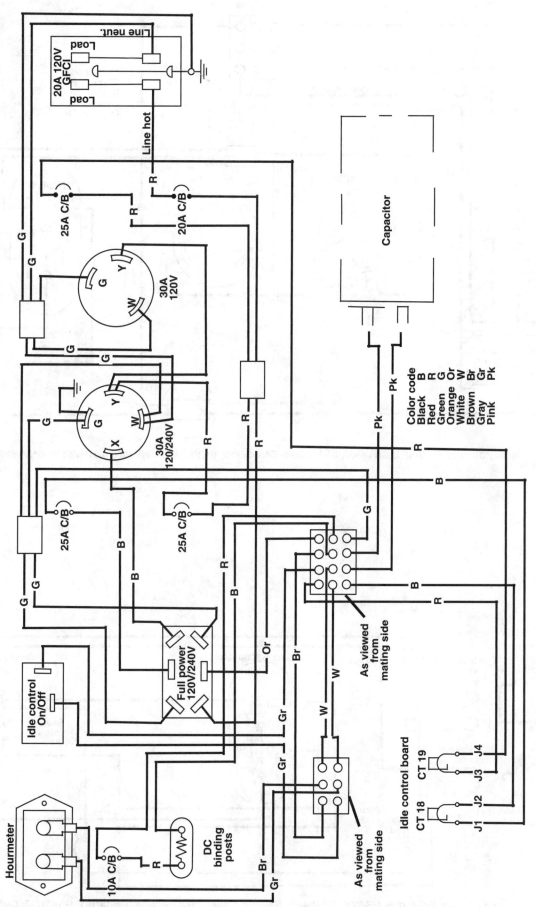

Fig. PC110 – Wiring diagram for Models DGHC6510, GHC6510, and H650CS.

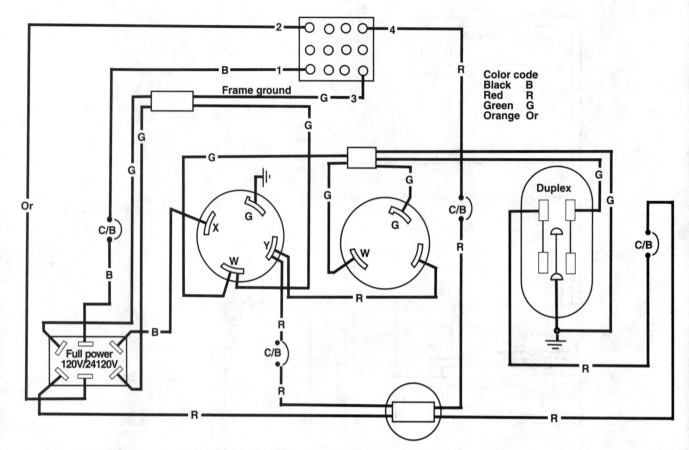

Fig. PC111 – Wiring diagram for Models GHV4500, GHV4500-1, GHV4500-2, GHV4500-3, GHV4500-4, GHV4510, GHV4510-1, GHV4510-2, and GHV4520.

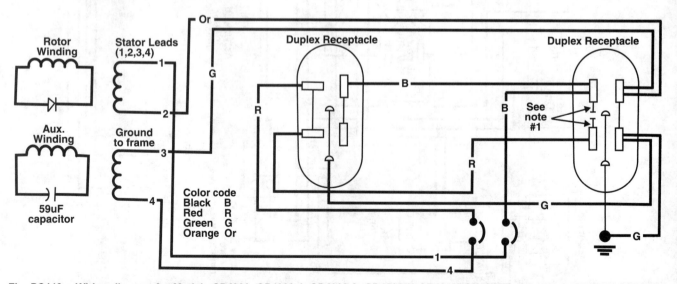

Fig. PC112 – Wiring diagram for Models GB4000, GB4000-1, GB4000-2, GB4000-3, GB4010, GB4010-1, GB4010-2, GB5000, GB5000-1, GB5000-2, GB5000-3, GB5000-4, GB5010,

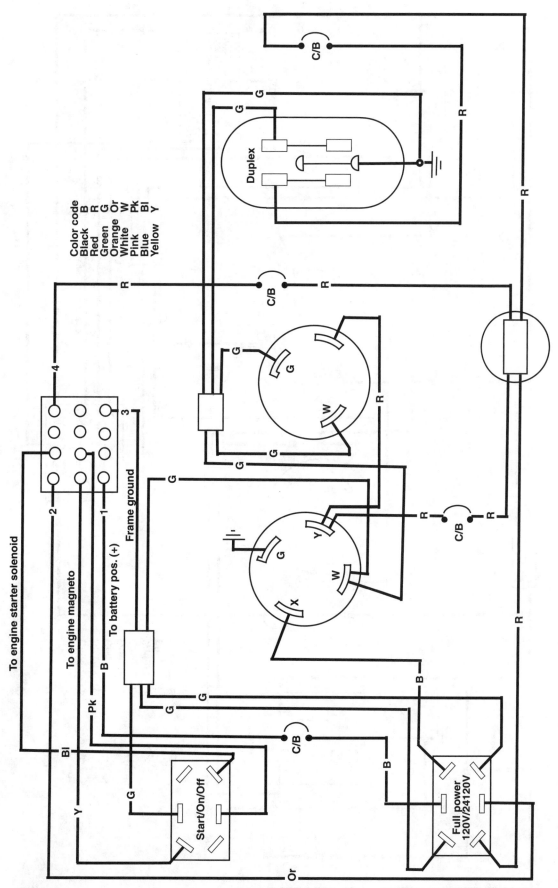

Color code
Black B
Red R
Green G
Orange Or
White W
Pink Pk
Blue Bl
Yellow Y

Fig. PC113 – Wiring diagram for Models GBE4010, GBV4600, GBV5000, GBV5000-1, GBV7000, GBV7000-1, GBV7000-2, GBV7000-3, GBV7010, GBV7010-1.

PORTER-CABLE

GENERATOR

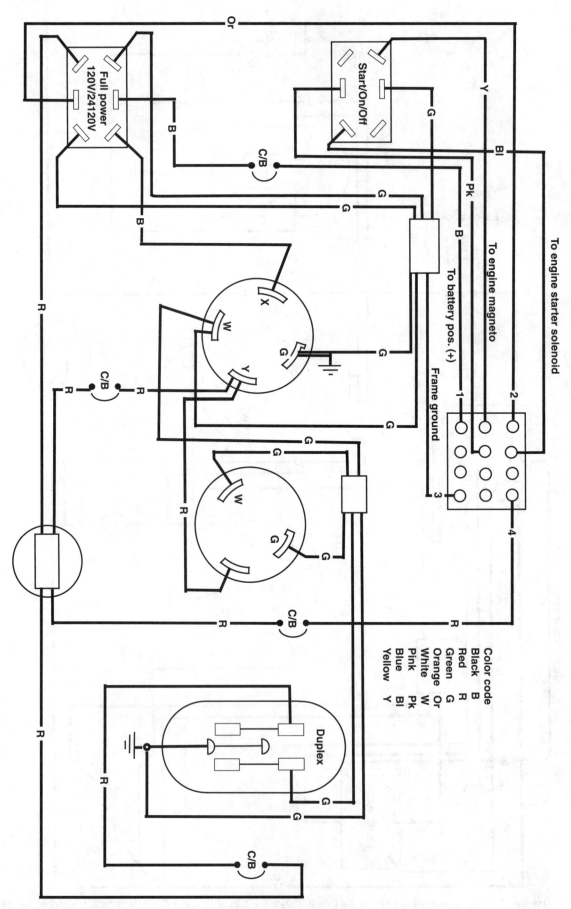

Fig. PC114 – Wiring diagram for Models BSV800, GBV7000-4, GBV7010-3, GBVE7010-3, and GBVE8000.

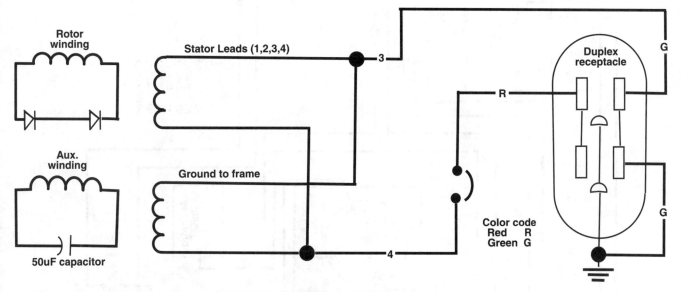

Fig. PC115 – Wiring diagram for Models CDGTP3010, CGTP3000, and CH250.

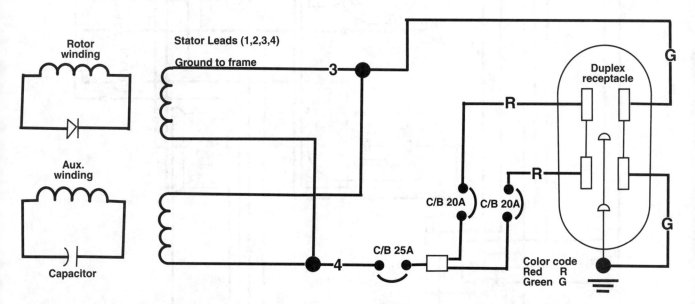

Fig. PC116 – Wiring diagram for Models CDGT3010, CDGT3010-1, CDGT3010-2,

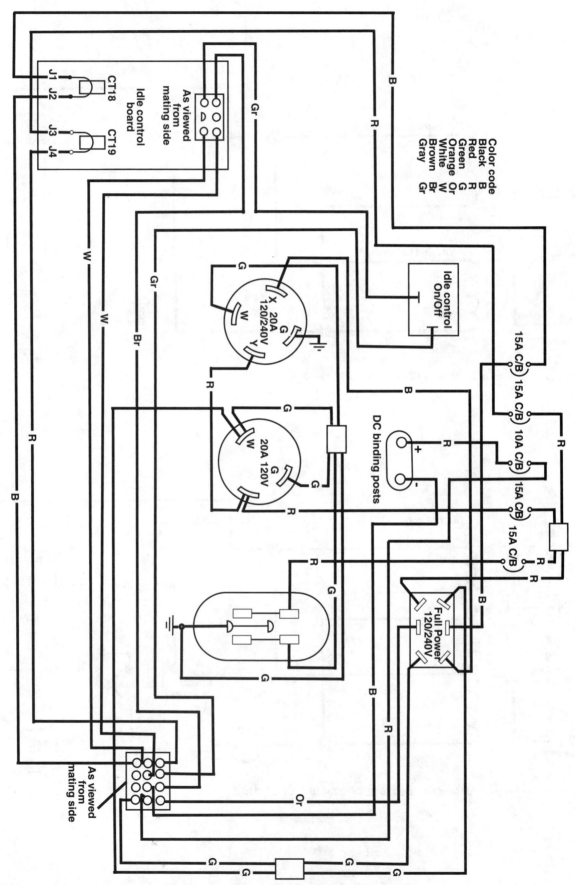

Fig. PC117 – Wiring diagram for Models CGBV4000 and CGBV4000-1.

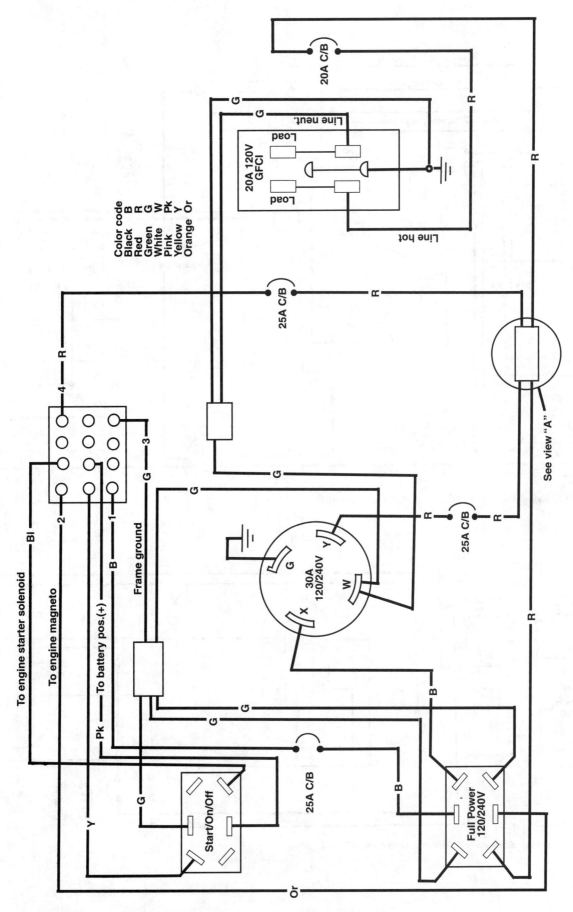

Fig. PC118 – Wiring diagram for Model GBFE6010.

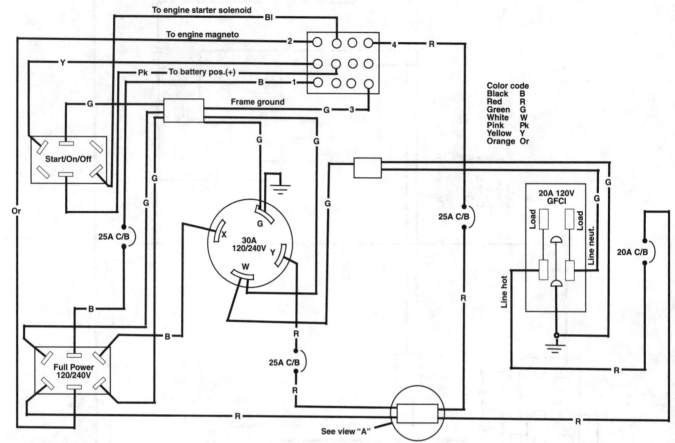

Fig. PC119 – Wiring diagram for Models BS600 and GBFE6010-1.

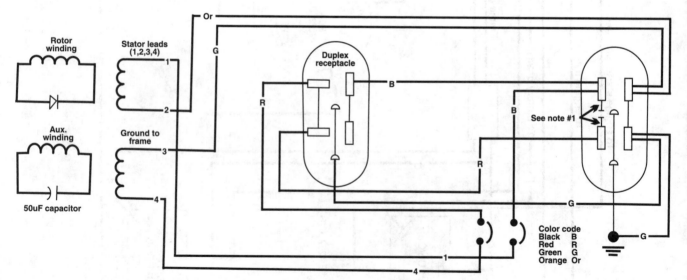

Fig. PC120 – Wiring diagram for Models GT5000 and GT5010.

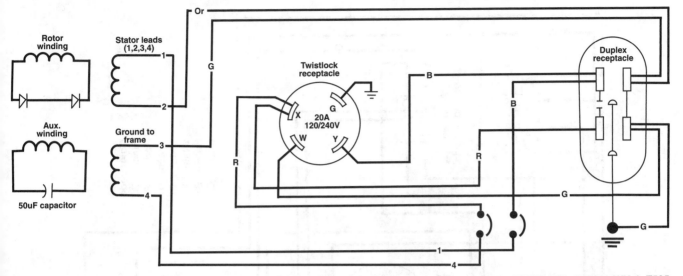

Fig. PC121 – Wiring diagram for Models BS500, DGT5010, DGT5010-1, GT5250, GT5250-1, GT5250-2, GT5250-WK, GT5250-WK-1, T525, and T550.

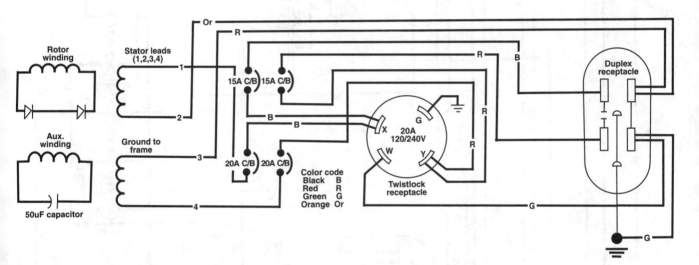

Fig. PC122 – Wiring diagram for Models GT5000C and GT5000C-1.

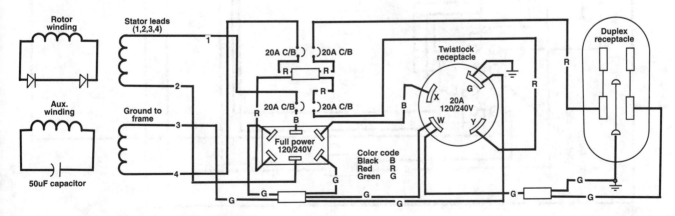

Fig. PC123 – Wiring diagram for Model GBVF5000.

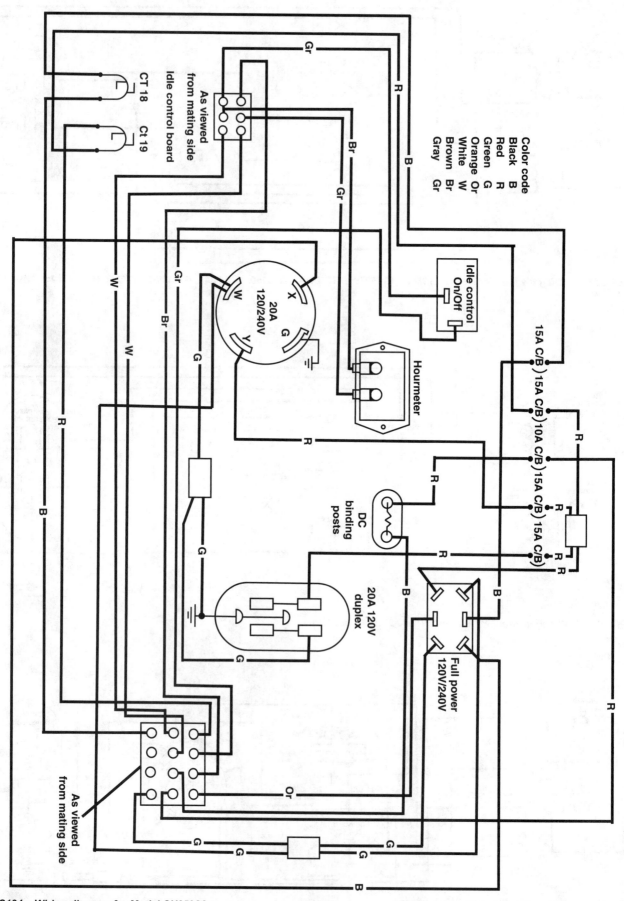

Fig. PC124 – Wiring diagram for Model CH350CS.

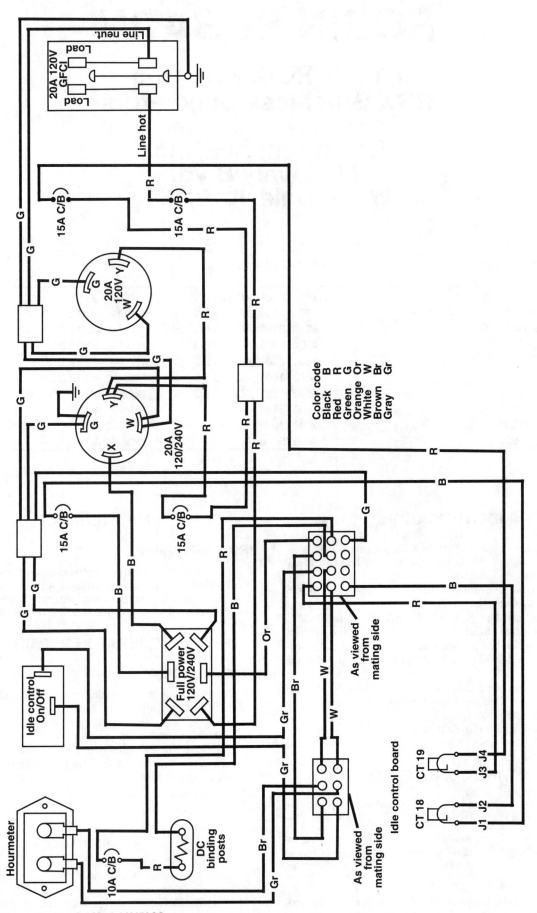

Color code
Black   B
Red   R
Green   G
Orange   Or
White   W
Brown   Br
Gray   Gr

Fig. PC125 – Wiring diagram for Model H450CS.

ROBIN SUBARU

R1300, RGD, RGV, and RGX Brushless Generators

Robin America, Inc.
940 Lively Blvd.
Wood Dale, IL 60191

Generator Model	Rated (Surge) Watts	Output Volts
R1300	1000 (1300) AC/100 DC	120 AC/12 DC
RGD2500	2200 (2500) AC/100 DC	120/240 AC/12 DC
RGD3300	3000 (3300) AC/100 DC	120/240 AC/12 DC
RGD5000	4500 (5000) AC/100 DC	120/240 AC/12 DC
RGV2800	2300 (2800) AC/100 DC	120/240 AC/12 DC
RGV4100	3600 (4100) AC/100 DC	120/240 AC/12 DC
RGV6100	4800 (5800) AC/100 DC	120/240 AC/12 DC
RGX2400	2000 (2400) AC	120/240 AC
RGX3500	3000 (3500) AC	120/240 AC
RGX5500	4800 (5500) AC	120/240 AC

IDENTIFICATION

These generators are direct drive, brushless, 60 Hz, single phase units with capacitor regulation and panel mounted receptacles.

The R1300 serial number plate is mounted on the side of the fuel tank, above the engine.

The serial number plate on the RGD2500, RGD3300, and RGX models is located on the generator end cover.

The serial number plate on the RGD5000 model is located on the back wall of the control box, facing the rewind starter.

The serial number plate on the RGV models is located on the outer side wall of the control box, above the generator end housing.

Refer to Figs. RB201-RB210 to view the generator components.

WIRE COLOR CODE

In places where Robin uses abbreviated color coding, the code is as follows:

Blk – Black	Blu – Blue	LBlu – Light blue
Brn – Brown	Grn – Green	Org – Orange
Gry – Gray	R – Red	W – White
Y – Yellow		

A wire with a slash separated ID is a 2-color wire. For example: R/W is a red wire with a white tracer stripe.

MAINTENANCE

ROTOR BEARING

The end housing bearing on these generators is a sealed, prelubricated ball bearing: Inspect and replace as necessary. If a seized bearing is encountered, inspect the rotor journal and the rear housing bearing cavity for scoring.

GENERATOR REPAIR

TROUBLESHOOTING

1. Before testing any electrical components, check the integrity of the terminals and connectors.
2. While testing individual components, also check the wiring integrity and fuse(s).
3. The use of the Dr. Robin generator tester #388-47565-08 is recommended for testing. The Dr. Robin tester has the following capabilities:

   a. AC voltmeter – 0-500 volts.

   b. Frequency – 25-70 Hz.

   c. Condenser capacity – 10-100 μF.

   d. Digital ohmmeter – 0.1-1,999 ohms.

   e. Insulation resistance – 3 MΩ.

An accurate engine tachometer will also be required.

4. Test specifications are given for an ambient temperature of 20° C (68°F). Resistance readings may vary with temperature and test equipment accuracy.

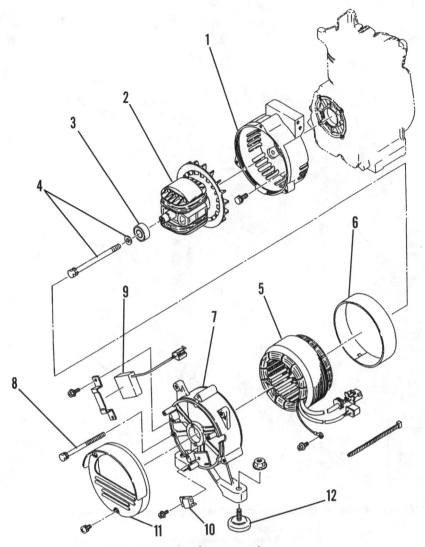

Fig. RB201 – Exploded view of model R1300 generator head components.

1. Engine adapter/front housing
2. Rotor assembly
3. Rotor bearing
4. Rotor through-bolt and washer
5. Stator assembly
6. Stator cover
7. Rear housing
8. Rear housing/stator through-bolt (4)
9. Capacitor/condenser
10. Diode rectifier
11. End cover
12. Rubber mount (2)

5. For access to the components listed in this **Trouble-shooting** section, refer to Figs. RB201-RB210 and the later **Disassembly** section.

6. All operational testing must be performed with the engine running at its governed no-load speed of 3750-3800 rpm on the R1300 generator, 3600 rpm on the RGD generators, or 3700-3750 rpm on the RGV and RGX generators, unless specific instructions state otherwise.

7. Always load test the generator after repairing defects discovered during troubleshooting.

Megger Test the Generator
System Insulation

1. Make sure that the circuit breakers are *on* and functioning.

2. Connect one megger test probe to any hot terminal of a receptacle.

3. Connect the other megger test probe to any ground terminal.

4. The megger should indicate a resistance of 1 MΩ or more. If less than 1 MΩ, megger test the rotor, stator, and control panel resistance individually. Refer to the subsequent Troubleshooting sections.

To Troubleshoot The Generator

1. Circuit breakers must be *on*. To test either the AC or DC (if equipped) breakers, access the breaker wiring and continuity test the breaker terminals with the generator shut down and at least one breaker feed wire disconnected. The breaker is faulty if it will not stay set, shows continuity in the off position or does not show continuity in the ON position.

2. Receptacles must be functioning. To test the receptacle(s):

    a. Access the receptacle connections.

    b. Disconnect the *return* feed wire from the receptacle being tested.

    c. Install a jumper wire between the *hot* and *return* outlet terminals.

    d. Test for continuity across the feed terminals. No continuity indicates a faulty receptacle.

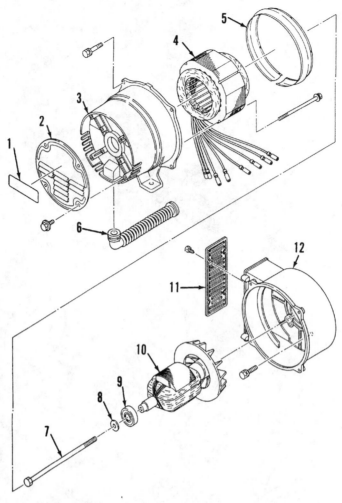

Fig. RB202 – Exploded view of models RGD2500, RGD3300, and RGX series generator head components.

1. Specification plate
2. End cover
3. Rear housing
4. Stator assembly

5. Support ring
6. Wiring-harness protective cover tube
7. Rotor through-bolt
8. Washer

9. Rotor bearing
10. Rotor assembly
11. Louvered plate
12. Engine adapter/front housing

e. Remove the jumper wire and retest for continuity across the feed terminals. No continuity this time indicates a good receptacle.

3. Start the generator, allow it to reach operating temperature, then load test the generator to check voltage output. A load circuit such as that shown in Fig. RB211 will work well for the AC output, using resistance heaters or incandescent lamps with a 1.0 power factor. A similar circuit can be constructed for the DC output, except that if a battery is used as the load, the voltage in Step 4 will test 1-2 volts higher.

4. Test the AC voltage at one of the 120V receptacles. Output should be 108-132 volts on all models except the RGD Series. On RGD generators, voltage should test 117-130 volts. AC voltage at the 240V receptacles should test 216-264 volts on all models except the RGD Series. On RGD generators, voltage should test 235-260 volts. DC voltage should test 6-14 volts.

5. Stop the generator.

6. Disconnect the stator winding wires. With the meter set to the R × 1 scale, apply the meter leads to the color coded wire terminals noted and test the stator winding resistance using the following specifications:

Model	Winding	Ohms
R1300	Brown to White AC	1.26-1.54
	Black to Black excitation/condenser	4.3-5.3
	Green to Green DC	0.56-0.68
RGD2500	Brown to White AC	0.66
	Blue to Light blue AC	0.66
	Black to Orange excitation/condenser	1.9
	Yellow to Yellow DC	0.12
RGD3300	Brown to White AC	0.44
	Blue to Light blue AC	0.44
	Black to Orange excitation/condenser	1.6
	Yellow to Yellow DC	0.11
RGD5000	Red to White AC	0.26
	Blue to Black AC	0.26
	Yellow to Yellow excitation/condenser	0.57
	Brown to Brown DC	0.13
RGV2800	Red to White AC	0.58
	Blue to Black AC	0.58
	Yellow to Yellow excitation/condenser	1.67
	Brown to Brown DC	0.25

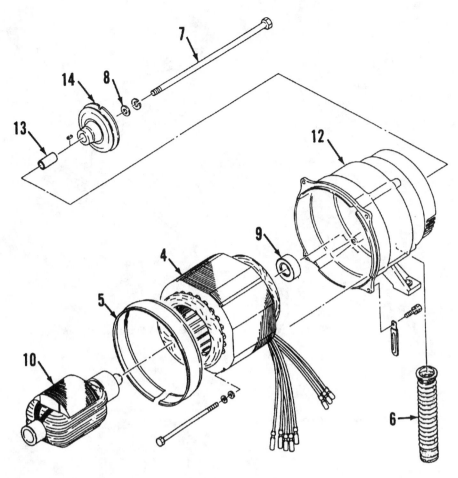

Fig. RB203 – Exploded view of model RGD5000 generator head components.

4. Stator
5. Support ring
6. Wiring harnes tube
7. Throughbolt

8. Washer
9. Rotor bearing
10. Rotor

12. Engine adapter housing
13. Spacer
14. Stator pulley

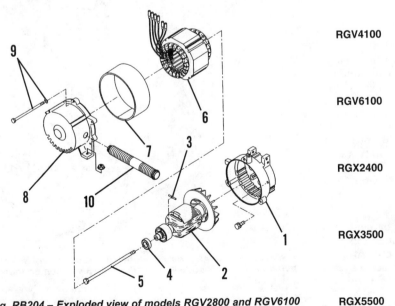

Fig. RB204 – Exploded view of models RGV2800 and RGV6100 generator head components.

1. Engine adapter/front housing
2. Rotor assembly
3. Rotor diode
4. Rotor bearing
5. Rotor through-bolt

6. Stator assembly
7. Stator cover
8. Rear housing
9. Throughbolt (4)
10. Wiring harness tube

Model	Connection	Value
RGV4100	Red to White AC	0.52
	Blue to Black AC	0.52
	Yellow to Yellow excitation/condenser	0.99
	Brown to Brown DC	0.18
RGV6100	Red to White AC	0.25
	Blue to Black AC	0.25
	Yellow to Yellow excitation/condenser	0.58
	Brown to Brown DC	0.13
RGX2400	Red to White AC	0.76-0.92
	Blue to Black AC	0.76-0.92
	Yellow to Yellow excitation/condenser	2.26-2.76
	Brown to Brown DC	0.27
RGX3500	Red to White AC	0.62-0.76
	Blue to Black AC	0.62-0.76
	Yellow to Yellow excitation/condenser	1.37-1.67
	Brown to Brown DC	0.22
RGX5500	Red to White AC	0.23-0.29
	Blue to Black AC	0.23-0.29
	Yellow to Yellow excitation/condenser	0.52-0.64
	Brown to Brown DC	0.14
ALL	Each wire to Ground	Infinity

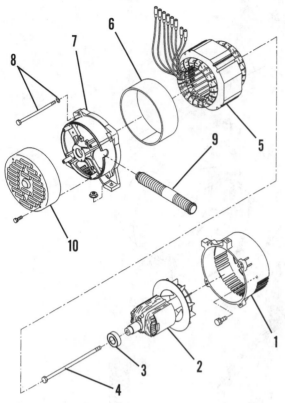

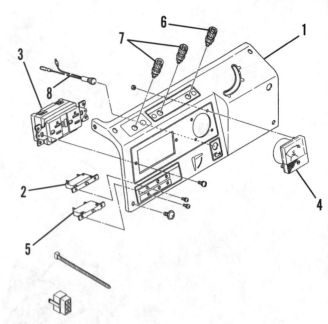

Fig. RB205 – Exploded view of model RGV4100 generator head components.

1. Engine adapter/front housing
2. Rotor assembly
3. Rotor bearing
4. Rotor throughbolt
5. Stator assembly
6. Stator cover
7. Rear housing
8. Rear housing/stator throughbolt (4)
9. Wiring harness protective cover tube
10. End cover

Fig. RB206 – Exploded view of model R1300 control panel components.

1. Control panel
2. AC circuit breaker
3. 120V AC GFCI receptacle
4. AC voltmeter
5. DC circuit breaker
6. DC (+) terminal
7. DC (-) terminal
8. Indicator lamp

If megger testing the stator, apply one megger lead to the White AC winding wire terminal, and apply the other lead to the lamination core. the megger should read a minimum of 1 MΩ. Repeat this step with at least one color coded wire terminal of every other winding.

The failure of any winding to test as specified indicates a faulty stator. Replace the stator.

7. The condenser is located in the rear housing on the R1300 Series, and in the control box on the RGD, RGV and RGX models. Some models have two condensers. They can be tested without removing them from the generator. To test each condenser:

NOTE: Even with the generator stopped, the capacitor could have a retained charge. Do not touch the capacitor terminals as electric shock may result. Always discharge the capacitor by shorting across the terminals with a screwdriver or similar tool with an insulated handle.

a. Unplug the condenser harness.
b. The recommended method of testing the condenser is by using the Dr. Robin tester, with the model capacitance as listed below.

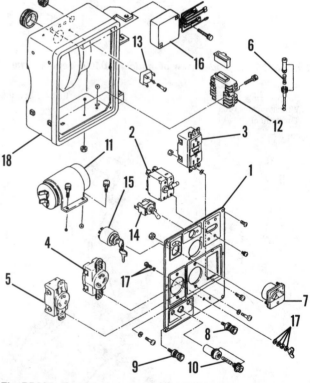

Fig. RB207 - Exploded view of model RGD3300 control box components. Model RGD2500 is similar, with fewer components.

1. Control panel faceplate
2. AC circuit breaker
3. 120V AC GFCI receptacle
4. 120V AC Twistlok receptacle
5. 125/250V AC Twistlok receptacle
6. AC fuse assembly
7. AC voltmeter
8. DC (+) terminal
9. DC (-) terminal
10. DC fuse assembly
11. Capacitor/condenser
12. Regulator
13. Diode stack assembly
14. Toggle switch
15. Ignition switch
16. Low-oil sensor
17. Ground terminal assembly
18. Control box

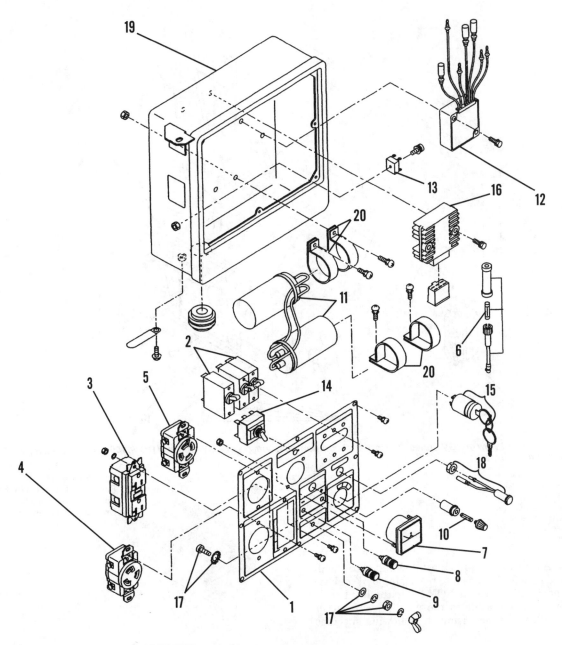

Fig. RB208 – Exploded view of model RGD5000 control box components.

1. Control panel faceplate
2. AC circuit breakers
3. 120V AC GFCI receptacle
4. 120V AC Twistlok receptacle
5. 240V AC Twistlok receptacle
6. AC fuse assembly
7. AC voltmeter

8. DC (+) terminal
9. DC (-) terminal
10. DC fuse assembly
11. Capacitors/condensers
12. Regulator
13. Diode stack assembly
14. Full-power switch

15. Ignition switch
16. Low-oil sensor
17. Ground terminal assembly
18. Indicator lamp
19. Control box
20. Clamp

R1300	10μF
RGD2500	10μF each
RGD3300	10μF each
RGD5000	30μF each
RGV2800	24μF
RGV4100	20μF each
RGV6100	28μF each
RGX2400	20μF
RGX3500	17μF each
RGX5500	28μF each

c. An acceptable alternate method of condenser checking is to replace the suspect condenser with a new or known good one. If the generator then performs to specification, the replaced condenser is faulty.

8. To service the rotor:

a. Test the winding resistance. Access the rotor winding connections. Disconnect/unsolder at least one winding end from one end of the diode and resistor, taking care (if unsoldering) to properly heat sink the diode and resistor to prevent burnout of good compo-

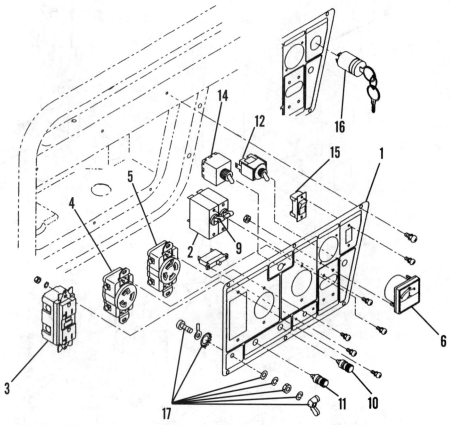

Fig. RB209A – Exploded view of RGV Series control panel components. View is of RGV4100 and RGV6100 components. RGV2800 is similar, with fewer components.

1. Control-panel faceplate
2. AC circuit breakers
3. 120V AC GFCI receptacle
4. 120V AC Twistlok receptacle
5. 240V AC Twistlok receptacle

6. AC voltmeter
9. DC circuit breaker
10. DC (+) terminal
11. DC (-) terminal
12. Idle-control switch

14. Full-power switch
15. Engine On-Off switch (recoil start)
16. Engine ignition switch (electric start)
17. Ground terminal assembly

nents. A small surgical hemostat clamped to the diode and resistor lead next to the solder joint makes a good heat sink. With the VOM set to the R × 1 scale, apply the meter leads to the winding connections and check resistance using the following specifications:

Model	Ohms
R1300	5.7-6.9
RGD2500	3.7
RGD3300	3.3
RGD5000	1.6
RGV2800	1.75
RGV4100	1.77
RGV6100	1.60
RGX2400	2.4-3.0
RGX3500	2.0-2.4
RGX5500	1.45-1.75

b. To megger test the rotor, apply one megger lead to one of the winding terminals, and apply the other lead to any ground on the rotor shaft or laminations. The megger should read a minimum of 1 MΩ.

c. To continuity test the rotor winding, apply one R × 1 VOM lead to one winding lead, and apply the other meter lead to any ground on the rotor shaft or laminations. There should be no continuity. Any continu-

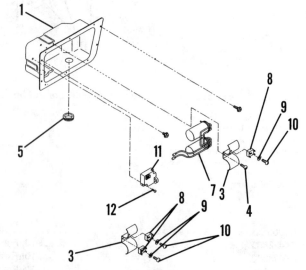

Fig. RB209B – Exploded view of RGV Series control box components.

1. Control-box
3. Clamps
4. Screw
5. Grommet
7. Capacitors/condensers

8. Diode stacks
9. Washer
10. Screw
11. Idle control switch
12. Screw

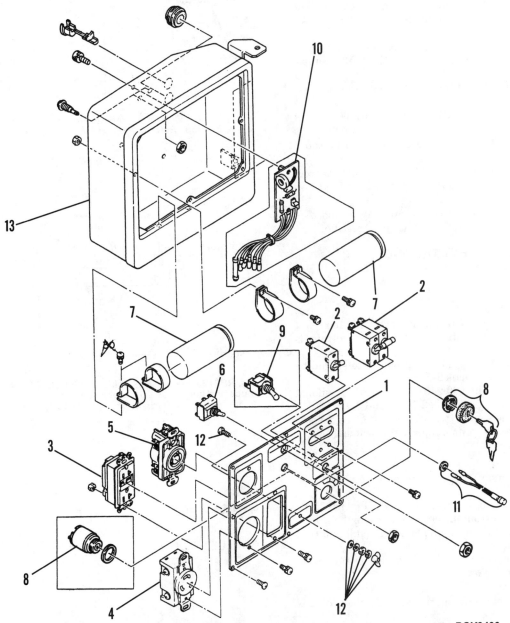

Fig. RB210 – Exploded view of RGX Series control box components. View is of RGX5500 components. RGX2400 and RGX3500 components are similar, except with smaller boxes and fewer components.

1. Control-panel faceplate
2. AC circuit breakers
3. 120V AC GFCI receptacle
4. 120V AC Twistlok receptacle
5. 240V AC Twistlok receptacle

6. Full-power switch
7. Capacitors/condensers
8. Engine ignition switch assembly
9. Idle-control switch

10. Idle-control unit
11. Pilot lamp
12. Ground terminal
13. Control box

ity reading in this test indicates a shorted rotor winding. replace the rotor.

d. To test the diode, connect the R × 1 meter leads to the diode terminals and note the reading, then reverse the leads, again noting the reading. A good diode will show continuity in only one direction. Note that some rotors have multiple diodes.

When replacing diodes on RGD5000 rotors, note the diode orientation prior to removal. place the white magnet mark to the left (Fig. RB212) and solder the

diode(s) in position so the cathode mark is at the bottom, then solder the coil ends to the terminal.

Repeated and frequent diode problems could indicate a faulty surge absorber resistor. proceed to the next test.

e. To test the resistor, first determine the resistors value from its markings. Test the resistor and compare its actual value to its rated value. Replace if beyond specification.

f. On rotors with permanent magnets attached, test the magnetism by holding a flat screwdriver blade approximately one inch (25 mm) from the magnet, letting the blade dangle while lightly holding the screwdriver handle. If the magnet attracts the blade, the magnet is good.

g. Failure of the rotor to test as specified indicates a faulty rotor. Replace the rotor or the faulty component.

9. If DC output is faulty, but the stator and rotor meet specification, test the diode rectifier:

a. Refer to Figs. RB213-RB215 for the correct rectifier diagram and test.

b. Using the R × 1 ohm scale, test the rectifier for continuity as noted.

c. If the test meter being used is reverse polarity from the recommended meter, the specified readings will be opposite those taken.

d. Failure of the rectifier to test completely as specified indicates a faulty rectifier. Replace the rectifier.

INDIVIDUAL COMPONENT TESTING

"Full Power" Voltage Selector Switch

To test the switch on units so equipped:

1. Access the switch terminals inside the control box.
2. Noting the wiring orientation, disconnect feed terminals 2, 5 and 8.
3. With the switch in the 120V position, there should be continuity between terminals 2-3, 5-6 and 8-9.
4. With the switch in the 120/240V position, there should be continuity between terminals 2-1, 5-4 and 8-7.
5. The failure of the switch to test as specified in Steps 3 and 4 indicate a faulty switch. replace the switch.

Oil Sensor

To test the sensor on gasoline-powered units so equipped:

1. Disconnect the sensor harness terminals (Fig. RB216).
2. Remove the sensor from the engine, then cap the sensor hole, making sure that the engine oil level is full and that the generator is sitting level.
3. Reconnect the sensor harness with the sensor resting on a rag or similar protector.
4. Start the engine.

a. If the engine stops within five seconds, the sensor is functioning properly.

b. If the engine fails to stop after 10 seconds, the sensor is faulty and must be replaced.

5. If the sensor met Step 4A specification, reinstall the sensor into the engine, then attempt to restart the engine: If the engine restarts, the sensor is OK. If the engine fails to restart, and the sensor wiring is OK, the sensor is faulty.

To test the system on diesel powered units so equipped:

1. Make sure that the engine oil level is full, and that the generator is sitting level.
2. Disconnect the two blue engine solenoid wires, and measure the resistance between the solenoid wire terminals (Fig. RB217). A good solenoid will test 235-290 ohms. Replace the solenoid if resistance is beyond these specifications.
3. Measure resistance between each blue wire terminal and the solenoid case ground. There should be infinity. Any continuity indicates a shorted solenoid winding.
4. Test the oil sensor unit:

a. Disconnect the black/yellow sensor lead.

b. Start the engine.

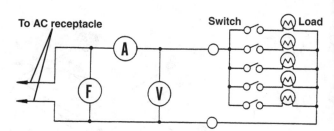

Fig. RB211 – View of a generator test load bank which can be made in shop. Refer to the text for construction and usage.

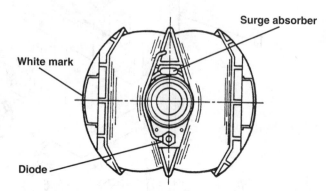

Fig. RB212 – End view sketch of model RGD5000 rotor showing diode and surge absorber placement relative to the white identification mark. Refer to the text for the full explanation.

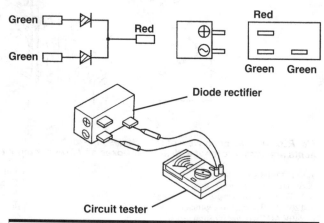

Analogue circuit tester		Apply black (-) needle of the circuit tester		
		Green	Green	Red
Apply red (+) needle of the circuit tester	Green		No continuity	No continuity
	Green	No continuity		
	Red	Continuity	Continuity	

Fig. RB213 – Diode rectifier continuity test for the model R1300 generator. Verify meter polarity or reverse meter leads.

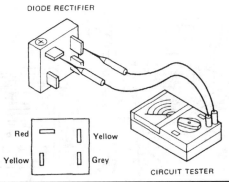

DIODE RECTIFIER

CIRCUIT TESTER

		Apply black (-) needle of the circuit tester			
		Yellow	Yellow	Red	Grey
Apply red (+) needle of the circuit tester	Yellow		No Continuity	No Continuity	Continuity
	Yellow	No Continuity		No Continuity	Continuity
	Red	Continuity	Continuity		Continuity
	Grey	No Continuity	No Continuity	No Continuity	

Fig. RB214 – Diode rectifier continuity test for models RGD2500 and RGD3300. Verify meter polarity or reverse meter leads.

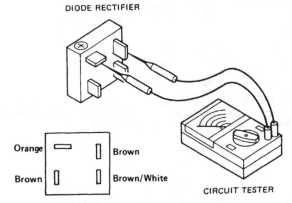

DIODE RECTIFIER

CIRCUIT TESTER

		Apply black (-) needle of the circuit tester			
		Brown	Brown	Orange	Brown/White
Apply red (+) needle of the circuit tester	Brown		No Continuity	No Continuity	Continuity
	Brown	No Continuity		No Continuity	Continuity
	Orange	Continuity	Continuity		Continuity
	Brown/White	No Continuity	No Continuity	No Continuity	

Fig. RB215 – Diode rectifier continuity test for model RGD5000, all RGV Series, and all RGX Series. Verify meter polarity or reverse meter leads.

c. Ground the black/yellow sensor lead to the engine crankcase.

d. If the Step C grounding activated the solenoid and stopped the engine, the oil sensor unit is functioning properly.

5. Test the pressure switch:

a. Disconnect the black/yellow sensor lead.

b. With the engine stopped and the ignition switch *off*, measure resistance between the engine oil pressure switch side of the black/yellow terminal and engine ground. Resistance should be infinity.

c. Start the engine, again measuring resistance. With the engine running, there should be continuity (0 ohms).

d. Any readings other than those specified in Steps B and C indicate a faulty engine switch.

Idle Control, RGV Series

1. Check the generator load start wattage. An overloaded generator will not return to rated rpm from idle. A motor powered tool, although within limits on running wattage, may require more starting wattage than the generator is rated to produce.

2. Inspect the angle of the solenoid mount bracket. The bracket should be positioned so that the solenoid plunger points directly at the spring hole in the throttle arm (Fig. RB218).

3. Disconnect and resistance test the two white solenoid wire terminals. Resistance should test 25-31 ohms. A resistance reading beyond this specification indicates a faulty solenoid.

4. Resistance test between each white solenoid wire terminal and the solenoid case. Resistance should be infinity. Any continuity reading in this test indicates a shorted solenoid winding.

5. Check the idle speed. With all loads disconnected from the generator and the idle control switch activated, idle speed should be 1900-2100 rpm on RGV2800 units or 2000-2200 rpm on RGV4100 and RGV6100 units. Idle speed should be adjusted with the throttle arm adjusting screw (Fig. RB219). If there is insufficient adjusting range with the screw, move the solenoid backwards.

6. Make sure that the main winding wires are passing through the idle control ZCT loop correctly (Fig. RB220).

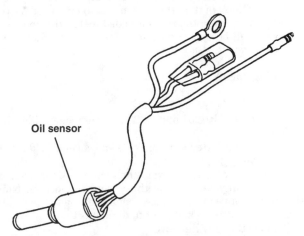

Oil sensor

Fig. RB216 – Typical oil sensor switch on gasoline powered generators. The number of wires may vary.

a. On 120V units, any one main winding wire can pass through the loop.

b. On 120/240V units, the red and the black main winding wires *must* pass through the loop from the same direction, or the idle control unit will not function.

7. Using the diagram and chart in Fig. RB221, resistance test the five terminals on the back of the idle control unit. Failure of any terminal pair to test to specification indicates a faulty control unit. Note that if the test meter being used is reverse polarity from the recommended meter, the specified readings will be opposite those taken. Reverse the meter leads and retest.

Idle Control, RGX Series

1. Check the generator load start wattage. An overloaded generator will not return to rated rpm from idle. A motor powered tool, although within limits on running wattage, may require more starting wattage than the generator is rated to produce.

2. Inspect the angle of the solenoid mount bracket (Fig. RB222). bend as necessary to correct.

3. Disconnect and resistance test the two light blue solenoid wire terminals. Resistance should test 235-290 ohms. A resistance reading beyond this specification indicates a faulty solenoid.

4. Resistance test between each light blue solenoid wire terminal and the solenoid case. Resistance should be infinity. Any continuity reading in this test indicates a shorted solenoid winding.

5. Start the generator (idle control switch *off*) and check the idle speed. With all loads disconnected from the generator and the idle control switch activated, idle speed should drop to 3150-3200 rpm on the RGX2400 and RGX 3500 units or 2700-2800 rpm on the RGX5500 units. Idle speed should be adjusted with the throttle arm adjusting screw (Fig. RB219). At the specified rpm, the AC voltmeter should indicate 75-85 volts.

6. Make sure that the main winding wires are passing through the idle control ZCT loop correctly (Fig. RB223).

a. On 120V units, any one main winding wire can pass through the loop.

b. On 120/240V units, the red and the black main winding wires *must* pass through the loop from the same direction, or the idle control unit will not function.

7. Using the diagram and chart in Fig. RB224, resistance test the four idle control unit leads. Failure of any terminal pair to test to specification indicates a faulty control unit. Note that if the test meter being used is reverse polarity from the recommended meter, the specified readings will be opposite those taken. Reverse the meter leads and retest.

DISASSEMBLY

R1300 Series

Refer to Figs. RB201 and RB206 for component identification.

1. Remove the slotted rear and side end cover shields from the generator.

2. Remove the control lever knob. Remove the control panel, carefully unplugging the wiring connectors. Note that the 6-pin stator harness connector has a lock tab which must be unlocked prior to disconnection.

3. Remove the air filter cover.

4. Remove the front cover shield fasteners, then pull the shield a short distance from the generator.

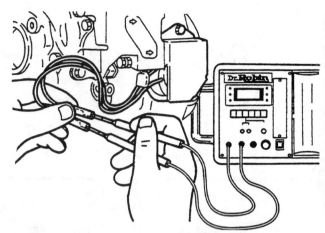

Fig. RB217 – *Testing the oil sensor solenoid on diesel powered generators. Refer to the text for the correct test procedure.*

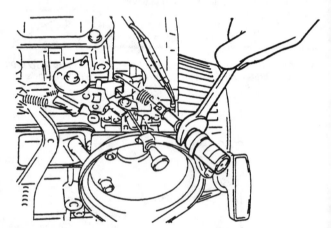

Fig. RB218 – *Inspecting the idle down solenoid mount angle and adjusting the solenoid position on the RGV Series generators. Refer to the text for the correct procedures.*

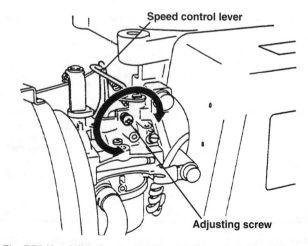

Fig. RB219 – *Adjusting the RGV and RGX series idle speed with the carburetor throttle lever screw.*

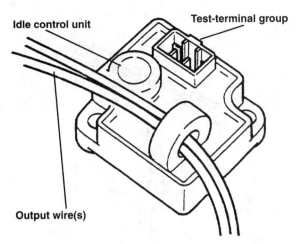

Fig. RB220 – Main winding wire(s) must pass through the ZCT loop on the RGV Series idle control unit. On dual voltage units, the two wires (black and red) must pass through in the same direction or the idle control will not function.

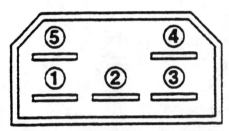

Terminal number of the Idle Control Unit

Circuit tester (with battery power source 1.5V)		Apply black (-) needle of the circuit tester				
		1	2	3	4	5
Apply red (+) needle of the circuit tester	1		∞	250 kΩ	250 kΩ	75 kΩ
	2	∞		∞	∞	∞
	3	250 kΩ	∞		250 kΩ	75 kΩ
	4	250 kΩ	∞			75 kΩ
	5	8.5k Ω	∞	7.8 kΩ	7.8 kΩ	

Fig. RB221 – Terminal resistance test for the RGV Series idle down control unit.

5. With the fuel valve *off*, remove the fuel strainer cup, then open the fuel valve and drain the fuel into an approved container. Unclamp the hoses from the fuel valve. Remove the front shield.

6. Disconnect the breather tube from the fuel tank fitting, then remove the fuel tank.

7. Remove the X-shaped fuel tank/shield bracket from the generator rear housing.

8. Remove the end cover.

9. Noting the position of the cable tie securing the rear housing harness, cut the tie, then disconnect and remove the condenser and the diode rectifier from the rear housing. Disconnect the harness ground cable from the rear housing.

10. If necessary, remove the two nuts holding the rear housing feet to the rubber mounts.

11. Place a temporary support, such as a wooden block, under the engine adapter/generator front housing.

12. Remove the three rear housing to front housing bolts.

13. Gently and evenly tap the rear housing with a plastic or rubber mallet to break the housing loose from the stator. remove the housing. If necessary, remove the rubber mounts.

14. Gently remove the stator cover, then lightly tap the stator lamination core with a plastic or rubber mallet to loosen the stator from the front housing. Remove the stator.

15. Remove the rotor throughbolt, then apply a small amount of penetrating oil through the rotor shaft onto the engine crankshaft taper.

16. Using a slide hammer with the proper threaded adapter, break the rotor free from the engine crankshaft. remove the rotor.

17. If necessary, unbolt and remove the front housing from the engine.

RGD and RGX Series

Refer to Figs. RB202, RB203, RB207, RB208, and RB210 for component identification.

1. On electric start models, disconnect the battery cables. Remove the battery.

2. Drain and remove the fuel tank.

3. Remove the harness protector tube ends from the generator and control box. Disconnect the control panel harness. Remove the control panel and control box.

4. Remove the exhaust system.

5. If necessary, remove the nuts holding the mounting feet to the rubber mounts, then lift the engine/generator assembly and remove it from the cradle frame. Otherwise, proceed to Step 6.

6. Remove the two nuts holding the generator rear housing feet to the rubber mounts.

7. Place a temporary support, such as a wooden block, under the engine adapter/generator front housing so that the rear housing feet clear the rubber mount studs.

8. On models equipped with an end cover plate, remove the plate. On RGD5000 models, remove the recoil starter assembly from the rear housing.

9. Remove the rotor throughbolt. On the RGD5000, also remove the recoil starter pulley assembly.

10. Remove the three or four rear housing-to-front housing bolts, carefully noting the position of any harness clamps or guides.

11. On all except RGV models, the recommended method of rear housing removal is by using a puller such as that pictured in Fig. RB225. On RGV models, and on all other models, if not using the puller, alternately tap on the bottom feet and top casting boss with a plastic mallet to loosen the rear housing from the stator. Remove the housing.

12. On RGD and RGX models, remove the four stator retaining bolts. With a wood block inserted vertically through the stator as shown in Fig. RB226, tap the rear housing bearing support on the block until the stator and stator support ring break free from inside the rear housing. Hold your fingers under the stator, as shown, to prevent the stator from dropping suddenly, damaging the stator windings. Note the stator wiring orientation and then remove the wires.

13. Using a slide hammer with the proper threaded adapter, break the rotor free from the engine crankshaft or driving shaft. Remove the rotor.

14. If necessary, unbolt and remove the front housing from the engine.

RGV Series

Refer to Figs. RB204, RB205, and RB209 for component identification.

1. On electric start models, disconnect the battery cables. Remove the battery.

2. Remove the harness protector tube ends from the generator and control box. Disconnect the control panel harnesses from the generator and engine, then remove the control box assembly.

3. Drain and remove the fuel tank.

4. Remove the muffler assembly.

5. On the RGV4100, remove the end cover.

6. Remove the two nuts holding the generator rear housing feet to the rubber mounts.

7. Place a temporary support, such as a wood block, under the engine adapter/generator front housing so that the rear housing feet clear the rubber mount studs.

8. Remove the four rear housing to front housing bolts. Note the stator wiring orientation. Remove the rear housing from the stator by lightly tapping against the mount feet with a plastic mallet.

9. Slide the stator cover off the stator, then remove the stator.

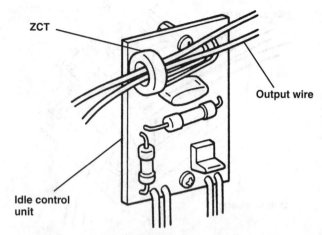

Fig. RB223 – Main winding wire(s) must pass through the ZCT loop on the RGX series idle control unit. On dual voltage units, the two wires (Black and Red) must pass through in the same direction or the idle control will not function.

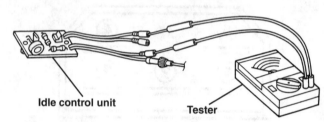

Tester Polarity (+) \ (-)	White	Light Blue (Fuse)	Light Blue	Red
White		Above 190 kΩ	Above 85 kΩ	Above 85 kΩ
Light Blue (Fuse)			∞	∞
Light Blue	20-50 kΩ	2-16 kΩ		0
Red	20-50 kΩ	2-16 kΩ		

Fig. RB224 – Terminal resistance test for the RGX Series idle down control unit.

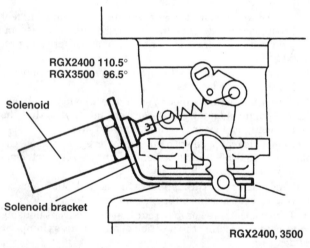

RGX2400 110.5°
RGX3500 96.5°

Solenoid

Solenoid bracket

RGX2400, 3500

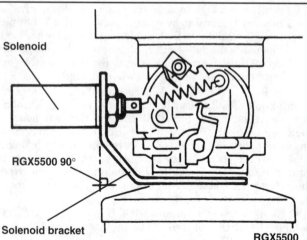

Solenoid

RGX5500 90°

Solenoid bracket

RGX5500

Fig. RB222 – View of idle down solenoid mount bracket angles on RGX series generators.

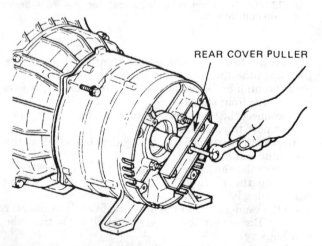

Fig. RB225 – On some models, a puller may be used to remove the rear housing

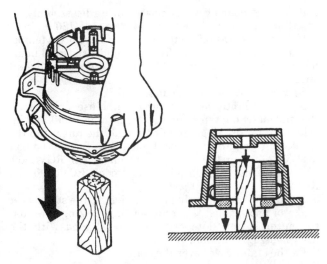

Fig. RB226 – Recommended stator removal method for RGD and RGX Series units is by lightly tapping the bearing support on a wood block inserted into the housing. Do not allow the stator to drop when it breaks free.

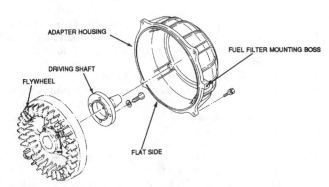

Fig. RB227 – Exploded view of flywheel and driving shaft used on RGD5000 units.

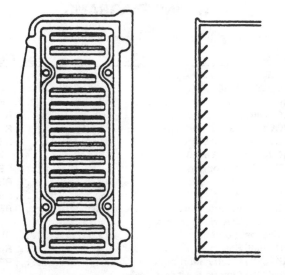

Fig. RB228 – Make sure that the louvered outlet panel on applicable RGD and RGX Series units is installed with the louvers facing upward and inward.

10. Remove the rotor throughbolt, then squirt a small amount of penetrating oil through the rotor shaft onto the engine crankshaft taper.

11. Using a slide hammer with the proper threaded adapter, break the rotor free from the engine crankshaft. remove the rotor.

12. If necessary, unbolt and remove the front housing from the engine.

REASSEMBLY

R1300 Series

Refer to Figs. RB201 and RB206 for component identification.

1. If the front housing was removed during disassembly, install it first. Torque the fasteners 12-14 N•m (105-120 in.-lb.).

2. Place a temporary support, such as a wooden block, under the front housing.

3. Service the rotor bearing, if needed.

4. Be sure the rotor magnets are clean. If used, make sure that the matching tapers on the engine crankshaft and inside the rotor are clean and dry. Install the rotor onto the crankshaft. Install the rotor throughbolt, and torque the bolt 12-14 N•m (105 120 in.- lb.).

5. Install the stator, stator cover, and rear housing assembly. Using a plastic or rubber mallet, lightly tap around the rear edge of the housing to seat the housing and stator against the front housing and over the rotor bearing. Insure that the stator wires are properly routed through the rear housing opening and are not pinched or kinked.

6. Install the three rear housing to front housing bolts. Torque the bolts to 5.5-7.5 N•m (50-65 in.-lb.).

7. Remove the engine spark plug and rotate the crankshaft to check for interference between the rotor and stator. There should be a consistent gap around the rotor laminations inside the stator. If any rubbing or scraping is evident, correct any misalignment prior to proceeding. Reinstall the spark plug.

8. Install the rear housing harness with the grommet in its proper slot.

9. Install and connect the condenser and the diode rectifier. Reconnect the harness ground cable.

10. Install the end cover.

11. Install the X-shaped fuel tank/shield bracket onto the rear housing.

12. Install the fuel tank, making sure to properly reconnect the breather tube to the tank fitting.

13. Reconnect the fuel hoses to the fuel valve inside the front cover shield, then install the shield.

14. Install the air filter cover.

15. Reconnect the control panel wiring, making sure that the lock tab on the 6-pin connector is properly secured. Install the control panel, making sure that no wires are pinched or kinked. Install the control lever knob.

16. Install the rear and side end cover shields to the generator.

RGD and RGX Series

Refer to Figs. RB202, RB203, RB207, RB208, and RB210 for component identification.

1. On RGD5000 units:

    a. If driving shaft (Fig. RB227) was removed, assemble the shaft to the flywheel. Torque the bolts to 55-70 N•m (40-50 ft.-lb., or 480-600 in.-lb.).

    b. Install the adapter housing with the flat side down and the fuel filter mounting boss on the air cleaner side of the engine. Torque the housing bolts to 20-23 N•m (180-205 in.-lb., or 15-17 ft.-lb.).

2. On all units except RGD5000:

    a. If the louver plate was removed, reinstall it with the louvers facing upward and inward (Fig. RB228).

    b. If the adapter housing was removed, install it and torque the bolts 12-14 N•m (105-120 in.-lb., or 9-10 ft.-lb.).

3. If the engine was left in the cradle frame, place a temporary support, such as a wood block, under the front housing so that the rear housing feet clear the rubber mount studs.

4. Service the rotor bearing, if needed.

5. Be sure the rotor magnets are clean. If used, make sure that the matching tapers on the engine crankshaft and inside the rotor are clean and dry. Install the rotor onto the crankshaft. Install the rotor throughbolt. Torque the bolt 11.5-13.5 N•m (100-115 in.-lb.) on models RGD2500 and RGX2400, and torque the bolt 22.6-24.5 N•m (200-215 in.-lb.) on models RGD 3300, RGX3500, and RGX5500. On RGD5000 units, install the rotor throughbolt finger tight at this time.

6. Set the rear housing on the workbench vertically, stator side up.

7. Feed the stator wires through the proper housing opening while sliding the stator down into the rear housing. Insert the four stator bolts into the housing to insure that the stator slots are aligned with the housing, then remove the bolts.

8. Place the support ring around the stator with the holes *up* and aligned with the flat sides of the stator laminations.

9. Using an aluminum or brass punch, tap around the support ring evenly to press the ring into the rear housing.

10. Install and torque the four stator bolts to 8-10 N•m (70-88 in.-lb.).

11. If the harness cover was removed, reinstall it, making sure that the small end fits into the rear housing. Feed the stator wires through the cover.

12. Install the stator/rear housing assembly over the rotor, aligned with the front housing. Use a plastic or rubber mallet to lightly and evenly tap the rear housing until it is snug against the front housing. Install the housing bolts. Torque the bolts to 5.0-6.0 N•m (40-55 in.-lb.).

13. On all models except RGD5000, install the end cover, then torque the screws to 5.0-6.0 N•m (40-55 in.-lb.).

14. On RGD5000 model, remove the rotor throughbolt. Install the starter pulley spacer, key, and pulley, then reinstall the throughbolt. Torque the bolt to 24-30 N•m (210-260 in.-lb., or 17.5-22 ft.-lb.). Install the recoil starter, torquing the bolts 4.0-6.0 N•m (35-55 in.-lb.).

15. Lift the generator rear housing slightly to remove the temporary front housing support. Lower the rear housing feet onto the rubber mounts. Install the mount nuts, torquing the nuts to 12-14 N•m (105-120 in.-lb.).

16. Install the exhaust system.

17. Install the control box and control panel. Feed the control box wires through the proper hole, connecting the control harness terminals to the generator terminals. Reconnect the end of the harness cover into the control box hole.

18. Install the fuel tank.

19. On electric start models, install and reconnect the battery.

RGV Series

Refer to Figs. RB204, RB205, and RB209 for component identification.

1. If the front housing was removed during disassembly, reinstall it. Torque the bolts to 12-14 N•m (105-120 in.-lb.).

2. If the engine was left in the cradle frame, place a temporary support, such as a wooden block, under the front housing so that the rear housing feet clear the rubber mount studs.

3. Service the rotor bearing, if needed.

4. Be sure the rotor magnets are clean. If used, make sure that the matching tapers on the engine crankshaft and inside the rotor are clean and dry. Install the rotor onto the crankshaft. Install the rotor throughbolt. Torque the bolt to 11.5-13.5 N•m (100-115 in.-lb.) on model RGV2800, 22.5-24.5 N•m (200-235 in.-lb.) on models RGV4100 and RGV 6100.

5. Noting the correct wiring orientation, fit the stator over the rotor and against the front housing, making sure that the outer stator slots are properly aligned with the front housing bolt holes.

6. Fit the stator cover over the stator.

7. Install the rear housing onto the rotor bearing and stator. Lightly and evenly tap the rear housing forward until it is snug against the stator, and the stator is snug against the front housing.

8. Install the three or four rear housing to front housing bolts. Torque the bolts to 4.5-6.0 N•m (40-55 in.-lb.).

9. Lift the rear housing. Remove the temporary front-housing support. Set the rear housing feet down onto the rubber mounts. Install the mount nuts, torquing the nuts to 12-14 N•m (105-120 in.-lb.). Make sure that the ground cable is properly reconnected between the left rear housing foot and the cradle frame.

10. On the RGV4100, install the end cover.

11. Install the muffler assembly. Torque the 6 mm fasteners 8-10 N•m (70-90 in.-lb.). Torque the 8 mm carbon steel fasteners to 19-25 N•m (165-215 in.-lb.). Torque the 8 mm stainless-steel fasteners to 22-28 N•m (190-245 in.-lb.).

12. Install and reconnect the fuel tank.

13. Install the control box assembly. Fit the harness cover tube over the stator wiring, fastening the small tube end into the rear housing hole. Connect the control box wiring to the stator wiring and the engine wiring. Then fit the upper end of the harness cover tube into the control box hole.

14. On electric start models, install and reconnect the battery.

WIRING DIAGRAMS

The wiring diagrams are listed as follows:

The symbols used on the wiring diagrams are decoded as follows:

Symbol	Component
MC	Main coil; AC winding
SC	Sub coil; Auxiliary or excitation winding
DC	DC winding
FC	Field coil (Rotor) winding
C	Condenser
D	Diode rectifier; diode stack
L_1	Pilot lamp
L_2	Oil sensor warning lamp
T	DC output terminal
F	Fuse
C.B.	Circuit breaker
NFB_1	No fuse breaker
NFB_2	No fuse breaker
VC SW	Voltage changeover switch
FP SW	Full power switch
S SW	Engine stop switch
OS	Oil sensor
OSC	Oil sensor controller
SIU	Solid state ignition unit
SP	Spark plug
MG	Magneto
IG	Ignition coil
MG. SW	Magnetic switch; relay
SOL	Solenoid
IC SW	Idle control switch
S.D.	Idle down control unit
E	Earth/ground terminal
ST. M	Starter motor
KEY SW	Key switch
BAT	Battery
V	Voltmeter
REC_1	AC output receptacle (Total 15A max.)
REC_2	AC output receptacle (240V)
REC_3	AC output receptacle (120V)

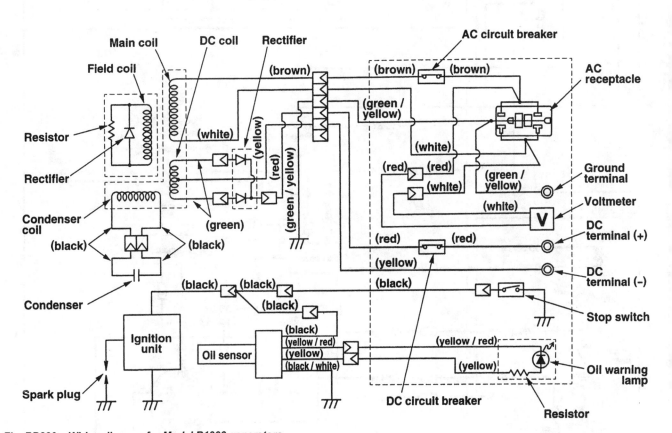

Fig. RB230 – Wiring diagram for Model R1300 generators.

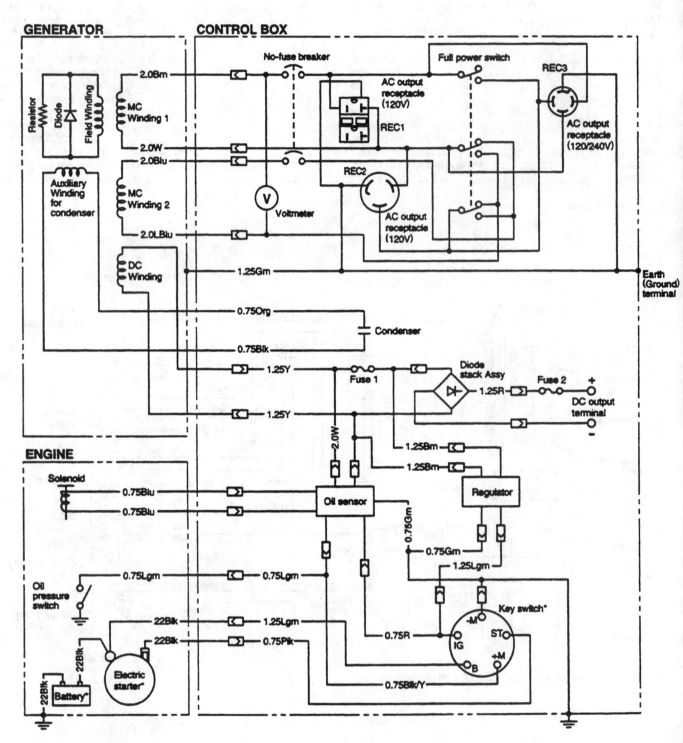

GENERATOR

CONTROL BOX

ENGINE

*Only for RGD3300.

Fig. RB229 – Wiring diagram for generator models RGD2500 and RGD3300.

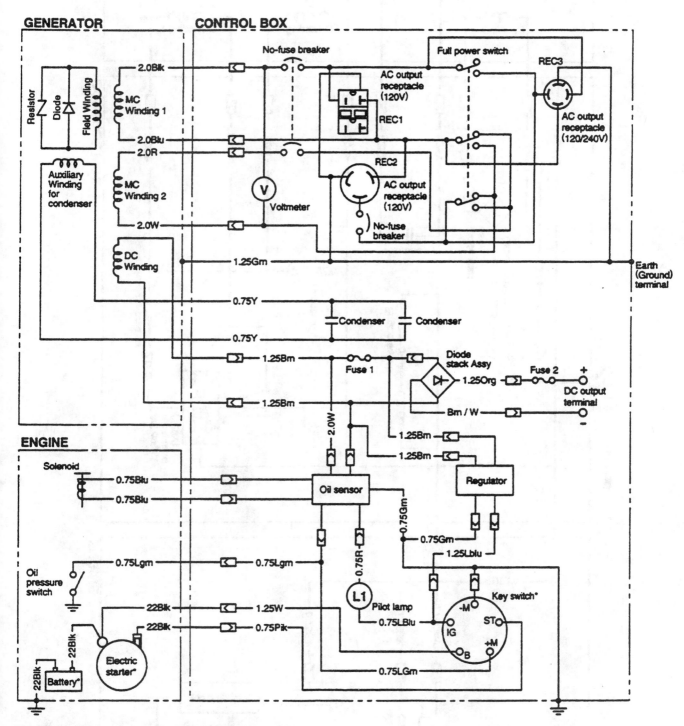

GENERATOR

CONTROL BOX

Resistor

Diode

Field Winding

MC Winding 1

2.0Blk

No-fuse breaker

Full power switch

REC3

AC output receptacle (120V)

REC1

AC output receptacle (120/240V)

2.0Blu

2.0R

MC Winding 2

REC2

AC output receptacle (120V)

Auxiliary Winding for condenser

Voltmeter

V

2.0W

No-fuse breaker

DC Winding

1.25Gm

Earth (Ground) terminal

0.75Y

Condenser

Condenser

0.75Y

1.25Bm

Fuse 1

Diode stack Assy

1.25Org

Fuse 2

+

DC output terminal

1.25Bm

Bm / W

−

ENGINE

1.25Bm

1.25Bm

Solenoid

0.75Blu

0.75Blu

Oil sensor

Regulator

0.75Gm

Oil pressure switch

0.75Lgm

0.75Lgm

0.75Gm

1.25Lblu

2.0W

0.75R

22Blk

1.25W

L1

Pilot lamp

Key switch°

−M

ST

22Blk

0.75Pik

0.75LBlu

IG

22Blk

Electric starter°

+M

B

22Blk

Battery°

0.75LGm

Fig. RB231 – Wiring diagram for Model RGD5000 generators.

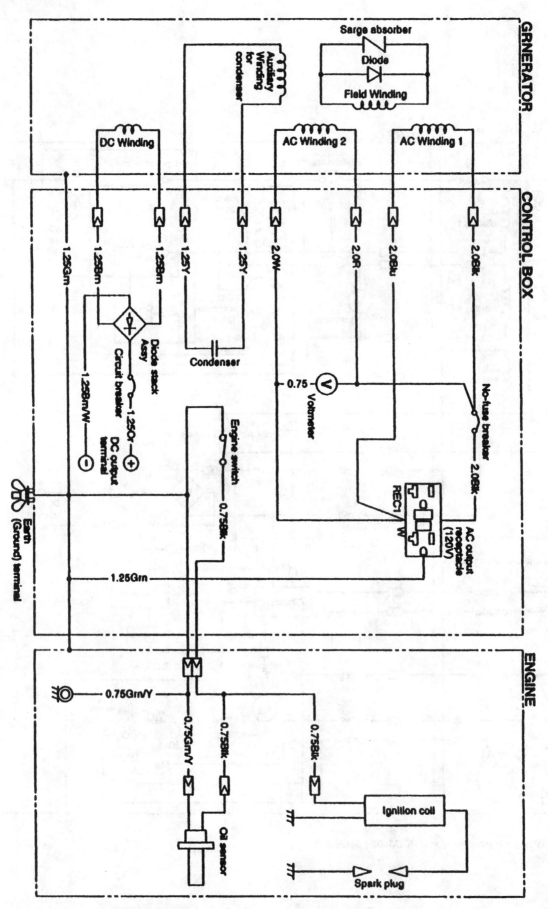

Fig. RB232 – Wiring diagram for Model RGV2800 generators.

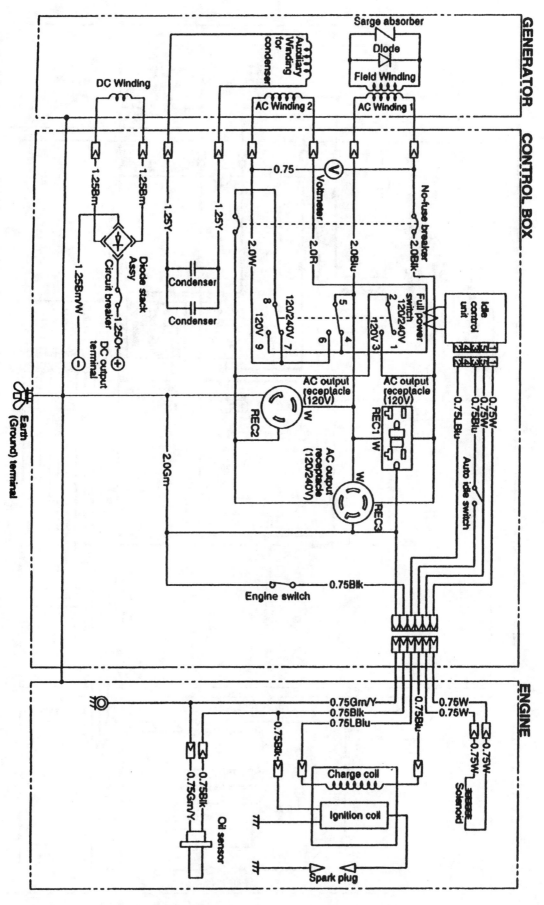

Fig. RB233 – Basic wiring diagram for Model RGV4100 generators.

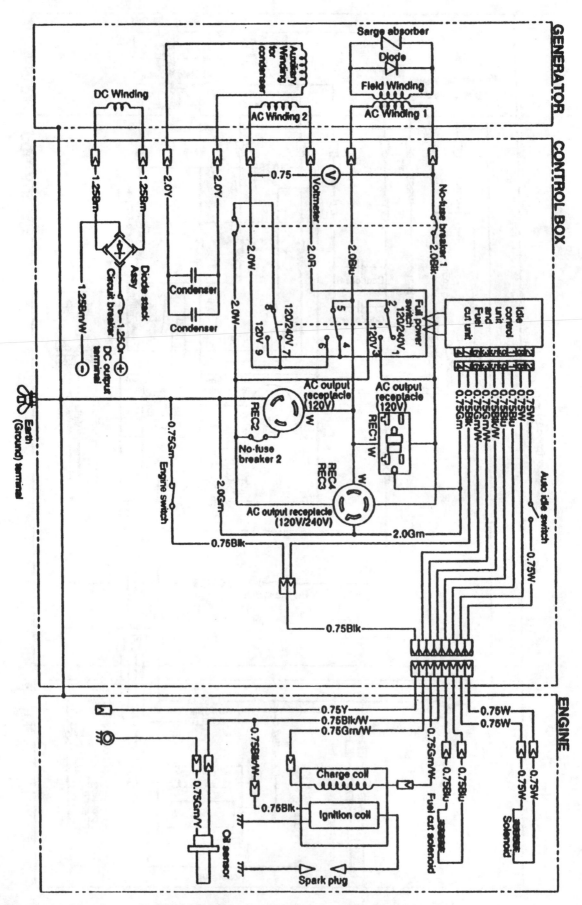

Fig. RB234 – Basic wiring diagram for Model RGV6100 generators.

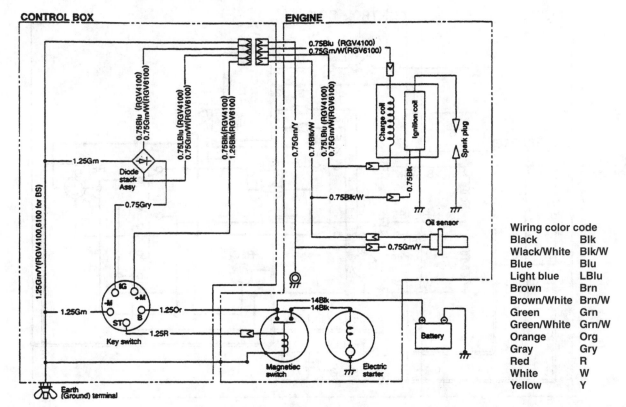

CONTROL BOX

ENGINE

Wiring color code
Black	Blk
Wlack/White	Blk/W
Blue	Blu
Light blue	LBlu
Brown	Brn
Brown/White	Brn/W
Green	Grn
Green/White	Grn/W
Orange	Org
Gray	Gry
Red	R
White	W
Yellow	Y

Fig. RB235 – Circuit wiring diagram for the electric start option on generator Models RGV4100 and RGV6100.

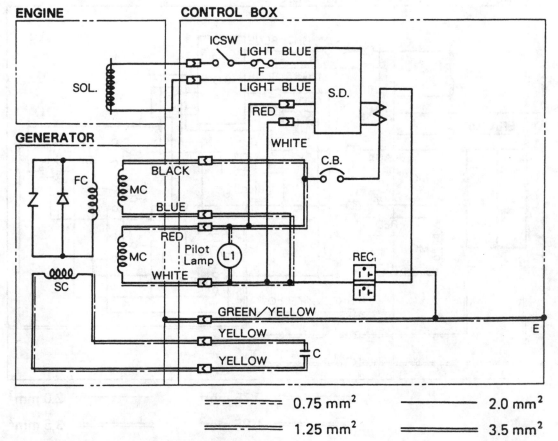

ENGINE

CONTROL BOX

GENERATOR

0.75 mm²   2.0 mm²

1.25 mm²   3.5 mm²

Fig. RB236 – Basic wiring diagram for Model RGX2400 generators.

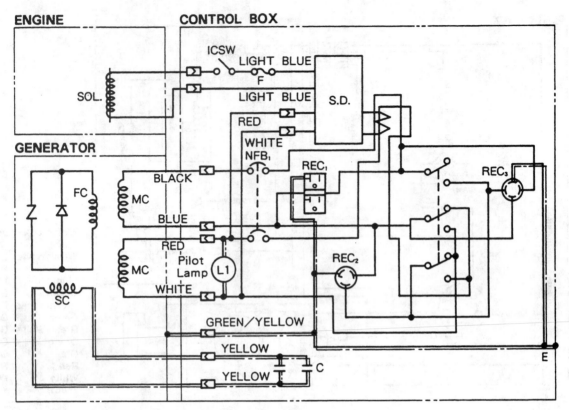

Fig. RB237 – Basic wiring diagram for Model RGX3500 generators.

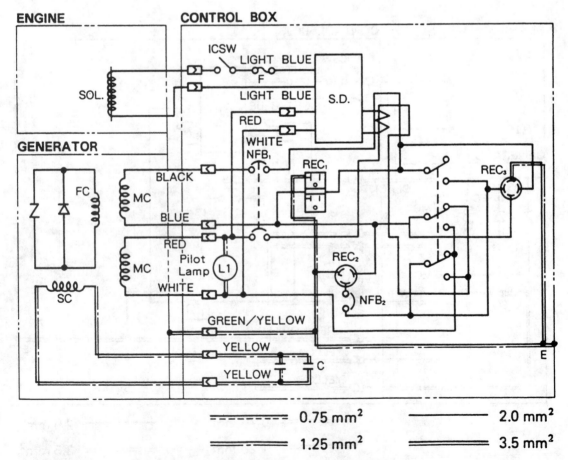

Fig. RB238 – Basic wiring diagram for Model RGX5500 generators.

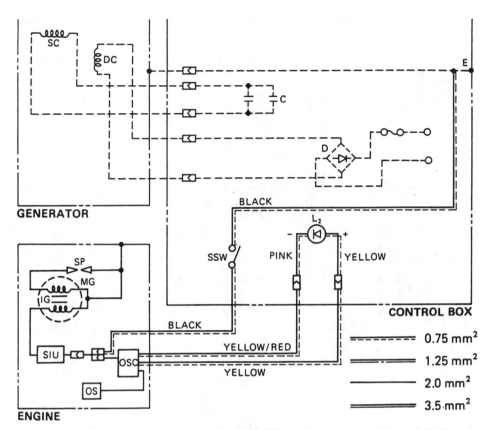

GENERATOR

CONTROL BOX

- - - - - - 0.75 mm²
———— 1.25 mm²
———— 2.0 mm²
———— 3.5 mm²

ENGINE

Fig. RB239 – Circuit wiring diagram for the oil sensor option on the RGX Series generators with recoil start.

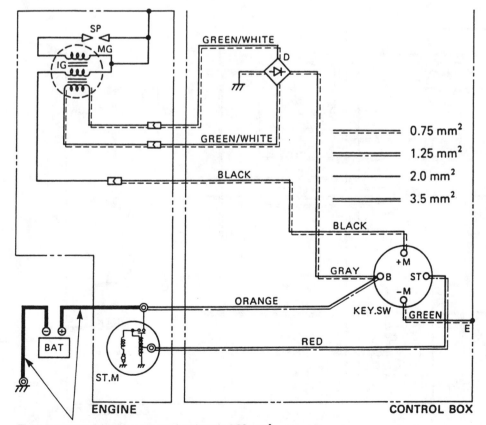

- - - - - - 0.75 mm²
———— 1.25 mm²
———— 2.0 mm²
———— 3.5 mm²

ENGINE

CONTROL BOX

The battery cords have a cross sectional area of 22 mm².

Fig. RB240 – Circuit wiring diagram for the electric start option on model RGX3500 generators without oil sensor.

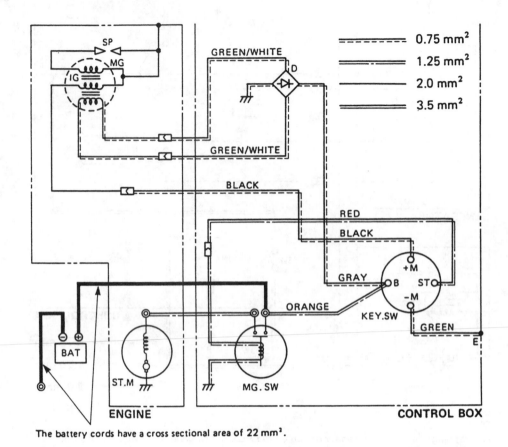

The battery cords have a cross sectional area of 22 mm².

Fig. RB241 – Circuit wiring diagram for the electric start option on model RGX5500 generators without oil sensor.

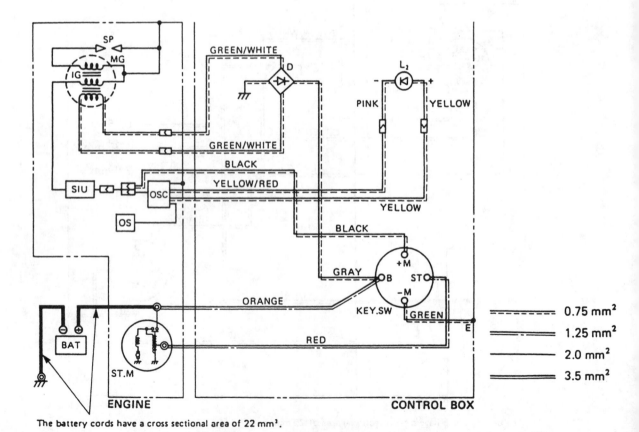

The battery cords have a cross sectional area of 22 mm².

Fig. RB242 – Circuit wiring diagram for model RGX3500 generators for the combined oil sensor and electric start options.

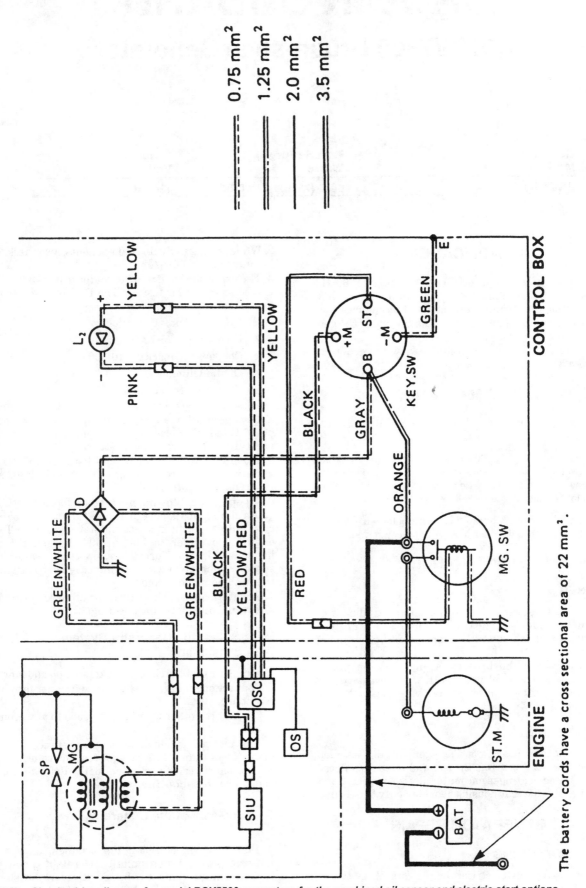

Fig. RB243 – Circuit wiring diagram for model RGX5500 generators for the combined oil sensor and electric start options.

ROBIN SUBARU
RGV7500 Brush-style Generator

Generator Model	Rated (Surge) Watts	Output Volts
RGV7500	6000 (7300) AC/100 DC	120/240 AC/12 DC

IDENTIFICATION

These generators are direct drive, brush style, 60 Hz, two pole single phase units with electronic automatic voltage regulation (AVR) and panel mounted receptacles.

The serial number plate is located on the end of the generator rear housing assembly.

Refer to Figs. RB301 and RB302 to view the generator components.

WIRE COLOR CODE

The wiring diagram color code for these generators is as follows:

Blk – Black	Blu – Blue	LBlu – Light blue
Brn – Brown	Grn – Green	Org – Orange
Gry – Gray	R – Red	W – White
LGrn – Light green	Y – Yellow	

A wire with a slash separated ID is a 2-color wire. For example: R/W is a red wire with a white tracer stripe.

MAINTENANCE

BRUSHES

Replace brushes when the brush length protruding from the brush holder reaches 5.0 mm (0.2 in.) or less. New brush length should be 15.0 mm (0.6 in.). Refer to Fig. RB303.

When servicing the brushes, always observe proper brush polarity: The light green AVR wire always goes to the positive (+) brush. The brown AVR wire goes to the negative (-) brush.

ROTOR BEARING

The end housing bearing on these generators is a sealed, pre lubricated ball bearing. Inspect and replace as necessary. If a seized bearing is encountered, also inspect the rotor journal and the rear housing bearing cavity for scoring.

GENERATOR REPAIR

TROUBLESHOOTING

1. Before testing any electrical components, check the integrity of the terminals and connectors.

2. While testing individual components, also check the wiring integrity and fuse(s).

3. The use of the Dr. Robin generator tester #388-47565-08 is recommended for testing. The Dr. Robin tester has the following capabilities:

   a. AC voltmeter – 0-500 volts.

   b. Frequency – 25-70 Hz.

   c. Condenser capacity – 10-100 µF.

   d. Digital Ohmmeter – 0.1-1,999 ohms.

   e. Insulation resistance – 3 MΩ.

An accurate engine tachometer will also be required.

4. Test specifications are given for an ambient temperature of 20° C (68° F). Resistance readings may vary with temperature and test equipment accuracy.

5. For access to the components listed in this **Troubleshooting** section, refer to Figs. RB301 and RB302 and the later **Disassembly** section.

6. All operational testing must be performed with the engine running at its governed no load speed of 3700-3750 rpm, unless specific instructions state otherwise. To adjust engine rpm, refer to Fig. RB304: Loosen the adjusting screw locknut. Turn the screw clockwise to decrease speed or counterclockwise to increase speed. Tighten the locknut and recheck rpm.

7. Always load test the generator after repairing defects discovered during troubleshooting.

Megger Test the Generator System Wiring Insulation

1. Make sure that the circuit breakers are *on* and functioning.

2. Connect one megger test probe to any hot terminal of a receptacle.

3. Connect the other megger test probe to any ground terminal.

4. The megger should indicate a resistance of 1 MΩ or more. if less than 1 MΩ, megger test the rotor, stator, and control panel resistance individually. Refer to the subsequent **Troubleshooting** sections.

To Troubleshoot the Generator

NOTE: When servicing the brushes, always observe proper brush polarity. The light green AVR wire always goes to the positive (+) brush. the brown AVR wire goes to the negative (-) brush.

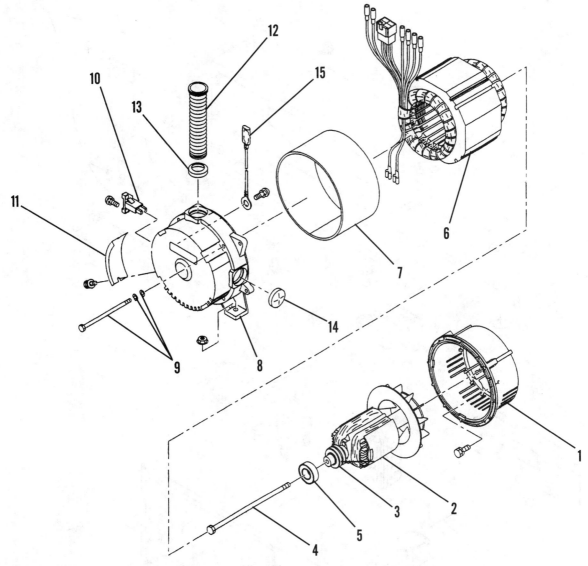

Fig. RB301 – Exploded view of Model RGV7500 generator head components.

1. Engine adapter/front housing
2. Rotor assembly
3. Sliprings
4. Rotor throughbolt
5. Rotor bearing
6. Stator assembly

7. Stator cover
8. Endbell/rear housing
9. Rear housing.stator
   throughbolt assembly (4)
10. Brush holder assembly
11. Brush cover

12. Wiring harness
    protective cover tube
13. Tube flange seal
14. Plug
15. Ground wire

1. Circuit breakers must be *on*. To test either the AC or DC breakers, access the breaker wiring and continuity test the breaker terminals with the generator shut down and at least one breaker feed wire disconnected. The breaker is faulty if it will not stay set, shows continuity in the OFF position or does not show continuity in the ON position.

2. Receptacles must be functioning. To test the receptacle(s):

   a. Access the receptacle connections.

   b. Disconnect the *return* feed wire from the receptacle being tested.

   c. Install a jumper wire between the *hot* and *return* outlet terminals.

   d. Test for continuity across the feed terminals. No continuity indicates a faulty receptacle.

   e. Remove the jumper wire and retest for continuity across the feed terminals. No continuity this time indicates a good receptacle.

3. Ground fault circuit interrupters (GFCIs) must be *reset* and functioning. To test the GFCI:

   a. Start the generator and allow a brief no-load warm up period.

   b. Press the *test* button. The *reset* button should extend. If it does, proceed to Step C. If the *reset* button does not extend, the GFCI is faulty or it is not receiving input from the generator. If the generator does not have proper output for testing the GFCI, disconnect the GFCI feed wires. Connect a wall socket jumper to the GFCI feed terminals, then test the GFCI.

   c. Depress the *reset* button. The *reset* button should then be flush with the *test* button. If the *reset* button

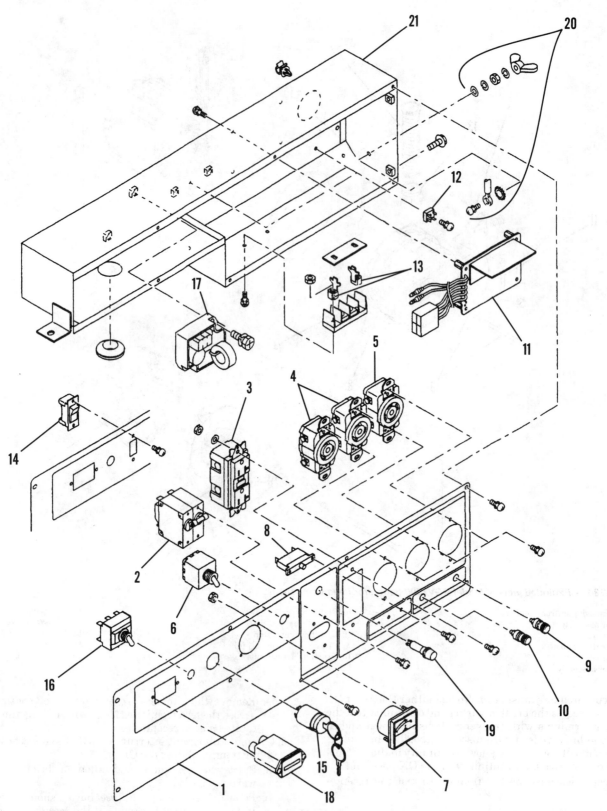

Fig. RB302 – Exploded view of Model RGV7500 control box components.

1. Control panel faceplate
2. AC circuit breaker
3. 120V AC GFCI receptacle
4. 120V AC Twistlok receptacle
5. 240V AC Twistlok receptacle
6. Full-power switch
7. AC voltmeter

8. DC circuit breaker
9. DC (+) terminal
10. DC (-) terminal
11. AVR
12. Diode stack
13. Terminal strip and cover assembly

14. Engine ON-OFF switch (recoil start)
15. Engine ignition switch (electric start)
16. Idle control switch
17. Idle control unit
18. Hourmeter
19. Pilot lamp
20. Ground terminal assembly

does not remain flush with the *test* button, the GFCI is faulty.

    d. Stop the generator.

NOTE: If the GFCI seems to trip for no apparent reason, check to make sure the generator is not vibrating excessively due to faulty rubber mounts or debris between the generator and frame. Excessive vibration can trip the GFCI.

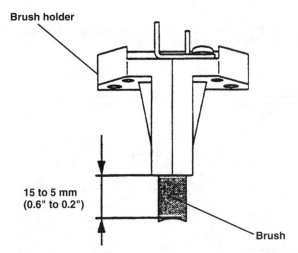

Fig. RB303 – View showing brush holder assembly with correct brush dimension limits.

Brush holder

15 to 5 mm (0.6" to 0.2")

Brush

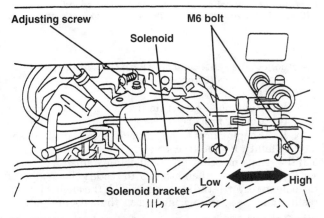

Fig. RB304 – View showing governor adjusting screw for bringing no load rpm within limits. Refer to the text for the correct adjustment procedure.

Adjusting screw

M6 bolt

Solenoid

Solenoid bracket

Low

High

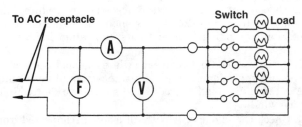

Fig. RB305 – View of a generator test load bank which can be made in shop. Refer to the text for construction and usage.

To AC receptacle

Switch

Load

4. Start the generator, allow it to reach operating temperature, then load test the generator to check voltage output. A load circuit such as that shown in Fig. RB305 will work well for the AC output, using resistance heaters or incandescent lamps with a 1.0 power factor. A similar circuit can be constructed for the DC output, except that if a battery is used as the load, the voltage in Step 5 will test 1-2 volts higher.

5. Test the output:

    a. AC voltage at the 120V receptacles should be 112-128 volts.

    b. AC voltage at the 240V receptacles should test 224-256 volts.

    c. DC voltage should test 6-14 volts.

6. Stop the generator.

7. Disconnect the stator winding wires. With the meter set to the R × 1 scale, apply the meter leads to the color coded wire terminals noted and test the stator winding resistance using the following specifications:

Mode	Winding	Ohms
RGV7500	Red to white AC	0.16
	Blue to black AC	0.16
	White to light green sub	0.11
	Brown to brown DC	0.61
	Each wire to ground	Infinity

    If megger testing the stator, apply one megger lead to the white AC winding wire terminal, and apply the other lead to the lamination core. The megger should read a minimum of 1 MΩ. Repeat this step with at least one color coded wire terminal of every other winding.

    The failure of any winding to test as specified indicates a faulty stator. Replace the stator.

8. To service the rotor:

    The slip rings are an integral part of the rotor assembly and cannot be replaced.

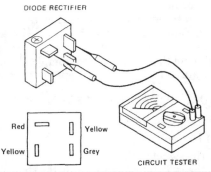

DIODE RECTIFIER

Red

Yellow

Yellow

Grey

CIRCUIT TESTER

		Apply black (-) needle of the circuit tester			
		Yellow	Yellow	Red	Grey
Apply red (+) needle of the circuit tester	Yellow		No Continuity	No Continuity	Continuity
	Yellow	No Continuity		No Continuity	Continuity
	Red	Continuity	Continuity		Continuity
	Grey	No Continuity	No Continuity	No Continuity	

Fig. RB306 – Diode rectifier continuity test for the RGV7500 generator. Verify meter polarity or reverse meter leads.

Clean dirty or rough slip rings with a suitable solvent, then lightly sand with ScotchBrite or sandpaper. Do not use steel wool, emery cloth or crocus cloth.

Cracked, chipped, broken, or grooved slip rings are not repairable. In these cases, the rotor will need replacement.

a. Disconnect the light green and the brown AVR brush wire connectors.

b. Measure field coil winding resistance through the brushes. Resistance should be 7.26 ohms.

c. Measure field coil winding resistance at the rotor slip rings. Resistance should be the same as Step B.

d. Measure field coil winding resistance from each slip ring to a ground on the rotor shaft or laminations. Resistance should be infinity/no continuity.

e. If test Steps B, C, and D met specification, the rotor and brushes are good.

f. If test Step B failed specification but Steps C and D met specification, the rotor is good. Inspect the brushes and test each brush for continuity through its terminal. If faulty, replace the brush holder assembly.

g. If both test Steps B and C failed specification, the rotor is faulty. Replace the rotor, then inspect and test the brushes.

h. If either Step D test read any continuity, the field coil winding is shorted. Replace the rotor.

i. To megger test the rotor, apply one megger lead to one of the winding terminals, and apply the other lead to any good ground on the rotor shaft or laminations. The megger should read a minimum of 1 MΩ. A rotor failing to meet this minimum has faulty winding insulation. Replace the rotor.

9. If DC output is faulty, but the stator and rotor meet specification, test the diode rectifier:

a. Refer to Fig. RB306 for the correct rectifier diagram and test.

b. Using the R × 1 ohm scale, test the rectifier for continuity as noted.

c. If the test meter being used is reverse polarity from the recommended meter, the specified readings will be opposite those taken.

d. Failure of the rectifier to test completely as specified indicates a faulty rectifier. Replace the rectifier.

INDIVIDUAL COMPONENT TESTING

"Full Power" Voltage Selector Switch

To test the switch:

1. Access the switch terminals inside the control box.
2. Noting the wiring orientation, disconnect feed terminals 2, 5, and 8.
3. With the switch in the "120V" position, there should be continuity between terminals 2-3, 5-6, and 8-9.
4. With the switch in the "120/240V" position, there should be continuity between terminals 2-1, 5-4, and 8-7.
5. The failure of the switch to test as specified in Steps 3 and 4 indicates a faulty switch.

Oil Sensor

To test the sensor on gasoline powered units so equipped:

1. Disconnect the sensor harness terminals (Fig. RB307).
2. Remove the sensor from the engine, then cap the sensor hole, insuring that the engine oil level is full and that the generator is sitting level.

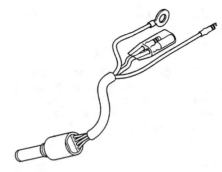

Fig. RB307 – Typical oil sensor switch on gasoline powered generators. The number of wires may vary.

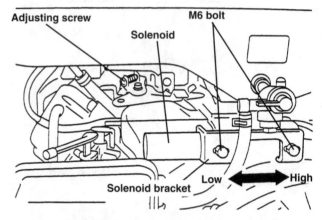

Fig. RB308 – To adjust the idle down rpm, move the slotted solenoid mount bracket as shown. Idle control speed should be 2200-2400 rpm with the idle switch on and no loads applied to

3. Reconnect the sensor harness with the sensor resting on a rag or similar protector.
4. Start the engine.

a. If the engine stops within five seconds, the sensor is functioning properly.

b. If the engine fails to stop after 10 seconds, the sensor is faulty. Replace the sensor.

5. If the sensor met Step 4A specification, reinstall the sensor into the engine, then attempt to restart the engine: If the engine restarts, the sensor is good. If the engine fails to restart, and the sensor wiring is good, the sensor is faulty.

Idle Control

1. Check the generator load start wattage: An overloaded generator will not return to rated rpm from idle. A motor powered tool, although within limits on running wattage, may require more starting wattage than the generator is rated to produce.

2. Disconnect and resistance test the two white solenoid wire terminals. Resistance should test 15-19 ohms. A resistance reading beyond this specification indicates a faulty solenoid.

3. Resistance test between each white solenoid wire terminal and the solenoid case. Resistance should be infinity. Any continuity reading in this test indicates a shorted solenoid winding.

4. Check the idle speed: With all loads disconnected from the generator and the idle control switch activated, idle speed should be 2200-2400 rpm. Idle speed should be ad-

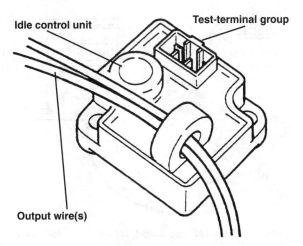

Idle control unit

Test-terminal group

Output wire(s)

Fig. RB309 – Main winding wire(s) must pass through the ZCT loop on the idle control unit.

justed with the slotted solenoid bracket (Fig. RB308): Loosen the bolts and move the bracket left to lower the idle speed or right to increase the idle speed.

5. Make sure that the red and the black main winding wires are passing through the idle control ZCT loop from the same direction, or the idle control unit will not function (Fig. RB309).

6. Using the diagram and chart in Fig. RB310, resistance test the nine terminals on the back of the idle control unit. Failure of any terminal pair to test to specification indicates a faulty control unit. Note that if the test meter being used is reverse polarity from the recommended meter,

the specified readings will be opposite those taken. Reverse the meter leads and retest.

AVR

To test the AVR system:

1. Test the engine mounted exciter coil.

   a. Access and disconnect the two yellow coil wire terminals from the AVR.

   b. Resistance test the two yellow wire terminals. the coil should test 2.4 ohms.

   c. If Step B met specification, the exciter coil is OK. If Step B failed, access and disconnect the two yellow wire terminals from the coil, itself, on the side of the engine crankcase. Repeat Step B.

   d. If Step C met specification, the coil to AVR wires are faulty. Repair or replace as necessary. If Step C failed, the exciter coil is faulty and must be replaced.

2. Inspect the AVR control unit. if it is burned, or has melted epoxy, it is normally faulty and should be replaced.

3. Test the AVR control unit:

   a. Access and disconnect the harness connectors.

   b. Using Fig. RB311, resistance test the color coded terminals as specified. Failure of any terminal pair to meet specification indicates a faulty control unit. Replace the unit.

   c. Occasionally, a control unit which appears normal and passes Step B test will have internal SCR or transistor damage which cannot be tested. If all other AVR components meet specification, the wiring

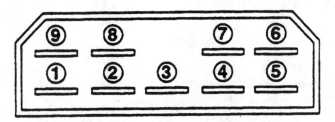

Terminal number of the Idle Control Unit

Circuit tester (with battery power source 1.5V)		Apply black (-) needle of the circuit tester								
		1	2	3	4	5	6	7	8	9
Apply red (+) needle of the circuit tester	1		110 kΩ	∞	110 kW	∞	∞	110 kW	50 kW	50 kW
	2	110 kΩ		∞	110 kW	∞	∞	110 kW	50 kW	50 kW
	3	∞	∞		∞	∞	∞	∞	∞	∞
	4	110 kW	110 kW	∞		∞	∞	110 kW	50 kW	50 kW
	5	∞	∞	350 kW	∞		∞	∞	∞	∞
	6	∞	∞	∞	∞	∞		∞	∞	∞
	7	110 kW	110 kW	∞	110 kW	∞	∞		50 kW	50 kW
	8	85 kW	85 kW	∞	80 kW	∞	∞	80 kW		0 kW
	9	85 kW	85 kW	∞	80 kW	∞	∞	80 kW	0 kW	

Fig. RB310 – Terminal resistance test for the RGV7500 idle down control unit. Refer to Fig. RB309 for terminal location.

integrity checks out, and the engine rpm is OK, but output is still questionable, the recommended procedure is to install a new AVR unit to see if output returns to specification.

Voltmeter

The recommended test for the voltmeter is to apply a load to the generator. If the voltmeter indicates voltage in the proper range, it is OK. If verification is necessary, the voltmeter leads can be accessed, and a test meter connected in place of the voltmeter.

DISASSEMBLY

Refer to Figs. RB301 and RB302 for component identification.

1. On electric start models, disconnect the battery cables and remove the battery.
2. Remove the harness protector tube ends from the generator and control box. Disconnect the control panel harnesses from the generator and engine, then remove the control box assembly.
3. Drain and remove the fuel tank.
4. Remove the muffler assembly.
5. Remove the brush cover from the generator end housing. Noting the wiring orientation, disconnect the brush wires, then remove the brush holder assembly.
6. Remove the two nuts holding the generator rear housing feet to the rubber mounts.
7. Place a temporary support, such as a wood block, under the engine adapter/generator front housing so that the rear housing feet clear the rubber mount studs.
8. Remove the four rear housing to front housing bolts. Noting the stator wiring orientation, remove the rear housing from the stator by lightly tapping against the mount feet with a plastic mallet.
9. Slide the stator cover off the stator, then remove the stator.
10. Remove the rotor throughbolt, then squirt a small amount of penetrating oil through the rotor shaft onto the engine crankshaft taper.
11. Using a slide hammer with the proper threaded adapter, break the rotor free from the engine crankshaft, being careful to prevent the rotor from falling. Remove the rotor.
12. If necessary, unbolt and remove the front housing from the engine.

REASSEMBLY

Refer to Figs. RB301 and RB302 for component identification.

The rotor must be installed before the stator. Exercise caution when handling the stator and rotor so the windings and winding coatings are not damaged.

1. If the front housing was removed during disassembly, reinstall it. Torque the bolts 12-14 N•m (105-120 in.-lb.).
2. If the engine was left in the cradle frame, place a temporary support, such as a wooden block, under the front housing so that the rear housing feet clear the rubber mount studs.
3. Service the rotor bearing, if needed.
4. Make sure that the matching tapers on the engine crankshaft and inside the rotor are clean and dry. Install the rotor onto the crankshaft. Install the rotor throughbolt. Torque the bolt to 22.5-24.5 N•m (200-235 in.-lb., or 16.5-18.5 ft.-lb.).

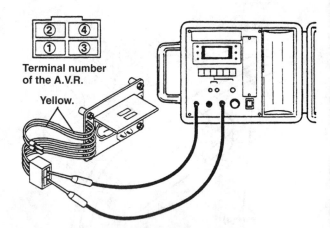

Fig. RB311 – Terminal resistance test for the RGV7500 AVR unit.

5. Noting the correct wiring orientation, fit the stator over the rotor and against the front housing, making sure that the outer stator slots are properly aligned with the front housing bolt holes.
6. Fit the stator cover over the stator.
7. Install the rear housing onto the rotor bearing and stator. Lightly and evenly tap the rear housing forward until it is snug against the stator, and the stator is snug against the front housing.
8. Install the four rear housing-to-front housing bolts. Torque the bolts to 4.5-6.0 N•m (40-55 in.-lb.).
9. Remove the spark plug and rotate the engine crankshaft to check for interference between the rotor and stator. There should be a consistent gap around the rotor laminations inside the stator. If any rubbing or scraping is evident, correct any misalignment prior to proceeding. Reinstall the spark plug and torque to 18 N•m (155 in.-lb.).
10. Lift the rear housing. Remove the temporary front housing support. Set the rear housing feet down onto the rubber mounts. Install the mount nuts while lifting slightly on the rear housing so the generator weight is not on the mount rubber. Torque the nuts to 12-14 N•m (105-120 in.-lb.). Make sure that the ground cable is properly reconnected between the left rear housing foot and the cradle frame.
11. Install the brush holder assembly. Reconnect the brush wires, making sure that the light green wire goes to the positive (+) brush. Install the brush cover.
12. Install the muffler assembly. Torque the 6 mm fasteners 8-10 N•m (70-90 in.-lb.). Torque the 8 mm carbon-steel fasteners to 19-25 N•m (165-215 in.-lb.). Torque the 8 mm stainless steel fasteners to 22-28 N•m (190-245 in.-lb.).
13. Install and reconnect the fuel tank.
14. Install the control box assembly. Fit the harness cover tube over the stator wiring, fastening the small tube end into the rear housing hole. Connect the control box wiring to the stator wiring and the engine wiring. Then fit the upper end of the harness cover tube into the control box hole.
15. On electric start models, install and reconnect the battery.

WIRING DIAGRAMS

Refer to Fig. RB312 to view the wiring diagram for the RGV7500 generator with recoil start.

Refer to Fig. RB313 to view the wiring diagram for the RGV7500 generator with electric start.

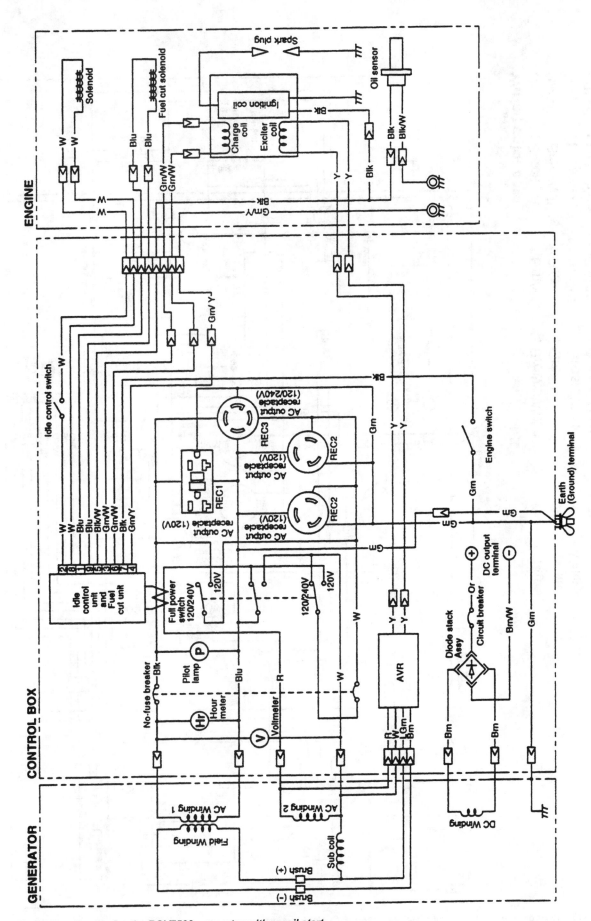

Fig. RB312 – Wiring diagram for the RGV7500 generator with recoil start.

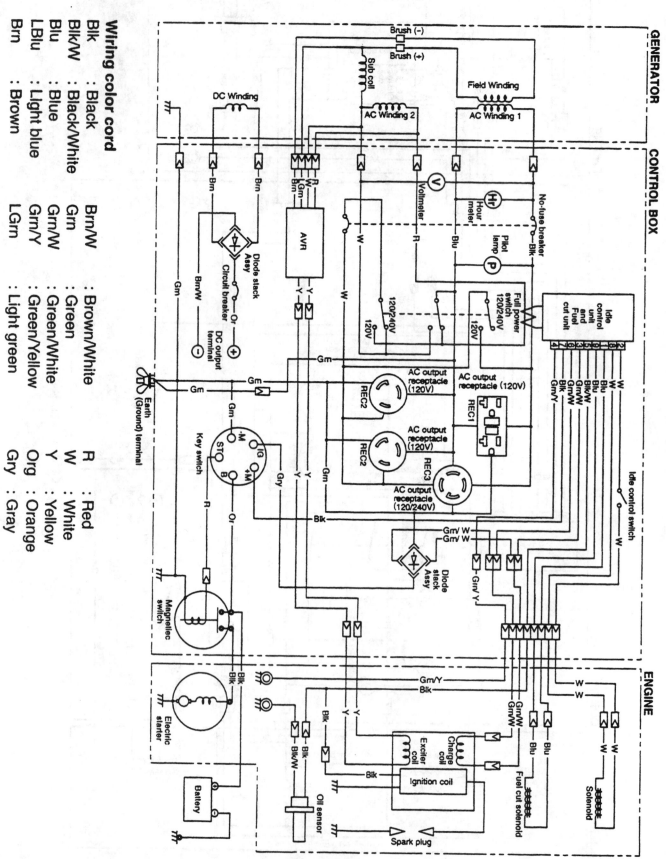

Wiring color cord

Blk	: Black	Brn/W	: Brown/White	R	: Red
Blk/W	: Black/White	Grn	: Green	W	: White
Blu	: Blue	Grn/W	: Green/White	Y	: Yellow
LBlu	: Light blue	Grn/Y	: Green/Yellow	Org	: Orange
Brn	: Brown	LGrn	: Light green	Gry	: Gray

Fig. RB313 – Wiring diagram for the RGV7500 generator with electric start.

ROBIN SUBARU
RGV12000 and RGV13000T
Brush-style Generators

Generator Model	Rated (Surge) Watts	Output Volts
RGV12000	10,000 (12,000) AC	120/240 AC
RGV13000T	10,000 (13,000) AC	120/208 AC

IDENTIFICATION

These generators are direct drive, self exciting, brush style, 60 Hz units with panel mounted receptacles.

Model RGV12000 is a two pole single phase unit with electronic automatic voltage regulation (AVR).

Model RGV13000T is a two pole 3-phase unit with diode rectified current transformer voltage regulation.

The serial number plate on the RGV12000 is located on the upper half of the face of the generator rear housing assembly.

The serial number plate on the RGV13000 is located on the upper circumference of the generator end cover, facing the bottom of the gas tank.

Refer to Figs. RB401 and RB402 to view the Model RGV12000 generator components. Refer to Figs. RB403 and RB404 to view the Model RGV13000 generator components.

WIRE COLOR CODE

The wiring diagram color code for these generators is as follows:

Blk – Black	Blu – Blue	LBlu – Light blue
Brn – Brown	Grn – Green	Org – Orange
Gry – Gray	R – Red	W – White
LGrn – Light green	Y – Yellow	Pik Pink

A wire with a slash separated ID is a 2-color wire. For example: R/W is a red wire with a white tracer stripe.

MAINTENANCE

BRUSHES

Replace the brushes when the brush length protruding from the brush holder reaches 5.0 mm (0.2 in.) or less. New brush length should be 15.0 mm (0.6 in.). Refer to Fig. RB405.

When servicing the brushes, always observe proper brush polarity.

NOTE: There are two styles of brush holders. When installing a brush holder, make sure that the mount screw holes align with the end housing holes, the brushes are perpendicular to the slip rings, and the brush radius aligns with the slip ring curvature.

Failure to do so could cause the brush holder to break either before or after the generator is started.

ROTOR BEARING

The end housing bearing on these generators is a sealed, prelubricated ball bearing. Inspect and replace as necessary. If a seized bearing is encountered, inspect the rotor journal and the rear housing bearing cavity for scoring.

The end housing must be removed for access to the bearing.

GENERATOR REPAIR

TROUBLESHOOTING

1. Before testing any electrical components, check the integrity of the terminals and connectors.
2. While testing individual components, also check the wiring integrity.
3. The use of the Dr. Robin generator tester #388-47565-08 is recommended for testing. The Dr. Robin tester has the following capabilities:

    a. AC voltmeter – 0-500 volts.

    b. Frequency – 25-70 Hz.

    c. Condenser capacity – 10-100 μF .

    d. Digital ohmmeter – 0.1-1,999 Ohms.

    e. Insulation resistance – 3 MΩ.

An accurate engine tachometer will also be required.

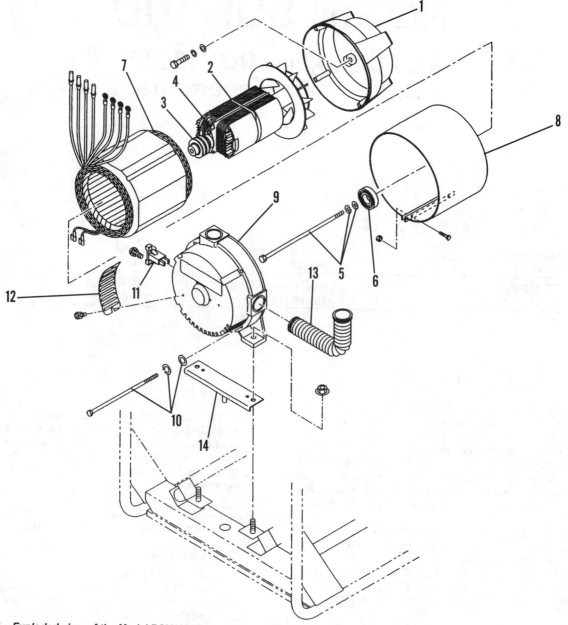

Fig. RB401 – Exploded view of the Model RGV12000 generator head components.

1. Engine adapter/front housing
2. Rotor assembly
3. Slip rings
4. Surge absorber
5. Rotor throughbolt assembly
6. Rotor bearing

7. Stator assembly
8. Stator cover
9. Rear housing
10. Rear housing/stator
   throughbolt assembly (4)

11. Brush-holder assembly
12. Brush cover
13. Wiring harness
   protective cover tube
14. Generator mount plate

4. Test specifications are given for an ambient temperature of 20° C (68° F). Resistance readings may vary with temperature and test equipment accuracy.

5. For access to the components listed in this Troubleshooting section, refer to Figs. RB401-RB404 and the later Disassembly section.

6. All operational testing must be performed with the engine running at its governed no load speed of 3700-3750 rpm, unless specific instructions state otherwise.

7. Always load test the generator after repairing defects discovered during troubleshooting.

Megger Test the Generator System Wiring Insulation

1. Insure that the circuit breakers are *on* and functioning.

2. Connect one megger test probe to any hot terminal of a receptacle.

3. Connect the other megger test probe to any ground terminal.

4. The megger should indicate a resistance of 1 MΩ or more. if less than 1 MΩ, megger test the rotor, stator, and control panel resistance individually. Refer to the subsequent Troubleshooting sections.

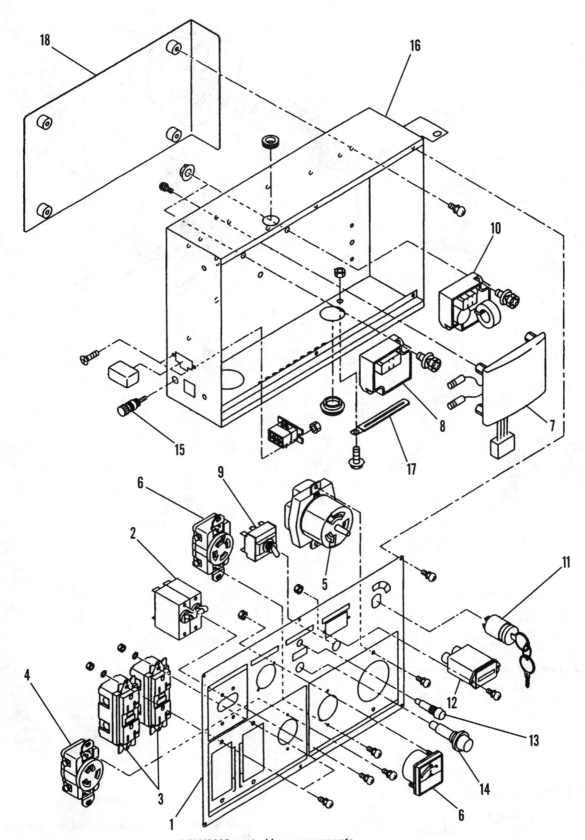

Fig. RB402 – Exploded view of the Model RGV12000 control box components.

1. Control-panel faceplate
2. AC circuit breaker
3. 120V AC GFCI receptacle
4. 120V AC Twitslok receptacle
5. 50A 125/250V AC Twistlok receptacle
6. 30A 240V AC Twistlok receptacle

7. AVR assembly
8. ECU
9. Auto-idle switch
10. Idle-control unit
11. Engine switch
12. Hourmeter

13. Pilot lamp
14. Low-oil light
15. Ground terminal
16. Control box
17. Clamp
18. Heat shield

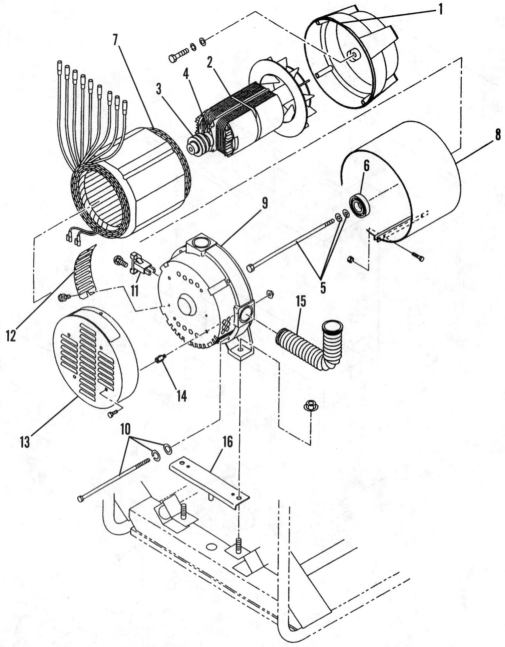

Fig. RB403 – Exploded view of the Model RGV13000T generator head components.

1. Engine adapter/front housing
2. Rotor assembly
3. Sliprings
4. Surge absorber
5. Rotor through-bolt assembly
6. Rotor bearing

7. Stator assembly
8. Stator cover
9. Rear housing
10. Rear housing/stator through-bolt assembly (4)
11. Brush-holder assembly

12. Brush cover
13. End cover
14. End cover mount studs (4)
15. Wiring harness protective cover tube
16. Generator mount plate

To Troubleshoot the Generator

1. Circuit breakers must be ON. To test the breakers, access the breaker wiring and continuity test the breaker terminals with the generator shut down and at least one breaker feed wire disconnected. The breaker is faulty if it will not stay set, shows continuity in the OFF position or does not show continuity in the ON position,

2. Receptacles must be functioning. To test the receptacle(s):

a. Access the receptacle connections.

b. Disconnect the *return* feed wire from the receptacle being tested.

c. Install a jumper wire between the *hot* and *return* outlet terminals.

d. Test for continuity across the feed terminals. No continuity indicates a faulty receptacle.

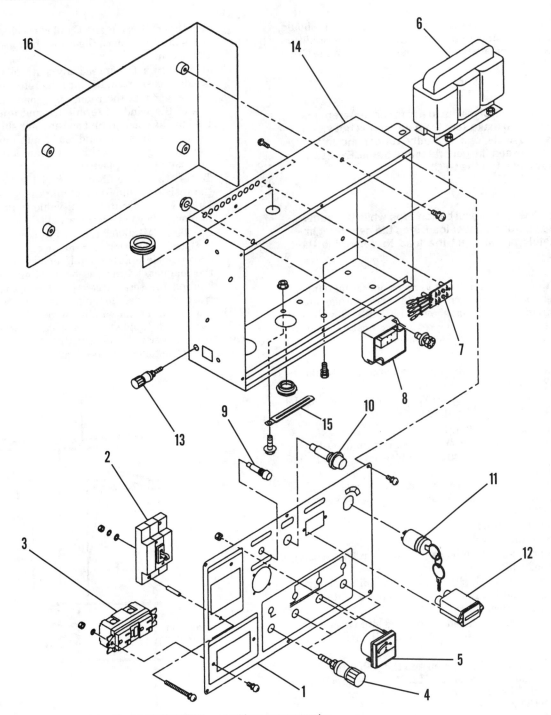

Fig. RB404 – Exploded view of the Model RGV13000T control box components.

1. Control-panel faceplate
2. AC circuit breaker
3. 120V AC GFCI receptacle
4. 3-phase AC outlet terminals (4)
5. AC voltmeter
6. Current transformer

7. Diode unit
8. ECU
9. Pilot lamp
10. Low-oil lamp
11. Engine ignition switch
12. Hourmeter

13. Ground terminal
14. Control box
15. Clamp
16. Heat shield

e. Remove the jumper wire and retest for continuity across the feed terminals. No continuity this time indicates a good receptacle.

3. Ground fault circuit interrupters (GFCIs) must be *reset* and functioning. To test the GFCI:

a. Start the generator and allow a brief no load warm up period.

b. Depress the test button. the *reset* button should extend. If it does, proceed to Step C. If the *reset* button does not extend, the GFCI is faulty or it is not receiving input from the generator. If the generator does not have proper output for testing the GFCI, disconnect the GFCI feed wires, connect a wall socket jumper to the GFCI feed terminals, then test the GFCI.

c. Depress the reset button. The *reset* button should then be flush with the *test* button. If the *reset* button does not remain flush with the *test* button, the GFCI is faulty.

d. Stop the generator.

NOTE: If the GFCI seems to trip for no apparent reason, check to make sure the generator is not vibrating excessively due to faulty rubber mounts or debris between the generator and frame. Excessive vibration can trip the GFCI.

WARNING: If the *reset* button trips while the generator is operating under load, stop the generator immediately and inspect the load to determine the fault.

4. Start the generator, allow it to reach operating temperature, then load test the generator to check voltage output. A load circuit such as that shown in Fig. RB406 will work well for the AC output, using resistance heaters or incandescent lamps with a 1.0 power factor.

5. Test the output:

a. AC voltage at the 120V receptacles should be 112-128 volts.

b. AC voltage at the 240V receptacles on Model RGV12000 should test 224-256 volts.

c. On Model RGV13000T:

1. AC voltage between 3-phase terminals O-U, O-V, and O-W (Fig. RB407) should each read 112-128 volts.

2. AC voltage between 3-phase terminals U-V, V-W, and U-W (Figs. RB408 or RB409) should each read 218-249 volts.

6. Stop the generator.

7. To service the rotor:

The slip rings are an integral part of the rotor assembly. they are not replaceable.

Clean dirty or rough slip rings with a suitable solvent, then lightly sand with ScotchBrite. Do not use steel wool, emery cloth or crocus cloth.

Cracked, chipped, broken, or grooved slip rings are not repairable. In these cases, the rotor will need replacement.

a. Disconnect the light green and the yellow brush wire connectors on Model RGV12000 or the green/white and the brown brush wire connectors on Model RGV13000T.

b. Measure field coil winding resistance through the brushes. Resistance on Model RGV12000 should be 4.11 ohms. On Model RGV13000T, 69.6 ohms.

c. Measure field coil winding resistance at the rotor slip rings. Resistance should be the same as Step B.

d. Measure field coil winding resistance from each slip ring to a good ground on the rotor shaft or laminations. Resistance should be infinity/no continuity.

e. If test Steps B, C, and D met specification, the rotor and brushes are good.

f. If test Step B failed specification but Steps C and D met specification, the rotor is good. Inspect the brushes and test each brush for continuity through its terminal. If faulty, replace the brush holder assembly.

g. If both test Steps B and C failed specification, the rotor is faulty. Replace the rotor, then inspect and test the brushes.

h. If either Step D test read any continuity, the field coil winding is shorted. Replace the rotor.

i. To megger test the rotor, apply one megger lead to one of the winding terminals, and apply the other lead to any good ground on the rotor shaft or laminations. The megger should read a minimum of $1\ m\Omega$. A rotor failing to meet this minimum has faulty winding insulation; replace the rotor.

j. To test the surge absorber resistor, first determine the resistor value from its markings. Test the resistor and compare its actual value to its rated value. Replace if beyond specification.

k. On the RGV13000T rotor, to test the magnetism of the attached permanent magnets, hold a flat screwdriver blade approximately one inch (25 mm) from the magnet, letting the blade dangle while lightly holding the screwdriver handle. If the magnet attracts the blade, the magnet is good.

7. Disconnect the stator winding wires. With the meter set to the R × 1 scale, apply the meter leads to the color coded wire terminals noted and test the stator winding resistance using the following specifications:

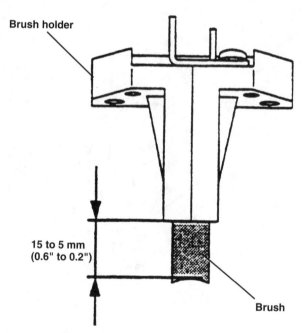

Fig. RB405 – View showing brush holder assembly with correct brush dimension limits.

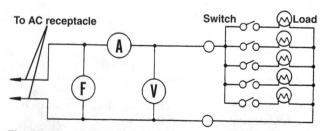

Fig. RB406 – View of a generator test load bank which can be made in shop. Refer to the text for construction and usage.

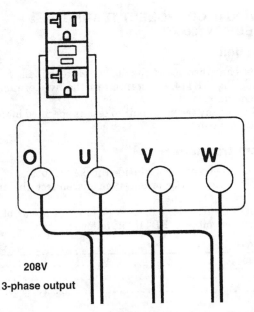

Model	Winding	Ohms
RGV12000	Red to white AC	0.11
	Blue to black AC	0.11
	White to light green auxiliary	0.37
RGV13000T	Red to white AC	0.23
	Red to black AC	0.23
	White to black AC	0.23
BOTH	Each wire to Ground	Infinity

If megger testing the stator, apply one megger lead to the white AC winding wire terminal, and apply the other lead to the lamination core. the megger should read a minimum of 1 MΩ. Repeat this step with at least one color coded wire terminal of every other winding.

The failure of any winding to test as specified indicates a faulty stator. Replace the stator.

8. Megger test the control panel by placing one meter probe on any live component and the other probe on any ground. the megger should read a minimum of 1 MΩ. Failure of the megger to read as specified indicates a component short. access the control box components and test each component individually to determine the fault.

INDIVIDUAL COMPONENT TESTING– BOTH MODELS

NOTE: If the meter being used is reverse polarity of the recommended meter when taking readings using the test chart figures, the test probes will need to be reversed to obtain the correct readings.

Oil Sensor Electronic Control Unit (ECU)

The ECU works in conjunction with the oil pressure switch to protect the engine, should oil pressure drop below 1.0 kg/cm (14-15 psi) due to insufficient crankcase oil. At that point, the ECU will ground the magnetos and stop the flow of fuel to the engine, as well as cause the warning lamp to blink for a two-minute period. The ECU also:

1. Kills the engine (grounds magneto and blocks fuel flow) when the ignition switch is turned *off.*

2. Deactivates the electric start circuit after three seconds if the starter is jammed.

3. Deactivates the electric start circuit if the engine does not start within 21 seconds.

4. Deactivates the electric start circuit if the remote control unit is connected but OFF, even if the ignition switch is turned to *start.*

5. Deactivates the electric start circuit once the engine starts and reaches 1050 rpm.

6. Momentarily activates the warning lamp when the ignition switch is turned *on,* as a periodic indicator that the lamp is good.

To test the ECU:

1. Access and unplug the ECU harness connector.

2. Referring to Fig. RB410, resistance test each numbered terminal pair as noted.

3. Failure of any terminal pair to test as specified indicates a faulty ECU.

Voltmeter and/or Pilot Lamp

The voltmeter and the pilot lamp cannot be tested with normal circuit testers due to the large internal resistances. Test these two items by applying rated voltage. Good items will function correctly.

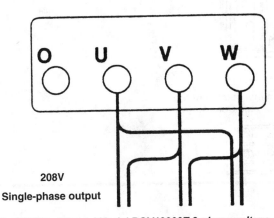

Fig. RB407 – View of Model RGV13000T 3-phase voltage output terminals wired to supply three single phase 120-volt AC circuits, each delivering 1/3 of the generator's rated output.

Fig. RB408 – View of Model RGV13000T 3-phase voltage output terminals wired to supply three single phase 208-volt AC circuits, each delivering 1/3 of the generator's rated output.

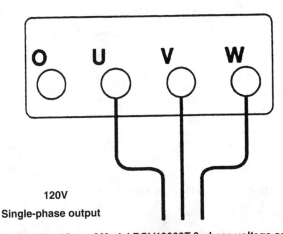

Fig. RB409 – View of Model RGV13000T 3-phase voltage output terminals wired to supply one 3-phase 208-volt AC 10kW continuous output circuit at a 1.0 power factor.

INDIVIDUAL COMPONENT TESTING – MODEL RGV12000

Automatic Voltage Regulator

1. Access and unplug the AVR terminal connectors.
2. Referring to Fig. RB411, resistance test each color coded pair of terminals as noted.
3. Failure of any terminal pair to test as specified indicates a faulty AVR.

Idle Control Unit

1. Access and unplug the control unit harness connector.
2. Insure that the black main winding wire is passing through the ZCT loop and is not pinched or kinked (Fig. RB412).
3. Using the diagram and chart in Fig. RB413, resistance test the five terminals on the back of the idle control unit.
4. Failure of any terminal pair to test as specified indicates a faulty control unit.

INDIVIDUAL COMPONENT TESTING – MODEL RGV13000T

Diode Unit

1. Access and disconnect the diode unit terminals.
2. Using Fig. RB414, resistance test the five terminals of the diode unit.
3. Failure of any terminal pair to test as specified indicates a faulty diode unit.

Current Transformer

1. Access the transformer inside the control box.
2. Noting the wiring orientation, disconnect the transformer wiring terminals.
3. Resistance test the paired numbered terminals of each transformer coil (Fig. RB415) as follows:

Terminals	Ohms
1-2	2.00
1-3	2.00
2-3	0.009

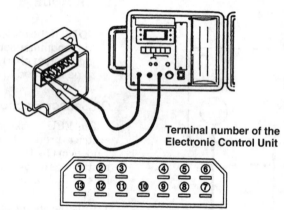

Terminal number of the Electronic Control Unit

		Tester (-)											
Pin No.	1	2	3	4	5	6	7	8	9	10	11	12	13
1		∞	∞	∞	∞	∞	∞	∞	∞	∞	∞	∞	∞
2	∞		∞	∞	∞	∞	∞	∞	∞	∞	∞	∞	∞
3	∞	18.4MΩ		10 MΩ	10.2 MΩ	∞	10.4 MΩ	11 MΩ	14.4 MΩ	5.3 MΩ	∞	∞	∞
4	∞	∞	∞		0.1 Ω	∞	∞	∞	∞	∞	∞	∞	∞
5	∞	∞	∞	0.1 MΩ		∞	∞	∞	∞	∞	∞	∞	∞
6	∞	∞	∞	∞	∞		∞	∞	∞	∞	∞	∞	∞
7	∞	∞	∞	∞	∞	∞		∞	2675 kΩ	∞	∞	∞	∞
8	∞	∞	∞	∞	∞	∞	∞		2590 kΩ	∞	∞	∞	∞
9	∞	∞	∞	∞	∞	∞	∞	∞		∞	∞	∞	∞
10	∞	5.3M Ω	∞	2392 kΩ	2389 kΩ	∞	2438 kΩ	2664 kΩ	4 MΩ		∞	∞	∞
11	∞	18M Ω	∞	10.3 MΩ	10.2 MΩ	∞	10.3 MΩ	11 MΩ	14 MΩ	5.3 MΩ		∞	∞
12	∞	∞	∞	∞	∞	∞	∞	∞	∞	∞	∞		∞
13	∞	∞	∞	∞	∞	∞	∞	∞	∞	∞	∞	∞	

Tester (+) applies to the Pin No. rows.

FIG. RB410 – Terminal resistance test for the 13-terminal electronic control unit. Verify meter polarity or reverse meter leads.

4. Resistance test between each numbered terminal on each coil and the transformer core. resistance should test infinity/no continuity

5. To megger test the transformer, apply one megger lead to one of the nine terminals. apply the other lead to the transformer core: The megger should read a minimum of 1 MΩ. Repeat this test with the other eight terminals.

6. Failure of any terminal pair on either of the three coils to test as specified in Steps 3-5 indicates a faulty current transformer.

DISASEMBLY

Refer to Figs. RB401-RB404 for component identification.

The rotor cannot be removed without first removing the stator. Exercise caution when handling the stator and rotor so the windings and winding coatings are not damaged. Exercise caution when handling a magnet equipped rotor–do not drop, heat, or allow debris to accumulate on the magnets. The rotor fan is part of the rotor assembly and cannot be replaced separately.

1. Disconnect the battery cables and remove the battery.
2. Remove the harness protector tube ends from the generator and control box. disconnect the control panel harnesses from the generator and engine, then remove the control box assembly.
3. Drain and remove the fuel tank.
4. Remove the muffler assembly.
5. On the RGV13000T, remove the end cover. On both models, remove the brush cover from the generator end housing. Noting the wiring orientation, disconnect the brush wires, then remove the brush holder assembly.
6. Remove the two nuts holding the generator rear housing feet to the rubber mounts.
7. Place a temporary support, such as a wood block, under the engine adapter/generator front housing so that the rear housing feet clear the rubber mount studs.
8. Remove the four rear housing to front housing bolts. Noting the stator wiring orientation, remove the rear housing from the stator by lightly tapping against the mount feet with a plastic mallet.
9. Slide the stator over the rotor and remove the stator.
10. Remove the rotor throughbolt, then apply a small amount of penetrating oil through the rotor shaft onto the engine crankshaft taper.
11. Using a slide hammer with the proper threaded adapter, break the rotor free from the engine crankshaft.

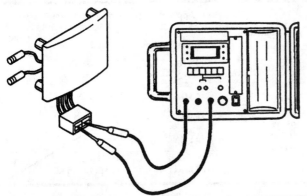

Analogue circuit tester	Apply black (-) needle of the circuit tester				
	Yellow	Red	White	Light Green	Brown/ Yellow
Apply red (+) needle of the circuit tester	**Yellow**		∞	∞	∞	∞
	Red	∞		100 kΩ	∞	200 kΩ
	White	∞	50 kΩ		∞	100 kΩ
	Light Green	16 kΩ	220 kΩ	125 kΩ		14 kΩ
	Brown/ Yellow	∞	185 kΩ	100 kΩ	∞	

Fig. RB411 – Terminal resistance test for the Model RGV12000 6-terminal automatic voltage regulator. "Yellow" test refers to (ital)each(end) of the two yellow wire terminals. Verify meter polarity or reverse meter leads.

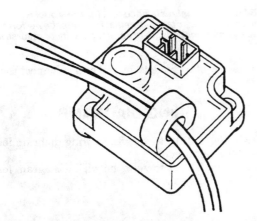

Fig. RB412 – Main winding wire(s) must pass through the ZCT loop on the RGV Series idle control unit. On dual voltage units, the two wires (black and red) must pass through in the same direction or the idle control will not function.

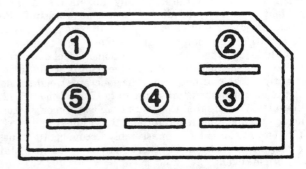

Analogue circuit tester (with battery power source 1.5V)	Apply black (-) needle of the circuit tester				
	1	2	3	4	5
Apply red (+) needle of the circuit tester	**1**		∞	65 kΩ	65 kΩ	30 kΩ
	2	∞		∞	∞	∞
	3	65 kΩ	∞		65 kΩ	30 kΩ
	4	65 kΩ	∞	65 kΩ		30 kΩ
	5	6.5 kΩ	∞	6.5 kΩ	6.5 kΩ	

Fig. RB413 – Terminal resistance test for the Model RGV12000 5-terminal idle control unit. Verify meter polarity or reverse meter leads.

Being careful to prevent the rotor from falling while removing the rotor.

12. If necessary, unbolt and remove the front housing from the engine.

REASSEMBLY

NOTE: When servicing the brushes, always observe proper brush polarity.

Refer to Figs. RB401-RB404 for component identification.

1. If the front housing was removed during disassembly, reinstall it. Torque the bolts 23-25 N•m (200-235 in.-lb., or 16.5-18.5 ft-lb).

2. If the engine was left in the cradle frame, place a temporary support, such as a wooden block, under the front housing so that the rear housing feet clear the rubber mount studs.

3. Service the rotor bearing, if needed.

4. Make sure that the matching tapers on the engine crankshaft and inside the rotor are clean and dry. Install the rotor onto the crankshaft. Install the rotor throughbolt. Torque the bolt 12-14 N•m (100-120 in.-lb.).

5. Noting the correct wiring orientation, fit the stator over the rotor and against the front housing, making sure that the outer stator slots are properly aligned with the front housing bolt holes.

6. Install the rear housing onto the rotor bearing and stator. Lightly and evenly tap the rear housing forward until it is snug against the stator, and the stator is snug against the front housing.

7. Install the four rear housing to front housing bolts. Alternately and incrementally torque the bolts to 12-14 N•m (100-120 in.-lb.).

8. Remove the spark plugs and rotate the engine crankshaft to check for interference between the rotor and stator. There should be a consistent gap around the rotor laminations inside the stator. If any rubbing or scraping is evident, correct any misalignment prior to proceeding. Reinstall the spark plugs and torque to 18 N•m (155 in.-lb.).

9. Lift the rear housing and remove the temporary front housing support. Set the rear housing feet down onto the rubber mounts. Install the mount nuts while lifting slightly on the rear housing so the generator weight is not on the mount rubber. Torque the nuts 12-14 N•m (105-120 in.-lb.). Make sure that the ground cable is properly reconnected between the generator and the cradle frame.

10. Install the brush holder assembly. Reconnect the brush wires, making certain to observe proper polarity. Install the brush cover.

11. Install the muffler assembly. Torque the 6 mm fasteners 8-10 N•m (70-90 in.-lb.). Torque the 8 mm carbon steel fasteners to 19-25 N•m (165-215 in.-lb.). Torque the 8 mm stainless steel fasteners to 22-28 N•m (190-245 in.-lb.).

12. Install and reconnect the fuel tank.

13. Install the control box assembly. Fit the harness cover tube over the stator wiring, fastening the small tube end into the rear housing hole. Connect the control box wiring to the stator wiring and the engine wiring. Then fit the upper end of the harness cover tube into the control box hole.

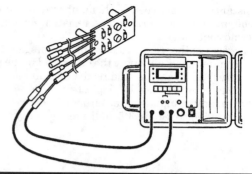

Diode unit		Apply black (-) needle of the circuit tester				
		Brown	Yellow 1	Yellow 2	Yellow 3	Light green
Apply red (+) needle of the circuit tester	Brown		∞	∞	∞	∞
	Yellow 1	16 kΩ		∞	∞	∞
	Yellow 2	16 kΩ	∞		∞	∞
	Yellow 3	16 kΩ	∞	∞		∞
	Light green	30 kΩ	16 kΩ	16 kΩ	16 kΩ	

Fig. RB414 – Terminal resistance test for the Model RGV13000T 5-terminal diode unit. Verify meter polarity or reverse meter leads.

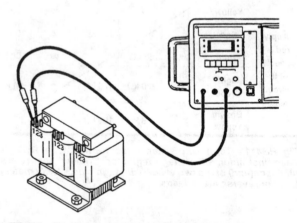

Fig. RB415 – Terminal resistance test for the Model RGV13000T current transformer. Each 3-terminal test must be performed on all three transformer coils. Verify meter polarity or reverse meter leads.

14. On electric start models, install and reconnect the battery.

WIRING DIAGRAMS

Refer to Fig. RB416 to view the wiring diagram for the RGV12000 generator.

Refer to Fig. RB417 to view the wiring diagram for the RGV13000T generator.

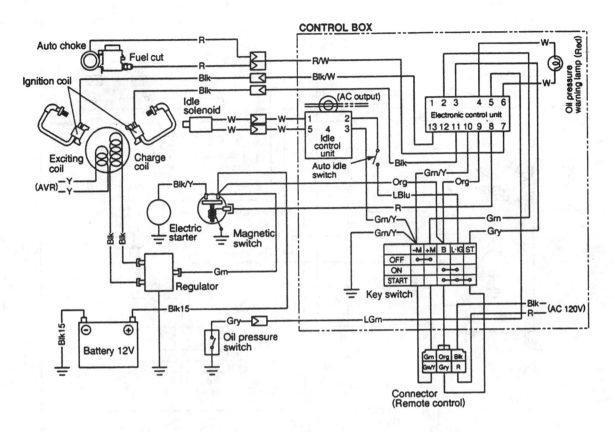

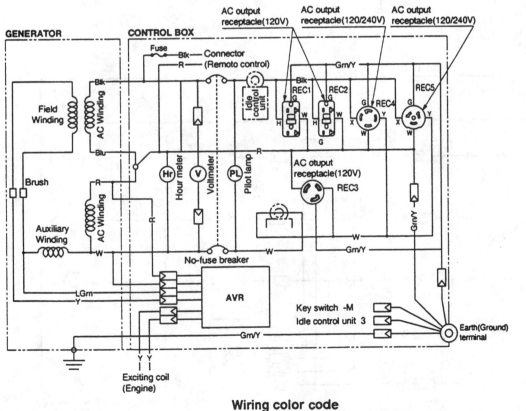

Wiring color code

Blk	: Black	Brn/W	: Brown/White	R	: Red	Blk/R	: Black/Red
Blk/W	: Black/White	Grn	: Green	W	: White	R/W	: Red/White
Blu	: Blue	Grn/W	: Green/White	Y	: Yellow	LGrn	: Light green
LBlu	: Light blue	Org	: Orange	Pik	: Pink		
Brn	: Brown	Gry	: Gray	Grn/Y	: Green/Yellow		

Fig. RB416 – Wiring diagram for the Model RGV12000 generator meter leads.

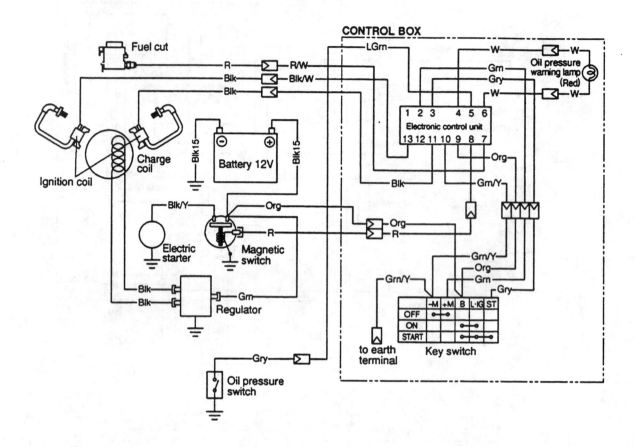

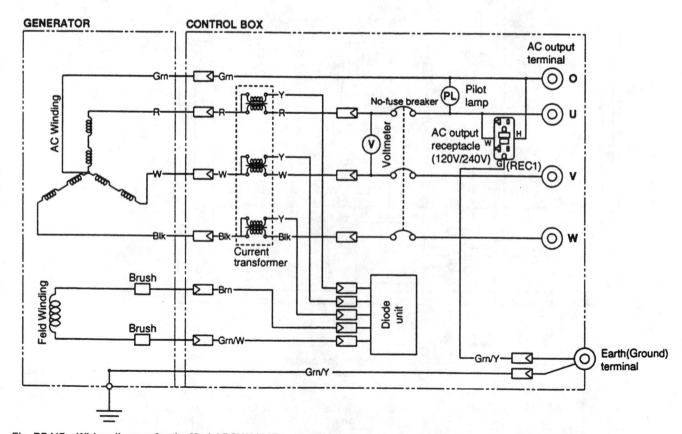

Fig. RB417 – Wiring diagram for the Model RGV13000T generator.

ROBIN SUBARU
RXW150 and RXW180 Welder/Generator

Generator Model	Rated Output: Watts	Volts	Amps	Duty Cycle
RXW150	2000 AC; 3440 DC	120 AC; 25.5 DC	135 DC	40%
RXW180	3000 AC; 3900 DC	120 AC; 26.0 DC	150 DC	50%

IDENTIFICATION

These welder generators are direct drive, self exciting, brush style units with panel mounted receptacles and controls.

AC output is provided by a 2-pole, 60 Hz, single phase stator winding.

DC current is generated by a 6-pole, 3-phase AC winding system, then converted by the 3-phase full wave diode rectifiers into DC.

On the Model RXW150, a 1/8 in. welding rod can be operated simultaneously with a 500-watt AC load. On the Model RXW180, a 1/8 in. welding rod can be operated simultaneously with a 1000-watt load. With separate and independent windings, inter system interference is avoided.

The Model RXW180 comes standard with electric start. Electric start is not available on the Model RXW150.

The serial number plate is located on the left side wall of the control box.

Refer to Figs. RB501 and RB502 to view the generator components.

WIRE COLOR CODE

The wiring diagram color code for these generators is as follows:

B – Black	L – Blue	Sb – Sky blue
Br – Brown	G – Green	Or – Orange
Gr – Gray	R – Red	W – White
Lg – Light green	Y – Yellow	P – Pink
Vl — Purple		

A wire with a slash separated ID is a 2-color wire. For example: R/W is a red wire with a white tracer stripe.

MAINTENANCE

BRUSHES

Replace brushes when they are excessively worn.

When servicing the brushes, always observe proper brush polarity.

NOTE: When installing a brush holder, make sure that The mount screw holes align with the end housing holes, the brushes are perpendicular to the slip rings and the brush radius aligns with the slip ring curvature.

Failure to do so could cause the brush holder to break either before or after the generator is started.

ROTOR BEARING

The end housing bearing on these generators is a sealed, prelubricated ball bearing. Inspect and replace as necessary. If a seized bearing is encountered, inspect the rotor journal and the rear housing bearing cavity for scoring.

The end housing must be removed for access to the bearing.

GENERATOR REPAIR

TROUBLESHOOTING

1. Before testing any electrical components, check the integrity of the terminals and connectors.
2. While testing individual components, also check the wiring integrity and fuse(s).
3. The use of the Dr. Robin generator tester #388-47565-08 is recommended for testing. The Dr. Robin tester has the following capabilities:

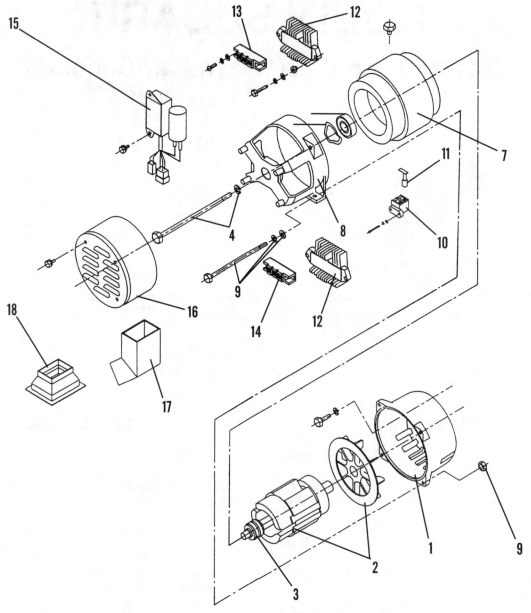

Fig. RB501 – Typical exploded view of the RXW Series generator head components.

1. Engine adapter/front housing
2. Rotor assembly with cooling fan
3. Sliprings
4. Rotor throughbolt
5. Compression ring
6. Rotor bearing
7. Stator

8. Endbell/rear housing
9. Rear housing/stator
   throughbolt assembly (4)
10. Brush holder
11. Brush (2)
12. Finned heat sink (2)

13. Rectifier 1
14. Rectifier 2
15. AVR unit
16. End cover
17. Rubber duct
18. Duct flange

a. AC voltmeter – 0-500 volts.
b. Frequency – 25-70 Hz.
c. Condenser capacity – 10-100µF.
d. Digital Ohmmeter – 0.1-1,999 ohms.
e. Insulation resistance – 3 MΩ.

An accurate engine tachometer will also be required.

4. Test specifications are given for an ambient temperature of 20° C (68° F). Resistance readings may vary with temperature and test equipment accuracy.

5. For access to the components listed in this Troubleshooting section, refer to Figs. RB501 and RB502 and the later Disassembly section.

6. All operational testing must be performed with the engine running at its governed no-load speed of 3700-3750 rpm, unless specific instructions state otherwise.

7. Always load test the generator after repairing defects discovered during troubleshooting.

To Megger Test the Generator System Wiring Insulation

1. Make sure that the circuit breakers are ON and functioning.

2. Connect one megger test probe to any hot terminal of a receptacle.

3. Connect the other megger test probe to any ground terminal.

4. The megger should indicate a resistance of 1 MΩ or more. If less than 1 MΩ, megger test the rotor, stator, and control panel resistance individually. Refer to the subsequent Troubleshooting sections.

Troubleshoot the Generator

1. Circuit breakers must be ON. To test the breakers, access the breaker wiring and continuity test the breaker terminals with the generator shut down and at least one breaker feed wire disconnected. The breaker is faulty if it will not stay set, shows continuity in the OFF position or does not show continuity in the ON position,

2. Receptacles must be functioning. To test the receptacle(s):

    a. Access the receptacle connections.

    b. Disconnect the *return* feed wire from the receptacle being tested.

    c. Install a jumper wire between the *hot* and *return* outlet terminals.

    d. Test for continuity across the feed terminals. No continuity indicates a faulty receptacle.

    e. Remove the jumper wire and retest for continuity across the feed terminals. No continuity this time indicates a good receptacle.

3. Ground fault circuit interrupters (GFCIs) must be *reset* and functioning. To test the GFCI:

    a. Start the generator and allow a brief no load warm up period.

    b. Depress the *test* button. the *reset* button should extend. If it does, proceed to Step C. If the *reset* button does not extend, the GFCI is faulty or it is not receiving input from the generator. If the generator does not have proper output for testing the GFCI, disconnect the GFCI feed wires, connect a wall socket jumper to the GFCI feed terminals, then test the GFCI.

    c. Depress the *reset* button. The *reset* button should then be flush with the 'Test' button. If the reset button does not remain flush with the test button, the GFCI is faulty.

    d. Stop the generator.

NOTE: If the GFCI seems to trip for no apparent reason, check to make sure the generator is not vibrat-

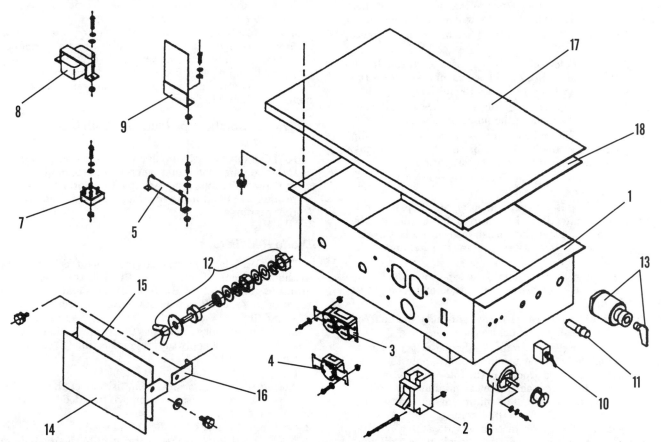

Fig. RB502 – Typical exploded view of the RXW Series control box components.

1. Control box and panel	7. Diode rectifier	13. Engine ignition switch
2. AC circuit breaker	8. Transformer	14. Terminal shield
3. 120V AC receptacle	9. Timer assembly	15. Rubber terminal-shield
4. 120V AC Twistlok receptacle	10. Idle-control switch	insulator plate
5. 200Ω resistor	11. Indicator lamp	16. Terminal shield hinge (2)
6. 300Ω resistor	12. Welding terminal assembly (2)	17. Control-box cover

ing excessively due to faulty rubber mounts or debris between the generator and frame. Excessive vibration can trip the GFCI.

WARNING: If the *reset* button trips while the generator is operating under load, stop the generator immediately and inspect the loads to determine the fault.

4. Start the generator, allow it to reach operating temperature, then load test the generator to check voltage output.

5. Test the output:

   a. AC voltage at the 120V receptacles should be 108-132 volts.

   b. DC voltage measured across the welding terminals under no load should be 70-75 volts on RXW150 model, or 60-65 volts for RXW180 model. While welding, with the current regulator dial turned to *maximum*, voltage should be approximately 25 volts.

6. Stop the generator.

7. To service the rotor:

   The slip rings are an integral part of the rotor assembly. they are not replaceable.

   Clean dirty or rough slip rings with a suitable solvent. Then lightly sand with ScotchBrite. Do not use steel wool, emery cloth or crocus cloth.

   Cracked, chipped, broken, or grooved slip rings are not repairable. In these cases, the rotor will need replacement.

   a. Disconnect the red and the black brush wire connectors.

   b. Measure field coil winding resistance through the brushes. resistance should be 30.6 ohms on RXW150 model, or 18.7 ohms on RXW180 model.

   c. Measure field coil winding resistance at the rotor slip rings. Resistance should be the same as Step B.

   d. Measure field coil winding resistance from each slip ring to a ground on the rotor shaft or laminations. Resistance should be infinity/no continuity.

   e. If test Steps B, C, and D met specification, the rotor and brushes are good.

   f. If test Step B failed specification but Steps C and D met specification, the rotor is good. Inspect the brushes and test each brush for continuity through its terminal. If faulty, replace the brushes and/or the brush holder assembly.

   g. If both test Steps B and C failed specification, the rotor is faulty. Replace the rotor, then inspect and test the brushes.

   h. If either Step D test read any continuity, the field coil winding is shorted. Replace the rotor.

   i. To megger test the rotor, apply one megger lead to one of the winding terminals, and apply the other lead to any good ground on the rotor shaft or laminations. The megger should read a minimum of 1 MΩ. A rotor failing to meet this minimum has faulty winding insulation. replace the rotor.

8. To service the stator, disconnect the stator winding wires. The welding coil winding wires will have individual terminals, the Main AC winding terminals will be in a two-terminal connector plug and the exciting coil and search coil winding terminals will be in a four-terminal connector plug.

   With the meter set to the R × 1 scale, apply the meter leads to the color coded wire terminals noted and test the stator winding resistance using the following specifications:

Model	Winding	Ohms
RXW150	U1-to-V1, V1-to-W1, and W1-to-U1 welding DC	0.033
	Sb1-to-Sb2 exciting DC	0.75
	White U to-white V main AC	0.24
	White-to-white/red search AC	0.36
RXW180	U1-to-V1, V1-to-W1, and W1-to-U1 welding DC	0.027
	Sb1-to-Sb2 exciting DC	0.55
	White U-to-white V main AC	0.19
	White-to-white/red search AC	0.68
BOTH	Each-wire-to-ground	Infinity

If megger testing the stator, apply one megger lead to the white AC winding wire terminal, and apply the other lead to the lamination core. The megger should read a minimum of 1 MΩ. Repeat this step with at least one color coded wire terminal of every other winding.

The failure of any winding to test as specified indicates a faulty stator. replace the stator.

9. Megger test the control panel by placing one meter probe on any live component and the other probe on any ground. The megger should read a minimum of 1 MΩ. Failure of the megger to read as specified indicates a component short. Access the control box components and test each component individually to determine the fault.

Individual Component Testing – Both Models

NOTE: If the meter being used is reverse polarity of the recommended meter when taking readings using any of the following test specifications, the test probes will need to be reversed to obtain the correct readings.

Diode Rectifiers

The RXW series generators use two diode rectifier units in the DC circuit. They are identified in the wiring diagrams as RX(A) and RX(B). Rectifier A is identified on the rectifier base by Part No. RM60SZ-6S, rectifier B as RM60SZ-6R.

Note that the following instructions apply to Rectifier A. To test Rectifier B, apply the test for Rectifier A, except reverse the meter leads.

To test Rectifier A:

1. Noting the wiring orientation, disconnect the rectifier wiring. It is not necessary to remove the diode unit from the finned heat sink.

2. Set the test meter to the R × 1 scale.

3. Apply the red/positive (+) meter lead to the diode base (Fig. RB503).

4. Apply the black/negative (-) meter lead to one of the three diode face terminals. The meter should read continuity.

5. Repeat Steps 3 and 4 with the other two diode face terminals.

6. Reverse the meter leads, then repeat Steps 3-5. The meter should read no continuity.

7. Failure of the rectifier to test any individual diode as specified in Steps 3-6 indicates a faulty rectifier.

Current Regulator

1. Noting the wiring orientation, access the regulator, disconnect the wiring, and remove the regulator, if necessary.

2. Set the test meter to the R × 1 scale.

3. Turn the regulator knob/shaft completely to the clockwise limit (viewed from the knob side of the regulator). Do not force.

4. With one meter lead attached to the center regulator terminal, attach the other lead to the terminal closest to the tip of the sliding contact arm. The meter should show little or no resistance.

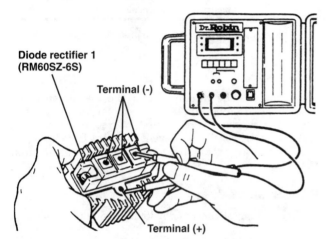

Fig. RB503 – View of the diode rectifier test. RXW Series units have two different diode rectifiers which appear identical. refer to the text for the correct test procedure.

5. Keeping both meter leads attached, slowly rotate the shaft to the other (counterclockwise) limit. The meter should show a consistent increase in resistance to approximately 200 ohm.

6. Failure of the regulator to test as specified indicates a faulty regulator.

Capacitor/Condenser

The capacitor is part of the AVR system. It is not serviceable separately.

Automatic Voltage Regulator (AVR)

To test the AVR:

1. Access the AVR unit inside the generator end cover and disconnect the AVR harnesses.

2. Set the test meter to a high resistance (1 MΩ) scale.

3. Using the wiring diagram and the chart in Fig. RB504, test continuity between the color coded wiring terminals as noted.

4. The failure of any terminal pair to test to specification indicates a faulty AVR assembly. Replace the AVR.

Idle Control System

1. To test the idle control solenoid:
   a. Disconnect the two white solenoid wires from the two yellow main harness wires (polarity is not critical).
   b. Resistance test the two white solenoid wires. Resistance should be 234-286 ohm. Failure of the solenoid to test in this range indicates a faulty solenoid. Replace the solenoid.

2. To test the idle control unit "timer":
   a. Unplug the 2-terminal and the 4-terminal connectors from the timer.
   b. Using the wiring diagram and the chart in Fig. RB505, test continuity between the color coded wiring terminals as noted.

A.V.R. wiring terminals		Black (-) meter probe							
		Red/White	Red/Black	Red	Black	Yellow	Yellow/Red	White	White/Red
Red (+) meter probe	Red/White		No continuity	Continuity	No continuity	No continuity	No continuity	No continuity	Continuity
	Red/Black	No continuity		No continuity	No continuity	No continuity	No continuity	No continuity	No continuity
	Red	Continuity	No continuity		No continuity	No continuity	No continuity	No continuity	No continuity
	Black	No continuity	No continuity	No continuity		No continuity	No continuity	No continuity	No continuity
	Yellow	No continuity	No continuity	No continuity	No continuity		No continuity	No continuity	No continuity
	Yellow/Red	No continuity	No continuity	No continuity	No continuity	No continuity		No continuity	No continuity
	White	No continuity	No continuity	No continuity	No continuity	No continuity	No continuity		No continuity
	White/Red	Continuity	No continuity	No continuity	No continuity	No continuity	No continuity	No continuity	

Fig. RB504 – Terminal pair continuity test for the AVR assembly. Reverse the meter probes if the test meter is reverse polarity of the recommended meter.

Idle control timer unit terminals		Black (-) meter probe					
		Red	Blue	Yellow	Yellow	Light blue	Light blue
Red (+) meter probe	Red		Continuity	Continuity	Continuity	No continuity	No continuity
	Blue	No continuity		No continuity	No continuity	No continuity	No continuity
	Yellow	No continuity	Continuity		No continuity	No continuity	No continuity
	Yellow	Continuity	Continuity	Continuity		No continuity	No continuity
	Light blue	No continuity	No continuity	No continuity	No continuity		No continuity
	Light blue	No continuity	No continuity	No continuity	No continuity	No continuity	

Fig. RB505 – Terminal pair continuity test for the idle control unit timer. Reverse the meter probes if the test meter is reverse polarity of the recommended meter.

c. The failure of any terminal pair to test to specification indicates a faulty idle control timer. Replace the timer.

DISASSEMBLY

Refer to Figs. RB501 and RB502 for component identification.

The rotor cannot be removed without first removing the stator. Exercise caution when handling the stator and rotor so the windings and winding coatings are not damaged. The rotor fan is part of the rotor assembly and is not replaceable separately.

1. On Model RXW180, disconnect the battery cables and remove the battery.
2. Drain and remove the fuel tank.
3. Remove the rubber harness duct from the generator end cover and remove the cover.
4. Remove the control box cover.
5. Note the wiring orientation and the location of all cable ties and restraints. Disconnect all wiring from the rear housing components.
6. Remove the control box assembly.
7. If necessary to remove the engine/generator assembly from the cradle frame:
   a. Remove the two upper side cradle support cross bars.
   b. Remove the nuts holding the engine and generator mounting feet to the rubber mounts.
   c. Lift the engine/generator from the cradle.
8. If not removing the engine/generator from the cradle, remove the two nuts holding the generator rear housing feet to the rubber mounts.
9. Place a temporary support, such as a wooden block, under the engine adapter/generator front housing so that the rear housing feet clear the rubber mount studs.
10. Remove the brush holder assembly.
11. Remove the four rear housing to front housing bolts.
12. Using a plastic mallet, carefully tap on the rear housing bolt ears to separate the housing from the stator. Remove the housing, carefully noting the wiring paths.
13. If necessary, carefully tap on the stator core to loosen it from the front housing. Remove the stator.
14. Remove the rotor throughbolt.

15. Using a slide hammer with the proper threaded adapter, or carefully tapping on the rotor lamination core with a plastic mallet, break the rotor free from the engine crankshaft. Be careful to prevent the rotor from falling.
16. If necessary, unbolt and remove the front housing from the engine.

REASSEMBLY

Refer to Figs. RB501 and RB502 for component identification.

Note that the rotor must be installed before the stator.
1. If the front housing was removed during disassembly, reinstall it. Torque the bolts to 12-14 N•m (100-120 in.-lb.).
2. If the engine was left in the cradle frame, place a temporary support, such as a wooden block, under the front housing so that the rear housing feet clear the rubber mount studs.
3. Service the rotor bearing, if needed.
4. Make sure that the matching tapers on the engine crankshaft and inside the rotor are clean and dry. Install the rotor onto the crankshaft. Install the rotor throughbolt. Torque the bolt 5.0-6.0 N•m (45-52 in.-lb...
5. Noting the correct wiring orientation (stator harness toward carburetor/muffler side of engine), fit the stator over the rotor and against the front housing.
6. Install the rear housing onto the rotor bearing and stator, making sure that the wave washer is installed and properly aligned against the rotor bearing, and the stator harness passes through the correct opening in the rear housing. If necessary, lightly and evenly tap the rear housing forward until it is snug against the stator, and the stator is snug against the front housing. Make sure that the stator wires are not pinched or kinked.
7. Install the four rear housing to front housing bolts. Alternately and incrementally torque the bolts to 12-14 N•m (100-120 in.-lb.).
8. Remove the spark plug and rotate the engine crankshaft to check for interference between the rotor and stator. There should be a consistent gap around the rotor laminations inside the stator. If any rubbing or scraping is evident, correct any misalignment prior to proceeding. Reinstall the spark plug and torque to 18 N•m (155 in.-lb.).

9. Install the brush holder assembly and any other components which may have been removed from the rear housing. Torque the M4 bolts to 1.0-1.5 N•m (8-13 in.-lb.). torque the M6 bolts to 5.0-6.0 N•m (45-52 in.-lb.).

10. Reconnect the main rectifier wires.

11. If the engine/generator assembly was removed from the cradle frame:

   a. Carefully fit the engine/generator into the frame.

   b. Install the nuts holding the engine and generator mounting feet to the rubber mounts. Torque the nuts to 12-14 N•m (100-120 in.-lb.).

   c. Install the two upper side cradle support cross bars. Torque the bolts to 5.0-6.0 N•m (45-52 in.-lb.)

12. If the engine/generator assembly was not removed from the cradle frame during disassembly:

   a. Carefully lift the rear housing.

   b. Remove the temporary front housing support.

   c. Set the rear housing feet down onto the rubber mounts.

   d. Install and torque the mount nuts to 12-14 N•m (100-120 in.-lb.).

13. Install the control box assembly.

14. Properly feed the generator-to-control box harness through the rubber harness duct. Referring to disassembly notes and the wiring diagram, reconnect all rear housing component wiring. Reinstall any cable ties and restraints removed during disassembly.

15. Install the control box cover, torquing the bolts to 5.0-6.0 N•m (45-52 in.-lb.).

16. Reconnect the rubber harness duct to the generator end cover. Install the end cover, torquing the bolts to 5.0-6.0 N•m (45-52 in.-lb.).

17. Install and reconnect the fuel tank. Torque the bolts to 3.0-4.0 N•m (25-35 in.-lb.).

18. On Model RXW180, install the battery. Reconnect the battery cables.

WIRING DIAGRAMS

Refer to Fig. RB506 to view the Model RXW150 wiring diagram.

Refer to Fig. RB507 to view the Model RXW180 wiring diagram.

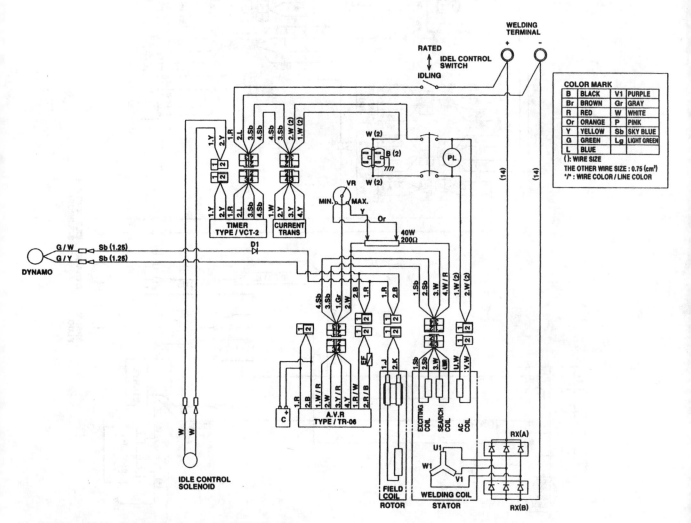

Fig. RB506 – Wiring diagram for the Model RXW150 generator.

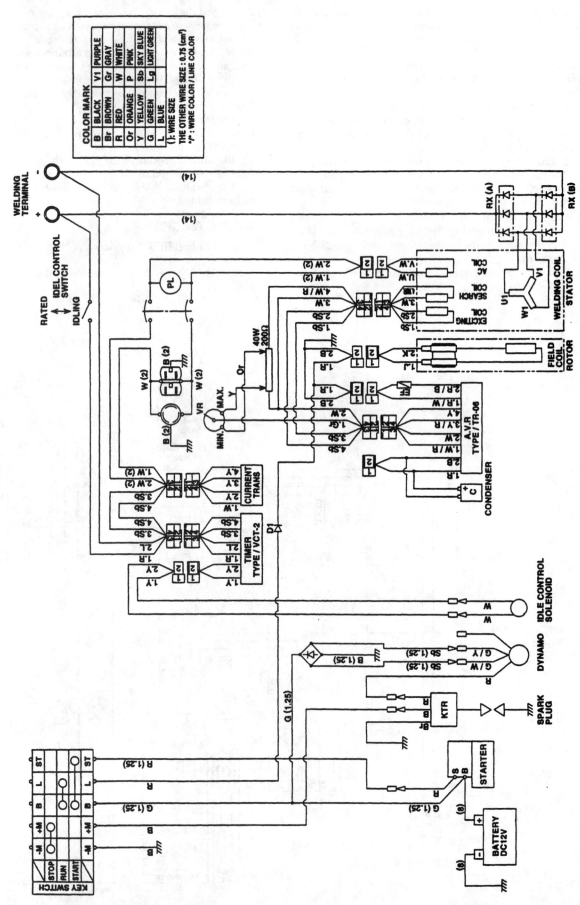

Fig. RB507 – Wiring diagram for the Model RXW507 generator.